Audel™ Guide to the 2005 National Electrical Code®

All New Edition

Paul Rosenberg

WILEY

Wiley Publishing, Inc.

Vice President and Executive Group Publisher: Richard Swadley
Vice President and Publisher: Joseph B. Wikert
Executive Editor: Carol A. Long
Editorial Manager: Kathryn A. Malm
Development Editor: Emilie Herman
Production Editor: Vincent Kunkemueller
Text Design & Composition: Wiley Composition Services

Published simultaneously in Canada

For general information on our other products and services please contact our Customer Care Department within the United States at (800) 762-2974, outside the United States at (317) 572-3993 or fax (317) 572-4002.

Wiley also publishes its books in a variety of electronic formats. Some content that appears in print may not be available in electronic books.

Library of Congress Control Number:

ISBN: 0-7645-7802-2

Printed in the United States of America

10 9 8 7 6 5 4 3 2

Contents

Foreword

I think that almost everyone who has been required to use the *National Electrical Code (NEC)** on a regular basis has often wished that it were easier to understand. Often, it seems that it lacks sufficient clarity and detail; other times, it seems to be overflowing with useless information. The purpose of this book is to help the reader sort through the voluminous code regulations and find the information he or she needs, with a minimum of effort. Perhaps it would help to understand where this code book comes from.

The *National Electrical Code* is one of many codes and standards published by the National Fire Protection Association (NFPA), a not-for-profit corporation. The code is revised every three years in order to keep up with new materials, tools, and methods that are constantly being developed. This work is performed by 21 separate committees, each consisting of approximately 10 to 15 persons, the majority of them engineers. Members of each committee meet several times, discuss all proposed changes, accepting some and rejecting others, and rewrite (as required) the sections of the *Code* that were assigned to their committee. Then, they circulate the changes among the various committees, coordinate the changes, and rewrite again. So, obviously, the updating of the *NEC* is no small chore. But the real difficulty is that it must remain applicable to all types of electrical installations, leaving no gaps. Because of this, it becomes rather difficult to interpret in many instances.

The purpose of this book is to arrange all of the pertinent requirements of the *National Electrical Code* in a manner that is user-friendly, allowing the reader to find the needed information painlessly and quickly. The challenge with the *NEC* is that many communities use it as law, and as such, it must be written accordingly. Every possible facet of every type of electrical installation must be covered. Because of this, the *NEC* is full of engineering requirements, installation requirements, and manufacturing requirements—all in engineering lingo and legalese. It's not hard to see why it is such a difficult document to comprehend. In order to make the *NEC* more easily understood and applicable, a number of guides have been written, most of which have a legitimate place. These guides serve to make all parts of the *NEC* understandable. They are written for engineers, designers, installers, and inspectors.

**National Electrical Code*® and (*NEC*®) are registered trademarks of the National Fire Protection Association, Inc., Quincy, MA.

The book you now hold in your hands is substantially different from standard *NEC* guidebooks. Rather than covering everything in the *NEC*, we concentrated only on the requirements for electrical installations. By omitting the engineering and manufacturing requirements, much of the confusion of the *NEC* is eliminated in one stroke. This leaves only the rules that actually apply to installing electrical wiring—which is the reason the *Code* is referred to 99 percent of the time.

This book is designed exclusively for the installer of electrical wiring, and is the result of many years of supervising and instructing electricians in the requirements of the *NEC*. Every effort has been made to make this book as easy to use as possible, both for the professional electrician and for the homeowner who wishes to do his or her own electrical work safely and efficiently, avoiding hassles with the local electrical inspector.

For actually installing electrical wiring, this book should be more useful than the standard *NEC* handbooks. For engineering questions, however, the *National Electrical Code* should be consulted.

2005

Throughout this book, you will see substantive changes for the 2005 *NEC* highlighted. Bear in mind that these changes will have the force of law once the 2005 *Code* is adopted in your jurisdiction.

As you go through both this book and the *Code*, you will find numerous references to other codes and standards. These various codes and standards are useful but must always be used in conjunction with the *NEC*, not separate from it. It is critical to remember that codes are generally adopted as law by local municipalities., while standards are not. So, codes contain mandatory requirements and standards contain suggested methods.

Finally, please remember that good workmanship and safety-consciousness are essential ingredients for any good electrical installation. Like fire, electricity can be the best of friends or the worst of foes. Without careful workmanship and an overriding concern for the safety of the installation and the installer, no electrical installation is worthwhile.

My sincere thanks go to all of the fine people I've worked with down through the years—I have had the good fortune of working with some of the finest people in the industry.

Paul Rosenberg

Introduction

The *National Electrical Code* is written as a minimum standard for electrical installation for the protection of life and property. It does not necessarily define the best installation methods, merely the minimum safety standards. Many purchasers of electrical installations will want to surpass the code.

When reading and interpreting the *NEC* there are certain words that you must pay attention to. These key words are:

Shall. Any time you see the word *shall* in the *NEC*, it means that you *must* do something a certain way. You have no choice at all; either you do it that specific way, or you are in violation of the code.

May. The word *may* gives you an option. You can do it the certain way that is stated, or you can do it another way; it is your choice.

Grounded Conductor. This is almost always the neutral conductor, although not necessarily. Take care not to let the word *grounded* confuse you; "grounded conductor" does not refer to a green wire.

Grounding Conductor. This is the green wire, more correctly called the "equipment grounding conductor," because it is used to connect equipment to ground.

You will find these ideas expressed in section 90.5 of the *NEC*, discussed below. They are defined as *Mandatory Rules* (shall), *Permissive Rules* (may), and *Explanatory Material* (Fine Print Notes). Special care must also be taken to differentiate between similar terms, such as "grounded conductor" (a neutral wire), and the "grounding conductor" (the green equipment grounding conductor). These terms are almost identical, and if you do not carefully examine each word, you could very easily make a wrong interpretation.

In addition to these terms, there are other, less-common terms (identification, listing, supervised, and so on) that can also be confusing. Remember that the *NEC* cannot be read casually. In order to make correct interpretations, every word must be considered. This requires extra work and effort.

Before getting to the main body of the *NEC* (starting with Article 100), it is important to cover two other sections that precede. The more important of these is Article 90, which explains what the Code is and what it applies to. The other, Article 80, is relatively new, and serves as a model local ordinance for the legal adoption of the *NEC*.Article 80—Administration and Enforcement.

Article 80 is a model ordinance for the administration and enforcement of the *NEC*.

Whether this section of the *NEC* will be adopted by most municipalities is still an unanswered question. You should definitely check with your local government to see whether these requirements have been adopted or not. Most municipalities have covered these concerns with local ordinances for a long time; some may choose to keep their own ordinances, and others may prefer simply to adopt the *NEC* rules as a package.

The rules of Article 80 should have no bearing on how you install electrical wiring, though it may mean slight changes in how your installations are inspected. So, while checking on its adoption is a good idea, don't expect it to change any of your installations.

Article 90—Introduction

This article lays the groundwork for the writing and application of the *National Electrical Code*. It begins by stating the purpose of the document, "the practical safeguarding of persons and property from hazards arising from the use of electricity," and goes on to explain that the *NEC* is written to provide safe installations, though not necessarily efficient ones.

Section 90.2 is especially important, as it identifies what sorts of installations are, or are not, covered by the *NEC*. Note that almost all wiring owned by utilities or mines, and in boats, aircraft, and automobiles are excluded.

90.1: Purpose

(1) Electricity can be dangerous if not used properly. The Code is written to provide a set of rules for the safe installation of electrical wiring.

(2) This Code's provisions are those essential for safety, and compliance with these rules may not always result in the most efficient, convenient, or least expensive installations; neither does it necessarily provide for the future expansion of electrical usage. It is however essentially free from hazards that may be encountered. Nonconformity to the rules of the *NEC* may result in hazards or overloading of wiring systems. Most of these problems result from not taking into consideration the increasing usages of electricity. If future needs are taken into consideration at the time of the original installation and adequate measures are taken to provide for the increased usage of electricity, these hazards and overloading may be greatly eliminated.

(3) In no manner is this *Code* intended to be used for design specifications or as an instruction manual for untrained persons. The rules of this *Code* will, however, add materially to proper design. It is also adopted as the regulations governing wiring installations by most government agencies. There may be additional requirements by the local agencies and these should be checked out.

90.2: Scope

(A) **Covered.** This *Code* covers:

(1) Electric conductors and equipment installed in or on: public or private buildings or other structures, mobile homes and recreational vehicles, floating buildings, and other premises, such as yards, carnivals, parking and other lots, and industrial substations.

Additional information concerning installations in multibuilding complexes or industrial buildings is found in the National Electrical Safety Code, ANSI C2-1997.

(2) The installation of conductors on the exterior of a premise is covered.

(3) The installation of conductors outside of a premise is covered.

(4) The installation of optical fiber cables and raceways. The inclusion of optical fiber cables in the *NEC* is odd, since these cables carry no electricity at all. They are included in the *National Electrical Code* for two primary reasons: (1) because they are usually installed by the same persons who install electrical wiring and (2) because optical fiber systems interact with, and depend upon, electrical and electronic systems.

The code's reference to "optical raceway" refers to special raceways whose use is dedicated to the optical cables they house. These are special inner ducts and possibly tubes associated with air-blown fiber. This is not defined clearly in the code, so check with your local inspector if you have any questions. Also, see 770.6 for details.

(5) Wiring in of offices, warehouses, or other buildings owned by electric utilities but not part a generating facility, substation, or control facility.

(B) Not Covered. This Code does not cover:

(1) Ships, watercraft, trains, aircraft, automobiles, or trucks, although mobile homes and recreational vehicles are covered.

(2) Installation of conductors is not covered in the *NEC* for underground mines. This does not exempt the above-ground installation of wiring, although self-propelled surface mining machinery and its trailing cables are excluded.

(3) Railroad generation, transformation, and transmission or distribution, if used only for signaling devices, and railroad trains are not covered in the *NEC*.

(4) Communication equipment located outdoors or indoors, if used exclusively by utilities, is not covered in the *NEC*.

(5) Electric utility wiring exclusively under the utility company's control, used for communication, metering, generation, transformation, and distribution of electricity, whether indoors or outdoors on property owned or leased by the utility, whether out of doors by established rights on private property and public highways, streets, or roads, are not covered by the *NEC*.

(6) Any metering, wiring, buildings or structures on any premise that is not owned or leased by the utility company is covered by the *NEC*. The *NEC* does cover all wiring other than utility metering equipment ahead of service equipment through building structures or any other place not owned or leased by the utility.

(C) Special Permission. Conditions and usages vary in different localities; therefore, the authority having jurisdiction for the enforcement of the *Code* must be able to grant exemptions for the installation of the wiring system equipment not under the control of the utilities. This occurs whenever utilities are connecting service-entrance conductors of the building or structure that they are serving. If such installations are outside the building or terminate just inside the building, special permission should be granted in writing.

There has been an abundance of work done by utilities, and often the work becomes a part of the *Code*. Should the installation of service laterals, for example, be deemed good engineering practice by utilities and acceptable by the enforcing authority, this practice may, by special permission, be permitted under the *Code*. This special permission does not eliminate the Special Permission under Article 100; it applies only to Section 90.2.

90.3: Code Arrangement

The *Code* is divided into an introduction and nine chapters. Chapters 1 through 4 deal with general applications of the *Code* to wiring and installations. Chapters 5, 6, and 7 supplement or amend the first four chapters, and deal with special occupancies and installations that involve special equipment or special conditions. Chapter 8 deals with communication circuits, and with the equipment and installation of radio and television. Chapter 9 deals with tables not included in, but to be used in conjunction with, the first eight chapters. Also included are examples for figuring requirements for installation. These examples are extremely valuable in the understanding of the preceding chapters.

Familiarity with the various *Code* chapters makes it easy to find what you want in the Code. Chapters 4 through 9 are special chapters and refer back to the first three chapters.

90.4: Enforcement

The *NEC* is written so that it can be enforced when adopted by agencies having the rights of inspection. The *Code's* enforcement and interpretation is placed in the hands of the enforcing agency or authority. These authorities are the ones who make the final decisions, hopefully using the good judgment that is essential in such interpretations. In many instances, the *Code* puts the entire responsibility of interpretation on the enforcing authority. For example, you will often find the phrase *by special permission;* this means special permission, in writing, by the *Code*-enforcing authority.

The enforcing authority is vested with the right to decide on the approval of equipment and materials. However, listings from the Underwriters' Laboratory, the CSA, or other independent testing laboratories are used for this purpose in many instances. One of the deterrents to Code understanding can be lack of communication between the inspector and the installer. Actually the inspector is the installer's friend, and all the inspector wants is a good safe job. The best advice to offer in this respect is to get acquainted with your inspector; he or she will be understanding and helpful in most cases.

Many industries have established procedures for installation and maintenance that are very effective and in many cases far more safety-oriented than the *Code* installations. This gives the enforcing authority the latitude to okay such installations.

90.5: Mandatory Rules, Permissive Rules and Explanatory Material

The *Code* includes both mandatory and advisory rules. The mandatory rules are characterized by the word "shall" This means that the rules must be strictly followed. Any time you see the word *shall* in the *NEC* it means that you must do something in a certain way. You have no choice at all; either you do it that specific way, or you are in violation of the *Code*. Permissive rules are characterized by the word *may*. The word *may* gives you an option. You can do it the specific way that is stated, or you can do it another way; it is your choice.itemizes the types of rules given in the *NEC*.

Explanatory material in the *NEC* is placed in Fine Print Notes (FPN). These notes are important for you to read, but *they are not enforceable*.

90.6: Formal Interpretations

An *NEC* committee is set up to render official *Code* interpretations when these are necessary. In the majority of questions arising on the *Code*, the interpretations are under the inspector's jurisdiction, as will be seen in the next section. However, there may be instances when official interpretations are required. No official interpretations will be made unless the Formal Interpretation Procedures outlined in the *Code* are followed.

90.7: Examination of Equipment for Safety

Most equipment and materials have been tested by electrical testing laboratories such as Underwriters' Laboratories (UL), and carry their label. However, the rates that UL charge equipment makers can be prohibitively high. (They are somewhat of a monopoly.) To work around this problem, some municipalities have experimented with allowing consulting engineers to certify the equipment as being safe. If UL rates remain as high as they are now (or possibly go even higher), this method may become far more common. Extreme care must be taken by any inspection authority or testing service in judging the safety of any equipment, device, or material. Care must also be taken to assure that the equipment, device, and so on, will be used only in the way intended. Section 110.3 and Article 100 cover examination of equipment and the meaning of "Listed."

90.8: Wiring Planning

This section is unusual in that it mentions planning for future expansion, but does not require anything specific. It has long been good trade practice to oversize electrical components. However, this is not required by the *NEC*. Oversizing is a design issue, not an installation issue. Nonetheless, responsible installers should oversize the electrical equipment they are responsible for providing, if at all possible. Conduits should not be filled to capacity, and distribution equipment should have plenty of empty space.

In the design of electrical systems by electrical engineers, ample provision should be made in the raceways for adequate wiring, as well as distribution and load centers which should be laid out in practical locations, keeping in mind their accessibility. The number of wires in enclosures and boxes should adhere to Code requirements in order to avoid fires and breakdowns and the inconveniences that accompany such troubles.

In reaching the goal of good wiring and installation, there is one requirement—good workmanship. Insulation damage, too many wires, and overfusing are points that must be carefully watched. Regardless of how good the design of the installation, cutting corners will defeat the intended product.

(A) **Future Expansion and Convenience.** Since the invention of the electric light, the amounts of electricity used in both home and industry have continually increased. Therefore, in designing wiring systems consideration should be given to large enough raceways and in some cases spare raceways to accommodate the changes—future uses of electricity or expansion of operations—that are certain to come. During the design phase, it would be a good idea to review Sections 110.16 and 240.24, which describe the necessary clearance distances and accessibility for future additions.

(B) **Number of Circuits in Enclosure.** You will find later in the *NEC* that there is a maximum number of conductors and circuits that you can put in a single enclosure such as raceways, boxes, and so on. These limitations for single raceways and boxes will reduce problems with short circuits and ground faults in a circuit.

Severe damage could be done to conductor insulation by pulling too many conductors in to raceways, or by pulling around too many bends. There are even times, when pulling large sizes of conductors, that the 360 degrees in total bends between pull boxes and the like could be too many. Since the *Code* is not intended to be a design manual, it is up to the designer and the inspection authority to watch for these things. The *Code* has taken into account (derated), as you will find in Article 310, certain numbers of current-carrying conductors in raceways to avoid overheating of conductors and raceways.

2005

90.9: Metric Units of Measurement

Metric units, together with our own units of measurement, are used in the *NEC*. In the 2005 edition, metric units are set in standard text, and English units are contained in parentheses. Horsepower, wire sizes, box sizes, and conduit sizes are generally set primarily in English units.

Chapter 1

General

Article 100—Definitions

The *National Electrical Code* (*NEC*) contains a great number of definitions, which are very important for interpreting the *Code*. If you have any doubt as to the exact meaning of a general term, refer to Article 100 and verify that meaning. You will also find that the definitions in this section are arranged in two categories—"General" and "Over 600 Volts."

But if you need the definition of a more specific term, you may have to find it in the article where it would be dealt with most directly. As you continue through the *Code*, you will find additional definitions scattered throughout other articles. These definitions are very specific to that article and are therefore included with that article and not in Article 100.

The following figures are useful in understanding the definitions. For a branch circuit, see Figure 100-1. For a multiwire branch circuit, see Figure 100-2. For an illustration of service drop, see Figure 100-3. Service-lateral and service-entrance equipment are illustrated in Figures 100-4 and 100-5, respectively.

Article 110—Requirements for Electrical Installations

Article 110 is by-passed in the study of the *Code* more often than any other article. It is short, but it is actually the foundation upon which the *Code* is written, as it contains provisions that are used throughout the entire *Code*.

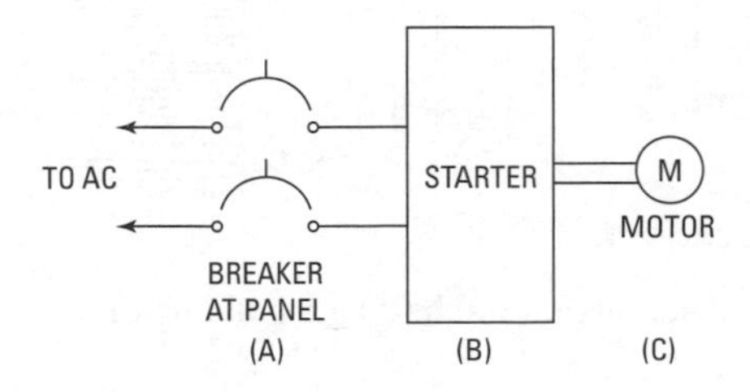

Figure 100-1 A motor circuit. The branch circuit extends from point A to point C.

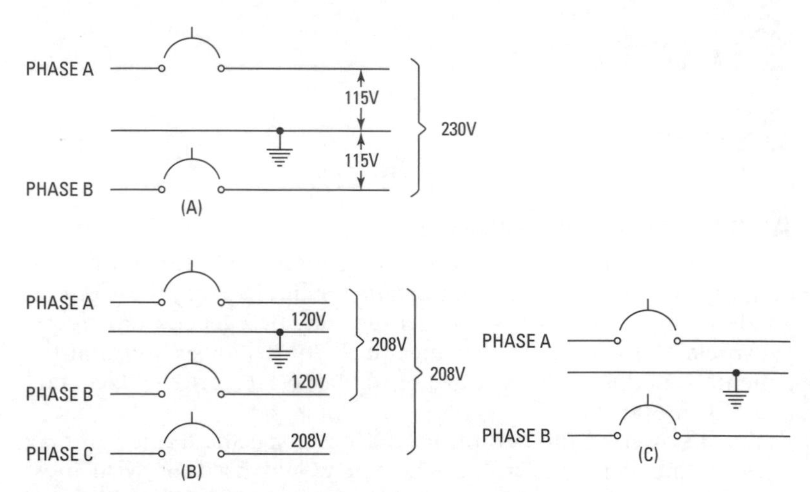

Figure 100-2 Variations of a multiwire branch circuit. Circuit C is not a multiwire branch circuit because it utilizes two wires from the same phase in conjunction with the neutral conductor.

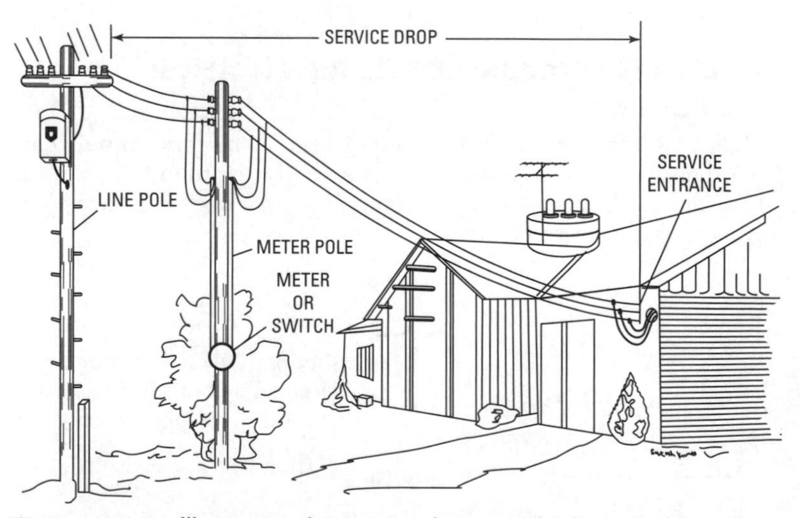

Figure 100-3 Illustrating the service drop attached to a building or other structure.

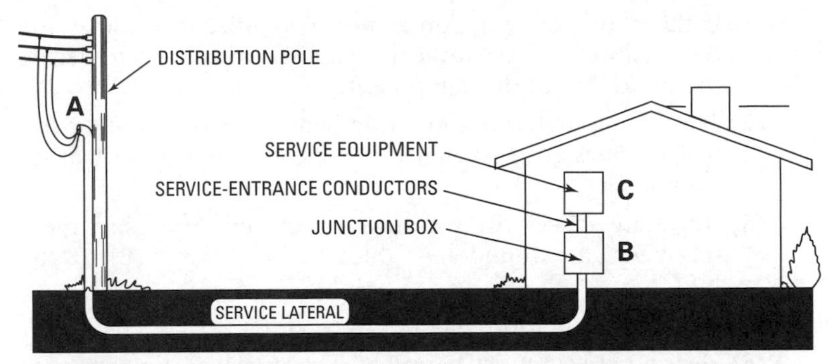

Figure 100-4 Illustrating the service lateral extending from point A to point B. The service entrance is from point B to point C.

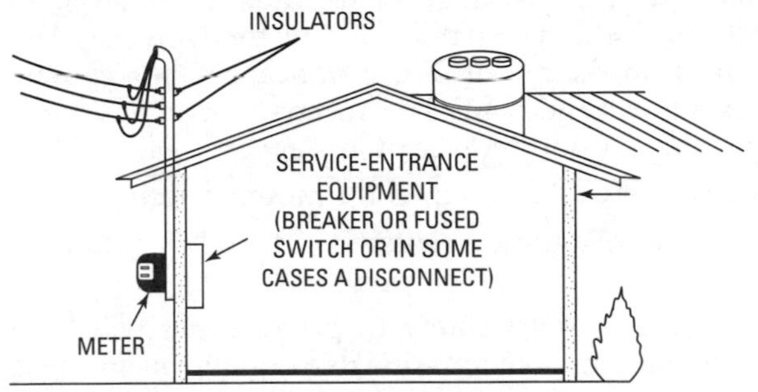

Figure 100-5 Showing the service-entrance equipment that will serve as the electrical disconnect supply.

I. General

110.2: Approval

See definition of *approved* under Article 100.

110.3: Examination, Identification, Installation, and Use of Equipment

(A) **Examination.** Observe the following considerations for the evaluation of equipment:

(1) Wiring devices and equipment that are suitable for use must be provided with identification of the product and of the use intended—environmental application. The identification, in most cases, is by labeling or listing.

If the above information is not available, it becomes the responsibility of the authority having jurisdiction to decide the suitability of the equipment.

(2) The wiring material and equipment must have their parts properly designed so that the enclosure will protect other equipment.

(3) Adequate splice-wire bending is required. The exact measurements are found in Tables 312.6(A) and (B) of the *NEC.*

(4) Electrical insulation may be checked.

(5) Heating effects must be taken into consideration on conductors. In Article 310, there are tables for reducing the ampacity of a conductor as ambient temperatures rise. The author finds that few are familiar with high-altitude rating of motors, which starts at 3500 feet above sea level. In higher altitudes, the air is thinner and therefore has less cooling effect on the motor. For instance, a 5-horsepower motor at a high altitude can't be expected to carry as much load as the same 5-horsepower motor at sea level.

(6) The equipment must be designed for minimal arcing.

(7) The use of voltages and currents must be taken into consideration.

(8) Other factors that affect safety to persons that will have occasion to come in contact with this equipment must be considered.

(B) Installation and Use. Labeling or listing will be effective only if the precautions noted on the installation and use instructions included with the labeling or listing service are followed. Alteration of equipment in the field voids any labeling or listing.

110.4: Voltages

The voltages referred to in the *Code* are the supply voltages, regardless of their source. The supply may be a battery, generator, transformer, rectifier, or a thermopile. When considering AC voltages, the voltage is the RMS voltage as explained in Article 100. There are really three general classifications of voltages in the *Code*—0 to 50 volts, 50 to 600 volts, and voltages that exceed 600 volts. Each is dealt with in separate parts of the *Code*. If wires having different voltages are run in the same raceway, there are specific rules to be followed. See Section 300.3(C).

No electrical equipment may be connected to a circuit that has a voltage higher than the equipment's rating.

110.5: Conductors

Unless the material of which the conductor is made is specifically identified, it is assumed to be copper. Any other material of which a conductor may be made, such as aluminum, shall be identified as such.

Copper and aluminum conductors have different ampacities and are covered in Article 310. Copper-clad aluminum conductors have the same ampacity as aluminum conductors.

110.6: Conductor Sizes

In dealing with wire sizes, the *Code* always refers to the American Wire Gage (AWG). At one time, this was known as the B&S Gage. Sizes of conductors larger than 4/0 are measured in circular mils.

110.7: Insulation Integrity

All wiring shall be installed free of shorts and grounds. This does not cover purposefully-grounded conductors, as covered in Article 250.

Shorts or grounds may be located before energizing circuits by the use of a megohm-type tester (available from several manufacturers).

Conductors of the same circuit and in the same raceway must be insulated with the same type of material. Therefore, insulation-resistance tests on each conductor should produce similar values. A case in point: Six 500 kcmil THHN conductors in the same conduit read approximately 1500 megohms on four conductors, and in the vicinity of 300 megohms on the other two conductors. While 300 megohms would have been a good value, the difference in the readings indicated problems. The low-reading cables were pulled out, and it was found that the insulation had been cut in many places. With time and condensation moisture, a fault would have occurred.

110.8: Wiring Method

Only recognized and suitable wiring methods are included in the *Code*. Basically, Chapter 3 covers approved wiring methods; Chapters 5 through 8 cover specific conditions and occupancies.

110.9: Interrupting Rating

Interrupting capacity is far different from the rating of the amperes that are required by a load. We are faced with what is known as fault currents. A fault current is the amount of current that might develop under a dead-short condition. This level of current is dependent upon the utility system supplying the current, the impedance of the system, and any fuses that may be up-line. At one time, this was not much of a problem, but with increased electrical usage and larger generating and distribution capacities, the problem of fault currents has increased. This is taken more into consideration now than in the past, and may become an increasingly important factor. If a piece of equipment is rated at X number of amperes, this does not necessarily mean that it can be disconnected under load or a fault condition without damage. Equipment is rated in carrying capacity as well as interrupting capacity. Sections 110.9 and 110.10 together require that all equipment be coordinated and protected from fault currents, not just from overcurrents. This requires the installer to get the cooperation of the utility company to verify available fault currents at the point of service.

110.10: Circuit Impedance and Other Characteristics

The fault currents are limited only by the capacity of the electrical supply, the impedance of the supplying circuits, and the wiring. As an example, the fault current will be much larger in circuits supplied from a large-capacity transformer supplying a heavily loaded city block than the fault current from a transformer serving a 5-horsepower irrigation pump in a rural area. The impedance of the supply to the 5-horsepower motor will be high in comparison to the impedance of the supply to the city block.

It is necessary to understand all coordinate fault currents, circuit impedances, and component short-circuit withstanding ratings. Fuse and breaker manufacturers have available easy-to-understand literature on fault currents and impedances, making it simple to check whether the equipment will withstand available fault currents.

It is also necessary to consider equipment that is connected to these circuits. In many cases, a wiring fault could spread its damage to these devices. This must be prevented. It is also important to understand that the requirements of the *Code*, especially in this section, will provide for a minimum level of safety; they don't guarantee that the equipment will not be damaged. Even with appropriately sized fault protection, damage to the equipment is possible, albeit without causing damage to other equipment or persons.

110.11: Deteriorating Agent

Environmental factors, such as wetness, dampness, fumes, vapors, gases, liquids, temperatures, and any other deteriorating effects, must also be noted; conductors and equipment used must be approved for the specific conditions of operation. The inspection authority is often faced with the responsibility of deciding in which category the installation belongs; it most certainly is beyond the scope of the *Code* to define and specify for every possible condition that will have to be met. The *NFPA National Fire Codes* will be of great value in this respect.

Protection shall be given to equipment, such as control equipment, utilization equipment, and busways, during construction if this equipment is approved for dry locations only. It shouldn't be permanently damaged by weather during the building construction. Section 300.6 further discusses protection from corrosion.

110.12: Mechanical Execution of Work

Electrical installers are required to install all electrical systems in a neat and workmanlike manner. Thus, the *Code* specifies that not just materials are important, but that workmanship is also extremely important.

This "neat and workmanlike manner" rule is actually one of the broadest in the *Code*. It can be applied to conduit bending, the trimming of panels, or to almost any aspect of an installation of electrical wiring. This gives the authorities having jurisdiction some discretion; they can invoke rulings based upon workmanship, which can be interpreted many ways. In actual practice, this rule can be applied either well or poorly, but is probably necessary. As expansive as the *Code* document is, human action is far more expansive, and no rule-book could address every possibility. This rule gives an inspector some latitude. The author has never seen it used in an overtly malicious fashion, though that does remain a possibility.

(A) **Unused Openings.** All openings in boxes, equipment, or enclosures of any kind must be effectively closed and must provide protection equal to that of the equipment or enclosure itself.

(B) **Subsurface Enclosures.** Conductors in underground enclosures (such as manholes) must be racked. This is necessary to provide for safe and easy access.

(C) **Integrity of Electrical Equipment and Connections.** All parts of electrical equipment must be kept free of paint, plaster, cleaners, and any other type of foreign material. This has long been a problem on construction sites, where plaster and paint end up in electrical panels and other items. All such contamination must be avoided.

110.13: Mounting and Cooling of Equipment

(A) Mounting. Mounting of equipment is an item directly related to workmanship. Wooden plugs driven into holes in masonry, plaster, concrete, and so on, will shrink and rot, thereby allowing the equipment to become loose. Thus, only approved methods of mounting and special anchoring devices may be used.

(B) Cooling. Electricity produces heat. Electrical equipment must be installed in such a way that circulation of air and convection methods of cooling are not hindered. Mounting equipment too close to walls, ceilings, floors, or other items will interfere with the electric equipment's designed means of cooling. Ventilation openings in the electric equipment must be kept free to permit natural circulation.

One should also watch the amount of total space in the room where the equipment is mounted. If it is inadequate to permit a low enough ambient temperature, means must be taken to permit the lowering of high ambient temperatures by natural or other means.

110.14: Electrical Connections

Because values of electrolysis (chemical decomposition caused by an electrical current) vary among metals, and because we are using copper or aluminum conductors, copper, being the more noble on the electrolysis series, will corrode the aluminum away. Therefore, you must be sure when making splices of terminations that the lugs or connectors are listed for the purpose for which you are using them. When using solder fluxes or inhibitors, make sure they are listed for the job you are doing. Wherever values for tightening torques are given, they must be adhered to.

The author has found very little available information on torquing values. Therefore, it might be appropriate to insert some torquing values in this book. Many breakdowns and possible fires might result from not adhering to proper torquing values, so Tables 110-1 through 110-3 are presented as guidelines for tightening connections. It might also be mentioned that dies on compression tools do wear, and to avoid breakdowns, the Biddle Co.'s Ducter can prevent this problem, as it will read down to one-half millionth of an ohm. This instrument has been invaluable to the author.

You will find additional torquing pressures in mechanical engineering handbooks. Loose connections can be a hazard, causing breakdowns and possibly fires. If the authority having jurisdiction so wishes, it may require torquing tests during inspections.

Table 110.1 Tightening Torque in Pound-Feet Screw Fit

Wire Size, AWG	*Driver*	*Bolt*	*Other*
18–16	1.67	6.25	4.2
14–8	1.67	6.25	6.125
6–4	3.0	12.5	8.0
3–1	3.2	21.00	10.40
0–2/0	4.22	29	12.5
AWG 200 kcmil	—	37.5	17.0
250–300	—	50.0	21.0
400	—	62.5	21.0
500	—	62.5	25.0
600–750	—	75.0	25.0
800–1000	—	83.25	33.0
1250–2000	—	83.26	42.0

Table 110.2 Screws

Screw Size, Inches Across Hex Flats	*Torque, Pound-Feet*
1⁄8	4.2
5⁄32	8.3
3⁄16	15
7⁄32	23.25
1⁄4	42

Table 110.3 Bolts

Size	*Duronze*	*Steel*	*Aluminum*
	Standard, Unlubricated		
3⁄8	20	15	16
1⁄2	40	25	35
5⁄8	70	50	50
3⁄4	100	90	70
	Standard, Lubricated		
3⁄8	15	10	13
1⁄2	30	20	25
5⁄8	50	40	40

(A) Terminals. Connections to terminals must ensure a good electrical and mechanical contact without injury to the conductors; connection must be by approved pressure connectors, solder lugs, or splices to flexible wires. The exception to the regulation is that No. 10 or smaller stranded conductors can be connected by means of clamps or screws with terminal plates having upturned lugs (Figure 110-1). Terminals for more than one conductor must be of the approved type for this purpose. When permitted to place a wire under a terminal screw, wrap it in such a direction that when you tighten the screw, the wire will not be squeezed out from under the head of the screw. On the smaller sizes of conductors, especially cord conductors, it is best to twist the conductor strands and apply some solder to them.

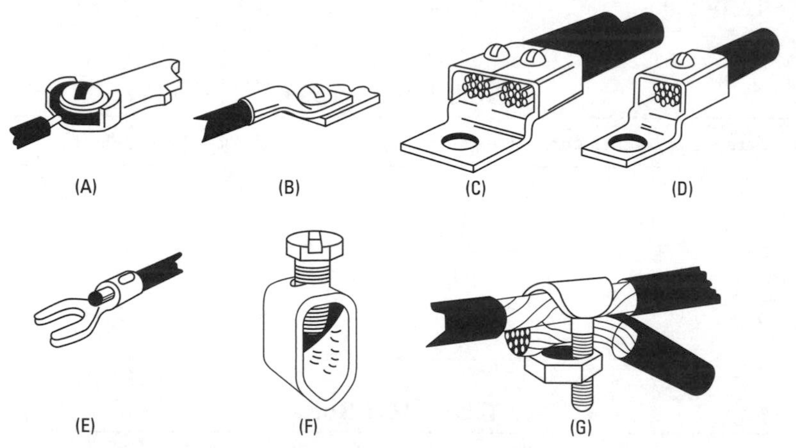

Figure 110-1 Various types of approved pressure connectors. (A) Terminal plate; (B) Soldered lug; (C) Double pressure-type lug; (D) Single pressure-type lug; (E) Open-end crimp-type lug; (F) Pressure-type connector; (G) Split-bolt clamp.

Compression-type connections are extremely good if the proper compression tool is used and it is in good shape. No. 10 or smaller conductors can be used for screws, studs, or nuts that have upturned lugs or equal design to keep the wire connection in place.

Any terminal or lug intended for use with aluminum must be so marked.

(B) Splices. Splices in wires are permissible in the proper places. When making a splice, the wires must be clean and a good electrical and mechanical connection must be made. The wires may then be soldered, provided a suitable solder and flux are used. The soldering temperature should be carefully controlled, because a cold solder joint is of no value; also, if the wires become too hot, the heat will damage the insulation. Remember that soldering is not permitted on conductors used for grounding. Approved connectors may also be used for splices, making sure the wires are clean and free from corrosion. After splicing, insulation at least equivalent to that on the wire must be applied to the splice. In general, this applies to all splices, but on high-voltage splicing, the specifications supplied with the high-voltage cables should be followed. When wire connectors are to be used on splices directly buried in the ground, they must be made with a type that is listed for that use.

This is extremely important. Many electrical connections fail because they are improperly made. Many troubles have been due to electrolysis between different metals, that is, the more-noble metal depleting the less-noble metal. Also, the oxidation of aluminum conductors (and this oxidation occurs practically instantly) creates a layer that has a very high resistance.

Another problem is the coefficient of expansion of different metals, creeping, and the difference in deformation of different metals. Be certain that you use connectors approved for use with this new product.

Inhibitors for use with aluminum are very important. Don't rely on the inhibitor alone, but thoroughly brush the aluminum conductor to remove the oxide film, and then immediately apply the inhibitor to prevent the recurrence of the oxide film.

(C) Temperature Limitations. The general principle of temperature limitations is that the operating temperatures of all circuit components (conductors, terminals, and equipment) must be coordinated so that no component is operated above its temperature rating. This section provides temperature limits for the termination of conductors. Terminations for circuits that are rated 100 amps or less and that use conductors from #14 through (and including) #1 are limited to 60°C. Conductors that have higher temperature ratings (such as the most common THHN conductors) can be used for these circuits, but the ampacity of such conductors must be determined by the "60°C" columns of Tables 310-16 through 310-19.

If the termination devices for the circuits mentioned above are listed for operation at higher temperatures, the conductors may also have their ampacity calculated at the higher temperatures.

Terminations for circuits that are rated over 100 amps, and that use conductors larger than No. 1, are limited to 75°C. Conductors that have higher temperature ratings (such as the most common THHN conductors) can be used for these circuits, but the ampacity of such conductors must be determined by the "75°C" columns of Tables 310-16 through 310-19.

Separately installed pressure connectors (such as a wire nut used between the termination points) must have temperature ratings equal to the temperature at which the conductor's ampacity was calculated. For example, if you are calculating the ampacity of a No. 8 conductor at 75°C, any splicing connector (such as a wire nut) that you use on those conductors must have a temperature rating of at least 75°C.

Design type B, C, D, or E motors are permitted to be terminated with conductors rated 75°C or higher, so long as the ampacity of the conductors will not heat them beyond 75°C. Remember in these situations that the supply source for the conductors must also be rated for the conductors.

II. 600 Volts, Nominal or Less

110.26: Working Space about Electric Equipment (600 Volts, Nominal or Less, to Ground)

Adequate space for safety must be maintained for easy maintenance of equipment. When equipment is located in locked rooms, it may still be considered accessible if the room is accessible to qualified personnel.

(A) Working Clearances. For working clearances, refer to Table 110.26(A)(1) in the *NEC*. Where enclosures are installed on each side of a workspace (whether or not either has live exposed parts), the amount of clear distance must be determined by Condition 3 in Table 110.26(A)(1).

In addition, the free space in front of electrical equipment must be at least 30 inches (762 mm) wide. This clear space must continue from the floor to the height specified in Section

110.26(E). Doors or panels on all electrical equipment must be capable of opening to at least a 90-degree angle. No equipment is permitted to extend more than 6 inches in front of another piece of equipment; for example, a large transformer may not be placed in front of a panelboard, even when the top of the transformer is lower than the bottom of the panel.

Condition 1: In this portion, insulated wire or bus bars are not considered live parts. If there are any exposed energized parts and parts that are grounded on the opposite side of the working space, or if there are exposed live parts on both sides of the equipment, suitable insulating materials must be installed for protection of only the live parts described above.

From this, we might conclude that a panel of this kind that will have to be worked on from time to time falls under Condition 1, and give a minimum 3 feet of clearance. This will also apply to bus bars and conductors.

Condition 2: In Condition 1, the panel was used as an example, but since the panel is usually contained in a metal enclosure, we must also look at Condition 2, which we find might be used under certain conditions.

Condition 3: Condition 1 might be an electrical closet, where panels are on two walls, in which case 3- and 4-foot conditions would prevail.

Exception

(a) If there are no renewable or replaceable parts on the back side of switchboards or motor control centers, and all parts of the unit are accessible from its front, working space is not required.

(b) The inspection authority has the right to make exceptions for smaller spaces where it seems appropriate. These judgments are applicable if the particular arrangement of the installation shows that it will provide sufficient accessibility or if no insulated parts carry more than 30 volts RMS, 42 volts peak, or 60 volts DC.

Concrete, brick, or tile walls should very definitely be considered grounds.

(c) Condition 2 working clearances are permitted between pieces of dead-front equipment that are located across an aisle from each other. However, this applies only in cases where written procedures ensure that pieces of equipment located across from one another will never be open at the same time. Also, this must be done in areas that are accessible to authorized personnel only.

(B) **Clear Spaces.** Clear spaces required around equipment can't be used for storage. If live parts are exposed, they must be guarded.

(C) **Access and Entrance to Working Space.** This portion is very important for persons working in the area discussed above. There shall be at least one entrance that is large enough to give adequate working space to the electrical equipment therein. Where switchboards and control panels are located with a rating of 1200 amperes or more and are 6 feet or more in width, it is required that one entrance be at least 24 inches in width and 6.5 feet in height at each end. Thus, in cases such as this, at least two entrances are required.

Exception

(a) This allows for a continuous unobstructed way of exit wherever switchboards or panelboards are located.

(b) Only one entrance is required if the working space around the various pieces of equipment in the room is doubled.

(D) **Illumination.** The equipment described in this article must be provided with a source of illumination.

(E) **Headroom.** The minimum ceiling height above the various pieces of equipment covered in this article is 6½ feet (1.98 m), except for residential service equipment or panelboards in existing dwellings rated 200 amps or less. (The requirements for equipment operating at higher voltages are given in Article 490.)

(F) **Dedicated Equipment Space.** Motor control centers and other equipment covered by Article 408 must be located in dedicated and protected spaces. An exception is made for control wiring that must be located adjacent to or near specific pieces of equipment.

For indoor locations, this dedicated space is required to be equal to the width and depth of the equipment from the floor up to a 6-foot level, or up to a structural ceiling if it is lower than 6 feet. (Suspended ceilings are not considered to be structural ceilings.) No piping or nonelectrical equipment may be located in this space. Sprinkler systems may be installed for these spaces so long as they are fitted with drip pans or other suitable protection.

Equipment located outdoors must be installed in enclosures that are adequate to the conditions, and must be protected from vehicles and accidental contact by unqualified persons. No other equipment is permitted in the dedicated space.

110.27: Guarding of Live Parts

This section applies to parts supplied with 600 volts or less.

(A) **Live Parts Guarded Against Accidental Contact.** This section covers the guarding or protecting of live parts of electrical equipment that are operated at 50 volts or more, so as to prevent accidental contact with them. Approved cabinets or enclosures shall be used, according to the requirements in other portions of the *Code*. The following are the means by which this shall be accomplished:

(1) Many references are made to only qualified persons having access to rooms, vaults, and so on. It is recommended that the reader refer to Article 100 and review the definition of qualified persons.

(2) So that only qualified persons may have access to live parts, suitable partitions or screens must be installed to keep away unqualified persons. Openings to live parts shall be of such a size that unqualified persons will be kept from accidentally contacting live parts. Again, qualified persons are mentioned. Their safety is thought of in making the equipment accessible without obstruction and in giving attention to the contact of conducting materials such as conduit or pipes.

(3) Balconies, galleries, or platforms must have sufficient elevations and be arranged such that unqualified persons have no access to live parts.

(4) Any live parts of equipment that are elevated a minimum of 8 feet or more above the floor or other accessible places are considered accessible to qualified persons only.

(B) **Prevent Physical Damage.** Many times electrical equipment is located in a work area where the activity around it might damage the equipment. In such a case, the enclosures or guards shall be of such strength as to prevent any damage to the electrical equipment.

(C) **Warning Signs.** Warning signs shall be posted at entrances to rooms or other guarded locations giving warning that only qualified personnel are permitted to enter. Although not specifically covered here, posting of dangers that might exist in any situation is always good safety practice.

Motors are covered in Sections 430.132 and 420.133, and parts supplied with over 600 volts are covered in Section 110.34.

110.18: Arcing Parts

Making and breaking of contacts usually causes sparking or arcing. Also, the white-hot filament of a lightbulb broken while in operation takes a little time to cool. Any parts that normally cause arcing or sparking are to be enclosed unless they are isolated or separated from combustible material. Lightbulbs are mentioned, but additional information is given in the articles covering hazardous areas, along with the specific requirements for switches, outlets, and other devices in hazardous locations.

Hazardous areas are covered in Sections 500 through 517.

110.19: Light and Power from Railway Conductors

It is not permissible to connect any circuits for light or power to any trolley wires that use a ground return signal.

The exceptions to this include car houses or any other freight station, and so on, that operate with electric railways.

110.21: Marking

All electrical equipment must be marked, showing the manufacturer's name and the electrical characteristics.

110.22: Identification of Disconnecting Means

It is essential that disconnecting means for appliances, motors, feeders, and branch circuits be properly identified as to what the disconnect serves. If overcurrent devices that have a series combination rating are used, they must be clearly marked to that effect. Such markings must be legible and durable. Panels usually have a card with the circuit numbers marked, which should be filled out in its entirety as a permanent record. This is one of the most frequent violations of the *Code*.

III. Over 600 Volts, Nominal

110.30: General

Since 1975, additions have been made at the end of various articles of the *Code* to cover over 600 volts, nominal. It is the intent that conductors and equipment used on volts higher than 600 volts, nominal, comply with this article and with all applicable articles. It is not intended that provisions of this article apply to equipment on the supply-side of the service conductors.

110.31: Enclosure for Electrical Installations

Areas where access is controlled by lock-and-key or other approved means, shall be considered accessible to qualified persons only. Examples of these areas include vault installations, room or closet installations, and areas surrounded by walls, screens, or fences.

The design and construction of enclosures shall be suitable to the nature and degree of hazard involved.

Any wall or fence less than 7 feet in height is not considered as preventing access. A 7-foot fence or wall is considered to be adequate. Fences or walls of lower height must have additional protection to the 7-foot limit. A fence made of no less than 6 feet of fence fabric and a 1-foot or greater extension, using three or more strands of barbed wire, is acceptable.

(A) **Fire Resistivity of Electrical Vaults.** Walls, roofs, floors, and doorways of vaults containing conductors operating at 600 volts or more must be fire-rated for a minimum of three hours. Equipment must also be marked with warning signs. Openings in equipment must be designed so that foreign objects inserted through such openings are deflected away from energized parts.

(B) **Indoor Installations.**

(1) **In Places Accessible to Unqualified Persons.** This section covers indoor installations to which unqualified persons might have access. The equipment shall be made with metal enclosures or a vault that is accessible only by lock and key.

Unit substations and any pull boxes or other means of connection associated with the equipment must be permanently marked with caution signs. Dry-type transformers must be ventilated so that they have openings in the equipment, but they shall be designed in such a manner that foreign objects inserted through the ventilating holes will have something to deflect them from the live parts.

(2) **In Places Accessible to Qualified Persons Only.** Section 110.34 and Article 490, Part III, are to be used for compliance when indoor electrical installations are considered accessible to qualified persons.

(C) Outdoor Installations.

(1) In Places Accessible to Unqualified Persons. Article 225 covers outdoor installations that are accessible to unqualified persons.

The *National Electrical Safety Code (ANSI) C2-2002* covers the clearance of conductors that are over 600 volts, nominal.

(2) In Places Accessible to Qualified Persons Only. Section 110.34 and Article 490, Part III, cover places of outdoor electrical installations where exposed live parts may be accessible to qualified persons. These sections deal with voltages over 600, nominal, and need not be repeated here.

(D) Enclosed Equipment Accessible to Unqualified Persons. Where equipment requires ventilation or other openings, the design of the equipment shall be such that foreign objects that might be inserted into ventilating openings will be deflected so as not to contact any live parts. Any such equipment that is in a position where it may be physically damaged from passing traffic must be protected by a suitable guard. Sometimes metal-enclosed equipment has to be located outdoors, where it might be damaged by the general public. If so, the design of such equipment shall be such that any exposed bolts, nuts, and so forth can't easily be removed by the public, and if such electrical equipment is located outdoors and is less than 8 feet from floor or ground, any doors or covers shall be hinged and capable of being locked. Manhole covers weighing more than 100 pounds need not be locked.

110.32: Work Space about Equipment

There shall be sufficient clear space around high-voltage equipment to permit ready and safe operation of such equipment.

If any energized parts are exposed, they shouldn't be less than 6½ feet measured vertically from any floor or platform, or less than 3 feet wide, measured parallel to the equipment. In all cases the width shouldn't be less than the space required for doors or hinge panels to open to a position of at least 90 degrees.

110.33: Entrance and Access to Work Space

(A) Entrance. The requirements for the entrance are to be not less than 6½ feet and not less than 2 feet in width. Adequate space must be provided for access to the working space around electrical equipment. If the switchboard or controller panels are more than 6 feet wide, entrance at each end will be required for both panelboards.

When only one entry is provided, it must be so located that the distance from switchboard to panelboard meets the minimum requirements for distance away from the equipment given in Table 110.34(A).

If bare or insulated parts of more than 600 volts, nominal, are located adjacent to such entrances, there must be suitable means taken to guard them.

(B) **Access.** When electric equipment is installed on platforms, balconies, mezzanine floors, or in attic or roof rooms or spaces, there must be permanent ladders or stairways installed for access. There is an OSHA regulation that requires ladders to extend 3 feet above the location to which they give access.

110.34: Work Space and Guarding

(A) **Working Space.** The minimum clear working space in front of electric equipment such as switchboards, control panels, switches, circuit breakers, transformers, motor controllers, relays, and similar equipment shouldn't be less than specified in Table 110.34(A) in the *NEC*, unless otherwise specified in the *Code*. Distances shall be measured from the live parts if such are exposed, or from the enclosure front or opening if such are enclosed.

(1) Insulated wire or insulated bus bars that are not supplied with over 300 volts shouldn't be considered live parts. If live parts are exposed on any one side and the parts on the other side are grounded in the working space, or if suitable guards made of wood or other insulating materials are in place, then the live parts shall be considered suitably protected.

(2) This section describes condition number two of Table 110.34(A). Note that masonry surfaces must be considered grounded surfaces.

(3) Describes condition number three of Table 110.34(A).

Exception

The deenergized parts are to be worked from the back on enclosed equipment. The required workspace is 30 inches nominal. If dead-front switchboards or control assemblies are in use, there are no fuses or breakers or adjustable parts on the back, and all connections are accessible from places other than the back, then the above 30-inch requirement will apply.

(B) Separation from Low-Voltage Equipment. When there is any low-voltage equipment in a vault, room or enclosure, such as switches, cutouts, or other equipment that operates at 600 volts, nominal or less, all exposed live parts or exposed wiring that operate at more than 600 volts, nominal, must be separated effectively from the low-voltage equipment and wiring by suitable partitions, screens, or fences.

Many utility companies will not permit low voltage in transformer vaults with high voltage, with the exception of low-voltage buses. This does not include lighting or other low voltage that might be required in the operation of the high-voltage equipment.

Exception

When 600 volts or less for switches or other equipment services only equipment within the high-voltage room, vault, or enclosure, such equipment in use in conjunction therewith at a voltage of 600 or less, nominal may be installed in the room that is accessible to qualified persons only.

(C) Locked Rooms or Enclosures. When there are live parts or exposed conductors that operate at over 600 volts, nominal, the entrances to any such building must be locked. There is an exception: Such locked entrances must be under the observation of qualified persons at all times. Permanent and conspicuous caution signs are to be installed where the voltage exceeds 600 volts, nominal, and must include the message "WARNING—HIGH VOLTAGE—KEEP OUT."

(D) Illumination. Adequate illumination must be provided to illuminate the high-voltage area properly for safe working, and the fixtures must be installed so that there will be no danger to anyone changing bulbs or working on the illumination system. All lighting outlets in these areas must be arranged so that no one making repairs or changing lamps will be exposed to live parts.

(E) Elevation of Unguarded Live Parts. See Table 110.34(E) for elevation of unguarded live parts.

(F) Protection of Service Equipment, Metal-Enclosed Power Switchgear, and Industrial Control Assemblies. Any pipes or ducts that are not related to the electrical installation and that require periodic maintenance, or whose malfunction would endanger the operation of the electrical system, must not be located in the vicinity of service equipment, metal-enclosed power switchgear, or industrial control assemblies. Protection must be provided where necessary to prevent damage from condensation, leaks, or breaks in foreign systems. If the pipes are installed for the protection of electrical equipment, they shouldn't be considered foreign objects.

110.36: Circuit Conductors

Circuit conductors may be installed in raceways, cable trays, as metal-clad cable, bare-wire cable, and buses, or as Type MV cables or conductors, as provided in Sections 300.39, 300.40, and 300.50. When bare live conductors are installed, they must conform to Section 490.24.

In the installation of conductors that carry high voltage, the sizing of bare conductors must be done with consideration for corona effects.

Insulators, their mountings, and conductor attachments, when used as supports for single conductors and bus bars, must be capable of safely withstanding the magnetic forces that will result between two or more conductors in the event of a fault current being imposed on them. The magnetic forces tend to push the conductors apart. The magnitude of the force depends upon the amount of short-circuit current involved, the spacing, and so on.

If open runs of lead-sheathed cables are used, they must be protected from physical damage and electrolysis of the sheath.

110.40: Temperature Limitations at Terminations

Conductors operating at over 600 volts are permitted to be terminated based upon a 90°C temperature rating and ampacity. See Sections 310.67 through 310.86 for the ratings of specific conductors.

IV. Tunnel Installations Over 600 Volts, Nominal

110.51: General

(A) Covered. This part applies to installation and use of high-voltage distribution and utilization power equipment that is portable and mobile, such as substations, cars, trailers, mobile shovels, draglines, hoists, drills, dredges, compressors, pumps, conveyors, underground excavators, and so on.

(B) **Other Articles.** The requirements of this part shall be in addition to other articles, such as Articles 100 through 710 of the *Code*. Grounding requires special attention; see Article 250.

(C) **Protection Against Physical Damage.** All conductors and cables used in tunnels must be located so that they are above the tunnel floor, and they must be thoroughly located and guarded so that they will not be subject to physical damage.

110.52: Overcurrent Protection

Article 430 covers the protection of motor-operated equipment from overcurrent. Transformers must be protected as covered in Article 450.

110.53: Conductors

Conductors installed in tunnels for high voltage may be installed (1) as in metal conduit or other metal raceways; (2) in Type MC cable; or (3) if approved, in other types of cable. Cables that are multiconductor, if approved for the service, may be a portable type supplying mobile equipment.

110.54: Bonding and Equipment-Grounding Conductors

Great care must be exercised when bonding or equipment-grounding, for personnel safety.

(A) **Grounding and Bonding.** Nonenergized metal parts of the electrical equipment, and metal raceways, and the cable sheath are required to be effectively grounded and bonded to all metal rails and pipes, not only at the portal, but at intervals not exceeding 1000 feet throughout the tunnel.

This is very important, as the best grounds will no doubt be encountered outside the tunnel, and thus extra paths are provided to this point.

(B) **Equipment-Grounding Conductor.** Inside the metal raceway, or inside the multiconductor cable jacket, an equipment-grounding conductor must be run with the circuit conductors. This equipment-grounding conductor can be insulated or bare.

110.55: Transformers, Switches, and Electrical Equipment

It is required that all transformers, switches, motor controllers, motors, rectifiers, and other equipment installed below ground be protected from physical damage by location or guarding.

110.56: Energized Parts

Any bare terminals or electrical equipment, including transformers, switches, motor controllers, and so on are to be enclosed to protect persons from accidental contact.

110.57: Ventilation Control Systems

When ventilation equipment is used, it shall be installed such that the ventilation may be reversed because during the summer, air will probably be going out of the tunnel, and in the winter, going into the tunnel.

110.58: Disconnecting Means

A switching device that conforms to the requirements of Articles 430 and 450 must be installed at each transformer or motor location for the purpose of disconnecting the transformer or motor. This switching device shall open all ungrounded conductors at the same time. A disconnect switch for a transformer must be rated no less than the ampacity of the transformer supply conductors.

110.59: Enclosures

Any enclosure used in a tunnel shall be drip-proof or submersible, as may be required by environmental conditions. Neither switch nor contactor enclosures are permitted to be used as junction boxes or raceways for conductors feeding through or tapping off at those locations to other switches, unless specially designed equipment is provided that is adequate for the purpose.

Chapter 2

Wiring and Protection

Article 200—Use and Identification of Grounded Conductors

200.1: Scope

Grounding is very important, and the conductors that are grounded intentionally must have identification to indicate for what grounding purpose they are used. This section also dictates that terminals to which grounded conductors are attached shall be identified. There are a number of definitions pertaining to grounding in Article 100. These seem to confuse some wirers. This will be covered further in Article 250.

200.2: General

All premise wiring shall have a grounded conductor, and this conductor shall be identified by white or gray insulation, or with insulation (not green) with three continuous white stripes, as covered in Section 200.6.

There are exceptions to all systems having a grounded conductor, and these exceptions are covered in Sections 210.10, 215.7, 250.21, 250.22, 250.162, 503.13, 517.63, 668.11, 668.21, and 690.41, Exception.

For identification, the conductors that are purposely grounded when they are insulated must have insulation of a different color from that of ungrounded conductors when the circuit voltage is 1000 volts or less. Circuits not rated less than 600 volts, up to 1000 volts and over, will be described in Section 250.184(A).

Be sure to read Section 200.6, which covers neutral conductors that are insulated. Section 200.6(B) takes into consideration that in sizes larger than No. 6, it is impossible to find white or gray insulation and makes provisions for this.

200.3: Connection to Grounded System

If there is a grounded conductor in the interior wiring system, it shouldn't be electrically connected to a supply system that does not have a corresponding grounded conductor. This condition could be dangerous and could cause many varied difficulties that most certainly would not be considered as safe wiring procedures. Electrically connected means that the connections are capable of carrying current. Conductors that are connected via electromagnetic induction, such as the windings on an isolating transformer, would not be considered electrically connected. Direct connections and electromagnetic induction must be separated in one's mind. Later, the text will deal with isolation transformers and give further clarification of how and what parts of the wiring system must be grounded.

200.6: Means of Identifying Grounded Conductors

Insulated conductors of No. 6 or smaller wire, when used as identified grounded conductors, shall have a white- or gray-colored insulation. Conductors of a color other than white, gray, or green that have three continuous white stripes along their entire length also meet the requirements of Section 200.6. If the insulated conductors are larger than No. 6, they shall be identified by a white- or gray-colored insulation, or by a distinctive marking (white) at the terminals while they are being installed. Type MI cable has bare conductors, so identification of the grounded conductors shall be marked generally by sleeving during the installation; the sleeving shall be white or gray, as indicated in Figure 200-1.

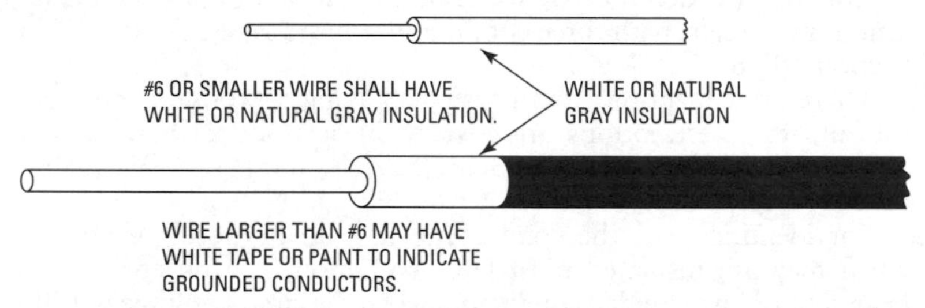

Figure 200-1 Method of identifying grounded conductors.

The insulated conductor that is intended to be used in flexible cords for the grounded conductor shall be white or gray insulation. Section 400.22 describes some other means of identification that are permitted.

Where maintenance supervision will be done by only qualified persons, grounded conductors in multiconductor cables may, at time of installation, be identified with a distinctive white marking, or any other equally effective means.

Section 690.31 allows for properly rated single conductors in photovoltaic systems to be identified by white markings only.

Grounded Conductors of Different Systems. It is sometimes necessary to run different systems with the grounded conductor for each system in the same raceway or other type of enclosure. Therefore, it becomes necessary for each grounded conductor for each of the systems involved to have its own identification. One grounded conductor may be white and others may be white with color tracers, provided the color tracer is not green.

200.7: Use of White or Gray Color or with Three Continuous White Stripes
White or natural gray is to be used only for the identification of grounded conductors. There are a few exceptions made necessary by the use of cords and cables, such as types AC, NM, and UF. These cables must always carry a white or gray conductor. If a cable with a white or gray conductor is used on a multiwire circuit in a circuit connected to the two *hot* legs of a three-wire circuit (as in Figure 200-2), or if three-wire cable is used on a three-phase system, the white or natural gray conductor shall be reidentified at its terminals (see Figure 200-3).

When two or three conductors in a cable assembly are used for a single-pole or three-way switch and a white conductor in the assembly is used as an ungrounded (hot or switch-leg) conductor, the white conductor must be reidentified by taping, painting, or other effective means wherever it is accessible.

When a flexible cord is permitted for connecting an appliance and the cord used has a white or natural gray conductor or other means permitted in Section 400.22, this conductor may be used even though it is not connected to a receptacle that has a grounded circuit.

If a circuit is less than 50 volts, a white or gray conductor (or one with three continuous white stripes) must be grounded if required by Section 250.20(A).

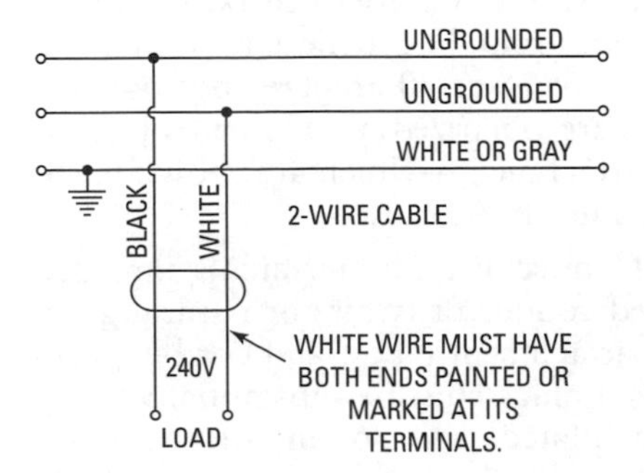

Figure 200-2 A two-wire cable connected to the two hot wires of a three-wire circuit.

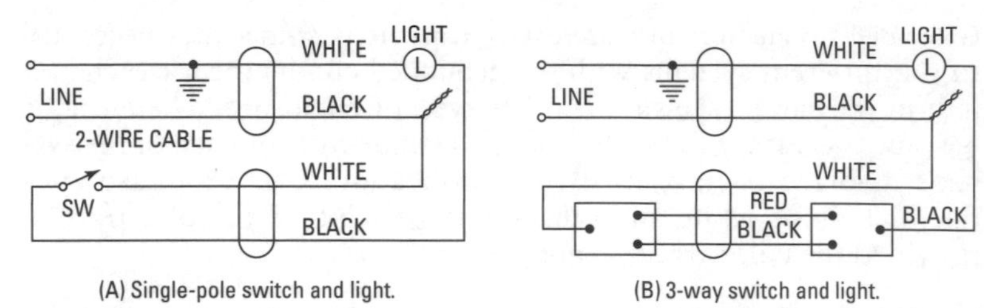

(A) Single-pole switch and light. (B) 3-way switch and light.

Figure 200-3 Methods of wiring common light switches with cable assemblies. Note that white wires must be reidentified when used as ungrounded conductors.

200.9: Means of Identification of Terminals

Terminals used by grounded conductors shall be substantially white in color. Other terminals not used for grounded conductors shall be colors other than white.

Where maintenance is done only by qualified persons, grounded conductors may be permanently marked at the time of installation by effective means; for instance, grounded conductors shall be marked with white or natural gray, and equipment-grounding conductors shall be bare or marked with green.

200.10: Identification of Terminals

(A) **Device Terminals.** All devices and equipment that have terminals with which to attach conductors to more than one conductor of a circuit shall be permanently marked so as to distinguish between the ungrounded conductor, the grounded conductor, and equipment-grounding conductors.

This would not apply to a single-pole toggle switch that opens or closes one conductor of a circuit. Terminals of lighting appliance branch-circuit panelboards are also an exception, because it is obvious where hot conductors, grounded conductors, and equipment-grounding conductors shall be attached. On devices that have over 30 amperes normal current capacity, unless they are polarized attachment plugs or polarized receptacles for attachment—which are shown in (B) below—the terminals need not be identified.

(B) **Receptacles, Plugs, and Connectors.** The terminals intended for attaching the grounded conductor (white or natural gray) on receptacles, polarized attachment plugs, and cord connectors for plugs and polarized plugs must be substantially white in color, such as chrome-plated, and so on, or the word "white" or the letter "w" shall be located by this terminal.

Should the terminal not be visible, the hole where the conductor is inserted must be either white in color or engraved with the word "white" or the letter "w". This indicates that the terminal is for use with either a white-colored or natural gray-colored conductor.

The terminal for attaching equipment-grounding conductors must be a green, hexagonal nonremovable screw or terminal nut, or a green pressure connector. If, instead of attaching the conductors under screws, the conductor is pushed into a hole with pressure connection, the entrance for the equipment-grounding conductor must be a distinctive green color.

(C) Screw Shells. For devices such as lamp sockets, the shells or thread part of the socket must be connected only to the grounded conductor. An example of this may be seen in Figure 200-4. This is true of all devices that have screw–shells, with the exception of screw-shell (edison base) fuseholders.

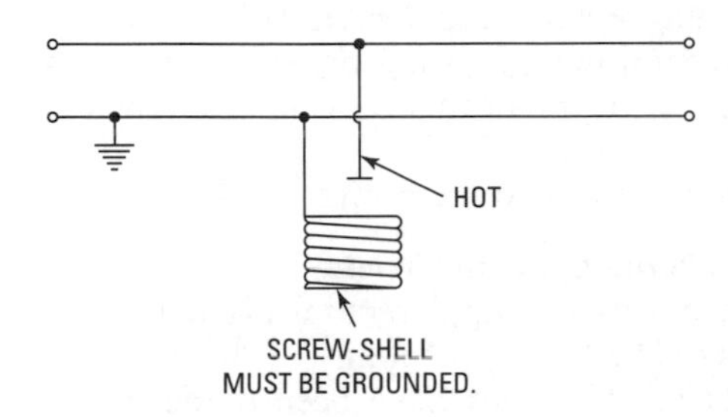

Figure 200-4 Illustrating the grounding of a screw-shellbase.

(D) Screw-Shell Devices with Leads. This part covers items commonly known as pigtail sockets; the lead that connects to the shell must be white or gray. The same will be true of lighting fixtures that have the leads attached.

(E) Appliances. Appliances to be connected (1) by permanent wiring methods or (2) by field-installed plugs and cords with three or more wires shall have markings to identify the terminal for the grounded circuit conductor. Appliances that have a single-pole switch or a single-pole overcurrent device in the line or any screw-shell lampholder shall be connected according to (1) or (2) of this paragraph, and an equipment-grounding conductor shall be included.

The last point should be called to your attention in more detail. An electric range or electric dryer circuit that is supplied from the service-equipment panel may have the grounded conductor (neutral) connected to the range or dryer frame as an equipment-grounding conductor also. However, if these circuits are from a feeder panel (as would be the case in a mobile home), the grounded conductor shouldn't be used as the equipment-grounding conductor and it shall be insulated, and a fourth conductor (either bare or of green color) used as the equipment-grounding conductor. Refer to Section 250.140, which also refers to Sections 250.134 and 250.138.

200.11: Polarity of Connections

Grounded conductors may not be connected in a way that reverses the designed polarity of a circuit or device.

Article 210—Branch Circuits

A. General Provisions

210.1: Scope

Branch circuits are defined and explained in Article 100. This article applies to branch circuits supplying lighting or appliance loads or combinations of such loads. Motor branch circuits are covered under Article 430.

See Section 668.3(C) for exceptions to the above.

210.2: Other Articles for Specific-Purpose Branch Circuits

There are a number of exceptions or supplemental provisions to this article on branch circuits. The listings of many of these exceptions are shown in Table 210.2. Changes have been made in the *NEC* amending or supplementing the provisions therein. They are not quoted here because the article and section that apply are in the *Code*. Refer to the *NEC* for a listing of additional systems and the sections in which they are to be found.

210.3: Classifications

In general, branch circuits will be classified by the maximum permitted ampere rating or setline of the overcurrent device. Otherwise they will be classified as 15, 20, 30, 40, and 50 amperes. If conductors of higher ampacity are used, the ampere rating or setting of the specified overcurrent device determines the classification of the branch circuit.

As an illustration, if there is a 15-ampere protective device in a circuit wired with No. 12 conductors, this will be a 15-ampere branch circuit; you can't make it a 20-ampere branch circuit merely because it uses No. 12 conductors. In all probability, the No. 12 wire was installed to handle a voltage drop that was too great, or, if installed in a raceway, it is possible that derating was required because of the fill. Derating is thoroughly covered in Article 310.

Exception

In industrial premises that are maintained and supervised by qualified persons servicing the equipment outlet circuits, greater than 50 amperes are permitted. Note that this exception applies only to equipment circuits, not to lighting circuits. See Section 210.23(D).

210.4: Multiwire Branch Circuits

The definition of a multiwire branch circuit is covered in Article 100. This is a critical term to understand because the *Code* so frequently refers to "a single circuit," or to similar terms. Also, it is important to note that a multiwire branch circuit may also be referred to as multiple circuits. As an example, the "two" small-appliance branch circuits required by Section 210.11(C)(1) are also clearly a multiwire branch circuit. So, keep in mind that there is some overlap in circuit definitions.

So that multiwire circuits will be thoroughly understood and the terminology used properly, diagrams are included in this section to explain them more fully. On a delta connected, four wire system, only two ungrounded conductors and the neutral will be considered the multiwire branch circuit. Figure 210-1 shows a multiwire branch circuit from a four-wire delta system; note that the ungrounded conductors extend from phases A and B. Phases A, B, and the neutral satisfy the condition of a multiwire branch circuit since there are 120 volts from A to the neutral, 120 volts from B to the neutral, and 240 volts from A to B. There is an equal potential

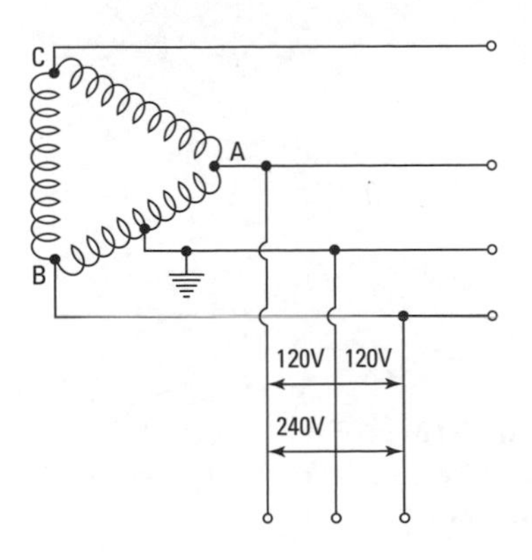

Figure 210-1 One type of multiwire circuit from a four-wire delta system.

difference between each phase wire and the neutral, and a difference of potential between the two phase wires.

A multiwire circuit shall be considered multiple circuits, and it is required that all conductors of such a circuit originate from the same panel.

Figure 210-2 does not satisfy the multiwire branch circuit conditions; both ungrounded phases are connected to the same phase, and so there is no difference of potential between the phase wires. Figure 210-3 does not satisfy the definition of a multiwire branch circuit; there is a voltage between the phase wires, but the same potential difference does not exist between each phase wire and the neutral. On the wye system, shown in Figure 210-4, the three phase conductors and the neutral satisfy the requirements for a multiwire branch circuit. Any two phase wires and the neutral satisfy the conditions; therefore, this is a multiwire branch circuit.

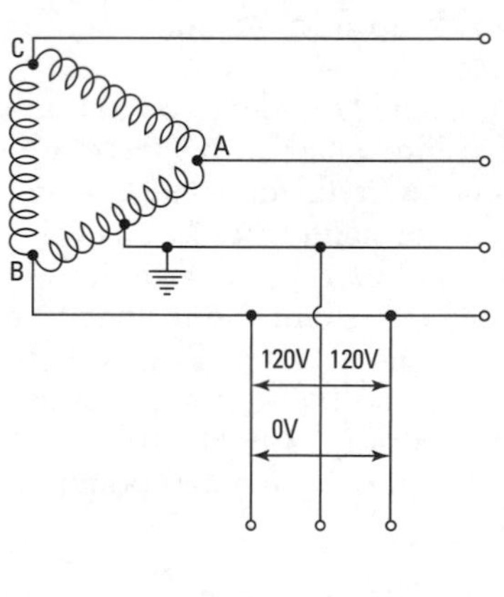

Figure 210-2 This is not a multiwire circuit from a four-wire delta system.

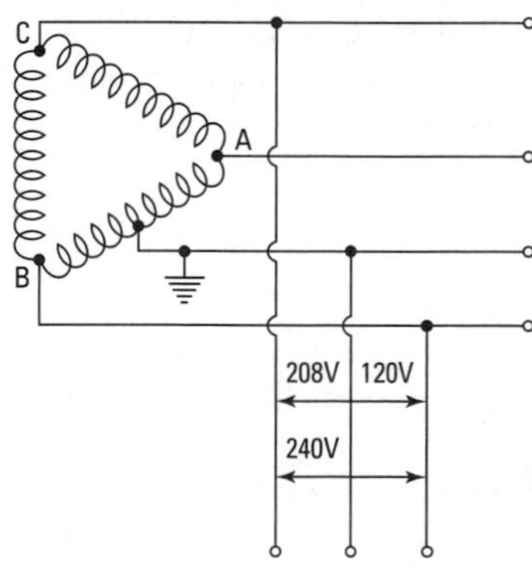

Figure 210-3 This is not a multiwire circuit from a four-wire delta system.

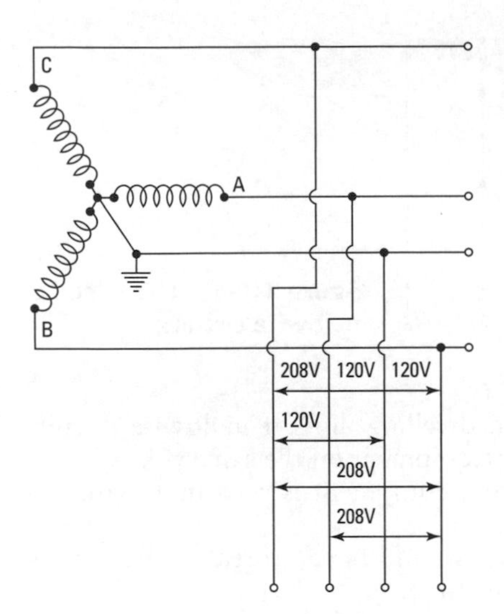

Figure 210-4 A multiwire circuit from a four-wire wye system.

Figure 210-5, which is a three-wire, 120/240-volt single-phase circuit, is also a multiwire circuit. Figure 210-6 does not satisfy the requirements of a multiwire branch circuit because there is no potential difference (zero voltage) between the two ungrounded phase conductors. Multiwire branch circuits are often misinterpreted. When the conditions of multiwire branch circuits are misapplied, the neutral may be forced to carry a heavy current, and this heavy current will result in heating and damage to the conductor insulation. No illustration is provided for a 277/480-volt wye system, but the same would apply as with the 120/208-volt wye system.

A device with a means of disconnecting simultaneously all ungrounded conductors at the panelboard where the multiwire branch

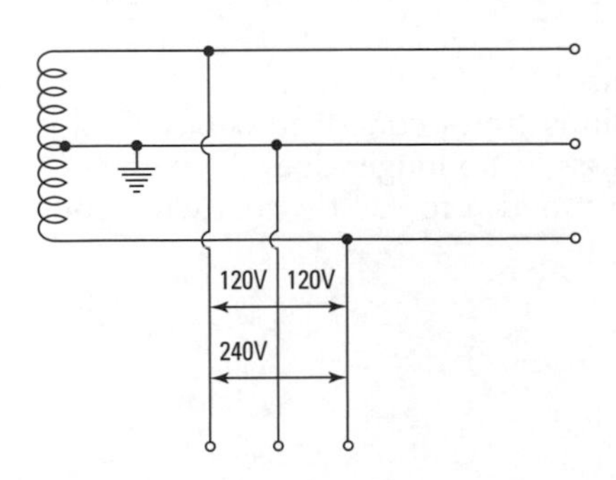

Figure 210-5 Identification for branch circuits.

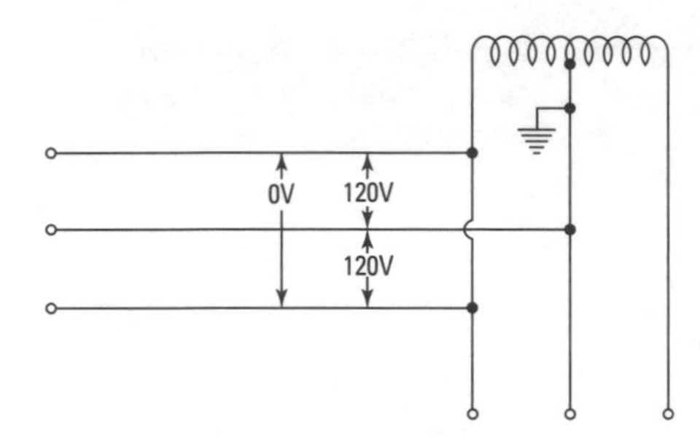

Figure 210-6 This is not a multiwire circuit.

circuit originates is required in dwelling units if multiwire circuits supply more than one device or equipment on the same yoke.

If only one piece of utilization equipment is on a multiwire circuit, the above does not apply.

All multiwire branch circuits should have a breaker that opens both hot conductors at the same time.

The above exceptions are for the reason that if one overcurrent device wire opened and a multiwire circuit supplied a circuit with only the hot conductors being used, a very severe safety hazard could result.

The continuity of the grounded conductor on multiwire circuits must be maintained. This will be covered later in Section 300.13(B).

When two or more systems having different voltages exist in the same building, every current-carrying conductor must be marked, showing its phase and system. This may be done by tags, tape, spray paint, or by some other acceptable means. The means of identification must be posted at any panelboard where such circuits originate.

Designs of three-phase, four-wire wye systems should take into account the possibility of high neutral currents due to third harmonics.

210.5: Color Code for Branch Circuits

Although the *NEC* has at various times contained a basic color code for branch circuit conductors, it no longer does. Nonetheless, it is still considered good trade practice to use the following code for most installations:

120/240/208-volt systems:

A phase	Black
B phase	Red
C phase	Blue

Neutral	White
Ground	Green

277/480-volt systems:

A phase	Brown
B phase	Orange
C phase	Yellow
Neutral	Gray
Ground	Green with yellow stripes

(A) Grounded Conductor. Grounded circuit conductors must be clearly identified. The requirements, formerly located in this section, have been moved to Section 200.6. Where there is a raceway, box, or gutter on any other type enclosure with conductors from more than one system and neutrals or grounded conductors involved, one system shall use a white or gray color or its neutral—a second system shall use the other color and more systems with neutrals will use a white conductor with a colored stripe for neutral. Thus, each system neutral may be easily identified. This colored stripe shouldn't be green since green always indicates equipment-grounding conductors. It is up to the authority that has jurisdiction to accept or reject the other means of identification (see Figure 210-7).

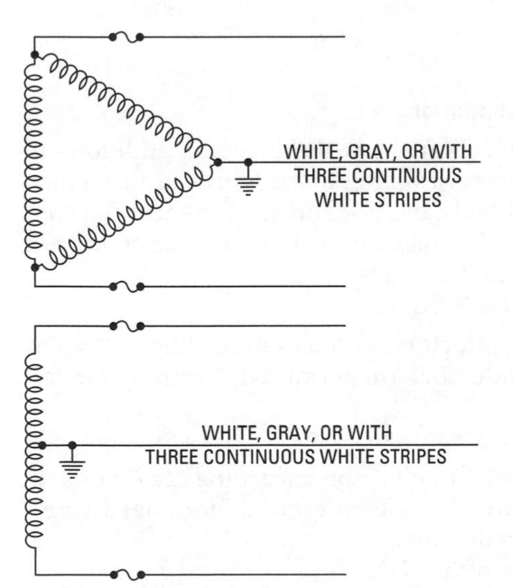

Figure 210-7 Showing a grounded conductor in a branch circuit.

Exception No. 1
Since mineral-insulated cable consists of bare conductors surrounded by insulating powder and the entirety is encased in a metal sheath, it becomes necessary at terminations to identify the neutral conductor. This is generally done by slipping the wide insulating sheath over one bare conductor.

Exception No. 2
More color-coded insulated neutrals will be covered in Section 200.6(A) and Section 200.6(B).

(B) Equipment-Grounding Conductor. Equipment-grounding conductors must be clearly identified. The requirements, formerly in this section, have been moved to Section 250.119.

2005

(C) Ungrounded Conductors. When there is more than one nominal voltage system (such as both 120/240 volt and 277/480 volt systems) in one premises, every ungrounded conductor must be permanently identified. Color-coding, tape, or tagging may be used. The circuit identifications must also be permanently posted in every branch circuit and distribution panel.

210.6: Branch Circuit Voltage Limitations

(A) Occupancy Limitation. There have been major additions to this section. It includes all dwelling units, which in turn could include guest rooms, hotels, motels, and so on, where the voltage shouldn't exceed 120 volts, nominal. When the conductor supplies terminals of the following:
See (1) and (2) in the *NEC*.

(B) 120 Volts Between Conductors. Voltages exceeding 120 volts, nominal, between conductors are permitted to supply the following:
See (1), (2), and (3) in the *NEC*.

(C) 277 Volts to Ground. Circuits not exceeding 277 volts to ground, but lower than 120 volts to ground, nominal are permitted to supply the following:
See (1) through (6) in the *NEC*.

(D) **600 Volts Between Conductors.** Circuits not exceeding 600 volts but more than 277 volts, nominal between conductors are permitted to supply the following:

(1) The alternative equipment supplies electric discharge fixtures if they are mounted according to one of the following: (a) on poles or similar structures used to light outdoor areas that include streets, roads, bridges, highways, athletic fields, or parking lots, provided they are mounted a minimum of 22 feet in height, or (b) a minimum height of 18 feet on structures such as tunnels. Permanently connected utilization of cord-and-plug equipment.

(2) Cord- and–plug-connected or permanently connected utilization equipment.

See *NEC* Section 410.78 for auxiliary equipment limitations.

Exception No. 1

Section 422.15(C) covers lampholders for infrared or industrial heating appliances. This exception applies to (B), (C), and (D) above.

Exception No. 2

This applies to railroad properties as described in Section 110.19 and covers (B), (C), and (D) above.

(E) **Circuits Over 600 Volts Between Conductors.** Circuits operating at more than 600 volts are permitted, but only where qualified persons service the installation.

210.7: Branch Circuit Receptacle Requirements

(A) The rules formerly contained in this section were moved for 2002 to Part III of Article 210, beginning with Section 210.50.

(B) Specific requirements for receptacles themselves (the wiring devices) are found in Article 406.

(C) If more than one circuit supplies receptacles (or other devices) on a single yoke (e.g., a duplex receptacle where both halves are wired to different circuits), a method of disconnecting the circuits simultaneously must be provided.

210.8: Ground-Fault Protection for Personnel

(A) **Dwelling Units.** Ground-fault circuit protection may be used in any location, circuits, or occupancies and will provide additional protection from line-to-ground shock hazards.

(1) For personnel protection, ground-fault circuit-interrupter protection is a requirement for all 125-volt, single-phase, 15- and 20-ampere receptacle outlets installed in bathrooms.

(2) Ground-fault circuit interrupter (GFCI) protection for personnel is required on all 125-volt, single-phase, 15- or 20-ampere receptacles installed in garages. Garages usually have cement or dirt floors, which are always considered ground potential.

If a receptacle is not readily accessible, a GFCI is not required. Exceptions to *NEC* Section 210.8(A)(5) are not to be considered as meeting the requirement of Section 210.52(G). In a garage where there are receptacles without GFCI protection for specific purposes, electric drills or other tools could be easily plugged into one of the receptacles. Since the floors of garages are usually concrete or dirt and the walls brick, these locations could be hazardous unless a GFCI is installed in those outlets.

(3) The *Code* requires that all 125-volt, single-phase, 15- and 20-ampere receptacles that are installed outdoors have GFCI protection.

Grade-level access means located not more than 8 feet above ground level of dwelling units and readily accessible.

(4) All 125-volt, single-phase receptacles installed in basements or crawl spaces are required to be GFCI receptacles, except if they serve only laundry circuits, sump-pump circuits, or specific-use circuits (such as to a freezer). See *NEC* for two exceptions.

(5) Any kitchen receptacle serving countertop areas shall have a GFCI for protection of the people.

Note

Receptacles for refrigerators and freezers are exempt from GFCI protection for people. However, if these are within the 6-foot limit from the sink, a shock hazard might still be present.

(6) All 125-volt, single-phase, 15- or 20-amp countertop receptacles within 6 feet (1.83 m) of a sink must have GFCI protection.

(7) All 125-volt, single-phase, 15- or 20-amp countertop receptacles within 6 feet (1.83 m) of a sink in or on a boathouse, dock, or seawall must have GFCI protection.

(B) Other Than Dwelling Units.

(1) In nondwelling occupancies, such as industrial or commercial places of business, all 125-volt, 15- or 20-amp receptacles that are installed in bathrooms must have GFCI protection. This includes such receptacles installed in hotel rooms and similar locations.

Bathroom—A *bathroom* is an area including a basin and at least one of the following: a tub, a toilet, or a shower.

It would appear to the author that a ground-fault circuit interrupter would be the least expensive and the most positive protection for the overall picture.

(2) All 125-volt, 15- or 20-amp receptacles installed on roofs must have ground-fault protection for personnel. Note that this does not apply to receptacles installed on the roofs of dwelling units and that the *NEC* does make an exception for dedicated receptacles serving snow-melting equipment.

2005

(C) Boat Hoists. Outlets in dwellings that supply boat hoists must be GFCI-protected if they are connected to 125-volt, 15- or 20-ampere circuits.

210.9: Circuits Derived from Autotransformers

An autotransformer is a transformer with only one winding and tap to give different voltages. That is, an autotransformer is a transformer whose windings are common to both primary and secondary. An autotransformer may not be used to supply branch circuits, unless the circuit being supplied by the autotransformer has a grounded conductor that is connected to a grounded conductor of the circuit that supplies the autotransformer. Improperly-connected autotransformers can result in a wiring system being exposed to voltages that it was not designed for. This would be a potentially dangerous situation.

Figure 210-8 shows three autotransformer connections. Figure 210-8A and B are both approved connections; Figure 210-8C is not approved since there is no common ground between the supply and the output of the autotransformer. Figure 210-9A represents an autotransformer dimmer circuit and is approved by the *NEC*.

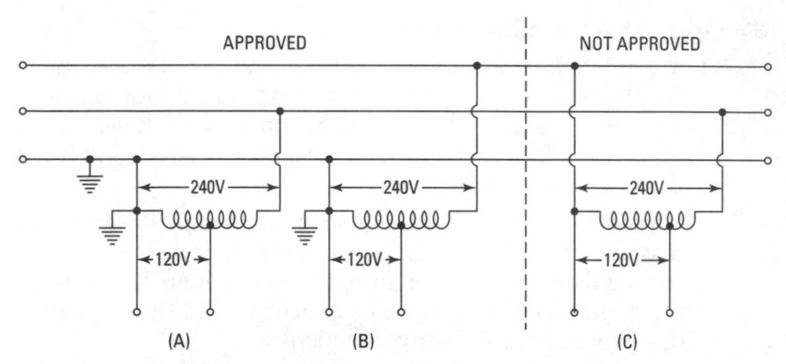

Figure 210-8 Autotransformer connections.

The above does not apply if the autotransformer is supplied from a system where the ground is connected to the ground supplying the autotransformer.

Autotransformers may be used to extend or add on to a branch circuit in an installation for which equipment is already installed and does not have a connection to a grounded conductor. An example would be changing a 208-volt, nominal supply to a 240-volt, nominal supply, or the reverse.

See Figure 210-9(B). An exception permits autotransformers to be used as booster transformers from 208 to 240 volts, or reducers from 240 to 208 volts. The neutral or grounded conductor that is ordinarily required when autotransformers are used can be omitted in these cases.

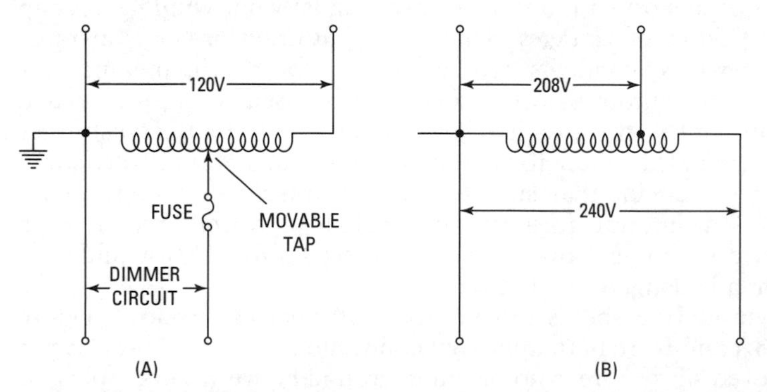

Figure 210-9 An autotransformer dimmer circuit (A) and a voltage booster (B).

210.10: Ungrounded Conductors Tapped from Grounded System

This section permits two or more ungrounded conductors to be tapped from ungrounded conductors of circuits that have a grounded neutral. One must be careful that these taps are not being made from a multiwire circuit because of feedbacks. It should be recalled that under multiwire circuits, this was prohibited.

Two-wire DC circuits and AC circuits of two or more ungrounded conductors may be thus tapped. Switching devices in these tapped circuits shall have a switching pole in each ungrounded conductor and all poles of multipole switching devices shall be manually switched together where such switching devices serve as disconnecting means as covered in Sections 410.48 (for two-pole switched lampholders), 410.54(B) (electric-discharge lamp switches), 422.31(B) (appliances), 424.20 (electric space heaters), 426.51 (snow-melting equipment), 430.85 (motor controllers), or 430.103 (motors).

210.11: Branch Circuits Required

Section 220.3 covers branch circuits for lighting and for appliances, including motor-operated appliances. Other branch circuits that are not covered in Section 220.3 are required elsewhere in the *Code*. Small appliance loads are covered in (B) below; laundry loads are covered in (C) below.

(A) Number of Branch Circuits. The total computed load and the size or rating of the branch circuits used determines the number of branch circuits required. The maximum load on any branch circuit shouldn't exceed the maximum specified in Section 210.22 and it is good practice to leave some spare capacity, as in most cases additional load will be needed at some time.

(B) Load Evenly Proportioned Among Branch Circuits. When computing on a volt-amperes-per-square-foot (0.093 sq m) basis, the load should be distributed as evenly as possible between the required circuits and according to their capacities.

Examples of computations are found in Chapter 9 of the *NEC*.

(C) Dwelling Units.

(1) Small Appliance Branch Circuits. Besides the branch circuits required by Section 220.4(A), two or more 20-amp small-appliance circuits are required to accommodate the connection of portable small appliance loads.

(2) Laundry Branch Circuits. In addition to the number of branch circuits required, including the small-appliance circuits, there shall be at least one 20-ampere branch circuit to supply laundry receptacle outlet(s), as required by Section 210.52. There shall be nothing else on this circuit.

(3) Bathroom Branch Circuits. Section 210.52(D) covers outlets in bathrooms. There must be at least one ground-fault receptacle installed near any wash-basin.

GFCIs can be obtained as receptacles or as breakers covering an entire circuit.

Receptacles in bathrooms must be supplied by 20-amp circuits and must be dedicated to those outlets only or, by exception, to an exhaust fan or similar equipment in that same bathroom.

Note that the circuit can serve one bathroom only.

As in kitchen counters, receptacles in bathroom countertops may not be installed face-up. If they were (in either area), water or other liquids could very easily pour into the receptacle and outlet box, creating a dangerous situation.

210.12: Arc-Fault Circuit Interrupter Protection

A number of manufacturers have designed arc-fault circuit interrupters (AFCIs). These have been required by the *NEC* since the 2002 edition, but there is still an ongoing argument as to their importance and effectiveness.

(A) Definition. *Arc-fault current interrupters* (AFCIs) are designed to interrupt branch circuits when they sense an arc current. (They do their sensing by means of a pre-programmed microchip.) The intent is to prevent the ignition of fires due to arcing.

2005

(B) Dwelling Unit Bedrooms. All branch circuits serving 125-volt, 15- or 20-amp receptacles in the bedrooms of dwelling units must be protected with a combination- type AFCIs. However, the Branch/Feeder type of AFCI will be permitted until January 1, 2008, in lieu of the combination device requirement.

Remember that the AFCI protection does not replace other types of circuit protection. It is not a GFCI and does not provide similar protection. AFCIs only interrupt circuits that arc. It is also important to note that AFCIs should not be used to protect circuits that feed devices such as smoke detectors. It is better that the smoke detector should be energized at all times. The principle here is the same as for not requiring overloads at fire pumps.

AFCIs must be located at the source of the branch circuit, unless it is installed within 6 feet (1.8 m) from the circuit's overcurrent device, or where the conductors between the AFCI and the overcurrent device are contained in a metal conduit or a metal-sheathed cable.

2005

210.18. Guest Rooms and Guest Suites
Guest rooms and suites that have permanent provisions for cooking must be wired with branch circuits and outlets that meet the requirements for dwelling units.

B. Branch-Circuit Ratings

210.19: Conductors—Minimum Ampacity and Size

(A) **General.** The branch-circuit rating is governed by the breaker size. Conductors shouldn't be less than the rating of the overcurrent protection, and not less than the maximum load to be served. This also applies to conductors serving branches with more than one receptacle that are to be used for cord-and-plug loads. These shall have an ampacity that is at least as great as the branch-circuit protection. Cable assemblies, if the neutral conductor is smaller than the ungrounded conductors, shall be legibly marked.

The ampacity ratings of conductors are covered in Tables 310.16 through 310.86. Motorbranch-circuit conductor sizing is covered in Part II of Article 430. Chapter 3 covers temperature limitations. The conductors for branch circuits shall be sized so that the voltage drop to the farthest outlet used for power, heating, lighting, or combinations of any of these, does not exceed 3 percent. The maximum allowable voltage drop for combinations of feeder and branch circuits shouldn't exceed 5 percent. Special noting of this is necessary since it is the first time that branch circuits have been taken into consideration when calculating voltage drop. Voltage drop on feeder conductors is covered in Section 215.2.

The required calculated loads may not be less than 100 percent of the noncontinuous load plus 125 percent of the continuous load.

A continuous load is a load where the maximum current is expected to continue for three hours or more. Branch-circuit conductors that supply continuous loads must be rated for 100 percent of the noncontinuous load and 125 percent of any continuous load.

(B) Multioutlet Branch Circuits. When branch-circuit conductors supply more than one receptacle-serving portable device via cord and plug, the ampacity of those conductors must be at least equal to the ampacity rating of the circuit. In other words, you may not string together outlets with No. 14 wire and connect them to a 20-amp circuit.

(C) Household Ranges and Cooking Appliances. There is a table covering range loads in Article 220. However, for range loads of 8¾ kW or more, the minimum branch-circuit rating shall be 40 amperes. This may have to be increased for ranges of larger capacity. The above requirement covers conductors supplying household ranges or other cooking equipment that are wall-mounted or counter-mounted.

In Table 220.19, you will find demand factors for ranges. Most inspection authorities believe that the branch circuit shouldn't be loaded in excess of 80 percent, as covered elsewhere in the *Code*, and since this table covers derating, they are very careful as to the ampacity of conductors serving ranges.

Exception No. 1

Branch-circuit conductors of 8¾ kW or more to ranges and the like ¾ are permitted to be smaller than the ungrounded conductors of the circuit. Column A of Table 220.19 computes the ampacities. Neutrals shouldn't be less than 70 percent of the branch-circuit rating, and shall never be smaller than No. 10 AWG.

Exception No. 2
On taps supplying wall-mounted ovens or cooking-top units that are supplied by 50-ampere branch circuit, the taps to the additional units shouldn't be less than 20-ampere capacity, but shouldn't be less than sufficient to cover the load the taps serve, and these taps shall be no longer than necessary to provide servicing of the equipment involved.

(D) Other Loads. When branch-circuit conductors supply cooking appliances other than the ones covered in (C) above and listed in Section 210.2, the ampacity of the conductors shall never be less than No. 14 AWG copper or No. 12 aluminum. Of course, the branch-circuit overcurrent protection shouldn't be sized greater than the ampacity of the conductor it serves.

Exception
Tap conductors for 15-ampere circuits from a circuit rated less than 40 amperes; if the tap conductor comes from a circuit rated at 40 or 50 amperes the tap conductors shouldn't be less than 20 amperes. This applies to tap conductors that supply any of the following loads:

(a) Shouldn't be longer than 18 inches where they serve individual lampholders or fixtures with this tap to the lampholder or fixture.

(b) Tap conductors covered by Section 410.67 to a fixture.

(c) Shouldn't be over 18 inches long to outlets served by the tap.

(d) Industrial heating appliances serving infrared lamp.

(e) De-icing and snow-melting cables or mats may have smaller tap conductors serving only the nonheating modes.

Section 240.4, covering fixtures and cords, is also an exception.

210.20: Overcurrent Protection
All branch-circuit conductors are required to be protected with overcurrent devices. The ratings of these devices must conform to the following:

(A) Continuous and Noncontinuous Loads. A continuous load is a load where the maximum current is expected to continue for 3 hours or more. Branch-circuit conductors that supply continuous loads must be rated for 100 percent of the noncontinuous load and 125 percent of any continuous load.

Exception
If the overcurrent devices are rated for continuous operation at 100 percent of their ratings, then the circuits they supply will follow suit.

(B) Conductor Protection. Branch-circuit conductors must be protected according to Section 240.3. Exceptions are made for tap conductors under the provisions of Section 210.19 and for fixture wires and flexible cords under the provisions of Section 240.4.

(C) Equipment. Equipment connected to branch circuits must be protected according to Section 240.2.

(D) Outlet Devices. These devices are covered in Section 210.21.

210.21: Outlet Devices

See Article 100 for the definition of outlet. Outlet devices shall conform to (A) and (B) below, and the ampere rating shouldn't be less than that of the load being served.

(A) Lampholders. Branch circuits of over 20 amperes shall have heavy-duty lampholders, and heavy-duty lampholders shall be rated at not less than 660 watts for medium-base types and not less than 750 watts for other types.

(B) Receptacles. A single receptacle is a single contact device with no other contact device on the same yoke (e.g., a common single-pole switch.) A multiple receptacle is a single device containing two or more receptacles (e.g., a common duplex receptacle.) A single receptacle on an individual branch circuit must have a rating of at least the same ampacity as the circuit, although certain single receptacles installed on individual motor branch circuits are exempted from this requirement. (See Section 430.109.)

- **(1)** Where a branch circuit serves a single receptacle that is the only receptacle in the branch circuit, the receptacle shall have a rating of not less than that of the branch circuit.
- **(2)** For two or more receptacles or outlets on the same branch circuit, Table 210.21(B)(2) tells us that receptacles shouldn't supply a total cord- and plug-connected load greater than that specified in the table.
- **(3)** You are referred to Table 210.21(B)(3) for branch circuits that supply two or more receptacles or outlet receptacles. If the branch circuit is larger than 50 amperes, the branch-circuit rating applies to the rating of the receptacle. The rating of the receptacle shouldn't be less than the branch-circuit overcurrent protection for over 50 amperes.

210.23: Permissible Loads

The branch-circuit overcurrent rating shall never be exceeded. Branch circuits shall carry only the load for which they are rated. Those consisting of two or more outlets shall supply loads according to the sizes of branch-circuit overcurrent protection that is covered in (A) through (C) below and is summarized in Section 210.24 and Table 210.24.

(A) **15- and 20-Ampere Branch Circuits.** Lighting units, other utilization equipment, or a combination of both may be supplied from 15- or 20-ampere branch circuits. The branch circuit supplying cord- and plug-connected utilization equipment shouldn't be loaded to more than 80 percent of the branch-circuit rating.

Where lighting units, cord- and plug-connected utilization equipment, or both are supplied by a branch circuit, the total rating of utilization equipment fastened in place on that circuit shouldn't exceed 50 percent of the branch-circuit rating.

Exception

Dwellings shall have a minimum of two small-appliance circuits. These are 20-ampere circuits and are covered in Section 210.1(C)(1), which will specify where they are to be located. They are to be used for no other purpose.

(B) **30-Ampere Branch Circuits.** Fixed lighting units with heavy-duty lampholders in anything other than dwelling unit(s) or utilization equipment in any occupancy may be supplied from 30-ampere branch circuits. The rating of any cord- and plug-connected utilization equipment shouldn't, however, exceed 80 percent of the branch-circuit ampere rating.

(C) **40- or 50-Ampere Branch Circuits.** This category includes fixed or stationary cooking appliances. In units other than dwelling units, it might also include fixed lighting units with heavy-duty lampholders, infrared heating units, or other utilization equipment. An electric range may be a fixed appliance or a stationary appliance, depending upon whether or not it was built-in, permanently connected to the branch circuit, or plugged in. It is recommended that a minimum of 40 amperes be used to supply clothes dryers, as too many won't meet the requirements of a 30-ampere branch circuit.

(D) **Branch Circuits Larger Than 50 Amperes.** Branch circuits 50 amperes or larger must never supply lighting loads. They are designed to supply only nonlighting loads.

210.24: Summary of Branch-Circuit Requirements

Table 210.24 summarizes branch circuits in dwelling units, and a branch circuit shouldn't be connected to serve more than one unit. The above requirements are for circuits having two or more outlets, but don't include the special circuits covered in 210.11(C)(1) and (2).

Branch circuits in dwelling units can supply only equipment in that unit, or loads associated only with that dwelling unit. All circuits such as central alarm systems, common area lighting, and so on, must come from a house panel, and not from a dwelling unit panel. This applies even if the dwelling unit has a building manager in occupancy.

Refer to the *NEC* for Table 210.24, "Summary of Branch-Circuit Requirements."

210.25: Common-Area Branch Circuits

This section deals with multi-unit residential buildings and specifies that circuits for a residential unit (one apartment) shouldn't be used to supply power to common areas. This forbids the practice of connecting a load in a common area to one of the apartment panels.

C. Required Outlets

210.50: General

Sections 210.52 through 210.63 cover the installation of receptacle outlets.

- **(A) Cord Pendants.** A cord connected and permanently installed should be considered the same as a receptacle outlet.
- **(B) Cord Connections.** Where flexible cords with a receptacle outlet are used, a receptacle outlet shall be installed. In the event that flexible cords are permitted, receptacle outlets for such cords are not necessary. This might be interpreted to mean that extension-cord lamps, with receptacles in the lamp head, must have a receptacle outlet for plugging in the extension cord. The author suggests that long cords should be plugged into a receptacle outlet, not solidly connected. The authority that has jurisdiction may enforce this interpretation.
- **(C) Laundry Outlet.** Where outlets are installed in dwellings for specific appliances such as laundry equipment, there shall be a maximum of 6 feet or less from the appliance they serve.

210.52: Dwelling-Unit Receptacle Outlets

(A) **General Provisions.** For all practical purposes, this covers all rooms in a residence that are used for living space. It does not cover bathrooms or hallways. The installation of receptacle outlets shall be such that no point along the floor line in any wall space is more than 6 feet, measured horizontally, from an outlet in that space. To clarify this, many outlets may be 12 feet apart.

It also includes any isolated wall space that is 2 feet or more in width, and wall space occupied by sliding panels, railings, or panel doors on exterior walls. If a room is divided by a fixed room divider, such as a standing bar counter, the room divider is also included in the 6-foot rule (see Figure 210-10).

The wall space includes all wall space that is unbroken along the floor line. This would eliminate counting doorways, fireplaces, and similar openings as wall space. Any wall space 2 feet or more in width shall be considered separately. The wall space will include corners of rooms along the floor line. The exclusion of fireplaces and so forth will prevent the use of cords across such potentially hazardous openings.

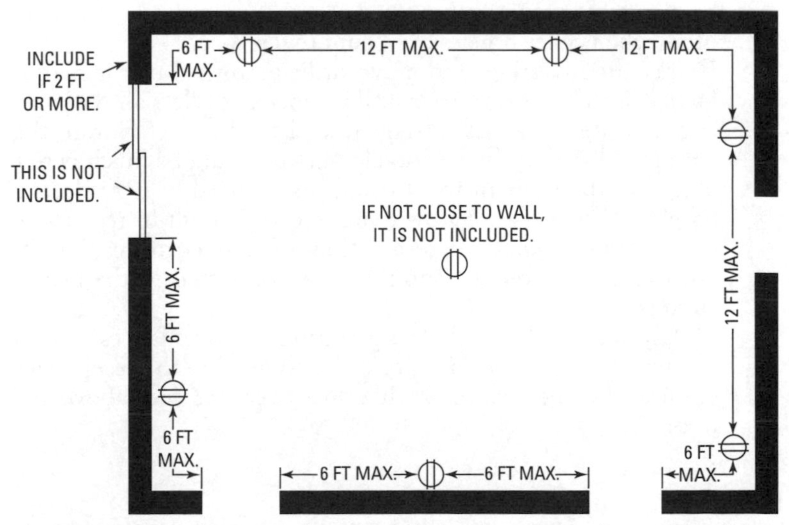

Figure 210-10 Proper wall receptacle outlet spacing for residential housing.

Sliding panels in exterior walls (such as in sliding glass doors) are excluded from being included in the wall space requirements, but any fixed-glass panels must be included as wall space.

Floor outlets that are not within 18 inches of the wall shouldn't be counted as fulfilling the requirements. Practical receptacles should be equally spaced along the wall line.

There has been considerable confusion in inspections on these matters, and the above should clarify many problems that formerly existed.

(FPN): The purpose of this requirement is to minimize the use of cords across doorways, fireplaces, or similar openings.

The pertinent points to remember are that at no point around the room shall a receptacle outlet be more than 6 feet from what is to be plugged into it, and no cords shall cross doorways, fireplaces, or similar openings placed in wall spaces.

By no means does this mean that you are required to start measuring 6 feet from a door and 12 feet between receptacle outlets. The intent is that, insofar as is practical, receptacle outlets be spaced equal distances apart so that there is no more than 6 feet from a receptacle outlet to the appliance plugged into the outlet. Also, floor-installed receptacles may be counted if they are located close to a wall. This occurs often when older houses are being rewired.

It is not advisable, and may not be permitted by the manufacturer's instructions, to install receptacle outlets above baseboard heaters, because the heat will tend to deteriorate the cord installation. The receptacle outlets would be much better placed in the floor in front of any baseboard heater. Electric baseboard heaters are now available with built-in receptacle outlets to eliminate deterioration of the cord insulation. Such receptacle outlets shouldn't be connected to the heater circuits.

Any receptacles in lighting fixtures or appliances within cabinets or cupboards are not to be counted in the receptacles required by the section if they are over 5½ feet above the floor.

Exception

Where there is baseboard heating, this heating can be obtained already equipped with factory-installed outlets or separate assemblies provided by the manufacturer. These outlets fulfill the outlet-and-wall-space rule. These special outlets shouldn't be connected to the heater circuit itself. The purpose is to prevent outlets from being installed above heating and prevent the cord from being placed over the hot baseboard heater, causing the deterioration of the cord.

(B) Small Appliances.

(1) Section 210.11(C) requires at least two small-appliance branch circuits in kitchens, pantries, dining rooms, and similar rooms. These are provided to power refrigerators, cooking appliances, and the like. These circuits are required to serve the receptacle outlets that are covered by Sections 210.52(B) and (F). However, these small-appliance branch circuits can supply only one outlet each. Other circuits can, of course, serve multiple outlets, but not these.

Exception No. 1
In addition to the two small-appliance branch-circuit outlets required in these areas, switched receptacles taken from general purpose branch circuits are allowed.

Exception No. 2
If one of these small-appliance branch circuits supplies a refrigerator only, it may be rated at 15 amperes.

Additional exceptions allowing the connection of other equipment to small appliance branch circuits:
(1) A receptacle that supplies only an electric clock (a very small load) may be tapped from a small-appliance branch circuit. (2) Range hoods can be powered by the small-appliance branch circuits whether connected with a cord and receptacle or directly connected (hard-wired). (3) If a gas range, oven, or countertop cooking device requires power for lighting or similar uses, that power may be taken from a small-appliance branch circuit.

(2) Countertop receptacle outlets that are installed in the kitchen shall be connected to not less than two small-appliance 20-ampere circuits, covered previously, supplying the kitchen and other specified rooms. This does not mean that only two small-appliance circuits may be used in wiring a dwelling; in fact, more than two will certainly prove an extra asset.

(C) **Countertops.** Countertops in kitchen and dining areas are of various widths and are often cut up by sinks, ranges, refrigerators, etc. When this is the case, any portion of the countertop space (countertops against walls only) that is 12 inches (305 mm) or wider shall have a receptacle outlet installed. No point on the countertop can be more than 24 inches horizontally from a receptacle outlet. Island or peninsula countertops 12 inches by 24 inches or greater must have receptacles installed so that no point on the perimeter of the countertop is more than 24 inches from a receptacle. These receptacles must be mounted above the countertop, or within 12 inches below the countertop. Note that no receptacles are allowed to be installed face-up in countertops. Any receptacle covered by appliances fastened in place or appliances occupying dedicated space adjacent to the basin shall have at least one receptacle outlet installed. This is to be used for electric appliances such as electric razors, toothbrushes, and hair dryers. Further information is given in Section 210.8(A)(1).

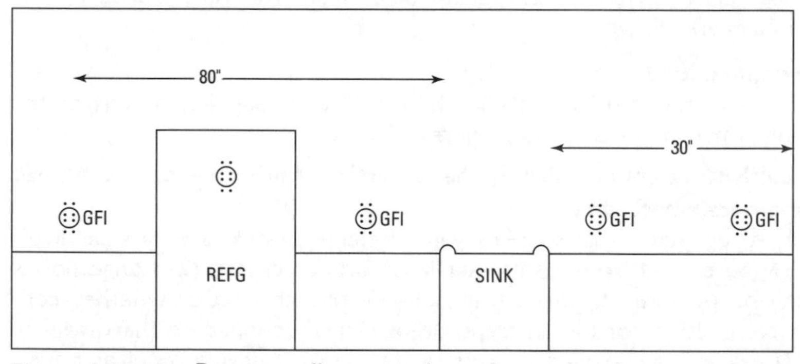

Figure 210-11 Proper wall receptacle outlet spacing for countertops in kitchen or dining areas.

(D) Bathrooms. There must be at least one ground-fault receptacle installed within 3 feet (900 mm) of any wash basin. GFCIs can be obtained as receptacles or as breakers and the service-entrance equipment covering an entire circuit.

As in kitchen counters, receptacles in bathroom countertops may not be installed face-up. If they were (in either area), water or other liquids could very easily pour into the receptacles and outlet box creating a dangerous situation.

2005

If a receptacle is installed in the side of the basin cabinet, no more than 12 inches (300 mm) below the countertop, it may be substituted for a receptacle in the wall or partition.

(E) Outdoor Outlets. See Section 210.8(A)(3). All one-family and two-family dwellings must have two outdoor receptacles, one near the front door, and the other near the back door. These receptacles must be ground-fault protected and must be no more than 6½ feet (1.98 m) above the finished grade. They must be easily accessible.

(F) Laundry Areas. Laundry areas shall have at least one receptacle outlet installed.

Exception

In apartments or multifamily dwellings that share laundry facilities in a common location on the premises and are available to all occupants, a laundry receptacle is not required in each dwelling occupancy.

In other than one-family dwellings, laundry facilities are often not permitted. When this is the case, no laundry receptacle need be installed.

(G) Basements and Garages. In one-family dwellings, each attached garage, unattached garage that has electrical wiring, or basement must have at least one receptacle. The laundry receptacle typically found in a basement can't be counted as this receptacle.

(H) Hallways. Any hallway 10 feet long, or longer, must have a receptacle. The length of the hallway is measured along its centerline, between doorways.

210.60: Guest Rooms

Hotels, motels, and similar occupancies have been interpreted as coming under this category, and the receptacle outlet spacing is the same as for dwelling units covered in Section 210.52. However, since many hotels and motels have beds, dressers, and other furniture permanently attached to the walls, receptacles by necessity must be located so that they will be convenient for the permanent fixtures that have been installed. At least two receptacles must be readily accessible, and receptacles placed behind beds must be placed or guarded so that cords connecting to the receptacle are not crimped.

210.62: Show Windows

Screw-shell lampholders have been used for many years to attach extension cords for floodlighting and the like. The *Code* now requires that at least one receptacle outlet be installed directly above the window. The receptacle outlets shall be installed for each 12 feet (3.66 m) of linear length or major portion thereof. It also requires that the show window be measured horizontally at its maximum width.

210.63: Heating, Air-Conditioning, and Refrigeration Equipment Outlet

These items require servicing from time to time. Therefore, 125-volt, single-phase, 15- or 20-ampere-rated receptacle outlets are a necessity to plug in service equipment. This receptacle shall be readily accessible to any unit located on a rooftop, or in an attic or crawl space. This receptacle must be on the same level, and within 25 feet of the equipment. Such receptacles on rooftops must be ground-fault protected (see Section 210.8(B)(2). Never connect the service receptacle to the load side of the disconnecting means for the equipment; the service receptacle must remain hot, even when the equipment is entirely disconnected. The above does not apply to one- and two-family dwellings.

210.70: Lighting Outlets Required

Lighting outlet installation is covered in (A) and (B) below.

(A) **Dwelling Units.** This requires at least one wall-switched lighting outlet to be installed in every habitable room: in hallways, bathrooms, stairways, attached garages, and at outdoor entrances. Along interior stairways, receptacles must be installed at each level that is six steps or more from the next level. There was some confusion in earlier versions of the *NEC* about installing a switched light at the vehicle door entrance to the garage. This is clarified: Because this is not considered an outside entrance, no switched light is required.

Also, at least one lighting outlet, controlled by a switch located at the entrance to the attic, basement, crawl space, or utility room, shall be installed.

Exception No. 1

In habitable rooms one or more receptacles may be controlled by a switch—where they control lighting outlets such as where a lamp will be turned on. This does not apply to kitchens and bathrooms. This will permit controlling of floor lamps and the like by means of a switch for entering the room, thus providing illumination.

Exception No. 2

Remote control or automatic control in lighting is permitted for hallways, stairways, and outdoor entrances. Remember, this does not cover lighting by a garage door. This will allow the prescribed lighting in these areas, if properly installed.

Exception No. 3

Occupancy sensors shall be permitted instead of switches to control lighting outlets. Typically these are small, passive infrared sensors that fit in a switch box and turn the light or lights on only when a person enters the area.

(B) **Guest Rooms.** This requires at least one wall switch-controlled lighting outlet or receptacle for a lamp in guest rooms of hotels, motels, and so on.

(C) **Other Locations.** One or more switch-controlled lighting outlets must be installed near equipment requiring service, such as any heating or air conditioning equipment in attics or crawl spaces. The switch must be installed at the entry point to the attic or crawl space.

Article 215—Feeders

215.1: Scope

This article covers the minimum size of feeder conductors supplying branch circuits. A feeder conductor here would have to go to a panel covering branch circuits, and loads to the feeder circuit shall be computed in accordance with Article 220.

Exception

Section 668.3(C) covers feeders at that point for electrolytic cells.

215.2: Minimum Rating and Size

(A) **General.** Parts II, III, and IV of Article 220 cover the calculations for feeder ampacities, and they shouldn't be of lower ampacity than thus computed, with minimum sizes as covered in (B) and (C), below. In the calculation of feeder sizes, voltage drops must be considered.

Feeder conductors need not be larger than the service entrance conductors when serving dwellings or mobile homes. By referring to Table 310.15(B)(6), the conductor size can be determined. Feeders must be sized, at a minimum, for 100 percent of the noncontinuous load, plus 125 percent of the continuous load connected to it. An exception is made for assemblies with overcurrent devices that are listed for operation at 100 percent of its rating. In these cases, the 125 percent of load requirement is dropped.

(B) **For Specified Circuits.** Conductors with a minimum ampacity of 30 are the minimum size permitted for loads consisting of the following types and numbers of circuits: (1) two or more two-wire branch circuits supplied by a two-wire feeder; (2) more than two 2-wire branch circuits supplied by a three-wire feeder; (3) two or more three-wire branch circuits supplied by a three-wire feeder; and (4) two or more four-wire branch circuits that are supplied by a three-phase, four-wire feeder.

(C) **Ampacity Relative to Service-Entrance Conductors.** Where feeder conductors handle the entire load supplied by service-entrance conductors with an ampacity of 55 or less, the feeder conductors shouldn't be smaller than the service-entrance conductors.

Unlike ordinary houses, mobile homes never have service-entrance conductors attached to them. Service-entrance conductors are mounted away from the mobile home, and feeders run from there to the mobile home.

Voltage drop is mentioned in Section 210.19(A). Feeders shall have no more than a 3 percent voltage drop at the furthest outlet, and combined branch circuits and feeders shouldn't exceed a 5 percent voltage drop. This, to a certain degree, is covered in Section 210.19(A), which discusses voltage drop for branch circuits. Further examples of this may be seen in examples 1 through 8 in Chapter 9.

The formula for voltage drop is as follows:

$$V_d = \frac{2L \times 12 \times 1}{CM}$$

where V_d = voltage drop

L = length of the circuit, feet, one way

I = current, amperes

CM = area of the conductors, circular mils

To arrive at the voltage drop in percent, the voltage supplied to the circuit is divided into the voltage drop.

Branch-circuit voltage drop was also covered in Section 210.19(A).

215.3: Overcurrent Protection

Part I of Article 240 covers how feeders shall be protected from overcurrent.

215.4: Feeders with Common Neutral

A common neutral feeder may be used for two or three sets of three-wire feeders or two sets of four- or five-wire feeders, providing that the neutral is large enough to take care of the unbalanced current that it may be required to carry, and that, when these are installed in a metal raceway or enclosure, all conductors are enclosed within the same raceway as required in Section 300.20. This is necessary to counteract induction that might be set up, which will cause heating and possibly imbalance.

Caution should be taken when current-carrying conductors and the neutral are brought into a large panel. The conductors should be run side-by-side (parallel) whenever possible, since a voltage imbalance might be caused by induction. Running the conductors side-by-side tends to cancel out the induction. This explanation does not appear in the *Code*, but is of importance.

At a *Code* panel session of questions and answers, a question arose: If you have conductors parallel of the same size and length, why is it required that they have the same type of insulation? The answer to this is that different types of insulation have different insulating qualities at different temperatures. As an example, if you had USE-RHW insulation on two conductors and THW on the third, the USE-RHW insulation resistance is much more stable under fluctuations of heat than the THW is. Therefore, you could possibly cause an imbalance of the impédance of the circuit to the lower insulation resistance of the THW, which would in turn cause a mismatch of the impedances at the end of the circuit.

215.5: Diagrams of Feeders

The enforcing authority may require diagrams that show feeder details, and they should show the area in square feet, total connected load before applying demand factors, demand factors selected, computed load after applying the demand factors, and size and type of conductors to be used. A wiring job of any size should have the feeder diagram supplied to the owner since it is a great assistance in determining where the feeders go and what they serve.

215.6: Feeder Conductor Grounding Means

If grounding conductors are required in branch circuits fed from feeders to a branch-circuit panel, there shall be a grounding means provided to which the grounding conductors may be attached. Also, in service-entrance equipment, the grounding bus must be grounded to the enclosure. In feeder panels, the grounding bus shall be grounded to the enclosures. This often brings up the question of where to attach the grounding conductor when it is required. The grounding conductor shouldn't be connected to the neutral bus of a feeder panel, but shall be connected to a separate grounding bus for the grounding conductors, or to the enclosure by approved lugs. Don't confuse a grounded conductor and a grounding conductor.

If the feeder panel is supplied by a metallic raceway that serves as a grounding conductor, the grounding conductors are attached to the enclosure and not the neutral bus since the neutral bus is isolated from the enclosure. If there is a grounding conductor but no metallic raceway, it is attached to the enclosure, as are the grounding conductors that leave the feeder panel. The latter part is not in the *Code* under this section, but is very appropriate.

215.7: Ungrounded Conductors Tapped from Grounded Systems

There are places where it will be necessary to tap into two-wire DC or AC circuits of two or more grounded conductors. Such a place might be a 240-volt, single-phase motor to be tapped from a three-wire, 240/125-volt system. It can easily be seen that you would have no use for the grounded conductor going to the motor, so such a two-wire tap on ungrounded conductors is permitted. Any switching device on this tap circuit shall open both ungrounded conductors, and one should be able to search the two ungrounded conductors simultaneously. However, there should be an equipment-grounding conductor in either the metal raceway of the separate green color or a bare conductor installed with the two ungrounded conductors, so that the motor would be at ground protection.

215.8: Means of Identifying Conductor with the Higher Voltage to Ground

There are many four-wire, delta-connected systems, the fourth wire being the tap of one phase to ground. In Figure 215-1A, you can see that two ungrounded conductors of the delta system will give 120 volts to ground, but notice that the third ungrounded conductor gives us a voltage of 208 volts to ground. In the field, this may be referred to as the wild leg, the high leg, or the stinger. This leg of the delta must have some kind of clear and permanent identification. The conductor can be orange in color, tagged, or permanently and effectively colored orange at any point where it could be used with the neutral. This identification shall be placed at any point where there might be connection made to the neutral or grounded conductor when it is present.

Figure 215-1B shows the voltage relationships on a grounded, four-wire wye system. On this system, 120 volts to neutral can be obtained from any ungrounded conductor and neutral.

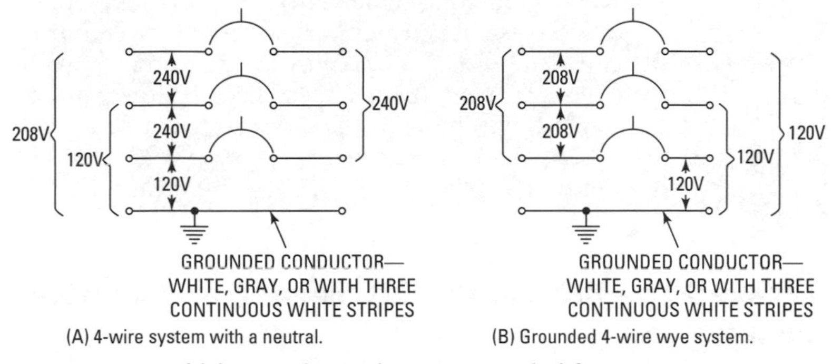

Figure 215-1 Voltage relationship on grounded four-wire systems.

215.9: Ground-Fault Protection for Personnel

There is nothing to prevent using GFCIs on a feeder supplying 15- or 20-ampere receptacle branch circuits. Section 210.8 covers some specifics, and Article 527 covers others. The point is that nothing prevents them from being used on 15- or 20-ampere circuits.

215.10: Ground-Fault Protection of Equipment

At a feeder disconnect rated 1000 amps or more, connected to 277/480-volt systems, ground-fault protection must be provided. See Section 230.95. If ground-fault protection is provided on the supply side of the feeder, no additional protection need be installed.

215.11: Circuits Derived from Autotransformers

Feeders may not be supplied from autotransformers unless the system supplied by the autotransformer has a grounded conductor and that grounded conductor is connected to the grounded conductor of the system that supplies that particular autotransformer. See the *NEC* for two exceptions to this rule.

2005

215.12: Identification for Feeders

(A) Grounded Conductor. These conductors (usually neutrals) must be identified according to Section 200.6.

(B) Equipment Grounding Conductor. Feeder equipment grounding conductors must be identified according to Section 250.119.

(C) Ungrounded Conductors. When there is more than one nominal voltage system (such as both a 120/240 volt and a 277/480 volt system) in one premises, every ungrounded conductor must be permanently identified. Color coding, tape, or tagging may be used. The circuit identifications must also be permanently posted in every branch circuit and distribution panel.

Article 220—Branch-Circuit, Feeder, and Service Calculations

I. General

220.1: Scope

In this article, the basic calculations for feeders, branch circuits, loads, and the method of determining the number of branch circuits will be discussed. This article might well be called "Fundamentals of Design," because it contains the minimum requirements for a particular design or application. This does not mean that in designing a wiring system one shouldn't take into consideration whether or not the minimum will be sufficient or whether allowances for future expansion should be provided for. This, of course, is very hard to anticipate with much accuracy, but definitely should be considered. A good and proper design always provides for at least the immediate future requirements that can be foreseen.

The ability to perform these calculations is far more of a design skill than an installation skill. It is very seldom that an electrician would need to perform such calculations for his daily work. Nonetheless, questions about these calculations are included in most electricians exams.

220.2: Computations

The following voltages shall be used in calculations involving branch circuits and feeders. This will hold true unless other voltages are specifically specified. The voltages to be used are 120, 120/240, 208Y/120, 240, 480Y/277, 480, 600Y/347, and 600, nominal. In addition, amperages may be rounded-off to the nearest amp. Fractions less than 0.5 ampere may be dropped.

220.3: Computation of Branch Circuits

This section of the *Code* is used to calculate the number and types of branch circuits that are required in any installation. These are minimum standards and may be exceeded at any time.

(A) Lighting Load for Listed Occupancies. The calculations for loads in various occupancies are based on watts (volt-amperes) per square foot. This is a minimum basis and consideration should be given to the ever-increasing trend toward higher levels of illumination. Each installation should be examined and not figured entirely on the watts-per-square-foot basis, but on the anticipated figure demands upon the system.

In figuring the watts per square foot (0.093 sq m), the outside dimensions of the building should be used. They don't include the area of open porches and attached garages with dwelling occupancies. If there is an unused basement, it should be assumed that it will be finished later; thus, it should be included in the calculations so that the capacity of the wiring system will be adequate to serve at a later date. Conduit or EMT should be installed in concrete basement walls during construction since the cost at that time will be much lower than at the time of finishing the basement.

The unit values given are based on 100-percent factor for the minimum requirements, so any low power factors should be taken into account in the calculations. If high power-factor discharge lighting is not used, allowance should be made for the increased amperage due to the lower power factor.

See Table 220.3(A) in the *NEC*. The unit load calculation shouldn't be less than shown in this table.

In calculation for dwellings, when figuring the area to be covered, porches, garages, and—if not adaptable for future use—unused or unfinished spaces should not be included.

Author's Note

If unfinished spaces are adaptable for future use, the area should be included in your calculations.

The load values used here are considered at 100-percent power factor. If less than 100-percent power factor, equipment is installed with sufficient capacity figured in to take care of the additional current values.

(B) **Other Loads**—All Occupancies. For any lighting other than general illumination, and for appliances other than motors, a minimum unit load per outlet is given in the table listed under this section of the *Code*.

In the *NEC* under this part you will find 11 categories that will be necessary when making calculations. Receptacle outlets are not to be considered as less than 180 volt-amperes (VA) for each receptacle. If more than one receptacle is contained in an outlet, each receptacle must be independently rated at 90 VA. For instance, an outlet with two receptacles in it would be rated at 180 VA (90 VA for each of the two receptacles). This will cover duplex receptacles or receptacles on the same mounting strap in an outlet box. Note, however, that this rule does not apply to the small-appliance and laundry circuits required by Section 210.11(C).

Multiple outlet assemblies are available, and if the multiple outlets are 5 feet or less from each other, then the continuous length of the multiple outlet assembly shall be considered as one outlet and figured at 180-volt-amperes capacity. Where a number of appliances will be used simultaneously, each foot (305 mm) should be considered as an outlet of not less than 180-volt-amperes capacity. These requirements shouldn't apply to either dwellings or guest rooms in hotels. One location where 1 foot might be considered as 180-volt-ampere capacity could be an appliance sales floor where a number of appliances might be connected for demonstration.

Another location might be in a school lab where a number of experiments using electrical apparatus would be conducted at the same time.

Household electric ranges are not subject to this regulation, but their minimum loads may be determined by using Table 220.19.

For show-window lighting, a minimum of 200 volt-amperes per linear foot (305 mm) of show window, measured horizontally along the base of the window, shall be used. Section 220.18 describes an acceptable method of computing electric clothes dryer loads.

(C) Loads for Additions to Existing Installations. Additional installations to existing electrical systems shall conform to the following:

(1) Dwelling Units. When computing additional loads for structural additions onto an existing dwelling unit, (B) or (C) (preceding) shall be used in the computation of the additional new load. Loads for structural additions to an existing dwelling unit or to a previously unwired portion of an existing dwelling that exceeds 500 square feet (46.5 sq m) shall be computed in accordance with (A) and (B) above.

It is necessary that the existing electrical systems be checked to make certain that the circuit or circuits being added have sufficient current-carrying capacity to take care of the additional load.

(2) Other Than Dwelling Units. When adding new circuits or extensions of circuits in other than dwelling units, either volt-amperes per square foot or amperes per outlet may be used, as covered in 220.3(B) and (C).

II. Feeders

220.10: General

Ampacity and Computed Loads

In no case shall the ampacity of the feeders be smaller than required to serve the load, and also in no case shall it be smaller than the sum of the loads of the branch circuits that it supplies, as computed in Part I of this article. This statement is subject to applicable demand factors that may apply in Parts II, III, and IV of this article.

See examples in Appendix D, and see Section 220.4(B) for maximum load permitted at 100-percent power-factor for lighting units.

220.11: General Lighting

The demand factors listed in Table 220.11 cover that portion of the total branch-circuit loads that are computed for lighting loads. These demand factors are only for the purpose of determining feeders to supply lighting loads, and are not for the purpose of figuring the number of branch circuits. The number of branch circuits has been covered in the preceding part and is also covered in Appendix D of the *Code,* as well as in following parts of this coverage of Article 220.

Each installation should be specifically analyzed, since the demand factors listed in this section of the *Code* are based only on the minimum requirements of load conditions and for 100-percent power-factor conditions. There will be conditions of less-than-unity power factor and conditions where these demand factors would be wrong. With the trend to higher illumination intensities and the increased use of fixed and portable appliances, the loads imposed on the system are very likely to be greater than the minimums. Also, electric discharge lighting should be of the high power-factor type; if not, additional provisions should be made for the low power factor involved.

Demand factors for small appliances for laundry equipment in dwellings are covered in Section 220.16.

Author's Note

See *NEC* Table 220.11, "Lighting Load Feeder Demands."

220.12: Show-Window and Track Lighting

Show-window lighting is a separate load from that calculated on the volt-ampere-per-square-foot basis. This lighting is to be figured at a minimum of 200 volt-amperes per linear foot (305 mm), measured horizontally along the base.

150 volt-amperes of load must be added to all load calculations for every 2 feet of track lighting (or any fraction thereof). This is accepted for dwellings or for the guest rooms of hotels or motels.

220.13: Receptacle Loads—Nondwelling Units

For installations other than dwelling units, receptacle loads calculated at 180 VA or less can be added to lighting loads, and derated in accordance with Tables 220.11 or 220.13.

220.14: Motors

The computing of motor loads is covered in Sections 430.24, 430.25, and 430.26.

Author's Note

See the *NEC* Table 220.13, "Demand Factors for Nondwelling Receptacle Loads."

220.15: Fixed Electric Space Heating

With one exception, the computed load on a feeder that serves fixed electrical space heating shall be equal to the total of the electrical space-heating load on all of the branch circuits. There is no demand factor. The feeder load current rating shall never be computed at less than the rating of the largest branch circuit supplied.

Exception No. 1

The inspection authority enforcing the *Code* may grant special permission to issue a demand factor for electrical space heating where they have duty-cycling or where all units won't be operating at the same time. The feeders are required to be of sufficient current-carrying capacity to carry the load as so determined.

220.16: Small Appliance and Laundry Loads—Dwelling Unit

(A) **Small Appliance Circuit Load.** The requirements for small-appliance receptacle outlets in single-family dwellings, multi-family dwellings with individual apartments that have cooking facilities, and in hotels or motels that have serving-pantry facilities or other cooking facilities, shall be a minimum of not less than two-wire, small-appliance circuits, as required in Section 210.11(C)(1). The calculated load for each circuit shouldn't be less than 1500 volt-amperes.

These loads may be included with the general lighting load, which makes it subject to the demand factors set forth in Table 220.11 in the *Code*.

Exception

Section 210.52(B)(1), Exception 2 allows for a branch circuit to power supplemental equipment for gas ranges and similar equipment. This circuit can be eliminated from load calculation required by section 220.16.

(B) **Laundry Circuit Load.** A minimum of one 20-ampere laundry circuit is required in the laundry room, but a feeder load of not less than 1500 volt-amperes shall be included for each two-wire laundry branch circuit installed, as required in Section 220.4(C). These circuit(s) may be included with the general lighting load and subjected to the demand factors provided in Section 220.11.

220.17: Appliance Load—Dwelling Unit(s)

A demand factor of 75 percent of the nameplate rating is permitted where there is a load of four or more fixed appliances served by the same feeder in a single-family dwelling or a multifamily dwelling.

220.18: Electric Clothes Dryers—Dwelling Unit(s)

When the wiring for a clothes dryer is installed, the clothes dryer has often not yet been purchased. Because of this, the electrician who installs the wires is often unable to check the nameplate rating to ensure that the feeder circuit that will feed the dryer is sized properly.

Electric clothes-dryer units shall be calculated at 5000 watts (volt-amperes) or the nameplate rating on the dryer, whichever is larger. When a house is being wired, the wirer can't be certain what the rating of the clothes-dryer to be purchased will be. This is why the *Code* has placed the 5000-watt minimum. Use of the demand factors in Table 220.18 in the *NEC* is permitted.

When two or more single-phase dryers are connected to a three-phase, four-wire service or feeder, the load must be calculated based on two times the total number connected between any two phases.

Many inspectors have insisted on No. 8 copper or the equivalent for dryers, as most dryers draw 30 amperes or more. A branch circuit shouldn't be loaded to exceed 80 percent of its capacity.

220.19: Electric Ranges and Other Cooking Appliances—Dwelling Unit(s)
When calculating feeder loads for electric ranges or other cooking appliances in dwelling occupancies, any that are rated over 1¾ kW shall be calculated according to Table 220.19 in the *NEC*. The notes following Table 220.19 are a part of the table and are very important in the calculation of feeder and branch-circuit loads.

Due to the increased wattages being used in modern electric ranges, it is recommended that the maximum demands for any range of less than 8¾ kW rating be figured using Column A in Table 220.19.

Three-phase, four-wire systems are often used. When calculating the current in such systems, it is necessary to use a demand of twice the maximum number of ranges that will be connected between any two-phase wires. An example of this is shown in Appendix D of the *NEC*.

Refer to the *NEC* for Table 220.19, "Demand Loads for Household Electric Ranges, Wall-Mounted Ovens, Counter-Mounted Cooking Units, and Other Household Cooking Appliances over 1¾ kW Rating." Column A is to be used in all cases except as otherwise permitted in Note 3.

There are five notes following Table 220.19 in the *NEC*. These are very clear and must be adhered to in some calculations of load. Please don't overlook these items. The *Code* panel has done an excellent job on these items and the examples in Appendix D.

For loads computed under this section, kVA shall be considered equivalent to kW. Heating loads are 100-percent power factor.

220.20: Kitchen Equipment Other Than Dwelling Unit(s)
Table 220.20 in your *NEC* may be used for load computation for commercial electric cooking equipment, dishwashers, booster heaters, water heaters, and other kitchen equipment. The demand

factors shown in that table are applicable to all equipment that is thermostatically controlled or is only intermittently used as part of the kitchen equipment. In no way do the demand factors apply to the electric heating, ventilating, or air-conditioning equipment. In figuring the demand, note that it shall never be less than the sum of the two largest kitchen equipment loads.

Author's Note

Refer to *NEC* Table 220.20, "Feeder Demand Factors for Kitchen Equipment—Other Than Dwelling Units."

220.21: Noncoincidental Load

When adding the branch-circuit loads, the smaller of two loads may be omitted if it is not likely that they will be used at the same time.

220.22: Feeder Neutral Load

The maximum unbalanced load controls the feeder neutral load ampacity.

Neutral feeder load must be considered wherever a neutral is used in conjunction with one or more ungrounded conductors. On a single-phase feeder using one ungrounded conductor and a neutral, the neutral will carry the same amount of current as the ungrounded conductor. A two-wire feeder is rare, so when considering the neutral feeder current, assume that there is a neutral and two or more ungrounded phase conductors. If there are two ungrounded conductors that are connected to the same phase, and a neutral, the neutral would be required to carry the total current from both phase wires, which would not be acceptable practice.

On three-wire, two-phase systems or five-wire, two-phase systems, the neutral shall carry 140 percent of the unbalanced load. On feeders that supply electric ranges, wall-mounted ovens, counter-mounted cooking tops, and so on, the maximum unbalanced load shall be considered to be 70 percent of the load on the ungrounded conductor. Therefore, the neutral supplying these may be 70 percent as large as the ungrounded conductors, providing that the neutral is no smaller than No. 10. The capacity of the ungrounded conductors is figured by Table 220.19 for ranges and Table 220.18 for dryers. For three-wire DC or single-phase AC, four-wire three-phase, three-wire two-phase, and five-wire two-phase systems, a further demand factor of 70 percent may be applied to that portion of the unbalanced load in excess of 200 amperes. There shall be no reduction of the neutral capacity for that portion of the load that consists of electric-discharge lighting, electronic computer/data processing, or similar equipment, when supplied by four-wire, wye connected, three-phase systems (this includes two- or three-wire circuits with neutrals). See Figure 220-1.

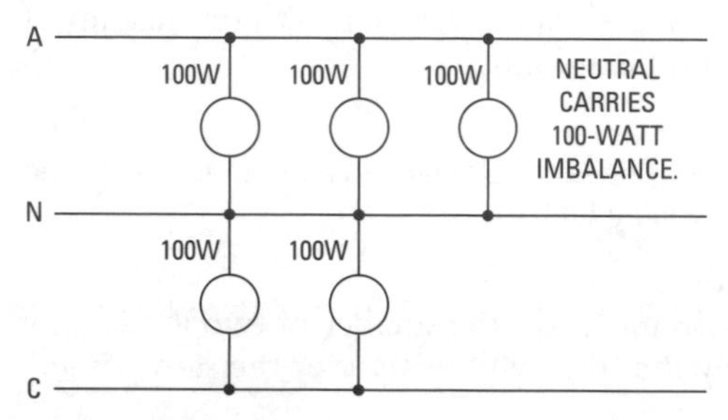

Figure 220-1 Unbalanced loads.

The question is often asked why a neutral that serves discharge lighting can't be derated. This is because of a third harmonic frequency produced by the discharge lighting and its ballast. This third harmonic may load the neutral to the maximum or higher allowable current. Figure 220-2 shows the effect of the third harmonic on a three-phase system, how the harmonics are in phase and will thus be added together.

As an example in figuring the neutral feed load, a four-wire, three-phase wye system of 120/208 volts will be used. Assume the loads shown in Table 220-1.

Table 220-1 Assumed Loads in Four-Wire, Three-Phase Wye System of 120/208 Volts

Discharge lighting	150 amperes per phase
Electric ranges	200 amperes per phase
Other loads [incandescent lighting, motor (3ϕ), miscellaneous]	295 amperes per phase
Total	**645 amperes per phase**

On 3ϕ motor loads a neutral is not involved, so it won't enter into the calculation of the neutral sizing.

The neutral current is calculated as shown in Table 220-2.

Table 220-2 Neutral Current

Ranges, 200 amperes @ 70%	140 amperes
Other loads	295 amperes
	435 amperes
200 amperes @ 100%	200 amperes
235 amperes @ 70%	164.5 amperes
	364.5 amperes
Discharge lighting	150 amperes
Total calculated neutral current	**514.5 amperes**

In the preceding case, it may be seen that the phase currents are 645 amperes per phase, while the neutral current may be derated to 514.5 amperes. These figures are based on demand-factor ratings.

As another example, assume a 50-unit apartment house using a four-wire, three-phase, 120/208-volt wye system, with fifty 3-wire electric ranges and a gross area of 1000 square feet per apartment, as shown in Table 220-3.

For the 50 ranges, this will put 16 ranges on one phase and 17 ranges on each of the other two phases. From Table 220.19 in the *NEC*, the indicated demand will be as follows: There will be a maximum demand of 17 + 17 ranges on any phase, or 34 ranges connected at the same time. According to Table 220.19, however, this will be 15,000 + 34,000 volt-amperes, or 49,000 watts, as shown in the following table, Table 220-4.

Table 220-3 50-Unit Apartment House Using a Four-Wire, Three-Phase, 120/208-Volt Wye System

General lighting, 50 × 3 × 1000	150,000 volt-amperes
Small-appliance loads, 50 × 2 × 1500	150,000 volt-amperes
	300,000
3000 volt-amperes @ 100%	3,000 volt-amperes
117,000 volt-amperes @ 35%	40,950 volt-amperes
300,000–120,000 volt-amperes @ 35%	45,000 volt-amperes
Minimum feeder capacitor for general lighting and appliances	88,950

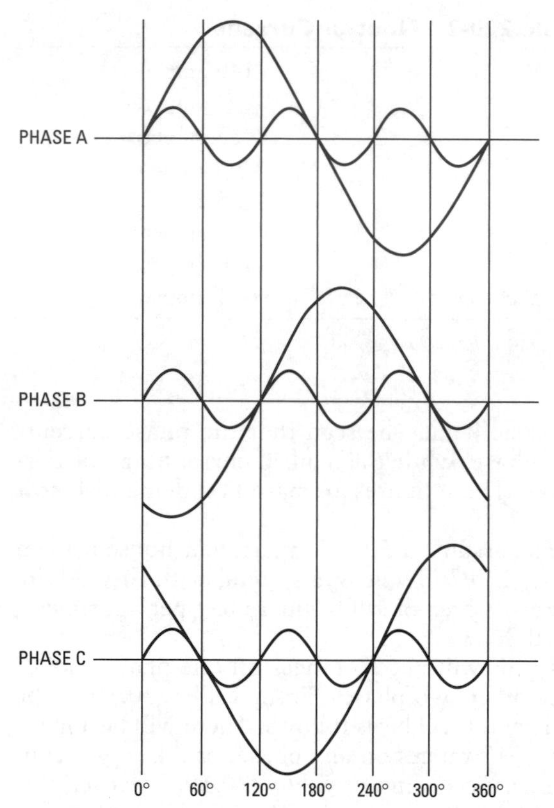

Figure 220-2 The effect of a third harmonic.

Table 220-4 Indicated Demand from Table 220-19 of the *NEC*

From Table 220-19	49,000 volt-amperes
General lighting and small appliance loads	88,950 volt-amperes
Total	**137,950 volt-amperes**

Maximum feeder demand:

$$\frac{137{,}950}{1.73 \times 208} = 384$$

To size the neutral, see Table 220-5.

Table 220-5 Sizing the Neutral

General lighting and small appliances	89,950 volt-amperes
Electric ranges, 49,000 @ 70%	34,300 volt-amperes
Total	**123,250 volt-amperes**

Total neutral current:

$$\frac{123{,}250}{1.73 \times 208} = 342$$

Table 220-6 shows the results.

Table 220-6 Total Amperes

200 amperes @ 100%	200 amperes
142 amperes @ 70%	99.4 amperes
Total	**299.4 amperes**

After figuring the phase load and neutral load in amperes, use the wire size necessary to carry that amount of current. In most cases, you will find that a conductor capable of carrying the exact number of amperes required won't be available, so use the next larger size.

III. Optional Calculations for Computing Feeder and Service Loads

220.30: Optional Calculation—Dwelling Unit(s)

(A) **Feeder and Service Loads.** This section takes into consideration the load calculations for single-family dwellings or an individual apartment of a multifamily dwelling served by a single three-wire, 120/240-volt or 120/208-volt set of service conductors or feeder conductors, with an ampacity of 100 amperes or more. It is permitted to use Table 220.30 instead of the method outlined in Part II of this article. If service-entrance conductors or feeder conductors are calculated by this section, the neutral load may be calculated using Section 220.22.

See Table 220.30 for optional calculation for dwelling unit loads in kVA.

(B) **Loads.** Take note of Table 220.30. Loads that are identified as "other loads" and "remainder of other loads" are shown below:

(1) Each 20-ampere small-appliance branch circuit and laundry circuit specified in Section 220.16 shall be calculated at 1500 volt-amperes.

(2) The same square footage, namely 3 volt-amperes per square foot, shall be used for general lighting and also for receptacles.

(3) For equipment that is permanently connected, fastened in place, or on a specific circuit (such as ranges, ovens, clothes-dryers, and water heaters), you must use the nameplate rating of the appliances.

(4) Lower power-factor loads, including motors, shall use the ampere rating on the load.

220.31: Optional Calculation for Additional Loads in Existing Dwelling Unit

If the present dwelling added onto is served by existing 120/240-volt or 208Y/120, three-wire service, you may use the table in your *NEC* for calculations. Also, 3 watts per square foot and 1500 volt-amperes for each 20-ampere small-appliance circuit shall be used. Ranges and wall-mounted or counter-mounted cooking equipment that is connected or fastened in place should use the nameplate rating.

Where space-heating equipment or air-conditioning is to be installed, at this point in your *NEC* you will find a listing of how to calculate these loads. Other loads shall include the same items that were covered above for small-appliance circuits and receptacles. Where there are other appliances fastened in place, including four or more separately controlled space-heating units, use the nameplate rating.

220.32: Optional Calculation—Multifamily Dwelling

(A) Feeder and Service Load. Table 220.32 may be used instead of Part II of this article to compute feeder or service load for multifamily dwellings that have three or more units. The following conditions shall be met:

(1) There can't be more than one feeder serving a dwelling unit.

(2) Electrical cooking equipment is supplied in each unit.

In Part II of this article, if the multifamily-dwelling load does not have electric cooking, and is more than the computed load in Part III for the identical load plus electric cooking using 8 kW for each dwelling cooking unit, the smaller of the two loads may be used.

(3) Each dwelling unit has electric space-heating equipment or air-conditioning or both.

When this optional method is used to compute the loads for feeders or services, the neutral load may be computed as covered in Section 220.22.

(B) House Loads. Part II of Article 220 shall be used for computing house loads. In addition, Table 220.32 shall be used for computing dwelling-unit loads.

(C) Connected Loads. Demand factors for the connected loads are covered in Table 220.32, and shall include the following:

(1) Each small-appliance branch circuit of two-wire, 20 amperes and its laundry branch circuit is covered in Section 220.16 and is figured at 1500 volt-amperes for each circuit.

(2) General lighting or general-use receptacles are figured at 3 volt-amperes per square foot.

(3) For appliances that are fastened in place or permanently connected on a specific circuit, the nameplate rating is to be used. Such appliances would include ranges and other cooking equipment, all wall-mounted or countertop units, clothes dryers, space-heating equipment, and water heaters.

Some water heaters are designed so that only one element at a time can come on, the top heating unit first and the bottom heating unit last. Here the maximum possible load shall be considered to be the nameplate load. See in the *NEC* for Table 220.32, "Optional Calculation—Demand Factors for Three or More Multifamily Dwelling Units."

(4) For motors and all low power-factor loads, the nameplate ampere or kVA rating shall be used.

(5) Where both air-conditioning or space-heating equipment are installed, the larger rating shall be used.

220.33: Optional Calculations—Two Dwelling Units

When the word feeder is used in these cases, it may also refer to the service. Where two dwellings are supplied by a single feeder or service and the load is computed under Part II of Article 220, and if this computed load exceeds the load specified under Section 220.32 for three identical units, you may use the lesser of the two loads.

220.34: Optional Method—Schools

Table 220.34 may be used in lieu of Part II of this article in the calculation of the service or feeder loads for schools if they are equipped with electric space heating or air conditioning or both. The demand factors in Table 220.34 apply to both interior and exterior lighting, power, water heating, cooking, other loads, and the larger of the space-heating load or the air-conditioning load.

When using this optional calculation, the neutral of the service or feeder loads may be calculated as outlined in Section 220.22. Feeders within the building or structure where the load is calculated by this optional method and calculated in Part II of Article 220 may use the ampacity as computed, but the ampacity of any feeder as calculated in Part II of Article 220 need not be larger than the individual ampacity for the entire building (this would mean the service entrance). Portable classrooms or buildings are not included in this section.

See the *NEC* for Table 220.34, "Optional Method—Demand Factors for Feeders and Service-Entrance Conductors for Schools."

220.35: Optional Calculations for Additional Loads to Existing Installations

When additional loads are added to existing facilities that have feeders and service as originally figured, you may use the maximum kVA figures in determining the load on the existing feeders and service if the following conditions are met:

(1) If the maximum data of the demand in kVA is available for a minimum of 1 year, such as demand meter ratings.

If demand data is not available, a recording ammeter may be used for 30 days, and the highest demand for any phase shall be used. This 30-day test must be conducted when the facility is in normal use.

(2) If the demand ratings for that period of 1 year are at 125 percent and the addition of the new load does not exceed the rating of the service. Where demand meters are used, in most cases the load as calculated will probably be less than the demand meter indications.

(3) If the overcurrent protection meets Sections 230.90 and 240.3 for feeder or service.

220.36: Optional Calculation—New Restaurants

When calculating the service or feeder load for a new restaurant, when the feeder carries the entire load, Table 220.36 can be used, rather than Part B of this article. Overload protection must be in accordance with Sections 230.90 and 240.3. Also, feeder or sub-feeder conductors don't have to be larger than service conductors, regardless of calculations.

IV. Method for Computing Farm Loads

220.40: Farm Loads—Buildings and Other Loads

(A) **Dwelling Unit.** Part II (but not Part III) of Article 220 shall be used for computing the service or feeder load of a farm dwelling.

(B) **Other Than Dwelling Unit.** The load for feeder, service-entrance conductors or service equipment should be computed in accordance with the farm-dwelling load and the demand factors that are specified in Table 220.40. This is for each farm building or load supplied by two or more branch circuits.

The conductors from the service pole to buildings or other structures are services, and this is covered in Section 230.21.

220.41: Farm Loads—Total

The farm-dwelling load and the demand factors in Table 220.41 should be used to compute the load on service-entrance conductors and service equipment. Remember, the dwelling load is computed as per computations for dwellings and added to the other loads. If there is equipment in two or more farm-equipment buildings or if loads have the same function, these loads should be calculated with Table 220.40, and they shall also be allowed to be combined as a single load for computing the total load. See the *NEC* for Table 220.41, "Method for Computing Total Farm Load," and Table 220.40, "Method for Computing Farm Loads for Other Than Dwelling Unit." Again, service drops and the like are calculated according Section 230.21.

Article 225—Outside Branch Circuits and Feeders

225.1: Scope

This article covers any electrical equipment or wiring that is attached to the outside or located on public premises. It also covers the conductors running between the buildings, structures, or poles that are involved with the premises being served. This article also covers electrical supply equipment that is attached to the outside of buildings, structures, or poles. It does not, however, cover services.

See Section 668.3(C) for outside branch circuits or feeders to electrolytic cells.

There is also the *National Electrical Safety Code* (*ANSI C2-2002*); it is used primarily by utility companies, but may also be involved in outdoor wiring of other property.

225.2: Other Articles

See Table 225.2 in the *NEC* for a listing of other article requirements for specific cases.

225.3: Calculation of Load

- **(A)** **Branch Circuits.** The provisions of Article 220, covering the calculation of loads on branch circuits, will apply to this article. In calculating loads for branch circuits, the ampacities allowable for single conductors in free air must be considered. These ampacities are given in Table 310.17 for insulated conductors and in Table 310.19 for insulated conductors and bare or covered conductors.
- **(B)** **Feeders.** The provisions covering feeders in Part II of Article 220 apply to outdoor feeders also, and Tables 310.17 and 310.19 apply for feeders for the ampacity of single conductors in free air.

225.4: Conductor Covering

This section covers the insulation of conductors and where such conductors are required. It must be remembered that these conductors are exposed to the elements and therefore must have an insulation that will withstand these conditions, including ultraviolet light.

Open conductors supported on insulators shall be insulated or covered when located within 10 feet (3.05 m) of a building or structure. Conductors that are in the form of a cable or are in raceways, with the exception of Type MI Cable, shall be of the rubber-covered or the thermoplastic type. In addition, where they are exposed to water or moisture, they shall comply with Section 310.8, which states that they must be resistant to moisture. There must be a W in the designation of the type of insulation. Festoon lighting conductors shall be rubber-covered or have a thermoplastic covering.

Exception

Equipment-grounding conductors and grounded circuit conductors need not be covered if other sections of the *Code* allow for the use of bare grounding conductors for the intended use.

225.5: Size of Conductors

The loads on the branch circuit and feeder conductors determine the ampacity of the conductor required. Loads will be determined according to Section 220.3 and Part II of Article 220. The ampacity will determine the size conductor to use, and this can be found in Tables 310-16 through 310-31.

225.6: Conductor Size and Support

(A) **Overhead Spans.** The following are the minimum sizes permitted for overhead conductors:

  (1) No. 10 copper or No. 8 aluminum of 600 volts, nominal or less for spans up to 50 feet (15.2 m) in length, and No. 8 copper or No. 6 aluminum for a longer span.

  (2) For over 600 volts nominal, No. 6 copper or No. 4 aluminum are the minimum sizes where open individual conductors are used, and No. 8 or No. 6 aluminum where in cable.

(B) **Festoon Lighting.** Conductors for festoon lighting that is not supported by messenger wires may be no smaller than No. 12. Any span of festoon lighting exceeding 40 feet must be supported by a messenger wire, and all messenger wires must be installed with strain insulators. Neither conductors nor messenger wires may be attached to fire escapes, downspouts, or other plumbing equipment.

225.7: Lighting Equipment Installed Outdoors

(A) **General.** Article 210 and (B) through (D) below cover branch circuits supplying lighting equipment out of doors.

(B) **Common Neutral.** The ampacity of a neutral wire must be calculated as the maximum load current between the neutral and any phase of the circuit.

(C) **277 Volts to Ground.** Branch circuits of more than 120 volts, nominal between conductors, and not over 277 volts, nominal to ground, are used to supply electric lighting where outdoor fixtures are located for illumination around industrial establishments, office buildings, schools, stores, and any other public or commercial buildings. The fixtures shall be 3 feet or more from any window, fire escape, stairs, platforms, and similar structures where a person may come in contact with the 277-volt conductors by ordinary means.

(D) 600 Volts Between Conductors. Section 210.6(D)(1) covers auxiliary equipment supplying electric discharge lamps, and equipment in which conductors exceeding 227 volts, nominal to ground, but not exceeding 600 volts, nominal between conductors are permitted.

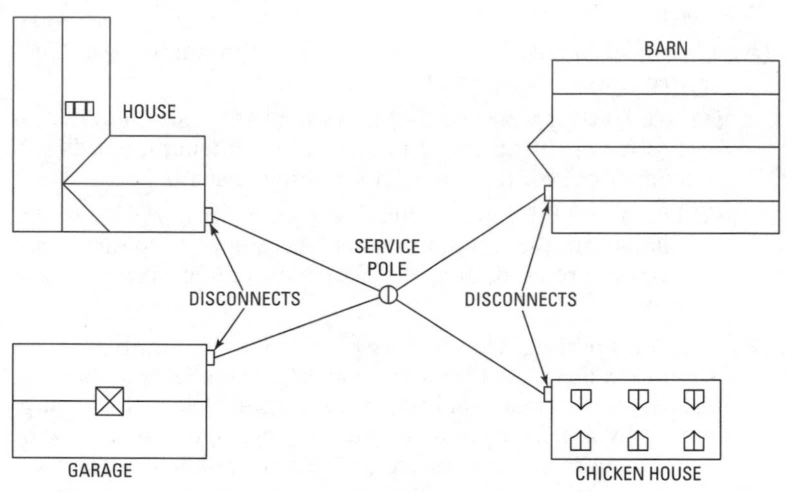

Figure 225-1 A group of farm buildings showing a service drop to each.

(E) Suitable for Service Equipment. Any disconnecting means covered specifically in (A) above must be listed for service equipment or be approved by the authority that has jurisdiction.

For garages and outbuildings on residential property, the disconnecting means may consist of a snap switch or a set of three-way or four-way snap switches suitable for use on branch circuits.

(F) Identification. If a building is fed by more than one feeder, branch circuit, or service (or any combination thereof), a permanent plaque must be placed next to each disconnect and must identify all power sources to the building.

Exceptions are made if the buildings are under the same management and proper switching procedures are established.

(G) Locked Feeder Overcurrent Devices. Branch-circuit overcurrent protection must be installed on the load side of the feeder if the feeder overcurrent device is locked up or somehow made inaccessible. This overcurrent device must be of lower ampacity than the feeder device.

225.9: Overcurrent Protection

The overcurrent protection of branch circuits is governed by Section 210.20. The overcurrent protection for feeders is covered by Part I of Article 240.

225.10: Wiring on Buildings

Wiring for circuits of 600 volts, nominal or less may be installed on the outside of buildings as open wiring on insulated supports, as multiconductor cable, as type MC or MI cable, in rigid metal conduit, messenger-supported wiring, cable trays, cable bus, wireways, flexible metal conduit, liquid-tight flexible metal conduit, liquid-tight flexible nonmetallic conduit, intermediate metal conduit, electrical metallic tubing, rigid nonmetallic conduit, and busways (see Article 368).

Circuits of over 600 volts, nominal are to be treated the same as the services in Section 230.202. Circuits for sign and outline lighting are covered in Article 600.

225.11: Circuit Exits and Entrances

Where outside branch and feeder circuits enter or leave a building, they are to be treated as service entrances as covered in Sections 225.23, 230.52, and 230.54. There is actually no difference except for classification of the purpose for which they are used.

225.12: Open-Conductor Supports

See the *NEC*.

225.14: Open-Conductor Spacing

(A) **600 Volts, Nominal, or Less.** Table 230.51(C) covers the spacing provided for conductors 600 volts, nominal or less.

(B) **Over 600 Volts, Nominal.** Conductor spacing over 600 volts is covered in Sections 110.36 and 490.24.

(C) **Separation from Other Conductors.** Open conductors shall be spaced not less than 4 inches apart. This applies to circuits (electrical) such as TV leads, telephone lines, etc. Of course, attention must be paid to the voltages involved and the separation spaced accordingly. There will be instances where 4 inches (102 mm) might not be sufficient due to higher voltages involved.

(D) **Conductors on Poles.** A 1-foot minimum spacing is required where racks or brackets are used. This again is determined by the voltages involved, with the spacing increased as required for higher voltages.

The horizontal climbing space for conductors on poles can be found in this section of the *NEC*.

225.15: Support Over Buildings

These require the same procedure as for service conductors covered in Section 230.29.

225.16: Point of Attachment to Buildings

This is covered in Section 230.26. Section 230.26 refers you to Section 230.24, where you will notice that (B) calls for a minimum of 10 feet (3.05 m) at the height of service entrance to building or at the drip loop of the building electric entrance. Please note there are two 10-foot (3.05 m) dimensions. Where there is a drip loop, this dimension should be used. Also remember that the dimension is measured from final grade level or above areas or sidewalks accessible only to people.

225.17: Means of Attachment to Buildings

This is covered in Section 230.27 of the *NEC*.

225.18: Clearance from Ground

The following shall be conformed to for open conductors not over 600 volts, nominal:

(A) If conductors don't exceed 150 volts to ground and are accessible to pedestrians only, they shall be a minimum of 10 feet above finished grade, sidewalks, or any other object from which they might be reached.

Here it would be a good idea to follow the instructions given in Section 225.16.

(B) If the voltage is no more than 300 volts to ground, the conductors may be a minimum of 12 feet above residential property, residential driveways, and commercial areas that are not subject to truck traffic.

(C) If the voltage exceeds over 300 volts to ground, the minimum height of the conductors shall be 15 feet instead of the 12 feet for voltages less than 300 volts to ground.

(D) An 18-foot minimum is required over public streets and alleys, roads, and parking areas that involve truck traffic. This also applies for driveways on other than residential driveways, and for areas such as forests, orchards and farmland, or cattle grazing areas.

(E) Clearances of conductors over 600 volts are not covered in the *NEC*. Refer to the *National Electrical Safety Code* (C2-2002), which the utility companies use.

For an illustration of the above clearances, see Figure 225-2.

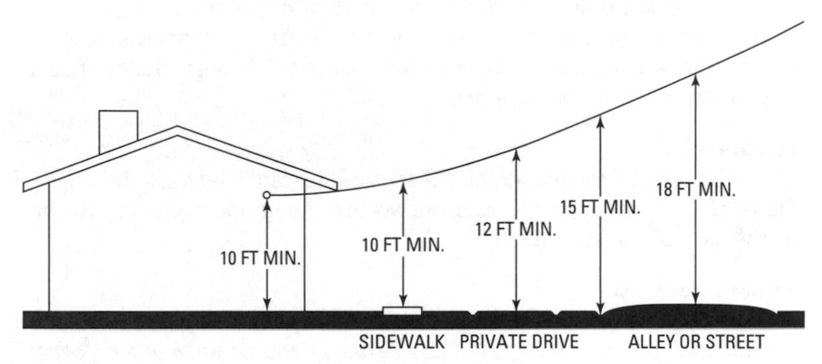

Figure 225-2 Minimum service-drop clearance.

225.19: Clearances from Buildings for Conductors of Not Over 600 Volts, Nominal

Many changes have been made from previous *Code*s for clearances over buildings for feeder and branch-circuit conductors. Ambient temperature affects conductor lengths; this should be considered because cold temperatures reduce the conductor length, giving more clearance than in the heat of summer when the conductors increase in length.

(A) **Above Roofs.** The vertical or diagonal clearance over roofs for conductors that are not fully insulated for the operating voltage shall have a clearance of not less than 8 feet (2.44 m) from the roof surface. This clearance must be maintained at least 3 feet beyond the edge of the roof.

Exception No. 1

Conductors over a roof surface subject to pedestrian or vehicular traffic must have a clearance from the roof as required by Section 225.18.

Exception No. 2

Where the voltages are less than 300 volts, the slope of the roof shall be taken into consideration. If it has a rise of 4 inches or more in 12 inches of the roof, the reduction of the clearance above the roof may be a minimum of 3 feet.

Exception No. 3

See Figure 230-4, which illustrates this exception; where the conductors don't exceed 300 volts to ground, the clearance is reduced for the overhanging portion of the roof only. This reduction is to be not less than 18 inches, provided that no more than 6 feet of conductors (with a maximum of 4 feet measured horizontally) pass over the roof overhang and where they are supported by a service mast or through-the-roof raceway or other approved support.

Exception No. 4

The requirement for maintaining 3 feet of clearance beyond the edge of the roof does not apply in situations where the conductors are attached to the side of the structure.

Author's Note

The through-the-roof raceway or other approved support must be capable of standing the strain of the service drop or be guyed or otherwise sufficiently supported.

(B) **From Nonbuilding or Nonbridge Structures.** This section involves clearance both vertically and horizontally from signs, chimneys, antennas, tanks, or anything that is a nonbuilding or nonbridge structure. The clearance for conductors can't be less than 3 feet (914 mm).

(C) **Horizontal Clearances.** Horizontal clearances can't be less than 3 feet (914 mm).

(D) **Final Spans.** Branch circuits and feeders to a building may supply from where they originate. You may attach these to the building, but they shall be kept a minimum of 3 feet from where they can be reached by persons, such as from windows, doors, and fire escapes.

There is an exception that will permit them to be less than 3 feet above a window.

(E) **Zone for Fire Ladders.** It is essential that when a building exceeds 3 stories or 50 feet in height, overhead lines shall leave at least a 6-foot clearance for fire ladders. When adjacent to buildings, an 8-foot clearance must be left for fire fighters.

This is followed by a fine-print note referring to the *National Electrical Safety Code*, which is used for high voltages or by utilities.

(F) Clearance from Building Openings. Feeder or branch conductors can't be installed where they will obstruct, or be installed underneath, openings in farm and commercial buildings through which materials can be moved.

225.20: Mechanical Protection of Conductors

Conductors on buildings, poles, and structures must be protected. Refer to Section 230.50.

225.21: Multiconductor Cables on Exterior Surfaces of Buildings

The requirements for multiconductor cables on exterior surfaces or buildings are the same as for service-entrance cables covered in Section 230.51. Also refer to Section 338.2, which covers service-entrance cable used for branch circuits and feeders. It is almost certain that it will be necessary to carry an equipment ground or otherwise satisfy the safety requirement of grounding to the satisfaction of the *Code*-enforcing authority.

225.22: Raceways on Exterior Surfaces of Buildings

Raceways mounted on the exterior of buildings and similar structures must be raintight and arranged to drain. Flexible metal conduit is the exception, per Section 398.12(1).

225.23: Underground Circuit

The requirements for underground branch circuits and feeders are the same as those covered in Section 300.5. However, remember that provisions must be made for an equipment- or grounding-conductor for feeders and branch circuits. This is important because the neutral of a feeder circuit is isolated from the cabinet and equipment, so the continuity of grounding must be provided or the purpose for this isolation will be defeated. If in doubt, check with the inspector before installation.

225.24: Outdoor Lampholder

With outdoor lampholders, connections to the conductors shall be staggered. Also, where the lampholders are of the pin type, which puncture the insulation, the conductors to which they are fastened shall be stranded-type conductors.

225.25: Location of Outdoor Lamp

For safety's sake, the location of lamps for outdoor lighting must be installed below all live conductors, electrical utilization equipment, transformers, and primary fuses. This will prevent persons changing the lamps from being injured by hot equipment.

Exception No. 1
Where relamping operations are provided with safeguards and proper clearances.

Exception No. 2
Where the equipment described above is controlled by a disconnect that can be locked, so that live parts can be killed, putting it in the open position.

225.26: Vegetation as Support
Trees can't be used for the support of any equipment or conductor span run overhead except for temporary wiring covered by Article 527.

From the above, one may realize that safety is of the utmost importance, and regulations are thus given for maintaining this safety.

II. More Than One Building or Structure

225.30: Number of Supplies
Where two or more buildings are under the same management and are located on the same property, each building or structure must be served by its own feeder or branch circuits. Remember that in these cases, multiwire branch circuits are considered to be one circuit only.

There are exceptions to this general rule in (A) through (E) below:

- **(A)** **Special Conditions.** Additional branch circuits or feeders may serve specific types of equipment: fire pumps, emergency systems, legally required standby systems, optional standby systems, and parallel power production systems.
- **(B)** **Special Occupancies.** Additional branch circuits or feeders may be installed with special permission (from the authority that has jurisdiction) to serve the following:
  - **(1)** Multioccupant buildings where there is no space available for supply equipment that is accessible to all occupants.
  - **(2)** A building or structure that is so large that two or more supplies are a practical necessity.
- **(C)** **Capacity Requirements.** When more than 2000 amperes at 600 volts or less are required, additional branch circuits or feeders may be installed.
- **(D)** **Different Characteristics.** Additional feeders or branch circuits are permitted when systems of different voltages or characteristics are required.

(E) **Documented Switching Procedures.** Additional branch circuits or feeders are allowed where documented safe switching procedures are established and maintained.

Safe switching procedures generally include documenting every switch and associated conductor. The primary concern is to ensure that the two different systems (from the branch circuits or feeders) are never connected. Such connection could cause any number of hazardous situations. Switching circuits are the most vulnerable to such problems, and that is presumably why they are singled out by the *NEC*. Note, however, that exactly what a safe switching procedure consists of is not mentioned by the *Code* and is, therefore, up to the discretion of the authority that has jurisdiction.

225.31: Disconnecting Means

There must be a means of disconnecting all ungrounded conductors that supply or pass through any building.

225.32: Location

The disconnecting means must be installed at the building it serves where the conductors pass through. This disconnect may be inside or outside, but must be located at the nearest readily accessible location relative to where it enters. The requirements of Section 230.6 (conductors in a raceway enclosed with 2 inches of concrete are considered to be outside the building, etc.) can be applied in these cases. Exceptions can be made:

(1) The disconnect may be located elsewhere if documented safe switching procedures are used.

(2) The disconnect may be located elsewhere if the installation qualifies under the provisions of Article 685 (Integrated Electrical Systems).

(3) Remote disconnecting means are allowed for poles or towers used for lighting.

(4) Remote disconnecting means are allowed for poles used only to support signs that are installed under the requirements of Article 600 (Signs).

225.33: Maximum Number of Disconnects

(A) General. As specified in Section 230.30, the disconnecting means may consist of no more than six switches or circuit breakers. These must be in a single enclosure, a group of enclosures, or a switchboard. Disconnects used solely for the control circuit of a ground-fault protection system are exempted from this requirement.

(B) Single-Pole Units. Two or three single-pole switches or circuit breakers may be used as a single disconnecting means for multiwire circuits, provided that they are equipped with a handle tie or master handle and disconnect all ungrounded conductors. The number of disconnecting means may not exceed six.

225.34: Grouping of Disconnects

(A) General. The disconnects allowed under Section 225.33 must be grouped together, and each must be marked, stating the load served. One exception exists: A disconnect serving only a water pump used for fire protection may be located elsewhere.

(B) Additional Disconnecting Means. The additional disconnecting means allowed under Section 225.30 must be far enough away from the other disconnecting means to minimize the possibility of simultaneous interruption.

225.35: Access to Occupants

In multioccupancy buildings, all occupants must have access to their own disconnecting means. This equipment may be accessible to management personnel only if the management is responsible for both supply and maintenance of the electrical system, and they are under constant supervision.

225.36: Suitable for Service Equipment

The disconnecting means mentioned in Section 225.31 must be a type that is suitable for use as service equipment. An exception is made for garages and other remote buildings on a residential property, where a switch or set of switches (including three-way and four-way) may serve as a disconnecting means.

225.37: Identification

A building that has a combination of feeders, services, or branch circuits passing through it must have a permanent plaque installed at the location of each feeder or branch-circuit disconnect. This sign must state the location of all other feeders, services, or branch circuits serving or passing through the building. See Section 230.2(E). Two exceptions are made:

(1) A plaque is not required for multibuilding industrial facilities under single management where it is assured that disconnecting can be maintained by using safe switching procedures.

(2) A plaque is not required at a garage or other remote structure on a residential property.

225.38: Disconnect Construction

(A) **Manually or Power Operable.** The disconnecting means may be either a manually operable or power-operable switch or circuit breaker. If a power-operable unit is used, it must be manually operable as well.

(B) **Simultaneous Opening of Poles.** The disconnecting means must simultaneously disconnect all ungrounded conductors.

(C) **Disconnection of Grounded Conductor.** If the primary disconnecting means does not disconnect the grounded conductor, some alternative means must be provided. This section does allow, however, for a pressure connector (lug) to serve as a disconnecting means.

In multisection switchboards, the disconnecting means for the grounded conductor may be in any section, provided that the section is so marked.

(D) **Indicating.** The disconnecting means must plainly identify whether it is on or off.

225.39: Rating of Disconnect

The rating of the disconnecting means must be at least equal to the load it will carry, determined under the provisions of Article 220. But, in no case shall it be less than what is required by each of the following:

(A) **One-Circuit Installation.** The disconnecting means for a single branch circuit must be rated at least 15 amperes.

(B) **Two-Circuit Installation.** The disconnecting means for two branch circuits must be rated at least 30 amperes.

(C) **One-Family Dwelling.** The feeder disconnecting means for a one-family dwelling must be rated at least 100 amperes.

(D) **All Others.** For other installations, the minimum feeder or branch circuit disconnecting means rating is 60 amperes.

225.40: Access to Overcurrent Protective Devices

If feeder overcurrent devices are not readily accessible, branch-circuit overcurrent devices (rated lower than the feeder device) must be installed on the load side in a readily accessible location.

III. Over 600 Volts

225.50: Warning Signs

Warning signs must be posted in any place where unauthorized persons may contact live parts. Those signs must read "WARNING—HIGH VOLTAGE—KEEP OUT."

225.51: Isolating Switches

Where oil switches or air, oil, vacuum, or SF^6 breakers are used as a disconnecting means, an isolating switch must also be installed on the supply side. This switch must have visible contacts and must meet the requirements of Sections 230.204(B), (C), and (D). An exception is made for disconnecting means installed on removable truck panels or certain metal-enclosed switchgear units. See the *NEC* for details.

225.52: Location

The disconnecting means must be located according to Section 225.32 or shall be electrically operated by a device that meets that location requirement.

225.53: Type

In addition to disconnecting all ungrounded conductors, each disconnect must have a fault-current closing rating no less than the available short-circuit current.

Article8
230—Services

I. General

230.1: Scope

This article gives a complete coverage of service conductors and equipment for the control and protection of all services. In Article 100, which covers definitions, these were the conductors and equipment from the source of power through the service-entrance equipment. The source might be distribution lines, transformers, or generators.

A thorough knowledge of this article is very important since a good service is the keystone of the entire installation. Services may be overhead or underground. A review of Article 100 and the following definitions pertaining to services is essential: service; service cable; service conductors; service drop; service-entrance conductors, overhead system; service-entrance conductors, underground systems; service equipment; service lateral; service raceway.

Although it is not a *Code* requirement, the local utility company should always be consulted as to the location for the service entrance.

230.2: Number of Services

Fundamentally, there is to be only one set of services to a building or premises. Should more than one service to a building or structure be required, as permitted in the exceptions that follow, some means, such as a plaque or directory clearly indicating locations where other services to the building or structure are, must be installed at each service drop or service lateral, or at each service-equipment location. This is essential to protect personnel, should one service be disconnected and people be led to believe that all electrical service to the building or structure have been disconnected.

See the *NEC* for Figure 230.1, "Services."

Exception No. 1

Separate services may be required for fire pumps. This would indicate an instance where a fire pump is being served from a separate transformer. It would accomplish nothing if the fire pump and the other service came from one transformer.

Exception No. 2

A separate service may be installed for emergency lighting and power. One would be a standby in the event of failure of the other service. An example of this might be a hospital where two separate services are installed, each being served from a different distribution feeder. In the event that one source fails, the other service might be fed from an emergency standby generator. This applies when there is a legal requirement for emergency services, standby or optimal standby systems.

Exception No. 3

By special permission (in writing), more than one service may be run to a building if there is no space available for service equipment that is accessible to all occupants.

Exception No. 4

Capacity Requirements. By special permission or where capacity requirements are in excess of 2000 amperes at a supply voltage of 600 volts or less, two or more services may be installed.

When we get into large ampacity services, it is felt that it is better to run two or more services than merely one service for the entire capacity requirement. Note "by special permission."

Exception No. 5
Buildings on Large Areas. For buildings covering large areas, by special permission, it may be necessary to have more than one service. This exception is not as prevalent as it was at one time, due to the increases in voltages and the use of transformer vaults and dry types of transformers.

Exception No. 6
Additional services may be required for different voltages, different frequencies, different phases, different classes of use, such as lighting rate and power rate, and controlled water-heater service.

The last exception is for *NEC* Section 230.40, Exception No. 2 only. If the ungrounded conductors are of size 1/0 or larger and are connected together at their supply end but go to the same location, and the service entrance end is not tied together so that it will become a parallel circuit, then they shall be considered one service lateral. This would mean that each set of individual service laterals could terminate in separate service entrances because they are in no way parallel.

Identification
If a building is fed by more than one service, a permanent plaque must be placed next to each disconnect and must identify all power sources to the building.

230.3: One Building or Other Structure Not to Be Supplied Through Another
Service drop or laterals are defined as the conductors that supply electricity to the point of usage. Service conductors that supply a building or structure must never be allowed to be run through one building to another unless the buildings are under single occupancy or management. Section 230.6 explains when conductors are considered to be outside a building. Even when under the same occupancy or management, one should be careful when permitting conductors to be run through the buildings (Figures 230-1 and 230-2).

In Figure 230-2, please note that the conductors to the meter and from the meter to supply the premises are separate conduits. You are never allowed to run metered and unmetered conductors in the same conduit. Some utilities that install these metering poles do, but that comes under the exception for utilities by the *Code*.

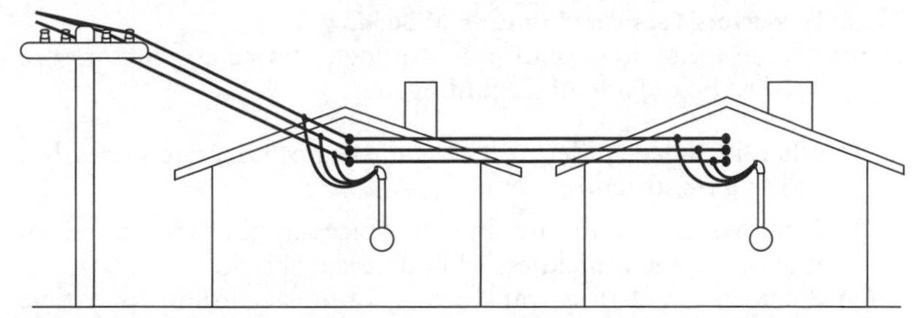

Figure 230-1 The service entrance for two buildings under the same ownership.

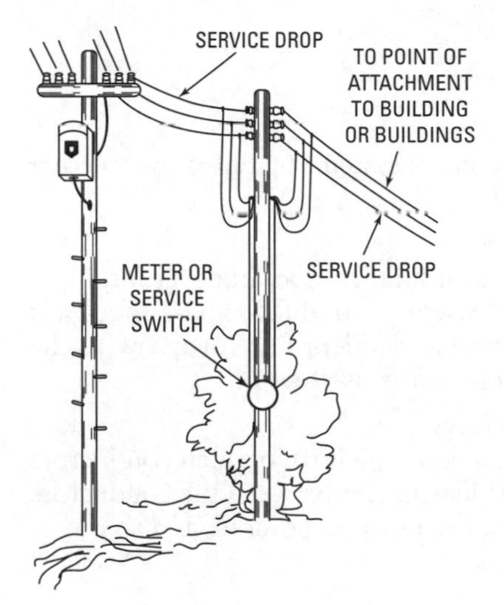

Figure 230-2 A metering pole.

(FPN): The minimum sizes for service conductors are given in the following references:

- Service Drops—See Section 230.23.
- Underground Service Conductors—See Section 230.31.
- Service-Entrance Conductors—See Section 230.41.
- Farmstead Service Conductors—See Part IV of Article 220.

230.6: Conductors Considered Outside of Building

If any of the following conditions are met, service conductors are considered to be outside of a building:

(1) When installed under at least 2 inches of concrete beneath a building or structure.

(2) When within a building, but in a raceway that is enclosed in at least a 2-inch thickness of concrete or brick.

(3) When installed in a transformer vault, according to Article 450, Part III.

230.7: Other Conductors in Raceway or Cable

Only service conductors may be installed in service conduits or cables.

Exception No. 1

Grounding conductors.

Exception No. 2

Conductors of load management control systems that have overcurrent protection.

230.8: Raceway Seal

All service conduits that enter a building from an underground distribution system must be sealed. See Section 300.5. Spare raceways must also be sealed. The sealant must be identified for use with the type of insulation on the conductors it is used with.

230.9: Clearances from Building Openings

When service conductors are installed in the form of open conductors or service cables, there must be at least 3 feet between the conductors or cable and any doors, windows, fire escapes, porches, and so on.

Exception

When conductors are run on the top of windows, they may be within 3 feet.

Service conductors can't be installed in a location where they might obstruct any openings through which materials can be moved. These openings are not uncommon in farm buildings and in some commercial buildings.

II. Overhead Services

230.21: Overhead Supply

See Figures 230-1 and 230-2. Overhead supplies are commonly referred to as service drops and are the conductors to a building or other structure. These drops could also be to a service pole where metering for a disconnect has been placed. On mobile home meters you will find that if the metering and disconnect are on a pole (because they may not be mounted on a mobile home), the service drop stops at that point, and the overhead conductors from the pole to the mobile home are called *feeders*. An example of this can be found not only in the section on mobile homes, but also in Part IV of Article 220, which covers loads on farms.

230.22: Insulation or Covering

Service conductors have to withstand exposure to atmospheric conditions or any other conditions that will cause leakage between conductors. The insulation for service conductors that are individually insulated is usually covered with extruded thermoplastic or thermosetting insulating material. There is an exception to the insulation. Where multiconductor cables such as duplex, triplex, or four-plex cables are used, the neutral conductor may be an uninsulated conductor.

See *NEC* tables in Article 310 for the ampacity of service drop conductors, run in cable or as open conductors. If covered with extruded thermoplastic or thermosetting insulation, the ampacity shall be the same as if they were bare conductors of the same size.

230.23: Size and Rating

The first requirement shall always be that service-drop conductors be of sufficient size to carry the load. They may not be smaller than No. 8 copper or the equivalent, except for limited loads, as covered in Section 230.42. An exception permits service-entrance conductors to be as small as No. 12 hard-drawn copper or No. 10 aluminum, providing that they furnish only a single circuit, such as a small polyphase motor, controlled water heaters, or similar loads, and providing they are hard-drawn copper or the equivalent.

Figures 100-1 and 230-2 show the service drops when the structure is a pole. The same condition could be a building with the meter on the building and/or a disconnecting means and other buildings or structures from said meter and/or disconnecting means. The grounded conductor of service drops shall be sized according to Section 250.23(B) in the *NEC*.

230.24: Clearances

As previously mentioned, high temperatures allow conductors to lengthen; therefore, clearances are given in the *Code* for the following conditions: 60°F (15°C), no wind; the potential amount of winter-time ice-load sag of the conductors must also be considered.

The *NEC* addresses clearance requirements in this section. This section pertains to services that are not readily accessible and don't exceed 600 volts, nominal; these services must conform to the following:

(A) Above Roofs. The basic requirement is that conductors shall have a vertical clearance of not less than 8 feet (2.44 m) from the highest point of the roofs above which they pass, which must be maintained for at least 3 feet (914 mm) in every direction from the edge of the roof.

Exception No. 1
Service conductors above roof surfaces that are subject to pedestrian or vehicular traffic must have a vertical clearance in accordance with Section 230.24(B).

Exception No. 2
The clearance may be reduced to 3 feet (914 mm) when the voltage between conductors does not exceed 300 volts, and the roof has a pitch of not less than 4 inches per 12-inch run.

Exception No. 3
No more than 6 feet of service conductors that operate at no more than 300 volts and that don't pass above anything except a maximum of 4 feet (1.22 m) of the overhang portion of the roof for the purpose of terminating at a (through-the-roof) service raceway or approved support may be maintained at a minimum of 18 inches (457 mm) from any portion of the roof above which they pass. This exception does not mention the slope of the overhang, but does specify the height of the point of attachment to the mast. Figure 230-3 shows where this exception is applicable.
Section 230.28 covers mast supports.

Exception No. 4
The requirement for maintaining 3 feet of clearance beyond the edge of the roof does not apply in situations where the conductors are attached to the side of the structure.

(B) Vertical Clearance from Ground. The following shall be conformed to for open conductors not over 600 volts, nominal:

(1) If the conductors don't exceed 150 volts to ground and are accessible to pedestrians only, they shall be a minimum of 10 feet above finished grade, sidewalks, or any other object from which they might be reached. Note, however, that only triplexed or quadraplexed cables (the typical modern service cables, with a bare messenger wire) may be used.

Here you would be well-advised to follow the instructions given in Section 225.16.

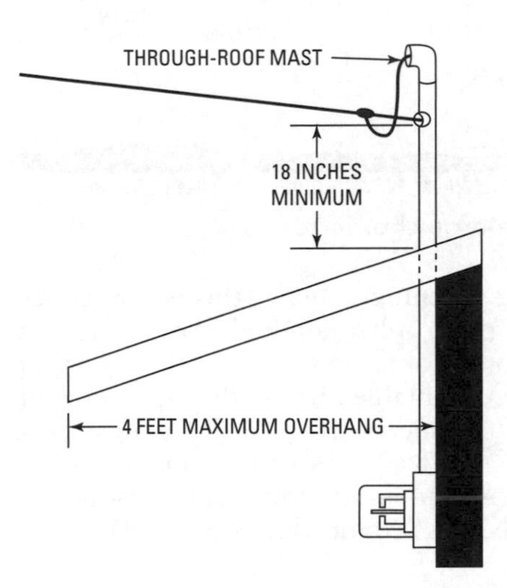

Figure 230-3 A service-drop mast mounted through the roof.

(2) If the voltage is no more than 300 volts to ground, the conductors may be a minimum of 12 feet above residential property, residential driveways, and commercial areas that are not subject to truck traffic.

(3) If the voltage exceeds over 300 volts to ground, the minimum height of the conductors shall be 15 feet instead of the 12 feet for voltages less than 300 volts to ground.

(4) An 18-foot minimum is required over public streets and alleys, roads, and parking areas when involved with truck traffic. This also applies for driveways on other than residential driveways, and for areas such as forests, orchards and farmland, or cattle grazing areas.

Clearances of conductors over 600 volts are not covered in the *NEC*. You are referred to the *National Electrical Safety Code*, which the utility companies use.

Refer to Figure 230-4 for ground clearances over the various areas where vehicles are used.

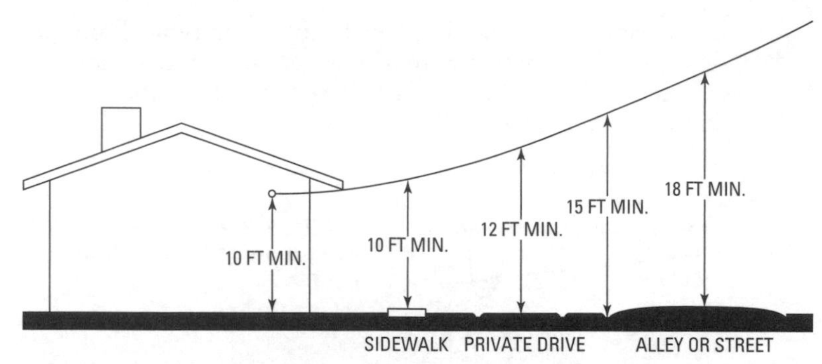

Figure 230-4 Minimum service-drop clearance.

(C) Clearance from Building Openings. Again, this is for service conductors not exceeding 600 volts, nominal. (For clearances of conductors of over 600 volts, nominal see the *National Electrical Safety Code,* available from the Institute of Electrical and Electronic Engineers [IEEE]). There must be a clearance of not less than 3 feet (914 mm) from windows, doors, porches, fire escapes, or similar locations. Conductors above windows may be less than the 3-foot (914-mm) requirement.

(D) Clearance from Swimming Pools. See Section 680.8.

230.26: Point of Attachment

The point of attachment of conductors (service drop) to a building or structure shall meet all of the requirements of Section 230.24. The minimum is 10 feet (3.05 m) above finished grade, but when going back to Section 230.24(B), note that the drip loop is mentioned, as well as service-drop cables supported and/or cabled together with a grounded bare messenger.

When sufficient attachment height for service drops can't be obtained due to the construction of the building, a mast type of riser (Figure 230-5) may be used, providing it is capable of withstanding the strain that might be imposed upon it. When considering the strain that might be imposed, the prevailing weather conditions should be considered. Each locality will no doubt have specifications that should be met.

230.27: Means of Attachment

In attaching multiconductor cables to buildings or other structures, only approved devices shall be used. In the attachment of open conductors, only approved, noncombustible, nonabsorptive types of fittings shall be used.

230.28: Service Masts as Supports

All fittings for service masts shall be properly identified, that is, listed for the purpose for which they are used. Any braces or guys supporting the mast shall be adequate for the purpose. See Section 230.26 in this book and Figure 230-5.

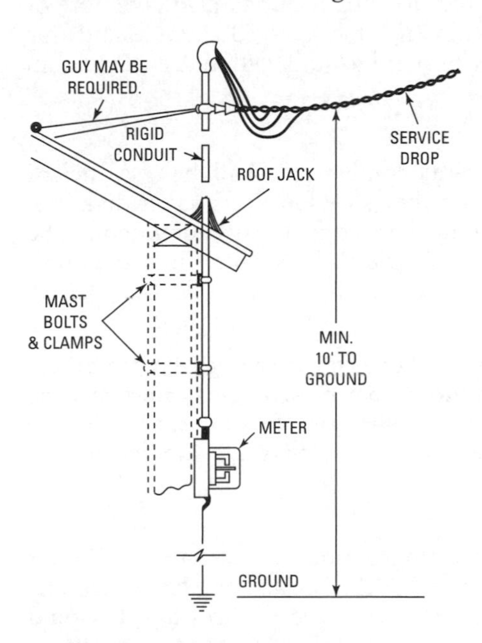

Figure 230-5 Mast installation for proper service-drop height.

230.29: Supports over Buildings

Substantial structures shall be used to securely support conductors that pass over roofs. They shall be of substantial construction and independent of the building where practicable.

III. Underground Service—Lateral Conductors

230.30: Insulation

The insulation on service-lateral conductors must be adequate for the voltage being carried, and the insulation must also be approved for direct burial so that there won't be detrimental leakage of current. In Article 300 you will find that service-lateral cables generally are noted by USE and most of the average conductors used in wiring carry a 600-volt rating, which in most cases will be more than adequate to cover the applied voltage. An exception for grounded uninsulated conductors is given; check your *NEC* for this. It is very important to know the electrolytic qualities of the earth in which they are being installed, if they are used. Also note that aluminum conductors or copper-clad aluminum conductors that are grounded should be used only when part of a cable that is listed for direct burial.

Take note of the bare copper for direct burial; also take note of the fact that the soil conditions must be judged suitable and this judgment is up to the authority that has jurisdiction, as provided in Section 90.4.

230.31: Size and Rating

As with all conductors, lateral conductors shall have sufficient ampacity to carry the load and they shouldn't be smaller than No. 8 copper or No. 6 aluminum or copper-clad aluminum. The grounded (neutral) conductor shouldn't be smaller than the minimum size permitted by Section 250.24(B)

Exception

When supplying loads with a limited demand and a single-branch circuit, such as for small polyphase motors, controlled water heaters, and the like, you may use conductors no smaller than No. 12 copper or No. 10 aluminum or copper-clad aluminum, provided they have the proper ampacity to carry the load.

230.32: Protection Against Damage

It is very important to see that ungrounded service laterals are not subjected to damage. See Section 300.5. The above refers to service-lateral conductors, not conductors enclosed in raceways. It would not be wrong, if they are enclosed in metallic or plastic raceways, to encase them in concrete of a red color; the author recommends 2000-pound concrete with a 7-inch slump and aggregate of no larger than an inch. Where service-lateral conductors enter a building, Section 230.6 applies, or raceway may be used as identified in Section 230.43.

IV. Service-Entrance Conductors

230.40: Number of Service-Entrance Conductor Sets

Only one set of service drops or service laterals shall supply the service-entrance conductors.

Exception No. 1

Buildings such as shopping centers that have more than one occupancy will be permitted to have more than one set of service-entrance conductors run to each occupancy, or they may be grouped so that more occupancies can be served by more than one set of service conductors by means of auxiliary gutters, and so on (see Figure 230-6).

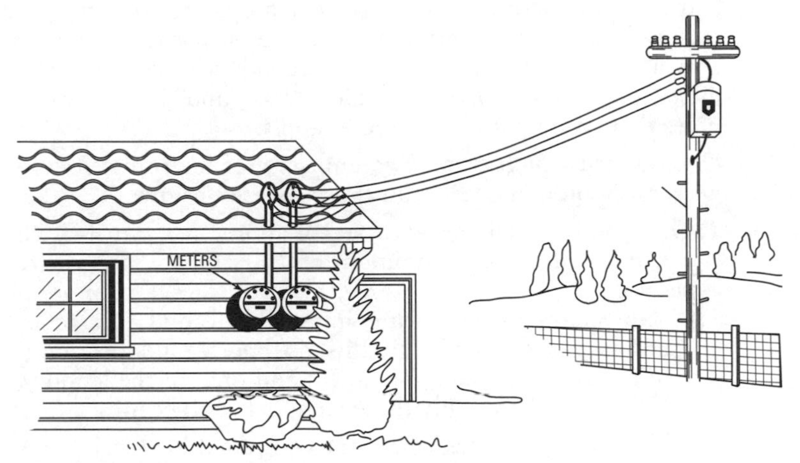

Figure 230-6 Two sets of service-entrance conductors tapped to one service drop.

Exception No. 2

This exception covers places where two to six disconnecting means are located in separate enclosures that must be grouped at one location supplying separate loads. From where we only have one service drop or service lateral, a separate set of service-entrance conductors is permitted to supply each or several of these service-equipment enclosures. Note that the disconnecting means shall be grouped.

230.41: Insulation of Service-Entrance Conductors

Service-entrance conductors entering buildings or other structures shall be insulated. Where only on the exterior of buildings or other structures, the conductors shall be insulated or covered; the attachment of service entrance conductors shall be installed so that water won't be siphoned into the service-entrance conductors.

Exception

Uninsulated grounded conductors may be used in service entrances as follows:

Check parts (A) to (D) in this section of the *NEC* for information about where ungrounded conductors are permitted or not permitted.

You would be well-advised to note the similarity of this section and Section 230.30; the same notation applies here.

(A) **General.** The first requirement, as always, is that the service-entrance conductors be of sufficient ampacity to carry the required load. The determination of the sizing is covered in Article 220, and the ampacity of the conductors can be determined by Tables 310.16 through 310.31, and the applicable notes that accompany these tables apply.

(B) **Ungrounded Conductors.** The ampacity of ungrounded conductors shouldn't be less than those covered below:

(1) One-family dwellings with six or more two-wire branch circuits shall have a minimum of 100 ampere, three-wire service.

Nowadays it is practically impossible to find a residence that has only six 2-wire circuits, as there would be a minimum of two for small appliances and one for the laundry, which would leave only three circuits for other uses.

(2) One-family dwellings where the computed load is 10 kVA or more shall have a minimum of 100-ampere, three-wire service.

(3) Other than those covered in (1) and (2), there may be loads that can have 60-ampere services.

Don't take the 100-ampere capacity as the maximum that is required. The computed load may require higher ampacity for the service conductors.

Exception No. 1
Where the load computed has not more than two 2-wire branch circuits, the service conductors may be No. 6 aluminum or No. 8 copper.

Exception No. 2
Some wiring is now done with demand-limiting equipment at the source of supply. If the maximum demand permits No. 8 copper, No. 6 aluminum, or No. 6 copper-clad aluminum.

Exception No. 3
If the load is limited to one single-branch circuit—such as on an illuminated advertising sign-No. 12 copper, No. 10 aluminum, or copper-clad aluminum may be used, but it may never be less than the branch circuits covered.

(C) Grounded Conductors. Section 250.23(B) sets the minimum size that will be permitted for the grounded (neutral) conductor, and this size shouldn't be less than as covered in that section.

230.42: Minimum Size and Rating

(A) General. The ampacity of service-entrance conductors, prior to derating, must be no less than specified by (1) or (2) of this section. Article 220 is used to determine load, and Section 310.15 is used to determine ampacity. Busway ampacity is determined by its labeling.

(1) 125 percent of continuous loads, plus 100 percent of noncontinuous loads.

(2) If the service-entrance conductors terminate in a device listed for operation at full load, 100 percent of all loads may be used as the ampacity.

(B) Ungrounded Conductors. The ampacity of ungrounded conductors must be at least the rating of the disconnecting means given in Section 230.79.

(C) Grounded Conductors. The minimum size of a grounded conductor must be as specified in Section 250.24(B).

230.43: Wiring Methods for 600 Volts, Nominal, or Less

The *Code* later covers not only what conditions in which service-entrance conductors may be installed, but the requirements of the *Code* limit wiring methods used. For these methods, check this section in your *NEC*. You will also find that it covers installation around flexible metal conduit according to the provisions of Section 250.79(A), (B), (C), and (E). See Section 348.2, as well.

Permission is granted with approved cable tray systems to support cables for use as service-entrance conductors. This is covered in Article 392.

230.46: Spliced Conductors

Service-entrance conductors may be spliced or tapped with clamped or bolted connections. Such splices must be made in enclosures or directly buried with the use of a listed underground splice kit. All splices must be made in accordance with Sections 110.14, 300.5(E), 300.13, and 300.15. (See Figure 230-7.)

On existing installations, a connection is permitted where service conductors are extended from a service drop to an outside meter and returned to connect to the service-entrance conductors of an existing installation.

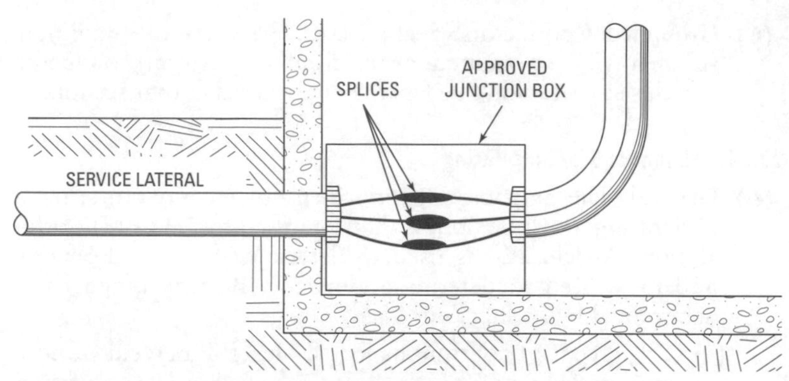

Figure 230-7 A splice in an underground service.

Where the service-entrance conductors are buses, the only method of getting continuity is by bolting the different pieces of bus bars to fit the various sections and fittings.

230.49: Protection Against Damage—Underground

Section 300.5 covers the protection required for underground conductors, which applies also to underground service-entrance conductors.

In areas where frost and freezing are problems, if the soil is of a rocky nature, frost heave can cause physical damage to the insulation of underground conductors, whether they are service conductors, feeders, or branch-circuit conductors. As a precautionary measure, one should lay a sand bed, install the conductors, and then cover them with a sand cover and back-fill with the earth that was removed. There should also be a loop at the point where they leave the ground, so that if the earth settles, the conductors won't be subjected to the pulling-out of terminations, and so on. Experience has also shown that if the conduit leaving the conductors at the point of their emergence from the earth is straight down, with an insulated bushing on the end of the conduit, frost won't cut the insulation as badly as it will if a 90-degree elbow is put on the end of the conduit. These are not *Code* requirements, but merely good procedures.

230.50: Protection of Open Conductors and Cables Against Damage—Aboveground

The following provides information about what is necessary to prevent physical damage to service-entrance conductors installed above-ground.

(A) Service-Entrance Cables. It is possible that service-entrance cables be installed where they may be exposed to physical damage, such as near driveways or coal chutes, near sidewalks or walkways, or where subject to contact with awnings, shutters, swinging signs, or similar objects; if so, they shall be protected by one or more of the following means: (1) by rigid metal conduit; (2) by intermediate metal conduit; (3) by rigid nonmetallic conduit suitable for the location; (4) by electrical metallic tubing; (5) by other approved means.

(B) Other Than Service-Entrance Cable. Open cables other than service-entrance cables and open conductors shall have a minimum clearance of 10 feet above grade level to protect them from physical damage.

Exception

MI and MC cables may be installed within 10 feet (3.05 m) from grade where they are not subject to physical damage, or are protected in accordance with Section 300.5(D).

230.51: Mounting Supports

The following covers mounting supports for individual open service conductors or cables.

(A) Service-Entrance Cable. Support service-entrance cables by straps or other approved means within 12 inches (305 mm) of the service head, gooseneck, or connection to a raceway or enclosure; otherwise, support the cables at intervals not to exceed 30 inches (763 mm).

(B) Other Cables. If the cables are not approved for mounting in contact with the building, they are to be mounted on insulators at intervals not to exceed 15 feet (4.57 m) and must have a minimum clearance of 2 inches (50.8 mm) from the surface over which they pass (Figure 230-8).

(C) Individual Open Conductors. Table 230.51(C) covers mounting of individual conductors. When exposed to weather, mounted on insulating supports, or attached to racks, brackets, insulators, or any other means, they shall be of a type approved by the authority that has jurisdiction or shall have a listing such as UL. If the conductors are not exposed to weather, you are permitted to mount them on glass or porcelain mounts. See Table 230.51(C), "Supports and Clearances for Individual Open Service Conductors," in the *NEC*.

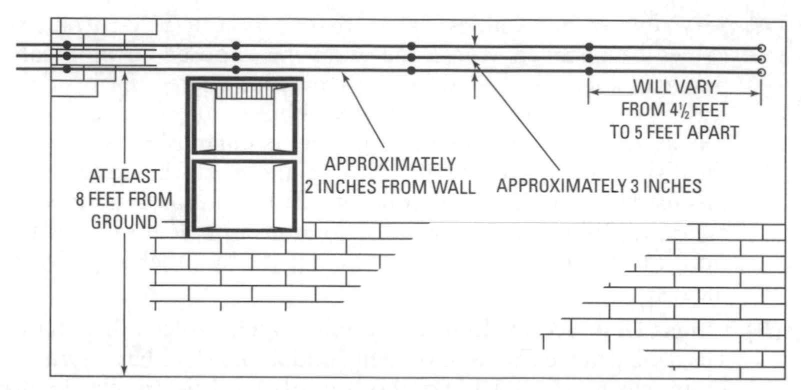

Figure 230-8 Service conductors run on an outside wall.

230.52: Individual Conductors Entering Buildings or Other Structures

Where individual conductors enter buildings, they are required to pass inward and upward through noncombustible and nonabsorptive insulating tubes, or through roof busings. The *Code* calls for drip loops. However, one should consider Section 230.54, which requires the service head to be above the service drop to prevent the entrance of moisture.

230.53: Raceways to Drain

Service-entrance raceways, when exposed to weather, shall be raintight and arranged to drain. If embedded in concrete they shall also be arranged to drain.

A particular incident is worth mentioning here: A large building housing turkeys had a service mast and service-entrance equipment. The service entrance was mounted in the enclosure where the turkeys were kept. The service mast had a rain-tight cap. However, the humidity was extremely high in the building and the temperature warm, so the moisture entered the service-entrance equipment, and as the warm air rose in the service mast, it came in contact with the cold. The moisture condensed and ran back down the service mast into the service equipment, causing some shorts. The author solved the problem by having duct seal placed at both ends of the service mast to stop the moisture from moving up the mast, and thus eliminated these shorting problems.

Exception

This exception allows for the use of flexible metal conduit (Greenfield), which is not rain-tight. The conduit must, however, be arranged to drain, as per Section 348.2.

230.54: Overhead Service Locations

(A) Rain-Tight Service Head. Always be sure that service-entrance raceways are equipped with approved rain-tight service heads.

(B) Service Cable Equipment with Rain-Tight Service Head or Gooseneck. Unless service cable is continuous from the meter enclosure or service equipment to the pole serving the same, rain-tight service heads shall be used or a gooseneck shall be formed. The connections shall then be taped and painted unless thermoplastic tape—that is, self-sealing tape-is used. This is to stop the siphoning of water down the cable.

(C) Service Heads above Service-Drop Attachment. To prevent water from entering the conductor, which causes some troubles due to siphoning between the conductor and its insulation, service entrance heads or goosenecks should always be located above the service drop.

There are, as usual, exceptions to this that might be required if it is impractical to meet the above requirements. Where necessary, to locate the service head below the service-drop conductors, this service head may be located at a point not to exceed 24 inches (610 mm) from the point of attachment.

There should also be a mechanical connector at the lowest point to prevent siphoning of water by the service-entrance conductors, as shown in Figure 230-9.

(D) Secured. It is essential that service cables be securely fastened in place to avoid possible damage.

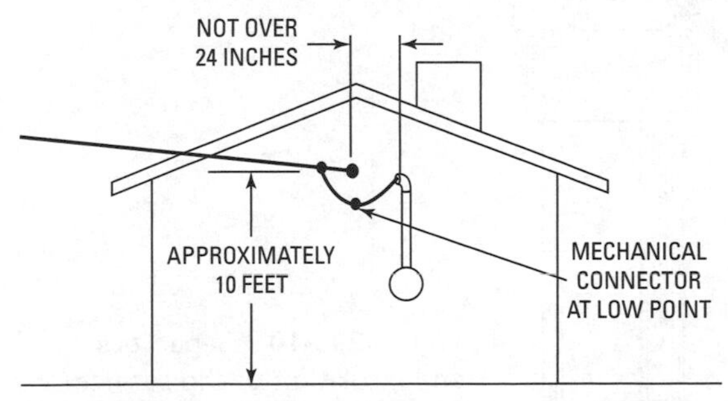

Figure 230-9 A connection to prevent siphoning of water into the service-drop head.

(E) Opposite Polarity Through Separate Bushed Holes. Service conductors brought through holes in the service-entrance head shall have the conductors of different polarities brought out through separate holes, thus avoiding a possible short between different phase conductors (Figure 230-10). This requirement is waived for jacketed service-entrance cables.

(F) Drip Loops. The drip loops required shall be formed on all conductors to prevent siphoning of moisture into the service equipment. Service conductors may be connected to service-entrance conductors in several ways. In one, the service-drop conductors shall be below the service-entrance head. In another, the service-drop conductors shall be below the termination of service-entrance cable sheath.

(G) Arranged So That Water Won't Enter Raceway or Equipment. In all cases, service-drop conductors must be arranged to prevent water entering the service entrance or the service-entrance equipment. This has been discussed just previous to this section.

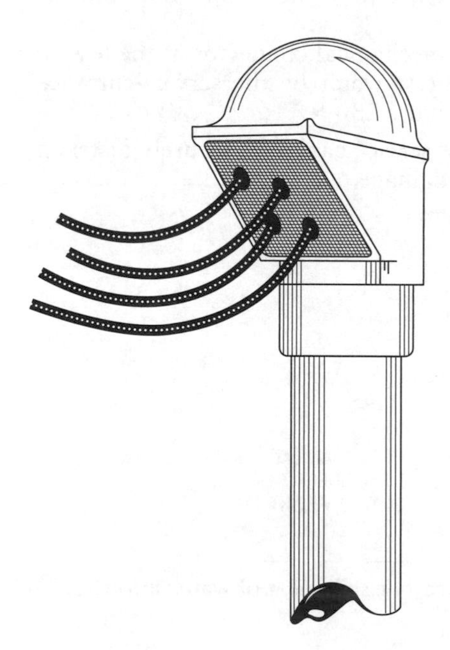

Figure 230-10 Separation and insulation of the service-entrance conductors to the service head.

230.55: Terminations at Service Equipment

Service-entrance raceway or cable must be terminated in service-entrance equipment such as boxes, cabinets, auxiliary gutters, or other effective means that enclose live parts to prevent easy access. An exception permits a raceway to terminate at a bushing when the service disconnecting means is mounted on a switchboard that has exposed bus bars on the back.

230.56: Service-Entrance Conductor with the Higher Voltage to Ground

If a four-wire delta with a midpoint of one grounded phase is used, one phase point to ground will give a higher voltage than the other phase of 208 volts, nominal, to ground. In the field we often refer to this as the wild leg, and every precaution must be taken to identify the wild leg on the outer finish, such as by orange color or by other effective means approved by the authority that has jurisdiction. This marking shall show up at every location or other connection place where it would be possible to connect the wild leg to the neutral.

V. Service Equipment—General

230.62: Service Equipment-Enclosed or Guarded

(A) **Enclosed.** With service equipment, the live parts shall be enclosed so as not to permit accidental contact. Thus service-entrance equipment usually has a dead front so that breaker panels only are exposed, and then a hinged door or removable cover to enclose these breaker handles; or as stated before, live parts in this equipment may be guarded as in (B) following.

(B) **Guarded.** Live parts not enclosed in such a way as on switchboards, panelboards, or control boards may be guarded as covered in Sections 110.18 and 110.27. Means shall be provided for locking or sealing doors that would give access to live parts.

230.66: Marking of Service Equipment

All service equipment must be marked by the manufacturer, noting that it is suitable for use as service equipment.

VI. Service Equipment—Disconnecting Means

230.70: General

A separate means shall be provided for disconnecting the service-entrance conductors from the building or structure wiring proper.

(A) Location. The service-disconnecting means should be located in a readily accessible point near the closest point of entrance of the service-entrance conductors. It may be a separate piece of equipment or a part of the service-entrance equipment panel where there are also branch circuits. The location of this disconnect may be either on the outside of the building or structure or inside the closest point where the service-entrance conductors enter.

Many inspectors require that the disconnecting means be set at a height of not to exceed 6½ feet above where a person would be standing when using the disconnecting means. Check with the inspection authority for the location. Bathrooms are not acceptable locations for service disconnects. Basements are generally acceptable locations if they are accessible.

(B) Marking. It is very important that the service disconnecting means be legibly marked so that it may be easily identified in case of electrical trouble or fire.

(C) Suitable for Use. It shall also be suitable for use for the prevailing conditions; for example, if it is outdoors, it shall be rain-tight. If service equipment is installed in hazardous (classified) locations, it shall comply with the requirements of Articles 500 through 517. (See Figure 230-11.)

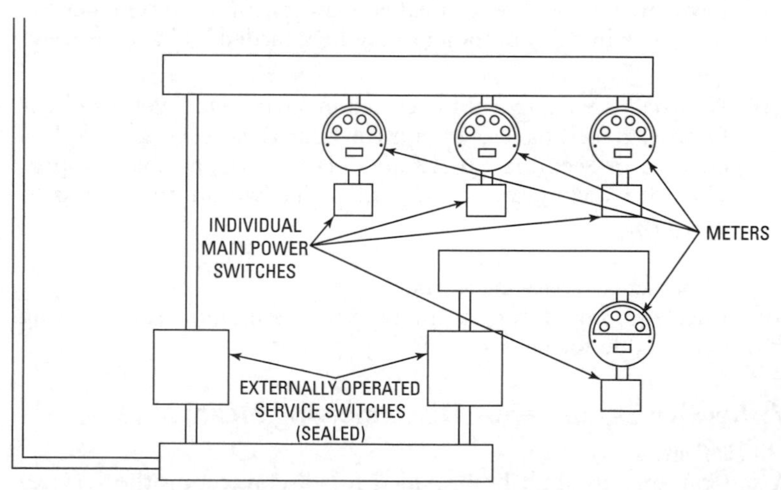

Figure 230-11 Typical service disconnect switches.

230.71: Maximum Number of Disconnects

(A) General. Section 230.2 covers disconnecting means for each service permitted under this section. Each set of service-entrance conductors that are permitted by Section 230.40, Exceptions No. 1, 3, 4, or 5 shall consist of not more than one disconnecting means in a single enclosure, but this is expanded to include also a group of separate enclosures for more disconnecting means. Remember, the maximum for disconnecting means that are grouped together or on a switchboard is six.

Exception

A disconnecting means that is used solely for control of ground-fault protection that has been installed as part of the equipment for ground-fault protection shouldn't be considered one of the not more than six disconnecting means.

(B) Single-Pole Units. On two- or three-pole services, single switches or breakers may be used, provided they have the capacity and are equipped with handle ties or master handle (this would be two or three breakers controlled by one handle) so that all disconnecting means to the service equipment open the ungrounded conductors with not more than six operations of the hand. The six operations of the hand are included to cover situations where you have up to six disconnecting means. More will be covered in Section 408.16(A), which covers service equipment that is in the panelboard.

This is commonly known as the six-operations-of-the-hand rule. Note the mention of multiwire circuits, and refer to the definition of a multiwire circuit in Article 100.

See Figures 230-12 through 230-14.

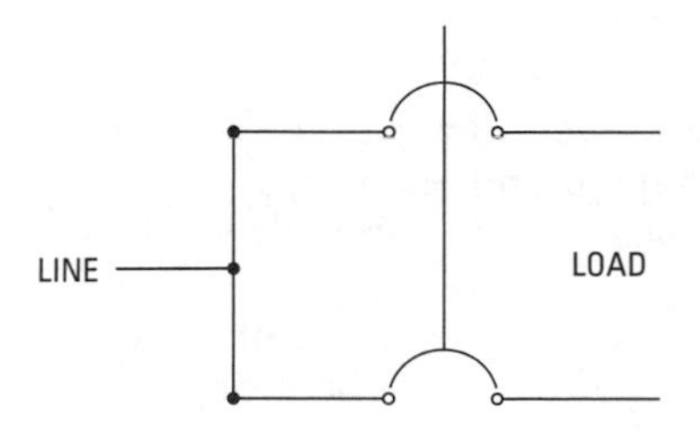

Figure 230-12 A two-pole circuit-breaker switch that cannot be used as a service disconnect.

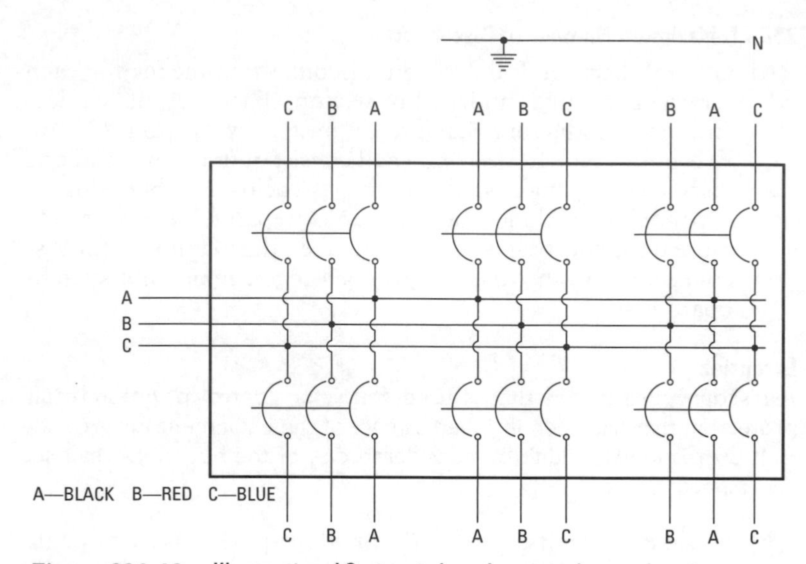

Figure 230-13 Illustrating 18 circuit breakers tied together in groups of three.

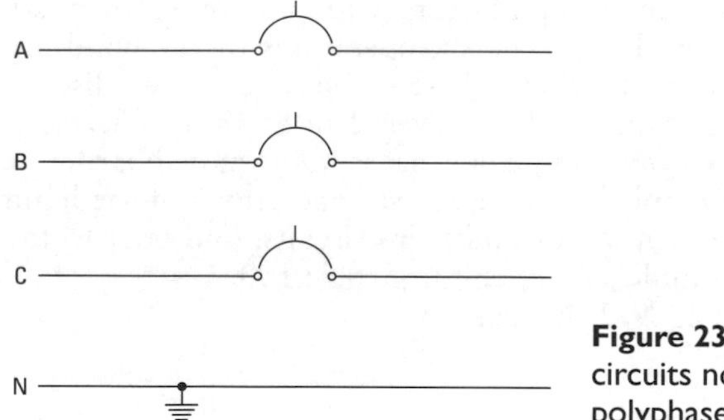

Figure 230-14 Multiwire circuits not supplying polyphase motors.

230.72: Grouping of Disconnects

(A) General. The two to six disconnecting means covered in Section 230.71 are to be grouped, and each separate one of the two to six disconnecting means shall be plainly marked as to what load each serves.

If one of the disconnecting means of Section 230.71 is used for fire protection only, it does not have to be grouped with the other disconnecting means, but may be in a remote place. As one can readily see, in the event of fire this disconnect shouldn't be as readily turned off as the rest of the disconnecting means. During a fire, this is the one disconnect that should be left alone.

(B) Additional Service-Disconnecting Means. To be compatible with Sections 700.12(D) and (E), which cover where one or more additional service-disconnecting means for fire pumps or for emergency services are installed, additional service-disconnecting means shall be sufficiently remote from the one-to-six service-disconnecting means for normal services, so that there will be no chance or at least minimal chance of simultaneous interruption of supply.

(C) Access to Occupants. In multiple-occupancy buildings it is essential that all occupants of the building have ready access to the disconnecting means serving their apartments.

Exception

In a multiple-occupancy building, if the building and electrical maintenance is provided by the management and this service is under continual supervision, it is not necessary that the disconnecting means be accessible to each occupant, because management has continuous service available.

230.74: Simultaneous Opening of Poles

The service disconnect must be capable of disconnecting all ungrounded conductors from the building or structure wiring.

230.75: Disconnecting Grounded Conductors

The disconnecting means may be designed so that it also opens the grounded conductor simultaneously with the ungrounded conductors. Where the disconnect does not accomplish this, there shall be other means provided in the service cabinet for disconnecting the grounded conductor from the interior wiring. This is usually accomplished by pressure-type connectors for the grounded conductor. The service-entrance ground should always be attached to the service side of the disconnecting means. This is required so that there will be ground on the service, even though the grounded conductors on the interior wiring have been disconnected. In multisection switchboards, a disconnecting means for the grounded circuit conductor may be in any of the switchboard sections, provided the section is so marked.

230.76: Manually or Power-Operable

A manually operable switch for disconnecting means of ungrounded service conductors or a circuit breaker with common trip or equipped with a handle that will operate the opening of all ungrounded conductors may be used; or, a power-operated switch or circuit breaker may be used as a disconnecting means, provided such equipment can also be operated by hand in case of trouble or power failure.

230.77: Indicating

Service disconnecting means must be plainly marked to show whether they are open or closed.

230.79: Rating of Disconnect

Article 220 covers calculation of load. The service-disconnecting means shouldn't be less than the load that is calculated, and in no case shall the service-disconnecting means be rated smaller than required in (A), (B), or (C) below:

(A) **One-Circuit Installation.** When limited loads are served by a single branch circuit, the rating of the service-disconnecting means shouldn't be lower than 15 amperes.

(B) **Two-Circuit Installations.** For installations that don't supply more than two 2-wire circuits, the minimum size of service-disconnecting means shouldn't be less than 30 amperes.

(C) **One-Family Dwelling.** Single-family dwellings shall have a service-disconnecting means of not less than 100 amperes, three-wire, where either of the following conditions exists: (1) If the calculated load is 10 kilo-volt-amperes or more, or (2) where six or more two-wire branch circuits are installed on the original installation.

The preceding conditions in no manner prohibit the installation of larger services and service disconnecting means to take care of any future added loads.

(D) **All Others.** Sixty amperes is the smallest service-disconnecting means that may be used for all other installations.

230.80: Combined Rating of Disconnects

Where more than one switch or circuit breaker is used as a disconnecting means in accordance with Section 230.71, the combined ratings of the switches or circuit breakers that are used shall be no less than what would be required had only one switch or circuit breaker been used.

230.81: Connection to Terminals

Soldered connections are never to be used for connecting service conductors to service disconnecting means. The conductors must be connected by means of pressure connectors, clamps, or other approved means.

230.82: Equipment Connected to the Supply Side of Service Disconnecting

Ordinarily, no circuits shall be connected ahead of the disconnecting means. There are, of course, exceptions to the above, as follows:

Exception No. 1

Any current-limiting device, including cable limiters.

Exception No. 2

Meter installations of less than 600 volts, nominal, and housed in a metal enclosure, if the metal enclosures are grounded as covered in Article 250.

Exception No. 3

Among the items that may be connected ahead of the disconnecting means are instrument transformers, consisting of current transformers and voltage transformers, and high-impedance shunts; in many places you will find that surge protection devices are also required. One such place is a grain elevator. These must be connected ahead of the disconnecting means. With surge protection, you will later find places where capacitors are used and fed through a fuse disconnecting means. These are connected on the load side of the service disconnecting means. Lightning arresters have to be connected ahead of any fusing or service disconnecting means. They are often mounted where the service drop and service conductors are spliced together or sometimes out of the disconnecting means. These take care of such surges as lightning surges, and so on.

Exception No. 4

This is an exception for connecting what may be called load-management devices, which consist of essential circuits for emergency systems. Among these emergency systems, fire pumps and fire sprinkler alarms, if supplied by the service equipment (many places, such as hospitals and places where electrical service must be maintained, have stand-by power systems that come on to pick up the load when power fails), are to be connected ahead of the disconnecting means so that they do not come on accidentally. See Figures 230-15 and 230-16 below.

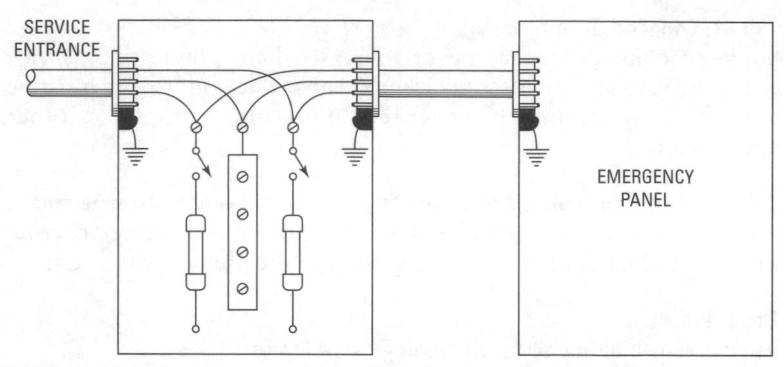

Figure 230-15 Unapproved means of connecting an emergency panel.

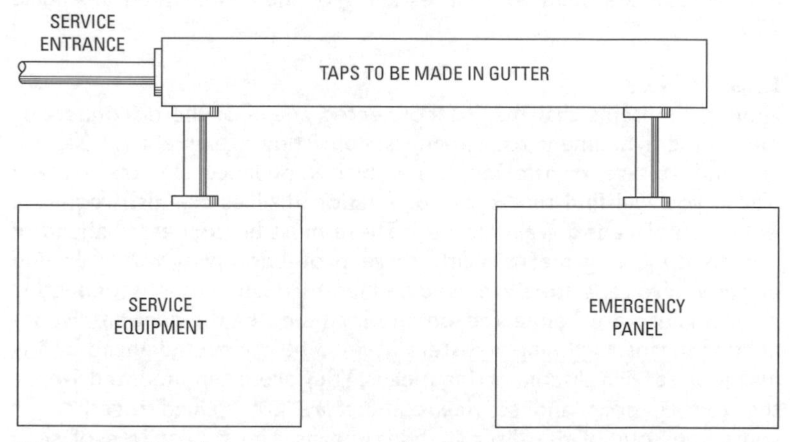

Figure 230-16 Approved means of connecting an emergency panel.

Exception No. 5
Solar electricity systems or interconnected power sources may be tapped ahead of service. Refer to Articles 690 and 705.

Exception No. 6
Where suitable overcurrent protection and disconnecting means may be provided and may be connected ahead of the regular service disconnecting means, if it is operated from the regular power. This exception covers control circuits for such.

Exception No. 7

If suitable overcurrent protection and disconnecting means are provided on ground-fault protection systems that are listed as part of the equipment, their installation may be connected ahead of the main service disconnecting means.

VII. Service Equipment—Overcurrent Protection

230.90: Where Required

Every ungrounded service conductor must have overcurrent protection. The various locations where this may be mounted were covered earlier.

(A) **Ungrounded Conductor.** This overcurrent protection is to be in series with the ungrounded service conductor it is protecting. The ampere rating or setting of adjustable overcurrent protection must never be rated over the ampacity of the service conductors it protects.

An example of the above would be that a 100-ampere service should have No. 3 THW conductors or larger, and a 300-ampere service should have 350 kcmil THW or larger conductors. Any derating that might be applicable in Article 310 must always be considered.

Exception No. 1

Motor-starting currents will have to be dealt with separately, and the ratings must be in conformity with Sections 430.52, 430.62, or 430.63. Although not stated in this section of the *Code*, you should always take into consideration what might be added in the future, and, if possible, make allowances for this at the time of the installation.

Exception No. 2

Where nonadjustable circuit breakers don't conform to the ampacity of the conductor, the next larger size may be used, provided that it is 800 amperes or less. If the breaker is adjustable, it shouldn't be set at a rating of more than 125 percent of the ampacity of the conductor. When fuses are used, if there is not a standard size to fit the ampacity of the conductor, the next larger size may be used, provided that it is 800 amperes or less. See Section 240.3, Exception No. 1 and Section 240.6.

Exception No. 3

As was stated in Section 230.71, there shouldn't be more than six circuit breakers or six sets of fuses for any service disconnecting means. Here, one must remember that a single-pole, a two-pole, or a three-pole breaker may count as one.

Exception No. 4

Fire Pumps. If we have a fire pump room for the service to the fire pump, it may be judged outside of the building, and the provisions above shouldn't apply. Locked-rotor current of the fire pump motor may be selected to carry the fire pump locked-rotor current indefinitely. Note: Here you are referred to another fire code, *NFPA 20-1983*, or to the ANSI standard, *Centrifugal Fire Pumps*. This is one point where other NFPA fire codes are recommended to be used with the *NEC*.

In considering the overload current protection for fire pumps, a set of fuses protecting all ungrounded conductors of the circuit is adequate. If single-pole breakers are used and grouped together, as previously covered, by a tie handle that opens all the breakers to the ungrounded protection device, they are also permitted. The grounded conductor must be connected to the service equipment and grounding electrode ahead of the fuse in the grounded conductor.

Exception No. 5

For 120/240-volt, three-wire services, overcurrent protection may be in accordance with Section 310.15(B)(6).

(B) Not in Grounded Conductor. There is to be no overcurrent device in the grounded service conductor. The exception to this is that the circuit breaker shall simultaneously open all conductors of the circuit. A little explanation of this might be in order. Should the grounded conductor overcurrent device open by itself and the ungrounded conductor overcurrent devices not open, an unbalanced voltage condition would exist that might cause damage to equipment (Figure 230-17).

230.91: Location of Overcurrent Protection

(A) Location. A no-fuse disconnect may be used, but the service overcurrent device shall be located immediately thereto. Doing it this way just means more expense. Section 230.70(A) indicates that the service disconnecting means must be located either inside or outside of the building or structure at the closest point of entrance.

(B) Access to Occupants. Occupants shall have access to the overcurrent protection devices in multiple-occupancy buildings.

An exception to this in Section 240.24(B).

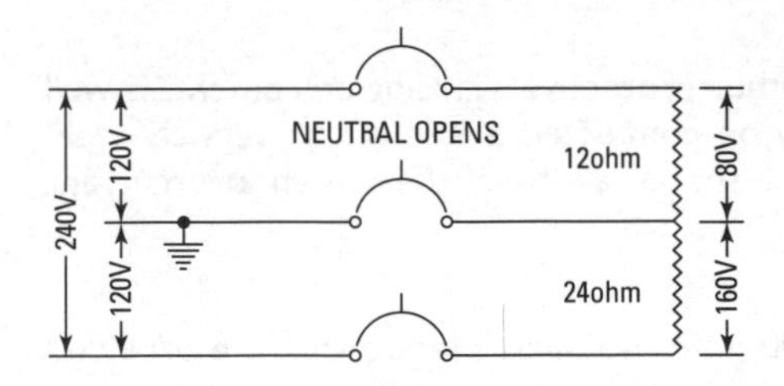

Figure 230-17 A breaker in the neutral. The neutral circuit breaker must be open simultaneously with the breaker contacts in the phase wires.

230.92: Locked Service Overcurrent Devices

If the service overcurrent devices are locked or sealed or otherwise not readily accessible, the feeder or branch-circuit overcurrent devices have to be located and readily accessible on the load side of the service overcurrent device. The circuits in this branch-circuit panel should be of lower rating than these service overcurrent devices. Since the branch-circuit devices will have a smaller rating than the inaccessible service overcurrent protection device, there will be less likelihood that it will trip.

230.93: Protection of Specific Circuits

To prevent tampering, it might become necessary to provide a locked or sealed automatic overcurrent device to serve some specific load, such as a water heater. Such a locked or sealed device must be in an accessible location.

230.94: Relative Location of Overcurrent Device and Other Service Equipment

The overcurrent devices are installed for protection and must protect all circuits and devices.

The service switch may be on the supply side. In fact, it has been repeatedly said that the service-disconnecting means must be on the supply side.

Exception No. 1

As permitted in Section 230.82, the following may be installed on the supply side of the service disconnecting means: high-impedance shunt circuits, lightning arresters, surge-protective capacitors, and instrument transformers (current and potential).

Exception No. 2

Surge suppressors, instrument transformers, and high-impedance shunts are allowed to be connected ahead of the service disconnect.

Exception No. 3
Circuits for fire alarms and other protective signaling equipment, as well as fire pump equipment, may be connected ahead of the service overcurrent protection, provided that they have their own overcurrent protection.

Exception No. 4
If the voltage is not over 600 volts, nominal, meters that are provided with metal housing and service enclosures are properly grounded, as described in Article 250.

Exception No. 5
If service disconnecting equipment is operated by power, the control circuit should be mounted ahead of the service equipment with proper overcurrent protection and a disconnecting means provided, so that if overcurrent devices in the service disconnecting means should open, current is still available from an electric control circuit.

230.95: Ground-Fault Protection of Equipment
Fault currents that are available, plus the damaging effects of ground faults, have made it necessary to do something about injury to personnel and damage to equipment.

Where service disconnecting means are rated 1000 amperes or more, ground-fault protection of equipment must be installed when the system is a solidly grounded wye system that has more than 150 volts to ground and does not exceed phase-to-phase voltages of 600 volts.

Exception No. 1
On industrial applications the disconnection in a nonorderly shutdown will cause increased hazards. This exception shouldn't apply to service-disconnecting means.

Exception No. 2
Service to fire pumps does not apply with the above provisions.

(A) Setting. The ground-fault maximum setting shall be 1200 amperes and the maximum delay shall be only one second where the ground-fault is equal to or greater than 3000 amperes. This ground-fault protection shall operate the service disconnecting means opening all ungrounded conductors of the faulted circuit.

(B) Fuses. Where fuses are used as the overcurrent protection, they shall be capable of interrupting any current higher than the interrupting capacity of the switch involved at any time when the ground-fault protection system does not cause a switch to be opened. Fuse manufacturers give the fuse rating and the ground-fault capabilities of the fuse interruption.

Note

A service-interrupting rating of the disconnecting means shall be based on the current rating of the largest fuse that could be installed therein, even though fuses of lower rating must be used. Circuit breakers that are adjustable for current ratings shall have disconnecting means installed based on the highest setting of the circuit breaker, even though lower rating of the circuit breaker is to be used.

Although 1000 amperes is the requirement in this section, it is recognized that it is desirable to have solidly grounded systems of less than 1000 amperes protected by ground-fault protection when the system is over 150 volts to ground and not exceeding 600 volts phase-to-phase.

As used in this section, the term solidly grounded means that the grounded conductor (neutral) is solidly grounded and not grounded through a resistive or impedance device.

Ground-fault protective equipment that opens the service-entrance equipment will protect said equipment and the equipment and conductors on only the load side of the service disconnecting means, and won't protect the service-entrance conductors.

The addition of ground-fault protective equipment at the service disconnecting means will by necessity require that the overcurrent devices used on the system elsewhere be coordinated by proper selection of the overcurrent devices downstream. Additional ground-fault protection will be needed downstream where the maximum continuity of service is required.

(C) Performance Testing. The *Code* requires that the ground-fault protection system be performance tested when first installed on site. Approved instructions are to be furnished with the equipment and the test run according to the test instructions.

You may be required to make a record of said test in writing and make this written record available to the authority that has jurisdiction.

VIII. Service Exceeding 600 Volts, Nominal

230.200: General

Service conductors and equipment used on circuits over 600 volts, nominal are covered in the following material, but previous sections also apply. Where the equipment is on the supply side of the point of service, the provisions of this article don't apply. There is a definition not found in Article 100 that appears in the *NEC* just after the first part of Section 230.200: "Service-Point."

Note

For circuits over 600 volts, you should also refer to the *National Electrical Safety Code (ANSI C2-2002)*.

230.201: Service Conductors

The service conductors are those conductors that run between the service point and the service disconnecting equipment.

230.202: Service-Entrance Conductors

The following parts shall be used to identify where service-entrance conductors to a building or separate enclosure are to be installed.

(A) **Conductor Size.** Conductors in a multiconductor cable shouldn't be less than No. 8, or if not in a cable, shouldn't be less than No. 6.

(B) **Wiring Methods.** Several methods of installing service-entrance conductors are listed in your *NEC* in Section 710.4.

Section 710.4(B) covers underground service-entrance conductors.

Approved cable tray systems may be used to support cables approved for use as service-entrance conductors. See Article 392.

Section 310.6 covers shielding on conductors.

(C) **Open Work.** See the *NEC*.

(D) **Supports.** In the event of short circuits, extra strain is placed on the supports due to magnetic fields. The supports, including the insulators, shall have sufficient strength to withstand this extra strain.

(E) **Guarding.** Qualified persons only shall be permitted access to open wires, and the necessary guarding to restrict access to such persons must be installed.

(F) **Service Cable.** Potheads or other suitable means shall be used to protect the conductors where they emerge from a metal sheath or raceway. This protection shall be against moisture and physical damage.

(G) **Draining Raceways.** Unless raceways have conductors that are identified for that use, they shall be arranged to drain where they are embedded in masonry, exposed to the weather, or in wet locations.

(H) **Over 15,000 Volts.** Refer to Sections 450.41 through 450.48. Conductors are required to enter either a metal enclosure for the switchgear or a transformer vault.

230.203: Warning Signs

High voltage warning signs must be posted where unauthorized persons might come in contact with live parts.

230.204: Isolating Switches

(A) **Where Required.** An air-break isolating switch must be installed on the supply side of the disconnecting means and on any equipment associated with the installation where oil switches or air or oil circuit breakers are used for the service disconnecting means. Air-break switches can be observed to determine if they are open, whereas oil switches and circuit breakers can't readily be checked.

Exception

Where such equipment is automatically disconnected from circuit breakers and all live parts (this applies to such equipment as is mounted on removable truck panels or metal-enclosed switch gear units that can't be opened unless the circuit is disconnected).

(B) Fuses As Isolating Switch. When fuses are such that they may be operated as a disconnecting switch, a set of fuses is acceptable as the isolating switch if the following conditions are met: (1) The oil disconnecting means is a nonautomatic switch, and (2) the set of fuses disconnects the oil switch and all associated service equipment from the service-entrance conductors.

(C) **Accessible to Qualified Persons Only.** Qualified persons only shall have access to isolating switches.

(D) **Grounding Connection.** See *NEC.* Take note of the grounding, as mentioned. It serves as a precaution should a switch accidentally become energized. Also with shielded-type, high-voltage cables of any great length, the cable and the shield become a fair-sized capacitor, and the grounding bleeds the charge away, thus preventing what might be a painful shock.

If a duplicate isolating switch is maintained by an electric supply company, a means for grounding need not be installed.

230.205: Disconnecting Means

(A) **Location.** Sections 230.70 or 230.208(B) give the location of the disconnecting means.

(B) **Type.** All ungrounded conductors shall be capable of being open at the service disconnect at the same time. They shall also be capable of being closed on ground faults that are equal to or greater than the maximum current available on such a ground fault.

(C) **Remote Control.** In multibuilding industrial facilities, the service disconnecting means is allowed to be located in a separate building or structure, but only if it is operated by an electric switch that is located at the generally specified location.

230.206: Overcurrent Devices as Disconnecting Means

Where the circuit breakers or fuses specified in Section 230.208 meet the requirements for service overcurrent devices that were covered in Section 230.205, they will constitute the service disconnecting means.

230.207: Equipment in Secondaries

See the *NEC.*

230.208: Overcurrent Protection Requirements

See the *NEC.*

230.209: Surge Arresters (Lightning Arresters)

Here the requirements in Article 280 for surge protection and lightning arresters or ungrounded overhead service conductors, if required by the authority that has jurisdiction, shall be placed on the supply side of this equipment.

This is not very specific, and decisions should be made based on the number of lightning disturbances.

230.210: Service Equipment—General Provisions

See the *NEC.*

230.211: Metal-Enclosed Switchgear

See the *NEC.*

230.212: Services Over 15,000 Volts

See the *NEC*.

Article 240—Overcurrent Protection

240.1: Scope

This article is arranged so that Parts I through VII provide the general requirements for overcurrent protection and overcurrent protective devices. These are general requirements; Part VIII covers supervised industrial installations operating at no more than 600 volts, and Part IX covers the overcurrent protection for over 600 volts, nominal.

Conductors have specific current-carrying capacities (ampacity) for different sizes, for different insulations, and for different ambient-temperature conditions. The purpose of overcurrent protection is to protect electrical systems in general, and the insulation of the conductors in particular, from damage caused by the current reaching too high a value.

Refer to Sections 110.9 and 110.10 for requirements covering interrupting capacity and protection against fault currents.

I. General

240.2: Protection of Equipment

Equipment must be protected by overcurrent devices. The list in this section of the *NEC* covers the equipment that is to be protected. See the *NEC*.

240.3: Protection of Conductors

It is the intent that conductors be protected according to their ampacities, as specified in Section 310.15, except as permitted by (A) through (G) below.

- **(A)** **Power-Loss Hazard.** If power loss will create a hazard, as in circuits handling material with a magnet, conductor overcurrent protection shouldn't be required. Short-circuit protecting devices shall be provided.
- **(B)** **Devices Rated 800 Amperes or Less.** There are standard ampere ratings for fuses and circuit breakers where the standard rating is not high enough for the ampacity of the conductors. You can use the next higher fuse or circuit breaker rating, or if the circuit breaker has adjustments higher than the standard rating, you may use a higher rating on the circuit breaker.

(C) Devices Rated Over 800 Amperes. When an overcurrent device is rated over 800 amperes, the conductors connected to it must be equal to or greater than the rating of overcurrent devices specified in Section 240.6.

(D) Small Conductors. Except where specifically permitted in (E), (F), or (G) that follow, the minimum overcurrent protection for No. 14 wire is 15 amperes, 20 amperes for No. 12 wire, and 30 amperes for No. 10 wire.

(E) Tap Conductors. Tap conductors may have overcurrent protection according to Sections 210.19(D), 240.21, 368.11, 368.12, and 430.53(D).

(F) Transformer Secondary Conductors. Section 450.3 covers secondary two-wire, single-phase conductors supplied by a transformer supplying only a two-wire secondary. The secondary is considered protected on the primary overcurrent protection of the transformer. If the primary overcurrent device does not exceed the value that is found by multiplying a secondary conductor ampacity by the secondary to primary transformer voltage ratio, then the secondary shall be considered to be protected by the overcurrent device of the primary.

(G) Overcurrent Protection for Specific Conductor Applications. See the *NEC* for specific exceptions.

240.4: Protection of Fixture Wires and Cords

In this section the overcurrent protection for tinsel cords and other flexible cords and their ampacities are covered in *NEC* Table 400-5. Supply overcurrent protection, as covered in Section 240.10, shall be permitted and be accessible for the cords.

Refer to the *NEC* for specifics on lamp cords, fixture wires, and extension cord sets.

240.6: Standard Ampere Ratings

Fuses and inverse time circuit breakers come in standard ratings. These are given in the *NEC* at this point.

Exception No. 1

There are additional standard ratings for standard Edison plug fuses and S-type plug fuses. More types are covered, but lower-than-30-ampere plug fuses (either type) are essential for proper protection of small loads.

In referring to standard ratings, note that when using adjustable circuit breakers, where the adjustment is readily accessible, the maximum setting of this overload adjustment shall be considered to be the maximum load resistance of this circuit breaker.

Exception No. 2

When current-adjustable circuit breakers are located behind sealable covers, bolted enclosure doors, or locked doors accessible to qualified personnel only, the setting does not have to be based upon the maximum setting of the breaker. This allows for the use of solid-state, adjustable-trip circuit breakers at a standard rating that is less than the maximum setting.

240.8: Fuses or Circuit Breakers in Parallel

Overcurrent devices consisting of fuses and/or circuit breakers shouldn't be arranged or installed in parallel.

Exception

If circuit breakers or fuses are assembled at the factory for paralleling and are listed for that purpose, they may be approved as being just one unit.

240.9: Thermal Devices

Thermally operated relays and other devices are operated by heat generation and a strip in the device that causes them to open. But remember, they are not designed to be used for short circuits. They may be used for motor protection as explained in Section 430.40. Fuses or circuit breakers that have a rating that does not exceed four times the rating of the motor shall be used ahead of the thermal device (Figure 240-1).

240.10: Supplementary Overcurrent Protection

Any supplementary overcurrent protection that is used in connection with lighting fixtures, appliances, or other utilization equipment, or for internal circuits and components of equipment, does not in any manner take the place of nor is deemed a substitute for the branch-circuit protection. Neither is it the intent that any such supplementary overcurrent protection device be subject to the accessibility required for branch-circuit protective devices. See Article 210.

240.11: Definition of Current-Limiting Overcurrent Protective Device

See the *NEC*.

240.12: Electrical System Coordination

The coordination of overcurrent devices has always been a problem, which too often has been overlooked. The *NEC* has taken this into consideration.

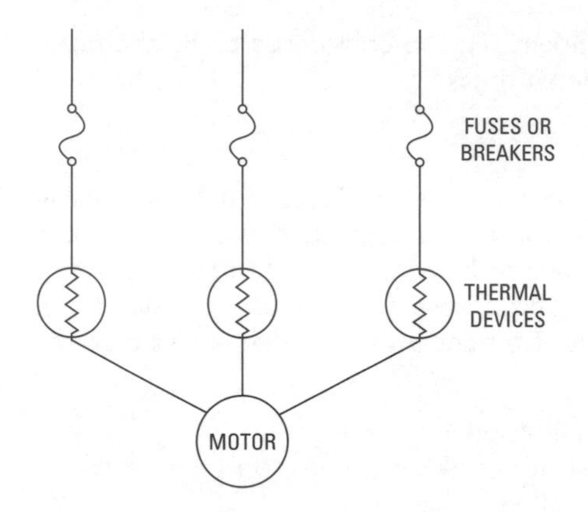

Figure 240-1 Thermal protection for motors.

Often, an orderly shutdown is vital to minimize hazard(s) to personnel as well as equipment problems. In this case, two conditions should be the basis of a system of coordination: coordinated short-circuit protection, and monitoring systems or devices that provide an indication of overload.

Note

Fault-current protection shall be so designed that in the case of a fault that is localized at one point of a system, it is taken care of by properly designed fault-current protection to open that point only, and thus not shut off the entire service to a building.

One example of such a system is a wye supply to motors with a resistance in the ground circuit, limiting ground-fault current to a predetermined value, such as possibly 10 amperes. A relay is connected across the ground resistor, or in series with it, which actuates an alarm.

A low-amperage ground fault in one phase won't cause a shutdown, but it also won't notify personnel that a ground fault has developed so that they can locate the ground and arrange for a proper shutdown for repairs. Two phases with ground faults at the same time, of course, would cause immediate shutdown.

240.13: Ground-Fault Protection of Equipment

Equipment must be provided with ground-fault protection in accordance with Section 230.95 when connected to a solidly grounded 277/480-volt wye system where the service disconnecting means is rated 1000 amperes or more. This rule does not, however, apply to the disconnecting means for industrial processes where a disorderly shutdown may be hazardous; it does not apply to fire pumps, and does not apply when the service has ground-fault protection that is not nullified by neutral connections to grounding electrodes.

II. Location

240.20: Ungrounded Conductors

(A) Overcurrent Device Required. An overcurrent device (fuse or overcurrent trip unit of a circuit breaker) shall be placed in each ungrounded conductor. An equivalent overcurrent trip unit may be a combination of a current transformer and a relay to open the overcurrent.

Note
See Article 430, Parts III, IV, VI, and XI.

(B) Any Circuit Breaker As Overcurrent Device. Any circuit breakers shall open all of the ungrounded circuit conductors.

Exceptions
Individual single-pole circuit breakers may be used for the protection of each ungrounded conductor of three-wire, direct-current or single-phase circuits, or for each ungrounded conductor of light or appliance branch circuits connected to four-wire, three-phase systems, or five-wire, two-phase systems, provided that lighting or appliance circuits are supplied from a system that has a grounded neutral and that no conductor in such circuits operates at a voltage greater than permitted in Section 210.6. See Figures 240-2 and 240-3.

240.21: Location and Circuit

Overcurrent devices of circuits shall be located at the point where the service to those circuits originates.

Smaller Conductor Protected. Tables 310.16 through 310.86 cover overcurrent protection for the larger conductor and provide for smaller conductors.

(A) Feeder and Branch-Circuit Conductors. Unless specified otherwise by specific Code sections, overcurrent devices are required for branch circuits and feeders at their point of supply.

(B) Feeder Taps Not Over 10 Feet (3.05 m) Long. If all of the following conditions are met, tap conductors from a feeder or a transformer secondary may be used.

(1) Not over 10 feet of tap conductor length.

(2) (a) Tap conductors supply a load computed not less than that for the circuit supplied, and (b) the taps are not less

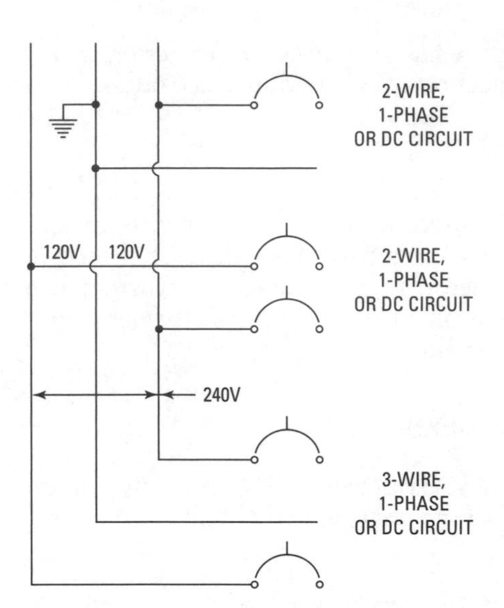

Figure 240-2 Overcurrent protection for a three-wire, single-phase system.

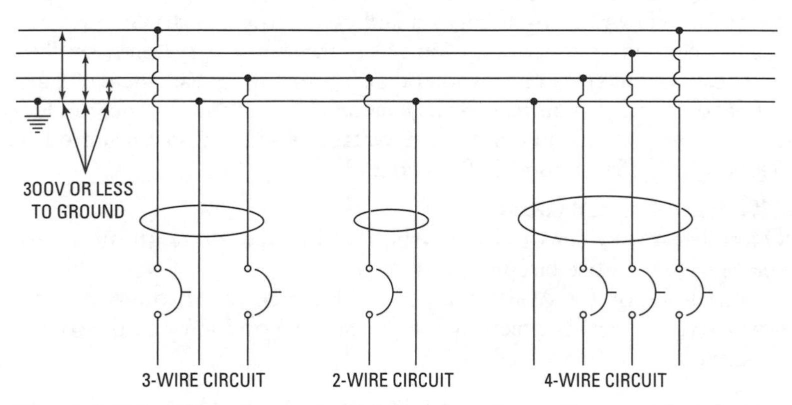

Figure 240-3 Overcurrent protection for a four-wire grounded system.

than the rating of the device involved, or (c) the tap conductors are not less than the overcurrent device where the tap conductors terminate.

(3) Tap conductors don't extend beyond any control devices or panelboards they supply.

(4) Where tap conductors supply switchboard, panelboard, or control devices, or the back of an open switchboard, they are enclosed in a raceway that extends from where the tap terminates. The only exception to this requirement is at the point of connection to the feeder.

Lighting and branch panelboards are covered in Section 408.16(A).

For explanation of some of the above points, see Figure 240-4. It appears that with transformers the only place this might apply is with unit load centers that have a transformer with a primary overcurrent device, short bus bars, and a series of breakers or fuses that don't serve as a lighting or appliance branch-circuit panelboard. This has been a question for some time and to all appearances is clarified here.

(5) The overcurrent device on the line side of the tap conductor doesn't exceed 1000 percent (ten times) the ampacity of the tap conductor.

(C) Feeder Taps Not Over 25 Feet (7.62 m) Long. The following conditions must be met for taps from a feeder where the tap conductors do not exceed 25 feet:

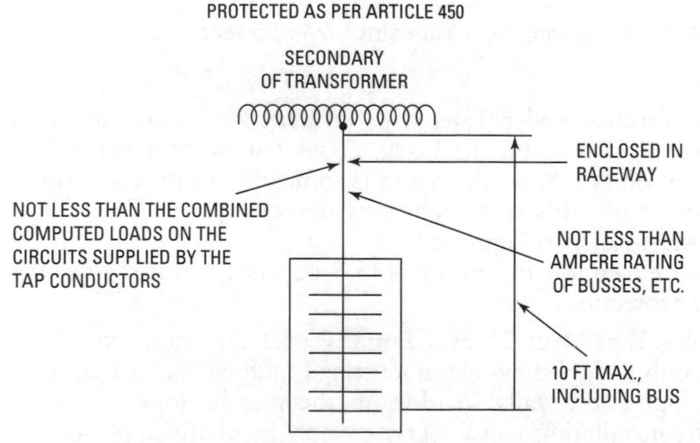

Figure 240-4 Illustrating a tap circuit not to exceed 10 feet in length.

(1) The ampacity of the taps are not less than one-third the ampacity of the feeder conductors or overcurrent protection from which they originate.

(2) The tap conductors terminate in a circuit breaker or a single set of fuses to limit the load to the ampacity of the tap conductors. Any such circuit breaker or set of fuses located at the tap load device may supply any number of additional loads from this tap conductor overcurrent device.

(3) The tap conductors are in a raceway suitable for protecting them from physical damage (Figure 240-5). Section 230.91 provides for service-entrance conductors where protected in the section just mentioned. See Figures 240-6, 240-7, and 240-8.

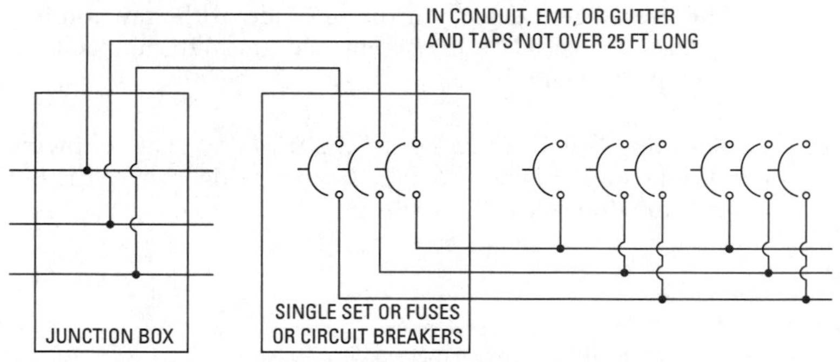

Figure 240-5 Illustrating tap circuits not over 25 feet long.

(D) Transformer Feeder Taps with Primary Plus Secondary Not Over 25 Feet (7.62 m) Long. A very important exception, Exception No. 8, is added and is somewhat misunderstood at times. It pertains to a separately derived system from transformers. Refer to Figure 240-9.

An exception in Section 445.5 covers generator overcurrent protection.

(E) Feeder Taps Over 25 Feet Long. Feeder taps may exceed 25 feet only in high-bay manufacturing buildings that are over 35 feet high at the walls. In addition, this may be done only if the entire installation will be serviced only by qualified personnel. In these cases, the horizontal length of the tap may be up to 25 feet, and the overall length of the run may be up to 100 feet.

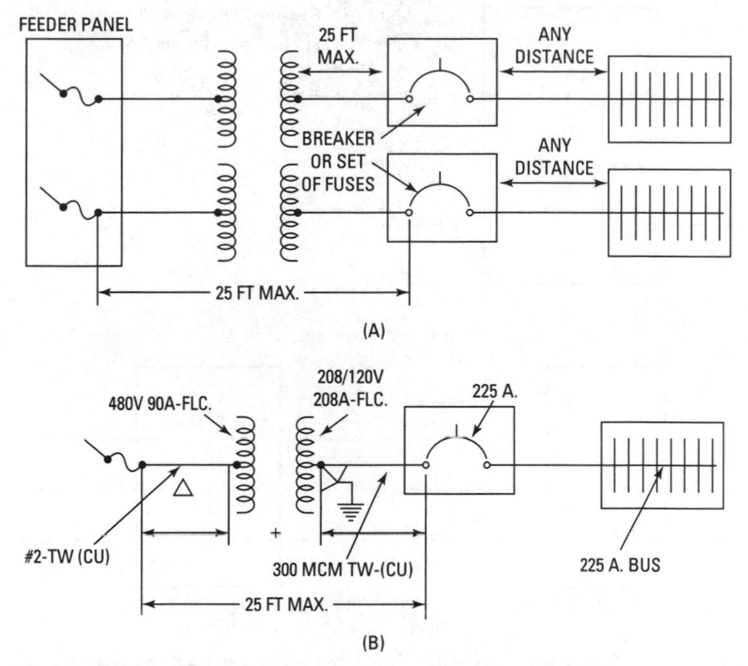

Figure 240-6 Transformer feed taps not over 25 feet long.

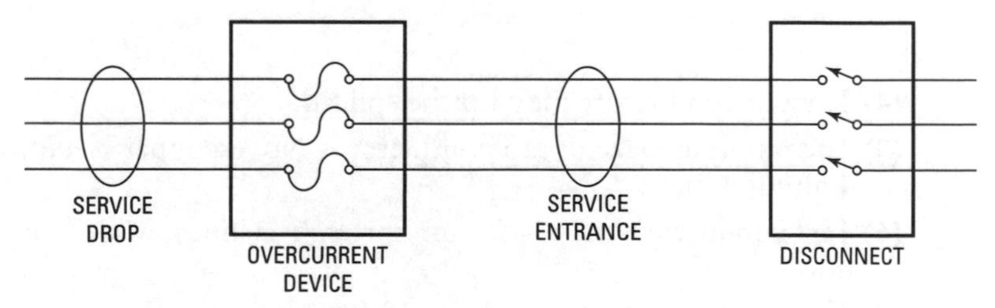

Figure 240-7 Separate disconnect and overcurrent devices that are separated because of service-entrance length.

Additional rules:

(1) The tap conductors must have an ampacity of at least one third that of the feeder overcurrent device.

(2) The tap conductors must end at a single overcurrent device (or set of fuses).

(3) The tap conductors must be physically protected.

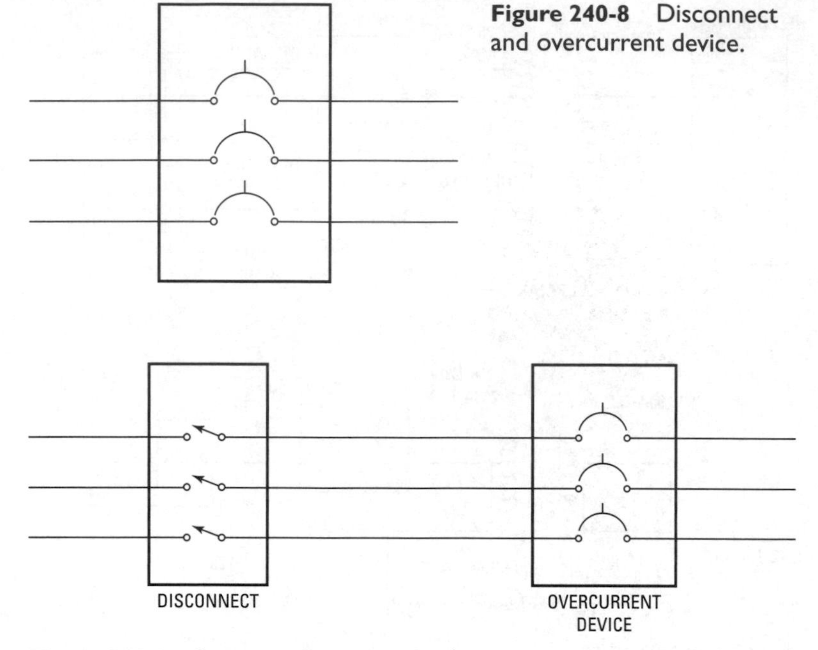

Figure 240-8 Disconnect and overcurrent device.

Figure 240-9 Separate disconnect and overcurrent devices that must be adjacent.

(4) The tap conductors may not be spliced.

(5) The minimum size of tap conductor is No. 6 copper or No. 4 aluminum.

(6) Tap conductors may not run through ceilings, walls, or floors.

(7) The tap must be made at least 30 feet above the floor.

(F) Branch Circuit Taps. In Sections 210.19, 210.20, and 210.24, taps and outlets for circuits that supply an electric range are considered as being protected by the branch-circuit overcurrent protection.

Basically, referring back to Section 210.19(B), Exception No. 2, tap conductors supplying electric ranges, and the like, from a 50-ampere branch circuit shall have an ampacity of not less than 20 amperes and the taps shouldn't be longer than necessary for servicing the equipment.

(G) Busway Taps. Busway taps as covered in Sections 368.10 through 368.14 are considered as protected.

(H) Motor Circuit Taps. Tap conductors for motors are discussed in Sections 430.28 and 430.53, and are considered protected as covered in these sections.

See the *NEC* for other rules regarding special types of taps.

240.22: Grounded Conductor

No overcurrent device is ever permitted in series with a conductor that is intentionally grounded.

An exception permits opening the neutral, provided all conductors involved are opened intentionally and the neutral or grounded conductor can't be opened independently.

Also, Sections 430.36 and 430.37 for motor overload protection permit the fusing of the grounded phase of a delta, but the grounded conductor from the delta must be connected ahead of the disconnecting means, overcurrent device, and proper ground protection.

240.23: Change in Size of Grounded Conductor

If the size of an ungrounded conductor is changed, a corresponding change in the grounded conductor is permitted.

240.24: Location in or on Premises

(A) Readily Accessible. Basically, the requirement that overcurrent devices be readily accessible is a must. They must be located so that they may be readily reached in emergencies or for servicing without reaching over objects, climbing on chairs, ladders, etc. There are four exceptions in the *NEC*: One covers busways, as provided in Section 368.12; the second covers supplementary overcurrent protection, as described in Section 240.10; the third refers to Section 230.92; and the fourth allows for the use of portable means (ladders and the like) to access overcurrent devices that are next to motors or other equipment.

(B) Occupant to Have Ready Access. This was added to the *NEC* to ensure that occupants have ready access to the supply conductors supplying their occupancy.

Exception No. 1

In a multiple-occupancy building where there is continuous management and supervision of electric service and electrical maintenance, overcurrent devices for services and feeders that supply more than one occupancy are permitted to be accessible only to the full-time management.

Exception No. 2

Guest room overcurrent devices may be accessible to management personnel only in hotels and motels.

- **(C)** **Not Exposed to Physical Damage.** Overcurrent devices shouldn't be subject to physical damage, and should thus be located with this in mind.
- **(D)** **Not in Vicinity of Easily Ignitable Material.** When an overcurrent protective device opens, there will be a spark and possible particles of hot metal that could cause a fire. This explains why most inspection authorities don't permit overcurrent panels in clothes closets and other similar locations.
- **(E)** **Not Located in Bathrooms.** No overcurrent device may be placed in the bathrooms of dwellings or in the guest rooms of hotels and motels. This requirement does not include GFCI receptacles and supplementary overcurrent protection.

III. Enclosures

240.30: General

Overcurrent devices shall be installed in an approved enclosure unless a part of a specially approved assembly that affords equal protection, or unless mounted on switchboards, panelboards, or controllers located in rooms or enclosures free from easily ignitable material and dampness.

Most branch circuit panelboards come with a cover over the hot portions of the panel so that the breaker handles can be switched without danger. Here another door is not necessarily required, but in most cases there will be an outside door.

240.32: Damp or Wet Locations

In damp places, as per Section 354.2(A), there must be a minimum of ¼-inch (6.35 mm) air space behind the enclosure, between the wall or other mounting space, and the enclosure shall be identified for the location. Many enclosures come with provisions built in to allow the ¼-inch (6.35 mm) air space. See Figure 240-10.

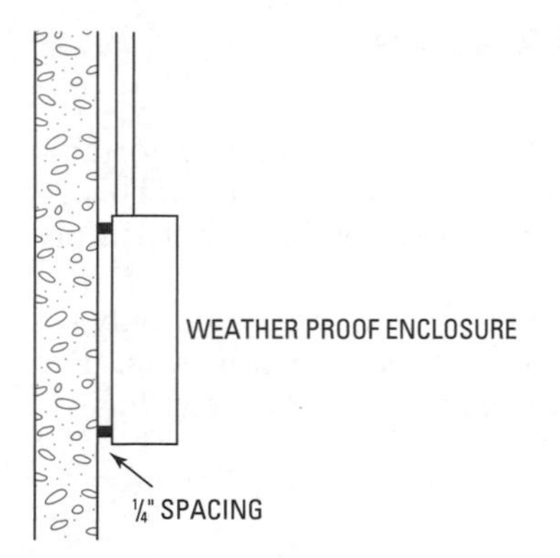

Figure 240-10 Spacing of enclosures for overcurrent devices in damp or wet locations.

240.33: Vertical Position

It is necessary to mount switches in a vertical position so that gravity can't close the switch unintentionally.

Exception

A case where this is impractical is covered in Section 240.81.

IV. Disconnecting and Guarding

240.40: Disconnecting Means for Fuses and Thermal Cutouts

In almost all cases, circuit opening devices are installed. There is an exception for thermal cutouts and fuses below 150 volts to ground and fuse cartridges of any voltage: Each circuit containing fuses or thermal cutouts shall have means of disconnecting them from the electrical supply current.

Recall that plug fuses come under the 125-volt classification and the screw-shell of the holder is connected to the load side. Thus, they are not required to have a disconnecting means. Service fuses (fuses at the outer end of the service entrance) may be installed ahead of the disconnecting means. See Section 230.82. A group of motors may be served by a single disconnecting means providing that the requirements of Section 430.112 are met. The same is permitted for fixed electric space-heating equipment in Section 424.22.

240.41: Arcing or Suddenly Moving Parts

See the *NEC*.

V. Plug Fuses, Fuseholders, and Adapters

240.50: General

(A) Maximum Voltage. Part V of Article 240 is quite important, as inspectors find considerable misuse of plug fuses. This in no manner suggests that they are to be avoided, but simply used properly.

Plug fuses and fuseholders are intended to be used only on circuits not exceeding 125 volts between conductors. For instance, circuits of 277 volts and two other conductors should definitely not involve plug-type fuses.

Exception

This approves the use of plug fuses and fuseholders on 120/208-volt wye systems.

(B) **Marking.** Ampere rating shall be plainly marked on every fuse, fuseholder, or adapter.

(C) **Hexagonal Configuration.** You will notice plug fuses with a mica glass on top that has a hexagonal configuration. The hexagonal mica indicates that the fuse is 15 amperes or less, and a round configuration indicates amperages from 15 to 30 amperes.

(D) **No Live Parts.** No live parts shall be exposed after the installation of plug fuses, fuseholders, or adapters.

(E) **Screw-shell.** The screw-shell of plug fuses shall always be connected to the load side of the circuit. The moment that the current is disengaged even partway, the shell has no voltage (Figure 240-11).

(F) **Classification.** Edison-base fuses shouldn't be rated at more than 125 volts and from 0 to 30 amperes.

(G) **Replacement Only.** This section states that plug fuses of the Edison-base type are recognized in the *Code* only as a

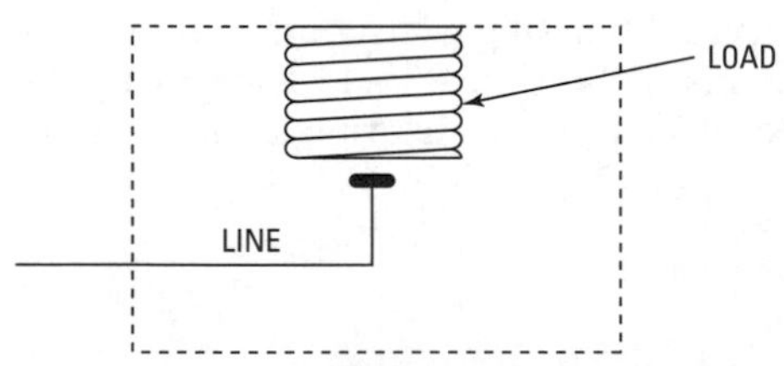

Figure 240-11 A screw-shell fuseholder.

replacement item in existing installations where there has been no evidence of overfusing or tampering. This means that all inspection authorities can condemn any new installations that use Edison-base fuses, regardless of how they are used or in what type of electrical device. The reason for this ruling is because of the dangerous practice of replacing a blown fuse with one of a higher rating than the ampacity of the conductor they are supposed to protect. This type of fuse also makes it easy to use a penny or other metal disk when a replacement fuse is not available. The proper plug fuse to be used on new installations is covered in Section 240.54.

240.52: Edison-Base Fuseholders

On all new installations requiring plug fuses, the Type S fuse is required to be used. Type S adapters should be installed in all Edison-base fuseholders in old installations to convert them to the new requirements.

240.53: Type S Fuses

Type S fuses are really screw-type fuses, and comply with (A) and (B) that follow:

(A) **Classificaton:** S-type fuses, like Edison-type fuses, are not to be used on circuits over 125 volts between conductors. We will see in (B) that the different sizes of S-type fuse are not interchangeable within 0 to 30 amperes. One S adapts from 0 to 15 amperes, another 16 to 20. Only a third S adapter takes 21 to 30. Note these classifications.

(B) **Noninterchangeable.** With reference to the above classifications, the 16- to 20-ampere and the 21- to 30-ampere classifications shouldn't be used in a fuseholder or adapter designed for lower amperages.

240.54: Type S Fuses, Adapters, and Fuseholders

(A) **To Fit Edison-Base Fuseholders.** S-type fuseholders are designed to screw into Edison-base fuseholders (Figure 240-12).

(B) **To Fit Type S Fuses Only.** Any fuseholders made only for S-type fuses or

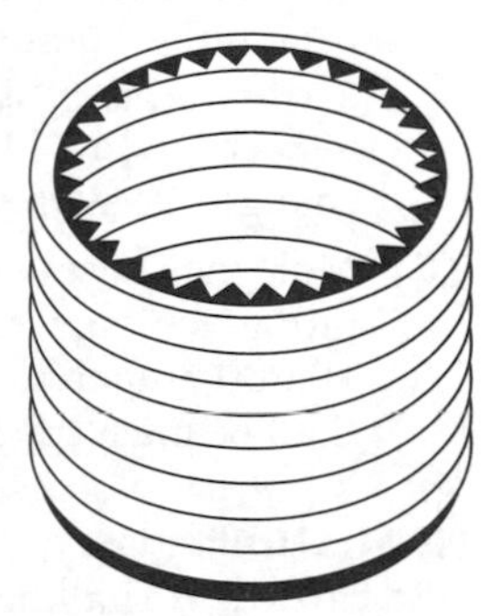

Figure 240-12 Type S plug-fuse adapter.

S-type adapters are designed so that the fuseholder itself or the adapter will accept only S-type fuses.

(C) **Nonremovable.** Type S adapters are made so that once screwed into an Edison base fuseholder, they can't be removed.

(D) **Nontamperable.** All precautions have been taken to prevent shunting around S-type fuseholders. Of course, anyone who is really determined to do it will occasionally succeed.

(E) **Interchangeability.** The designs of S-type fuses and fuseholders are standardized by the manufacturer.

VI. Cartridge Fuses and Fuseholders

240.60: General

(A) **Maximum Voltage—300-Volt Type.** The Type SC fuse for 300-volt systems became a necessity due to the extended use of 277/480-volt wye systems, which previously required the use of a 600-volt fuse. These are a noninterchangeable type of cartridge fuse and are made to fit certain devices the size of circuit breakers. This allows the use of either circuit breakers or fuses in the same panel.

Exception

This applies to conductors that have not over 300 volts to ground and are supplied with a grounded neutral.

(B) **Noninterchangeable—0–6000 Ampere Cartridge Fuseholders.** Adapters may be purchased for using lower-current-rated cartridge fuses in a higher-current-rated fuseholder. A 100-ampere-rated fuse switch may be adapted to use a 60-ampere fuse, but a 60-ampere fuse switch is not to be altered to adapt a 100-ampere fuse. Current-limiting fuses, which are covered in Section 240.11, are a special type of fuse and are not interchangeable with fuses that are not current-limiting.

(C) **Marking.** Fuses shall be plainly marked to show their ampere rating, voltage rating, and interrupting rating, if it is over 10,000 amperes. The name or trademark of the manufacturer shall be marked, as well as if they are the current-limiting type.

240.61: Classification

Fuse cartridges shall be marked with the current ampere rating, but if the voltage rating on the fuse is higher than the circuit rating, the voltage rating may be used.

VII. Circuit Breakers

240.80: Method of Operation

Circuit breakers shall be trip-free and capable of being opened and closed by hand without employing other sources of power, even though they are designed to be normally operated by electrical, pneumatic, or other sources of power.

Large circuit breakers that are opened and closed by means of electrical, pneumatic, or other power shall be capable of being closed by hand for the purpose of maintenance and shall also be capable of being opened by hand under load without the use of any other form of power.

240.81: Indicating

The open and closed position of a circuit breaker must be clearly marked so that it is easy to tell whether the breaker is on or off.

If circuit breakers are mounted in a position other than horizontal, such as vertical, the on position of the handle shall be up.

240.82: Nontamperable

It is required that any changing of a circuit breaker's top current also requires either the dismantling of the breaker or the breaking of a seal.

240.83: Marking

Circuit breakers must be clearly identified. The primary requirements are as follows:

(1) The marking must be durable, and visible after installation.

(2) Breakers under 800 volts and 100 amps must have their rating molded, etched, or stamped into their handles.

(3) Any circuit breaker with an interrupting rating over 5000 amps must have the rating shown. Breakers used only for supplemental protection are exempted. If the breaker is being used as part of a series combination, the combination rating must be marked on the equipment being fed.

(4) Circuit breakers listed for switching 120- and 277-volt fluorescent lights must be marked "SWD."

(5) Circuit breakers must be marked with the voltages for which they were designed.

240.84: Series Ratings

Series combination ratings can't be applied on systems where motors are supplied on the load side of the main overcurrent device, if the available short-circuit current is greater than the interrupting capabilities of the load-side circuit breakers.

240.85: Applications

The requirements of this section are simply that no circuit breaker should be applied beyond its rating. For example, a breaker rated 480 volts shouldn't be used on any circuit exceeding 480 volts.

In addition, this section reiterates that a two-pole breaker may not be used to protect a corner-ground delta system unless it is marked "1N–3N" by the manufacturer.

240.86: Series Ratings

The following rules apply when a circuit breaker is used on a circuit that has a higher fault current availability than its rating, but is on the load side of a second circuit breaker having a higher rating.

2005

(A) **Selected under Engineering Supervision in Existing Installations.** Series-rated combination devices may be specified by a professional engineer who is a specialist in electrical installations. The selection must be carefully documented and stamped by the engineer.

(B) **Tested Combinations.** A combination of a line-side overcurrent device and a load-side circuit breaker may be used. In this case, the combination must be tested and the details marked on the associated switchboard or panelboard.

(C) **Motor Contribution.** Series ratings may not be used when motors are connected between the two devices or if the sum of the full-load currents is more than 1 percent of the interrupt rating of the breaker with the lower rating.

VIII. Supervised Industrial Installations

240.90: General

This portion of Article 240, Part VIII, applies only to circuits and equipment in supervised industrial installations, which is defined in Section 240.91.

240.91: Definition of Supervised Industrial Installation

A *supervised industrial installation* is defined as the portions of an industrial facility that meet all of the following conditions:

- **(A)** Where maintenance and engineering supervision is sufficient to ensure that only qualified people service and monitor the system.
- **(B)** Where the load used for industrial processes (as calculated under Article 220) is 2500 kVA or more.
- **(C)** Where the facility is served by at least one service that is over 150 volts to ground and 300 volts phase to phase.

Offices, warehouses, garages, machine shops, and recreational facilities that are not part of an industrial facility do not qualify under this definition.

240.92: Location in Circuit

Overcurrent devices must be installed in each ungrounded circuit conductor as follows:

- **(A) Feeder and Branch Circuit Conductors.** These circuits must be protected at the point of supply, except as allowed under Section 240.21, or by (B) or (C) below.
- **(B) Transformer Secondary Conductors of Separately Derived Systems.** If the following three conditions are met, conductors may be connected to a transformer secondary without overcurrent protection:
  - **(1) Short-Circuit and Ground-Fault Protection.** Such conductors must be protected by one of the following: (a) The length of the secondary conductors does not exceed 50 feet, and the primary overcurrent device does not have a rating greater than 150 percent of the secondary ampacity times the secondary-to-primary voltage ratio; (b) the secondary conductor is no longer than 75 feet and is protected by a differential relay with a trip setting no more than the conductor ampacity; or (c) the secondary conductor is no longer than 75 feet, and calculations (performed under engineering supervision) determine that existing system overcurrent devices will provide sufficient protection.
  - **(2) Overload Protection.** Conductors must comply with one of the following: (a) They terminate in a single overcurrent device that limits load to the conductor ampacity; (b) they terminate in a group of overcurrent devices (no more than

six) that limits load to the conductor ampacity; (c) overcurrent relays are used to sense all of the secondary current and to limit the load by opening overcurrent devices; or (d) calculations (performed under engineering supervision) determine that existing system overcurrent devices will provide sufficient protection.

(3) Physical Protection. The secondary conductors must be protected from damage.

(C) Outside Feeder Taps. Outside conductors may be tapped to a feeder or transformer secondary without overcurrent protection and the tap point when all of the following conditions are met:

(1) The conductors are protected from damage.

(2) The overcurrent devices (no more than six) limit load to the conductor ampacity.

(3) The conductors are installed outdoors, except at the point of termination.

(4) The overcurrent device is part of a disconnecting means or located adjacent thereto.

(5) The disconnecting means is located in a readily accessible location outside the structure or immediately next to the point of entrance.

IX. Overcurrent Protection Over 600 Volts, Nominal

240.100: Feeders and Branch Circuits

(A) Feeder and branch circuit conductors operating at over 600 volts must have overcurrent protection in each ungrounded conductor at the point where the conductor is supplied or at another location determined by engineering supervision. Protection may be provided by one of the following:

(1) Overcurrent Relays and Current Transformers. A combination of overcurrent relays and current transformers is permitted. See the *NEC* for technical details.

(2) Fuses. A fuse connected in series with each ungrounded conductor.

(B) Protective Devices. These must be capable of interrupting all available current.

(C) Conductor Protection. The operating time of the protective device and the available short-circuit current must be coordinated with the conductor's characteristics to prevent damage under fault conditions.

240.101: Additional Requirements for Feeders

(A) **Rating or Setting of Overcurrent Protective Devices.** The continuous rating of a fuse may not be more than three times the ampacity of the conductors. The long-term trip setting of a circuit breaker or electronically actuated fuse may not be more than six times the ampacity of the conductors.

For fire pumps, see Section 695.4(B).

(B) **Feeder Taps.** Conductors tapped to a feeder may be protected by the feeder overcurrent device if that device also protects the tap conductors.

Article 250—Grounding

I. General

250.1: Scope

Grounding is very important. You will find some places that are not to be grounded, but in the majority of the cases, grounding takes precedence. The reasons for grounding are protection from lightning, safety to people, and safety to property. There is some confusion at times among the definitions covering grounding; be sure to check your definitions of the different groundings. Of the definitions most commonly causing confusion, the distinction between grounding of services and equipment grounding is the most frequent.

In the *NEC* you will find the type of conductors and sizes to be used for each of the above-mentioned groundings. To have equipment grounding, there must be service-entrance grounding. As we proceed, various methods of grounding and of bonding-equipment grounding will be discussed in detail. You will also find certain places where grounding is not required if proper guarding and isolation are provided. We have mentioned some of the purposes for grounding; it also protects if a circuit comes in contact with higher voltage or if a fault occurs to ground. Then it facilitates the opening of overcurrent devices. See Section 110.10.

250.2: Effective Grounding Path

Caution must be taken to ensure that the grounding path from circuit equipment and metal enclosures will be continuous and not subject to damage. They must be capable of safely handling fault currents that may be imposed on it. The impedance must be sufficiently low to keep the voltage ground minimum and facilitate the opening of overcurrent devices covering the circuit.

Grounding conductors can't use the earth as the only equipment-grounding conductor.

250.3: Application of Other Articles

Other articles in the *Code* list additional grounding requirements for specific conductors and equipment installations other than those covered in this article. A listing to which additional grounding requirements apply is given in this section. In Table 250.3 of the *NEC* you will find a list of many places in the *NEC* where additional information on grounding is given, along with the section numbers.

250.4: General Requirements for Grounding and Bonding

This section contains the primary principles of grounding and binding. In other words, this section explains the overall goals of grounding and bonding, as a prelude to the specific rules of the balance of Article 250.

(A) Grounded Systems.

- **(1) Electrical System Grounding.** Grounded electrical systems must be connected to the earth in such a way as to limit unexpected voltages that may be imposed by surges, lightning, and so on, and to stabilize voltages during normal use.
- **(2) Grounding of Electrical Equipment.** Conductive materials not intended to carry current must be connected to the earth so as to limit the voltages they may be exposed to.
- **(3) Bonding of Electrical Equipment.** Conductive materials (metal) that are not intended to carry current (boxes, raceways, etc.) must be connected together and connected at the supply source to make an effective path for ground-fault currents.
- **(4) Bonding of Electrically Conductive Materials and Other Equipment.** This section merely extends the preceding item, (3), from electrical equipment to other conductive objects that are likely to become energized.
- **(5) Effective Ground-Fault Current Path.** This section merely states that the effective ground-fault path mentioned in (3) must be permanent, have low impedance, and be properly connected to the earth.

(B) Ungrounded Systems.

(1) Grounding Electrical Equipment. In ungrounded systems, conductive materials not intended to carry current must be connected to the earth so as to limit the voltages they may be exposed to by unintended contact with energized parts or from lightning.

(2) Bonding of Electrical Equipment. Conductive materials (metal) that are not intended to carry current (boxes, raceways, etc.) must be connected together and connected at the supply source to make an effective path for ground-fault currents. These paths must be capable of handling the maximum ground current likely to be imposed upon them.

(3) Bonding of Electrical Conductive Materials and Equipment. This section merely extends the preceding item, (2), from electrical equipment to other conductive objects that are likely to become energized.

(4) Path for Fault Current. This section merely states that the effective ground-fault path mentioned in (2) must be permanent, have low-impedance, and be properly connected to the earth.

250.6: Objectionable Current over Grounding Conductors

(A) Arrangement to Prevent Objectionable Current. In the grounding of electrical systems and circuit conductors, all conducting noncurrent-carrying materials including surge arresters must be arranged so that objectionable flow of current will be prevented over all grounding paths and grounding conductors.

On occasion, metal-sided houses with metal water pipes have started electrical fires because of grounded current flowing through the metal siding to ground.

Sometimes persons are concerned about reading current on the grounding electrode conductor. This is natural, as it is a parallel path from the service equipment to the transformer ground. There are other cases that might appear to be of concern, but are really not. In meeting the requirements of this section, one must carefully analyze whether it is a problem or not.

(B) Alterations to Stop Objectionable Current. In using multiple grounding connections, if there is current flowing, check to see if it is a natural condition or objectionable. If objectionable, one or more of the following steps will help:

(1) Disconnect one or more (but not all) grounding connections.

(2) Alter the placement of the grounding connections.

(3) Eliminate any continuity between grounding connections.

(4) Take other steps that are acceptable to the authority that has jurisdiction.

(C) Temporary Currents Not Classified as Objectionable Currents. Grounding and grounding conductors have a job to do, and it is possible under certain conditions, such as a ground fault, that there will be current flowing while they are performing their duty. Such currents are not classified as being objectionable, as covered in (A) and (B) above.

(D) Limitations to Permissible Alterations. This article must not be interpreted to allow electronic equipment to be installed on AC circuits that are not grounded. Current that has electrical noise (electromagnetic interference) or voltage fluctuations (dirty power) can't be considered an objectionable current, as specified in this section.

250.8: Connections of Grounding and Bonding Equipment

The grounding conductors and bonding jumpers must be attached to circuits, conduits, cabinets, equipment, and the like that are to be grounded, by means of suitable lugs, listed pressure connectors, clamps, exothermic welding, or other approved means. Soldering is never permitted—neither is the strap-type of grounding clamp. Be certain that the device used is identified as being suitable for the purpose (as sheet metal screws are not).

250.10: Protection of Ground Clamps and Fittings

Unless approved for general use without protection, ground clamps and fittings shall:

(1) Be located so that they won't be subject to damage.

(2) Be enclosed in metal, wood, or equivalent protective covering.

With exothermic welding, because the resistance of the connection is checked with a high-voltage tester, you know that the problems of a high-resistance connection are eliminated.

250.12: Clean Surfaces

Good electrical continuity must always be made. If there are any nonconductive paints, lacquers, enamels, or the like on equipment that is to be grounded, they shall be removed from any contact surfaces or threads.

All corrosion shall be removed. If rebar is used as the grounding electrode, thoroughly clean the surface of the rebar by grinding to remove the oxidation. If an exothermic weld is used, heat the rebar to drive out the moisture in its pores.

II. Circuit and System Grounding

250.20: Alternating-Current Circuits and Systems to Be Grounded

Items (A), (B), (C), and (D) that follow cover AC circuits that will be grounded. Circuits not covered by these four items may be grounded.

(A) Alternating-Current Circuits and Systems Less Than 50 Volts to Be Grounded. Any AC circuit of 50 volts or less must be grounded if it meets any of the following conditions:

- **(1)** If the transformer supplying the 50 volts or less is supplied by voltages in excess of 150 volts to ground, the 50-volt side must be grounded.
- **(2)** When transformers are supplied from an ungrounded system.
- **(3)** If 50-volt AC conductors run overhead outside of the buildings, then one wire must be grounded.

(B) Alternating-Current Systems of 50 Volts to 1000 Volts. The conditions that follow will require that AC systems of 50 volts to 1000 volts that supply premises' wiring systems must be grounded:

- **(1)** If the voltage to ground on ungrounded conductors is 150 volts or less.
- **(2)** On a four-wire wye system, if the neutral is used as a circuit conductor. This would not include a 480/277-volt high-resistance grounded system where the midpoint of the Y is grounded through a resistor, but this connection is not used as a circuit conductor.
- **(3)** This is best described by Figure 250-1E, where the midpoint of one phase of a delta supply is grounded and this midpoint is used as a neutral.
- **(4)** See Sections 230.22, 230.30, and 230.41 for requirements for a service conductor that is uninsulated, such as a quadraplex.

Exception No. 1

An exception to the grounding of these systems covers electric furnaces, or any other means of heating metals for refining, melting, or tempering.

Exception No. 2
Grounding is not required for separately derived systems used for speed control in industrial plants or for rectifiers. Nothing else shall be connected to these.

Ground-fault detectors if used on ungrounded systems will provide additional protection.

Exception No. 3
If the preceding four special conditions are met, separately derived systems supplied by primary voltage ratings less than 1000 volts are not required to be grounded. The first is that only qualified persons maintain service or supervise the installation. For the other three conditions, see the *NEC*.

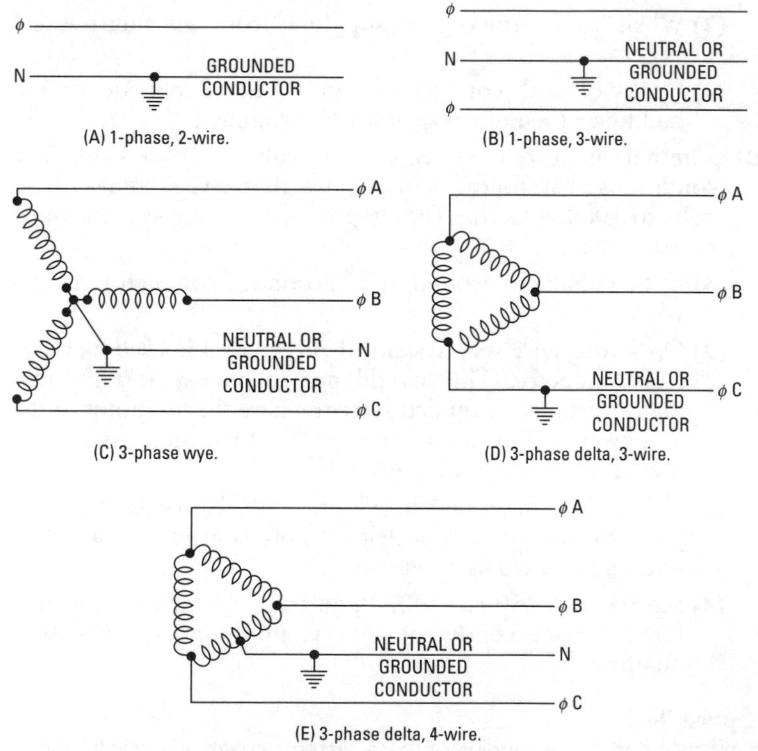

Figure 250-1 Grounding different types of circuits.

Exception No. 4
Article 517 does not permit a neutral ground in operating rooms; this system is an isolated system and uses a special transformer with a metal barrier between the primary and secondary windings.

Exception No. 5
As mentioned in item 2 above, a four-wire wye with the resistive ground on the center where the fourth wire is not used as a circuit conductor need not be grounded. This will also apply to impedance grounded neutral systems. The impedance may be a resistor, as mentioned before. The purpose of this is to limit the ground-fault current values. Such systems may be used where the voltages run from 480 volts to 1000 volts, three-phase if all of the following conditions are met: (a) Qualified persons only will be servicing and maintaining the installation. (b) Where the system being supplied must have continuity of service. (c) To identify grounded phase conductors, ground detectors must be supplied to indicate that there is a grounded phase. This permits operation in most cases until shutdown can occur or until repair. (d) In no case shall this high-impedance neutral be used to supply line-to-neutral loads.

(C) Alternating-Current Systems 1 kV and Over. It is required that systems that supply mobile portable equipment and are supplied by AC systems of 1 kV or over be grounded, as covered in Section 250.188. Where not supplying portable equipment, such systems may be grounded. When grounding systems (AC) 1 kV and over, the grounding shall meet the applicable parts of this Article 250.

(D) Separately Derived Systems. Where interior wiring systems are supplied by generators not connected to the supply system, such as emergency generators, or where transformers in the supply system are used for other than supply voltage and the primary and secondary are isolated (which also includes converter windings that have no direct connections to the supply source), grounding is required where stated above. Grounding of separate systems is explained in Section 250.30.

Note
On-site generators are not considered separately derived systems, if the neutral of the generator is solidly connected to the neutral of the supply service. Section 250.30 involves systems that are not separately derived systems and that aren't specifically required to be grounded. You are also referred to Section 445.5 for the minimum size of conductors required to carry the fault currents.

250.22: Circuits Not to Be Grounded

See the following for circuits that need not be grounded:

(A) **Cranes.** Electrically operated cranes operating over combustible fibers, as provided in Section 503.13, which covers Class III locations.

(B) **Health-Care Facilities.** Circuits in operating rooms, as per Article 517. This will also apply to health-care facilities, as provided in the same article, such as locations where anesthetics will be used.

250.24: Grounding Service-Supplied Alternating-Current Systems

(A) **System Grounding Connections.** The reason for grounding on the supply side of the service-entrance equipment is because if the supply side is ever disconnected, there will still be a ground on the supply system. In the event that the distribution system is ungrounded, there would, of course, be no ground from the primary supply. The ground at the service-entrance equipment shall be connected at an accessible point on the load side. It is preferred that this connection be made in the service-entrance equipment; thus, it is accessible without breaking the utilities meter housing seal (Figure 250-2). See Section 250.32 for two or more buildings supplied by a single service and Section 250.30 for separately derived systems.

Exception No. 1

Provisions of Section 250.30(B) require that separately derived systems meet the requirements of this section, and a grounding electrode conductor is to be connected to the separately derived system.

Exception No. 2

Section 250.32 covers the necessity of grounding conductor connection where there are separate buildings.

Exception No. 3

Section 250.142 covers the grounding meter enclosing all ranges, wall-mounted ovens, clothes dryers, and counter-mounted cooking units.

As an inspector, the author likes to see the grounding electrode conductor connected in the service equipment enclosures of any size of service. Here it is accessible for service problems.

Exception No. 4

A single grounding electrode connection to a tie point is permitted when services are dual-fed in a common enclosure, or are grouped together in separate enclosures but employ a secondary tie.

Exception No. 5

The grounding electrode conductor may be connected to the grounding bar (or bus) where the main bonding jumper is connected, if the main bonding jumper [see Sections 250.24(B) and 250.28] is a wire or bus bar connected between the neutral bar or bus and the grounding bar or bus.

Exception No. 6

Section 250.36 covers high-impedance grounded neutral systems and what grounding connections are required.

(B) Grounded Conductor Brought to Service Equipment. When an AC system is served by 1000 volts or less and is grounded at any point, you are required to run the grounded conductor to each service. The grounded conductor size is determined by and shouldn't be less than the requirements for grounded conductors covered by Table 250.66 in the *NEC*. The grounded conductor shall also be routed in with the phase conductors. If service conductors are larger than 1100 kcmil and made of copper, or 1750 kcmil made of aluminum, the table indicated above can't be used. The grounded conductor shouldn't be smaller than 12½ percent of the area of the largest phase conductor. If the phase conductors are paralleled, the grounded conductor size should be based on the total area of the paralleled conductors. Refer to Section 310.4.

Exception No. 1

The size of the grounded conductor is not required to be larger than the size of the phase conductors. This means that the conductors, if paralleled or single, are of an individual phase. Of course, remember that if they are paralleled, you use the combination of the sizes of the parallel conductors for determining the size of the grounded conductor.

Exception No. 2

Where high-impedance grounding neutral systems are used, as covered in Section 250.27 under connection requirements.

Exception No. 3

If a number of service disconnects are in a service-rated assembly (as in the case of many meter banks), one grounded conductor must bond the assembly.

250.26: Conductor to Be Grounded—Alternating-Current Systems

Article 200 covers the use and identification of grounded conductors. Ordinarily, the grounded conductor is commonly known as the white wire, although it will be recalled that a neutral gray color may also be used. Figure 250-2 illustrates the different grounding circuits used.

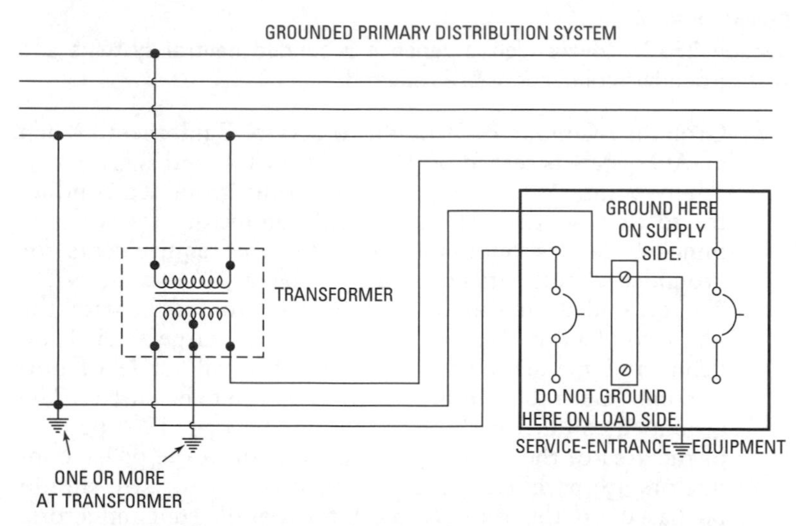

Figure 250-2 Grounding on the supply-side of the service-entrance equipment will provide a ground on the supply if it is ever disconnected.

For AC systems in premises, the following grounded conductors must be used:

(1) One conductor shall be grounded if the circuit is single-phase, two-wire.

(2) The neutral conductor shall be grounded if the circuit is single-phase, three-wire.

(3) On a wye system that has a center tap common to all phases.

(4) On a delta system that has one phase conductor grounded.

(5) In a delta where one phase is used as in 250.24(B), the mid-tap shall be grounded and be the neutral conductor. Remember not to connect to the wild leg through the neutral.

Article 200 tells how to identify the grounded conductor.

250.28: Main and Equipment Bonding Jumpers

An unspliced jumper shall be used to connect the equipment grounding conductor and the service-equipment enclosure to the grounded conductor. This shall be in the service-equipment or the service-conductor enclosure (Figure 250-3). Usually, in service-entrance equipment there is a grounding bar for attaching the neutral of the service and branch circuits. This is usually connected to the enclosure by means of inserting a screw to the bar into the enclosure. Then an additional bar is bolted or clamped to the service equipment for attaching the equipment grounding conductors. A case such as this constitutes the bonding.

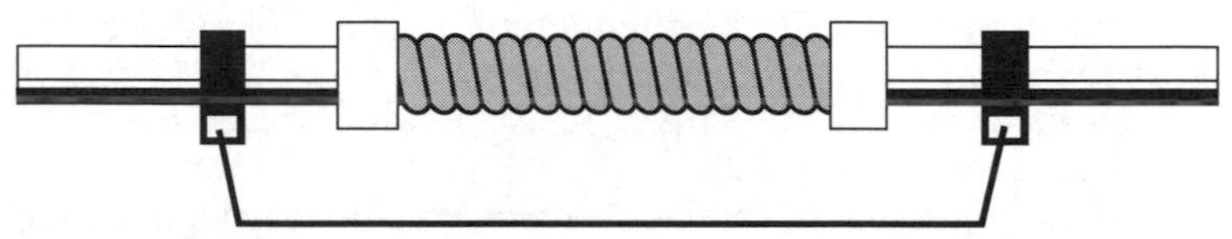

Figure 250-3 Bonding must extend around any flexible conduit used in conjunction with service-entrance equipment.

(A) Material. Copper or other noncorrosive material is to be used for main and equipment bonding jumpers. This automatically eliminates aluminum in many places. Remember that if a metal other than copper is used, it shall be sized equivalently to what would be required for copper jumpers.

(B) Construction. When a screw is used as the main bonding jumper, it must be identified by being colored green. It must also be visible after installation.

(C) Attachment. Section 250.8 specifies the manner in which main and equipment bonding jumpers shall be attached for circuits and equipment. Section 250.8 specifies the manner of attachment for grounding electrodes.

Both (A) and (B) should be reviewed, noting that soldering is not permitted. This question of aluminum for common grounding conductors always arises. This is explained in (A), and to my knowledge there is no grounding clamp approved by UL for grounding with aluminum. If it were to go to a made electrode, it could not be run closer than 18×0 to the earth; and if to a water pipe, the water piping is subject to sweating and would cause corrosion of the aluminum.

(D) **Size—Equipment Bonding Jumper on Supply-Side of Service and Main Bonding Jumper.** See Figure 250-3. Many people seem to be confused by Tables 250-66 and 250-122. By looking at the headings there should be no confusion. Table 250.66 describes grounding electrode conductors for AC systems. Table 250.122 describes minimum equipment-grounding conductors for grounding raceway and equipment. Raceways are commonly used to route grounding electrode conductors to a water pipe. If the metal raceway does not have a connection to the conductor, a bonding jumper must be used. The jumper must be as large or larger than the grounding electrode conductor.

Table 250.66 describes grounding electrode conductors, if the service-entrance phase conductors are larger than 1100 kcmil copper or 1700 kcmil aluminum. This will require that the bonding jumper circular mill area be less than 12½ percent of the size of the largest phase conductor. See the *NEC* for the remainder of (D).

Bonding jumpers for DC systems can be no smaller than the system grounding conductor size given in Section 250.166.

(E) **Size—Equipment Bonding Jumper on Load Side of Service.** Table 250.95 lists the size of the equipment bonding jumpers on the load side of the service overcurrent services. The bonding jumper is to be a single conductor sized according to Table 250.122 for the largest overcurrent device that is provided to supply the circuits. If the bonding jumper supplies two or more raceways or cables, the same rulings apply.

Exception

It is not necessary that the equipment bonding jumper be larger than the phase conductors, but in no case shall it be smaller than No. 14 AWG.

(F) **Installation—Equipment Bonding Jumper.** It is permitted to install the equipment bonding jumper either inside or outside of the raceway or enclosure. When it is installed outside of the raceway or enclosure, it shouldn't be over 6 feet in length and shall be routed with the raceway or enclosure. If the equipment bonding jumper is inside the raceway, refer to

Section 310.12(B) for the conductor identification requirements you are to comply with.

250.30: Grounding Separately Derived Alternating-Current Systems

Section 250.30 covers the grounding of separately derived AC systems. Grounded systems are covered in Part A, and ungrounded systems are covered in Part B.

(A) Grounded Systems. This section defines the rules for grounded separately derived systems. Note, however, that high-impedance grounded neutral systems may be made as specified in Sections 250.36 and 250.186.

(1) Bonding Jumper. If bonding jumpers are used, they should be sized according Section 250.28(A) through (D). This shall be used for connecting the derived phase conductors to the grounded conductor from the derived system to the equipment grounding conductors of said system. This connection is to be made at any point of the separately derived system to the grounding source of the first disconnecting means, which can include the overcurrent devices. As an alternative, the separately derived system ground may be made at the source of the separately derived system when it does not have either overcurrent device or disconnecting means. The point of connection must be the same as for the grounding electrode conductor. (See Section 250.30(A)(2).)

You would be well-advised to check back to Section 240.6(C), which explains where overcurrent devices are required to be mounted. This is frequently a factor for separately derived systems.

Exception No. 1

A bonding jumper may be installed at both the source and the first disconnecting means, if this does not establish a parallel path with the grounded circuit conductor. (Connection to the earth in this case is not considered a parallel path.) A grounded conductor used in this manner must be at least the size of the bonding jumper, though it does not have to be larger than the ungrounded conductors.

Exception No. 2

Also covered is the size of the bonding jumper on a system of not over 1000 volt-amperes that supplies a Class 1 remote-control or signaling system. The bonding jumper shouldn't be smaller than the derived phase conductors, but in no event shall it be smaller than No. 14 copper or No. 12 aluminum.

(2) Grounding Electrode Conductor. Section 250.66 covers the sizing of grounding electrode conductors. The grounding electrode conductor for the derived phase conductors shall be used to connect the grounded conductor of the derived system to the grounding electrode. Unless permitted by Section 250.24(A), the grounding connection may be made at the same point on the separately derived system as where the bonding jumper is installed.

Exception

A grounding electrode conductor is not required if the system is supplied by a transformer rated not more than 1000 volt-amperes and the system supplies Class I remote-control or signal circuit. However, the system grounding conductor must be bonded to the transformer framer or enclosure and the bonding jumper meet the requirements of the exception in (1) above, and the transformer frame enclosure has to be grounded as required by Section 250.134.

Where multiple separately derived systems are connected to a common grounding electrode conductor (see Section 250.30(A)(3)), that grounding electrode conductor must be sized according to Section 250.66, using the largest phase conductor from each system in the calculations.

(3) Grounding Electrode Conductor Taps. Taps from a separately derived system may be connected to a common grounding electrode system. Such tap conductors must connect the grounded conductor of the separately derived system to the common grounding electrode conductor. Tap conductors must be sized according to Section 250.66. Connections must be made at an accessible location, with listed compression connectors to a bus bar (at least ¼ inch × 2 inches), or by exothermic welding. The installation of the grounding electrode conductor and the taps must conform to the usual rules for installing grounding electrode conductors, found in Section 250.64. Any exposed structural steel or metal piping in the area served by the derived system must be properly bonded to the grounding electrode conductor. See Section 250.104.

(4) Grounding Electrode. The grounding electrode must be as close as practical and is preferred in the same area as the grounding conductor to the system. Where grounding electrodes are used, they should be (1) the nearest effectively grounded metal building or structure; (2) a metal water pipe within 5 feet (1.5 m) of the point of entrance into the building that is effectively grounded by sufficient metal water pipe buried in the earth; or (3) made electrodes, as specified in Sections 250.50 and 250.52, provided the electrodes that were specified in (1) and (2) above are not available. In industrial and commercial buildings where only qualified persons service the electrical system, and if the entire length of the water pipe is exposed, the connection may be made at any point on the piping system.

(5) Equipment Bonding Jumper Size. If the bonding jumper is run with the phase conductors to the disconnecting means, it must be sized according to Section 250.28(A) through (D).

(6) Grounded Conductor. If a grounded conductor is installed and the bonding jumper is not located at the source of the separately derived system, it must be routed with the phase conductors, and may not be smaller than the grounding electrode conductor (see Section 250.66), although in no case must it be larger than the largest phase conductor. If the derived phase conductors are in parallel, the grounded conductor must have its size based on the total circular mil area of the paralleled phase conductors. Grounded conductors for high-impedance grounded neutral systems must be installed per Section 250.36.

(B) Ungrounded Systems.

(1) Grounding Electrode Conductor. A grounding electrode conductor must connect the metal enclosures of the derived system to the grounding electrode. The conductor must be sized according to Section 250.66,and the connection may be made at any point between the source and the first disconnecting means.

(2) Grounding Electrode. The requirements here are the same as for 250.30(A)(4). An exception is made for vehicle-mounted generators, for which you should refer to Section 250.34.

250.32: Two or More Buildings or Structures Supplied from a Common Service

(A) Grounded Systems. This section applies to two or more buildings or structures served from one service. Part II describes the grounded system of each building or structure;

each of them shall have a grounding electrode that is also connected to the metal enclosure of the disconnecting means that actually would be a feeder, and a grounding electrode is to be provided connected to the disconnecting means, as well as the grounded circuit conductor of the AC supply on the supply side of the building or structure disconnecting means.

Exception No. 1

If the building or structure is supplied by only one branch circuit and there will be no equipment used in the building or structure that requires grounding, no grounding electrode is required. This is a little questionable, as it is inspected when wired, but can be changed by the occupant without the authority that has jurisdiction ever knowing about it.

Exception No. 2

In some cases a grounding electrode won't be required for the separate building or structure, if an equipment-grounding conductor is run with the current supply wires to the building or structure. This equipment-grounding conductor is for grounding noncurrent-carrying parts of equipment. It is also used for interior metal piping systems and for building or structural metal frames. It shall also be bonded to the disconnecting means of separate building or structure or grounding electrodes. This is further described in Part VIII of this article. When using the equipment-grounding conductor, and there are no other grounding electrodes, two or more branch circuits may be used. Be careful where livestock is involved. If the equipment-grounding conductor serving in such an area should be run underground in the earth to the disconnecting means, it must be insulated or covered.

(B) **Ungrounded Systems.** When two or more buildings or structures are supplied in an ungrounded system, you are required to install a grounding electrode at each building or structure. This is further described in Part VIII. The grounding electrode shall be connected to both metal enclosures of a building or structure, as well as to the disconnecting means.

Exception

The metal enclosure of a building or its structure disconnecting means does not have to be connected to the grounding electrode conductor, provided all of the following conditions are met: (1) The entire wiring system has an equipment grounding conductor (green wire); (2) No grounding electrodes already exist; (3) Only one branch circuit exists in the building; (4) Any equipment conductor run underground must be insulated if livestock is commonly in the building.

(FPN): There are special requirements for agricultural buildings. Refer to Section 547.8(A), Exception.

(C) Disconnecting Means Located in Separate Building or Structure on the Same Premises. When one or more buildings is under the same management and the disconnects are remotely located [see Sections 230.84(A) and 230.35(A)], the following conditions must be complied with:

(1) The neutral (grounded circuit conductor) is to be connected to the grounding electrode at the first building only.

(2) Any building that has two or more branch circuits must be provided with a grounding electrode, and an equipment grounding conductor from the first building must be connected to it. This equipment grounding conductor must be run with the circuit conductors.

(3) The connection of the grounding conductor to the grounding electrode conductor at the second building must be made in a junction box, panelboard, or similar enclosure located just inside or just outside the second building.

Exception No. 1

The second building is not required to have a grounding electrode if it has only one branch circuit that supplies no equipment that is required to be grounded.

Exception No. 2

Any equipment grounding conductor run underground must be insulated when livestock is commonly found in the building.

(D) Grounding Conductor. Grounding conductors must be sized according to Table 250.122 and installed in accordance with Sections 250.64(A) through (F). If a metal conduit is used to protect a grounding conductor, it must be bonded to the conductor at both ends.

Exception No. 1

The grounding conductor does not have to be larger than the biggest ungrounded conductor.

Exception No. 2

The part of the grounding conductor that is the only connection between the electrode and the grounded or grounding conductor or the metal enclosure of the disconnecting means does not have to be larger than No. 6 copper or No. 4 aluminum.

250.34: Portable and Vehicle-Mounted Generators

(A) **Portable Generators.** It is not required that the frame of a portable generator be grounded if the following conditions are met:

(1) Only equipment is mounted on the generator and/or cord- and plug-connected to outlets on the generator.

(2) Any of the metal parts of the equipment being supplied has equipment grounding conductor to the terminals of the outlet of the generator receptacles, which in turn are bonded to the generator frame.

It is to be noted that equipment is to be mounted on the generator, or that a receptacle(s) is to be mounted on the generator and cord-and-plug connections are to be made to the equipment for which the power is being supplied. Also, there is to be an equipment grounding conductor in the cord and the receptacles connected as on any system, with the equipment grounding conductor going back to the generator frame.

(B) **Vehicle-Mounted Generators.** A generator mounted on the frame of a vehicle shall be considered as serving as the grounding electrode for the system being supplied under the following circumstances:

(1) Tile generator frame shall be bonded to the frame of the vehicle.

(2) Current supplies only equipment mounted on the vehicle and/or cord- and-plug-connected equipment when supplied from outlets mounted on the vehicle or on the generator.

(3) Equipment grounding conductor is supplied from the equipment being served to receptacles on the vehicle or generator where they are properly bonded to the generator frame.

(4) The provisions of this article are also complied with.

(C) **Neutral Conductor Bonding.** Bonding of conductors other than the neutral to the frame is not required.

Refer to Section 250.5, especially (D), for grounding portable generators that supply fixed wiring systems.

250.36: High-Impedance Grounded Neutral System Connections

The following provisions, (A) through (F). are to be complied with when high-impedance grounded neutral systems are used. Such installations are allowed by Section 250.20(B).

(A) **Grounding Impedance Location.** When high-impedance grounding is used, it must be located between the grounding source neutral and the grounding electrode. Where a neutral does not exist, the high-impedance must be installed from a grounding transformer to the grounding electrode.

(B) **Neutral Conductor.** The neutral conductor from its connection point to the grounding impedance must be fully insulated, and must have an ampacity that is no less than the maximum current rating of the grounding impedance. The neutral may not, however, be smaller than No. 8 copper, or No. 6 aluminum.

(C) **System Neutral Connection.** The neutral conductor should only be connected through the grounding impedance to the grounding electrode.

Note

When a circuit is closed, there is a value of charging current that sometimes exceeds the overcurrent protection on that system. The impedance usually is not just selected for ground-fault current slightly greater than or equal to this capacity charging current; impedance grounding also aids in limiting any transient overvoltages to safe values. For more information, refer to ANSI/IEEE Standard 142, which gives recommendations for industrial commercial power systems. The design of impedance grounding in reality becomes an engineering and design problem.

(D) **Neutral Conductor Routing.** The impedance grounding system from the neutral point of a transformer or generator may be installed in a separate raceway. Thus, you can see it is not required to be run in the same raceways as the phase conductors to the disconnecting means or overcurrent device. The impedance grounding limits the ground current so it may be treated from a solidly grounded neutral.

(E) **Equipment Bonding Jumper.** An equipment bonding jumper (connected between equipment grounding conductors and the grounding impedance) must be an unspliced conductor and must run from the first disconnecting means to the grounded side of the impedance.

(F) Grounding Electrode Conductor Location. The grounding conductor is to be attached at any point from the grounded side of a grounding impedance to the equipment grounding connection that may be at the service equipment or disconnecting means of the first system. At first this might sound like paralleling grounding conductors, but the impedance grounding conductor originates from the phase neutral, and the equipment grounding conductor referred to here connects from the grounding electrode to the equipment disconnecting means of the first system.

III. Grounding Electrode System and Grounding Electrode Conductor

250.50: Grounding Electrodes

Grounding electrode methods have been constantly improved, and when you refer to Section 250.56, you find that the grounding electrode resistance actually has to be measured.

Table 250-1 may assist you in determining approximate resistivity of various soils. This table does not appear in the *NEC*.

In (A) through (D) below, all shall be bonded together from the grounding electrode service, if available at the building or structure being served. Section 250.64(A) covers the installation of the bonding jumpers, and Section 250.66 covers the size of the bonding jumper to be used and connected as specified in Section 250.70. The unspliced electrode conductor (note *unspliced*) may be run to any available grounding electrode. Internal metal water piping is not allowed to serve as the only grounding electrode; it must be supplemented by another type of electrode. In addition, the grounding electrode conductor must be connected to it within five feet of its point of entrance to the building.

Exception No. 1

It is allowed to extend the grounding electrode conductor by use of an exothermic weld (commonly called a *Cad-weld*) in industrial and commercial locations. This is also allowed using irreversible compression connectors.

Exception No. 2

Where only qualified persons will service metal water pipes in commercial and industrial buildings, such water pipes are allowed to be used as the only grounding electrode conductor. However, the entire length of the piping that is subject to ground currents must be exposed.

(A) Metal Underground Water Pipe. A buried metal underground water system may be used as a ground electrode if there is 10 feet (3.05 m) or more in direct contact with the earth (including any metal well-casing bonded to the water pipe). These shall be electrically continuous or made electrically continuous by bonding around installed joints or sections of insulating pipe to the point where the common grounding conductor connects to the pipe and any bonding connection. Water meters shouldn't be relied upon for the grounding path. The grounding should be on the street side of the water meter or the water meter should be effectively bonded around, as shown in Figure 250-11. Resistivities of different soils are shown in Table 250-1.

Table 250-1 Resistivities of Different Soils

		Resistivity OHM-CM	
Soil	***Average***	***Min***	***Max***
Fills—ashes, cinders, brine wastes	2370	590	7000
Clay, shale, gumbo, loam	4060	340	16,300
Same—with varying proportions of sand and gravel	15,800	1020	135,000
Gravel, sand, stones, with little clay or loam	94,000	59,000	458,000

You are required, when using a metal underground water-pipe system as a grounding electrode, to supplement it with at least one additional electrode such as those described in Section 270.81 or 250.52. This ensures some bonding electrode in the event that some outside water system is eventually replaced with nonmetallic water pipe. This supplemental electrode may be supplied from the grounded service-entrance conductor, a grounded service raceway, any grounded service enclosure, or the interior metal water piping at any point that is convenient.

If the building or structure has nonmetallic water piping and metal water piping within the building or structure, this will eliminate the use of the interior water piping as an electrode. However, the *Code* also requires bonding the equipment grounding electrode to interior metallic water piping, even though supplied by a nonmetallic underground water piping supply.

If the supplemental electrode is one that is described in Section 250.52(C) or (D), the size of the grounding electrode need not be larger than No. 6 copper or No. 4 aluminum.

Author's Note

Aluminum is mentioned, but recall that aluminum is not permitted closer than 18 inches to the earth or in contact with anything corrosive, and the service-entrance conductor is to be without a splice.

(B) Metal Frame of the Building. The metal frame of a building may be used as a grounding electrode, provided it is effectively grounded and would give the same protection as the other methods of grounding.

(C) Concrete-Encased Electrode. Concrete-encased steel reinforcing bar (must be bare, or have some type of electrically conductive coating such as zinc) of No. 4 AWG copper encased in concrete footings or direct contact with the earth. The criteria for this shall be: (1) not less than 20 feet (6.1 m) long and not less than ½ inch (12.7 mm) in diameter;in most cases more than 20 feet (6.1 m) will be available, making for a lower resistance ground; or (2) a minimum of 20 feet (6.1 m) of bare copper conductor not smaller than No. 4 AWG.

For a large industrial complex, a preferred method is to use metallic water-pipe bonding, accomplished by welding across the joints of the cast-iron pipe exothermically and tying the steel column anchor bolts by exothermic welding to the bolt and then to the rebar in the casson and the rebar in the concrete pudding. These are then all tied together as the grounding electrode for a building. The results of a ground test to this installation are usually very close to, or just a little above, zero-ohms resistance.

Concrete in contact with earth will always draw some moisture, which causes this type of grounding to be very effective.

(D) Ground Ring. The ring of copper conductor not smaller than 2 AWG, buried at a depth of not less than 2½ feet and in direct contact with the earth can be used as the grounding electrode for buildings or structures.

In encircling the building or structure, you will find that in most cases there will be much more than 20 feet (6.1 m) of conductor. The author has found that a better ground may be obtained by burying the conductor about 3 feet from the building and also by periodically driving ground rods and attaching them to the bare conductor by approved methods.

If the connection is made by exothermic welding, these welds often are deceptive. One sure method to be certain that they make good contact, if it is rebar, is to buff the rust off the rebar and heat it to drive the moisture out of the pores. Then, to verify the load resistance connection, a load-resistance-reading ohmmeter that will go down to at least ½-ohm resistance should be used to verify the proper weld. This would also apply to exothermic welding to ground rods, except that the buffing and preheating would not be required.

250.52: Made and Other Electrodes

If electrodes as specified in Section 250.50 are not available, you may use one or more electrodes covered in (A) through (D) below. If practical, these made electrodes should reach permanent moisture levels. If one electrode is not long enough to reach permanent moisture, an additional one may be added to lengthen it. There shall be no paint or enamel or nonconductive coatings on the electrode. If more than one made electrode is required (including those used for lighting protection), then they shall be spaced a minimum of 6 feet apart—a greater distance apart would be preferable. Where two or more made electrodes are used, they are considered and treated as a single electrode system.

(A) Metal Underground Gas Piping System. Underground metal gas piping systems may not be used as grounding electrodes.

(B) Other Metal Underground Systems or Structures. Other underground systems or structures such as underground tanks or piping.

(C) Rod and Pipe Electrodes. When rod or piping electrodes are given, they must be a minimum of 8 feet long and shall be installed as follows in items (1) through (3). One gross violation of this is the grounding electrodes used on antenna systems where the installer usually uses a 3-foot rod.

(1) When electrodes are pipe or conduit, they shouldn't be smaller than ¾-inch trade-size. When using iron or steel piping, the conduit used shall be galvanized-coated or otherwise coated with conductive noncorrosive material.

(2) Rods used as electrodes made of steel or iron shall be at least ⅝ inch in diameter. Nonferrous rods such as solid copper, stainless steel, or steel that is copperclad shouldn't be less than ½ inch in diameter.

(3) The electrodes that are driven shall be driven to the 8-foot minimum or more. In other words, an 8-foot rod shouldn't be left sticking out of the ground. In mountainous areas where bedrock is encountered close to the surface, the rod electrode may be driven at an oblique angle or may be buried in a trench that shall be at least ½ foot deep. The upper end of the ground rod is to be flush with the surface or below surface level. If this can't be done, the end above ground shall be protected, as described in Section 250.117, from physical damage.

(D) Plate Electrodes. Plate electrodes shall be a minimum of 2 square feet, and if made of iron or steel, they shall be at least ¼ inch thick. If made of a nonferrous metal such as copper, they may be .06 inch in thickness. These are often used by utilities at the bottom of a pole hole.

(E) Aluminum Electrodes. It is not allowable to use aluminum electrodes. See comments at Section 250.64 for an explanation.

IV. Grounding Conductors

250.54: Material

The material for grounding conductors is as specified as follows in (A), (B), and (C):

(A) Grounding Electrode Conductor. The grounding electrode conductor shall be of copper or other corrosive-resistant material, solid or stranded, insulated or bare, without splices or joints (except in the case of bus bars). Electrical resistance per foot (linear) shouldn't exceed that of the allowable copper conductors that might be used for this purpose. Thus, if aluminum (in cases where permissible) is used, the conductor must be larger than copper would have to be for the same purpose. If aluminum is used, see Section 250.64.

There is an exception permitting bus bars to be spliced.

There is another exception that covers instances where more than a single enclosure is permitted, as described in Section 230.40. Here you are permitted to connect taps to the grounding electrode conductor. These taps shall be run into each enclosure. The main grounding electrode conductor is to be sized according to Section 250.66, but the sizing shall be for the largest conductor serving each separate enclosure.

The tap conductors can't be attached to the grounding electrode conductor so as to cause a splice. Exothermic welding can be used to extend the grounding electrode conductor in industrial or commercial locations.

(B) Types of Equipment-Grounding Conductor. When the equipment-grounding conductor encloses the circuit conductors, it shall consist of one or more of the combinations listed in the *NEC.*

Exception No. 1

All of the fittings of a flexible metal conduit must be listed.This also applies to flexible metallic tubing, but the following conditions must be met: (a) Six feet is the maximum length that can be used of any combination of flexible metal conduit, flexible metallic tubing, and liquid-tight flexible metal conduit in any grounding run; (b) The circuit shall be protected by overcurrent devices not to exceed 20 amperes; and (c) The flexible metal conduit or tubing shall be terminated only in fittings listed for grounding.

Exception No. 2

Liquid-tight flexible metal conduit, in sizes up to ¼ inch, that have a total length not to exceed 6 feet from the ground return path and fitting for terminating the flexible conduit are also to be listed for grounding. Liquid-tight flexible metal conduit ⅜- and ½-inch trade-size shall be protected by overcurrent devices at 20 amperes or less, and sizes of ¾ inch through 1¼-inch trade-size, protected by overcurrent devices of 60 amperes or less.

(C) Supplementary Grounding. Separate grounding electrodes are permitted to augment grounding conductors, as specified in Section 250.54(B). The earth is not to be used as the sole equipment-grounding conductor.

It is required that a metallic equipment-grounding conductor be used; the earth resistance in almost every case is too high to cause the overcurrent devices to operate properly when a ground fault occurs.

250.56: Resistance of Made Electrodes

Made electrodes shall have a resistance to ground of 25 ohms or less, wherever practicable. When the resistance is greater than 25 ohms, two or more electrodes may be connected in parallel or extended to a greater length. It should be noted that a made electrode that measures more than 25 ohms must be augmented by one additional electrode of a type permitted by Section 250.52. The *Code* does not go into the mechanics of grounding, but good practice indicates that the electrode has a lower resistance when not driven close to a foundation. See Sections 250.50 and 250.52.

Continuous water-piping systems usually have a ground resistance of less than 3 ohms. Metal frames of buildings often make a good ground and usually have a resistance of less than 25 ohms. As pointed out in Section 250.50, the metal frame of a building (when effectively grounded) may be used as the ground. Local metallic water systems and well casings also make good grounds in most cases.

Grounding, when made electrodes are used, can be greatly improved by the use of chemicals, such as magnesium sulfate, copper sulfate, or rock salt. A doughnut-type hole may be dug around the ground rod into which the chemicals are put. Another method is to bury a tile close to the rod and fill it with the chemical. Rain and snow dissolve the chemicals and allow them to penetrate the soil (Figure 250-4).

It is recommended that the resistance of ground rods be tested periodically after installation. This is rarely done, however, except

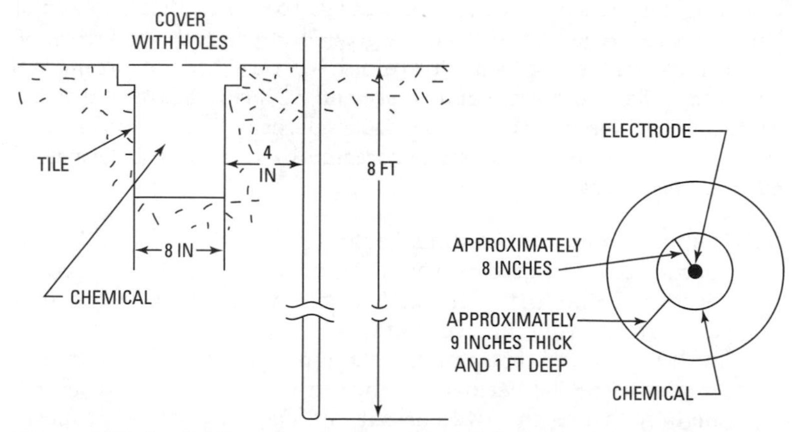

Figure 250-4 Adding chemicals to the soil to lower its ground resistance.

by utility companies who realize the importance of adequate ground. The testing of ground resistance is a mystery to many electricians. Never attempt to use a common ohmmeter for the testing; the readings obtained are apt to be almost anything due to stray AC or DC currents in the soil or DC currents set up by the electrolysis in the soil. There are many measuring devices on the market, such as the ground Megger—a battery-operated ground tester that uses a vibrator to produce pulsating AC current. In recent years, the transistor-type ground tester has appeared on the market.

An example of what you might expect to get by the paralleling of ground rods might be as follows (these figures are general and shouldn't be taken as being the results in every case): Two rods paralleled with a 5-foot spacing between will reduce the resistance to about 65 percent of what one rod would be; three rods paralleled with a 5-foot spacing between will reduce the resistance to about 42 percent; and four rods paralleled will reduce the resistance to about 30 percent.

In summation, there is only one method of telling if the grounding electrode meets these requirements—by testing the ground resistance by means of a good ground-resistance tester.

250.58: Common Grounding Electrodes

This section applies to alternating-current systems and requires that the same grounding electrode in or at the building be used to ground the conductor enclosure and equipment in or on that building. See Sections 250.24 and 250.32.

When two or more grounding electrodes are bonded together, they are considered one.

250.60: Use of Air Terminals

Lightning-rod conductors and made electrodes for the grounding of lightning rods shouldn't be used in place of made electrodes for grounding wiring systems and equipment. However, they may be bonded together; in fact, the *Code* recommends the bonding of these electrodes to limit the difference of potential that might appear between them. See Sections 250.50 to 250.52.

To assist in the above, see Sections 250.50, 800.31(H)(7), and 820.22(H) for required bonding.

A difference of potential between two different grounding systems on the same building can be eliminated if the two grounding systems are tied together (bonded). This is required in the case of lightning protection system conductors and grounding electrodes. To have potential differences between grounding systems can be quite dangerous, as large voltages can develop between any grounded metal objects, such as water pipes, sinks, and so on.

250.64: Grounding Electrode Conductor Installation

The metallic enclosures for the grounding conductor shall be continuous from the cabinet or equipment to the grounding electrode and shall be attached at both ends by approved methods. Articles 342, 344, 352, and 358 govern the installation of the enclosure for the grounding conductor. A common error is often made here—if the system or common grounding conductor is smaller than No. 6, it must always be in rigid metal conduit, intermediate metal conduit, rigid nonmetallic conduit, electrical metallic tubing, or armor cable. This is often confused with a conductor used for grounding equipment and enclosures only.

(A) Aluminum or Copper-Clad Aluminum Conductors. Due to corrosion, aluminum grounding conductors shouldn't be placed in direct contact with masonry, earth, or other corrosive materials. In addition, bare or insulated copper or aluminum conductors are forbidden to be in direct contact with masonry, earth, or anywhere subject to corrosive conditions. Also, where aluminum grounding conductors are used, they shouldn't be closer than 18 inches to the earth. Please note that this does not prohibit the use of aluminum for grounding conductors, but merely places certain restrictions on it. One sadly abused use of aluminum grounding conductors is in antenna grounding. Remember also that the grounding conductor is to be without splice.

(B) Securing and Protection from Physical Damage. No. 4 or larger grounding conductors may be attached to the surface—knobs or insulators are not required. Mechanical protection will be required only where the conductor is subject to severe physical damage. No. 6 grounding conductors may be run on the surface of a building if protected from physical damage and rigidly stapled to the building structure. Grounding conductors smaller than No. 6 shall be in conduit, EMT, intermediate metal conduit, or cable armor.

If a metal raceway is used merely as a physical protection for the common grounding conductor, it is often not attached to either the enclosure or the grounding electrode. If this is the case, you are required to bond the common grounding conductor to both ends of the metal protective raceway. This will lower the impedance of the circuit, which is very necessary upon fault.

It is recommended that magnetic metal enclosures, such as steel pipe or armor, not be used where protection from physical damage can be otherwise obtained, such as by the size of the conductor itself or by nonmetallic enclosures.

2005

(C) Continuous. The grounding electrode conductor may not be spliced, except by irreversible compression-type connectors listed for grounding, or by exothermic welding. Sections of bus bar are permitted. Also, jumpers from grounding electrodes and grounding electrode conductors to copper or aluminum busways may be used. Such busways must be at least ¼ inch by 2 inches (6mm x 50 mm), and the connections must be made with listed connectors.

(D) Grounding Electrode Conductor Taps. If there is more than one service enclosure (permitted by 230.40, Exception 2), taps may be made from the grounding electrode conductor. Such taps must be sized per Section 250.66, and must enter the service enclosures. The connections must be made so that the grounding electrode conductor remains without splice.

(E) Enclosures for Grounding Electrode Conductors. If a metal enclosure is used over the grounding conductor, it shall be electrically continuous from its origin to the grounding electrode. It shall also be securely fastened to the grounding clamp so that there is electrical continuity from where it originates to the grounding electrode. If a metal enclosure is used for physical protection only for the ground conductor and is not continuous from the enclosure to the ground-rod bonding, it shall be done from the grounding conductor to the metal enclosure at each end of the metal enclosure used for physical protection. The requirements of Article 342 shall be met for the installation where intermediate metal conduit is used. If rigid metal conduit is used for protection of the grounding conductors, the requirements of Article 344 shall be complied with. If rigid nonmetallic conduit is used for protection, it must meet the requirements of Article 352, and if EMT is used, it must meet the requirements of Article 358.

(F) To Electrode(s). The grounding electrode conductor may be run to a convenient electrode in a grounding electrode system, or to one or more grounding electrodes individually. The grounding electrode conductor must be sized according to the largest grounding conductor required by any of the grounding electrode it is connected to.

250.66: Size of Alternating-Current Grounding Electrode Conductor

Table 250.66 gives the size of the grounding electrode conductor when used on a grounded or ungrounded AC system. The size shouldn't be less than shown in the table.

Exception

Smaller grounding electrode conductors may be used in the following circumstances:

(a) If they are connected to made electrodes [see Section 250.52(C) and (D)], the part of the conductor directly contacting the electrode may be only No. 6 copper or No. 4 aluminum.

(b) If they are connected to concrete-encased electrodes [see Section 250.50(C)], the part of the conductor directly contacting the electrode may be only No. 4 copper.

(c) If they are connected to a ground ring [see Section 250.50(D)], the part of the conductors directly contacting the electrode is only required to be as large as the conductor in the ground ring.

250.68: Grounding Electrode and Bonding-Jumper Connections to Grounding Electrode

This section tells us that the grounding conductor shall be connected to the grounding electrode in such a manner that a *permanent and effective ground* will result. Most inspectors insist that this connection be accessible wherever possible, or exothermically welded or brazed. The connection to the grounding electrode shall be accessible. The bonding conductors must be long enough to allow for the removal of the equipment without harming the bond.

Where water piping is used as the grounding electrode, any joints that may be disconnected for repairs and replacement shall be bonded around. The same shall be done with insulated coupling, unions, and water meters. This applies to water meters in the house or other buildings where the grounding conductor is connected to the piping on the building side of the meter. It is a common practice to connect the grounding conductor to the street side of the piping ahead of the water meter. Most inspectors prefer it this way when practical to do so. If the meter is at the curb, and there are 10 feet or more of buried water piping (metallic), bonding is not required (Figure 250-11).

Exception

Buried connections or encased connections to a buried grounding electrode, concrete-encased grounding electrode, or driven grounding electrode shouldn't be required to be accessible.

250.70: Methods of Grounding and Bonding Conductor Connection to Electrodes

In connecting grounding conductors to grounding fittings by suitable lugs, use pressure connectors, clamps, or other listed means (including exothermic welding). Soldering must never be used. The ground clamps shall be of a material that is compatible for both the grounding electrode and the grounding conductor. No more than one conductor shall be connected to an electrode unless the connector is listed for the purpose. One of the methods below should be used and exothermic welding could be added:

- **(A) Bolted Clamp.** Listed bolted clamps may be used for connecting the conductor to the made electrode provided they are made of bronze or brass, or of plain or malleable iron. If plain or malleable iron is used and they are not accessible, the copper grounding conductor might cause some electrolysis on the iron.
- **(B) Pipe Fitting, Pipe Plug, and so on.** A pipe plug or similar device screwed into a pipe or fitting may be used for connecting the conductor to the made electrode.
- **(C) Sheet-Metal Strap-Type Grounding Clamp.** Ordinary sheet metal grounding clamps are not approved. There is a listed clamp that has a rigid metal base that clamps firmly to the electrode and the strap. Only this strap is of such material and heavy enough that it is very unlikely that it will stretch after being installed.
- **(D) Other Means.** Other approved methods.

Author's Note

In my experience, exothermic welding has proved to be an excellent method. However, with exothermic welding you can be fooled about the connections. Eyeballing and hammer testing don't always disclose defects

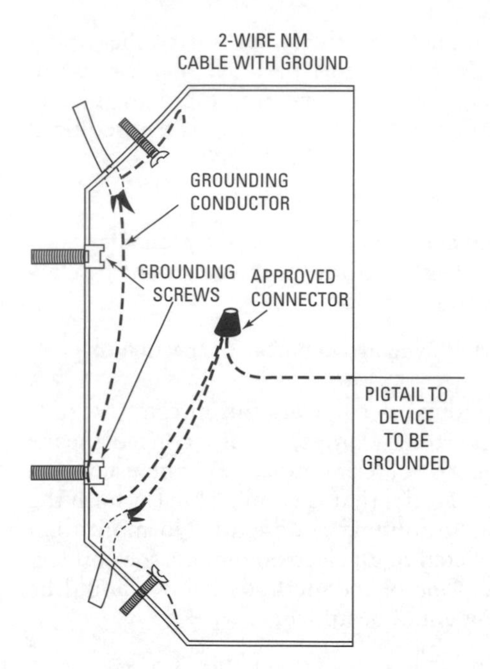

Figure 250-5 Illustrating the proper method for grounding boxes and carrying the grounding conductor on to the device to be grounded.

in high-resistance connections. This is a positive check on the resistance of the connection to the grounding electrode.

V. Enclosure and Raceway Grounding

250.80: Service Raceways and Enclosures

All metal service enclosures and service raceways must be grounded. A metal conduit elbow in a nonmetallic conduit run is not required to be grounded so long as it is no less than 18 inches below grade.

250.84: Underground Service Cable

Where an underground service is supplied by a continuous metal-sheathed or armored cable that is bonded to the underground system, it need not be grounded at the building and may be insulated from the interior conduit. Please note that this is quite different from an overhead system in which the cable sheath or armor and/or conduit shall be grounded at the service entrance.

250.86: Other Conductor Enclosures and Raceways

Other metal enclosures and raceways must be grounded, as well.

Exception No. 1
Grounding is not required for raceways used when extending open wiring, nonmetallic cables, and knob-and-tube wiring not containing equipment grounds. These runs must be no more than 25 feet long, must be kept free from contact with persons, and must be kept away from any contact with grounded surfaces.

Exception No. 2
Short sections of enclosures used to protect cable assemblies need not be grounded.

Exception No. 3
Enclosures that are not required to be grounded according to Section 250.43(I).

Exception No. 4
A metal conduit elbow in a rigid nonmetallic conduit run is not required to be grounded so long as it is at least 18 inches below grade or beneath a 4-inch thick concrete slab.

VI. Bonding

250.90: General

To ensure electrical continuity and also sufficient capacity to conduct the fault-current likely to occur safely, bonding shall be required to ensure that necessary electric continuity is provided. This section is very basic and important. Larger available fault currents demand proper means of safely handling this problem.

250.92: Service Equipment

This is an item that has been a part of the *Code* for many years, but is often overlooked. It is recommended that a careful understanding of the reason for bonding at this location be gained because of its importance. The way in which this bonding is to be done is covered in Section 230.94. An example of proper bonding is shown in Figure 250-6.

The parts of equipment required to be grounded effectively follow in (A) and (B):

(A) **Bonding of Service Equipment.** Noncurrent-carrying metal parts of equipment shall be effectively bonded as covered in (1), (2), and (3):

- **(1)** Section 250.84 gives the exception for service raceways, service cable armor or sheath, or cable trays.

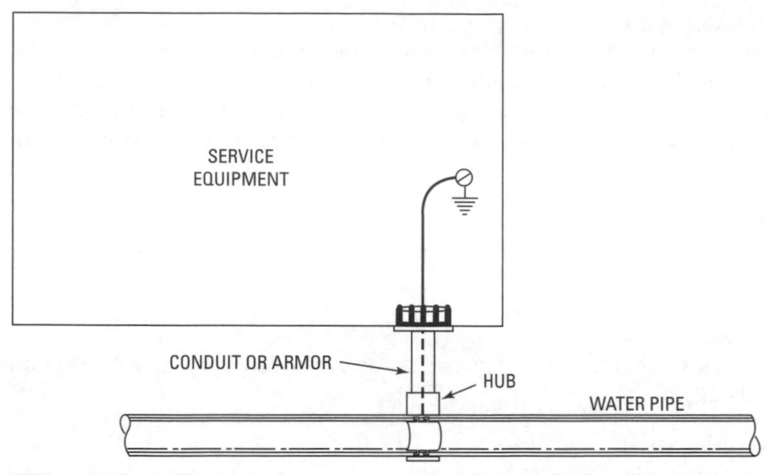

Figure 250-6 Conduit or armor used to protect the grounding wire shall be bonded to the grounding electrode and to the service-entrance enclosure.

(2) Bonding is required for all service-equipment enclosures when service entrance conductors are present. This includes any meter ratings, boxes, auxiliary gutters, etc. that may be connected in the service raceway or armor. Meter socket threaded hubs served by metal conduit will serve as a bond.

(3) Section 250.64(A) covers metal raceways or armor in which the grounding electrode conductor is contained. This section requires that all metal raceway or cable armor be bonded to the grounding electrode conductor at both ends and at every metal box, enclosure, etc. in between.

(B) Bonding to Other Systems. At the service, an accessible means for connecting bonding and grounding conductors must be provided, using at least one of the following methods:

(1) Exposed service raceways (metal raceways only).

(2) An exposed grounding electrode conductor.

(3) Other approved means.

Refer to Sections 800.40 and 820.40 for requirements for communications and CATV circuits.

250.94: Method of Bonding Service Equipment
The following is the procedure to be followed to ensure continuity at service equipment:

(A) **Grounded Service Conductor.** Section 250.8 requires bonding equipment to the grounded service conductor.

(B) **Threaded Couplings.** Threaded couplings or threaded bosses when used with rigid or intermediate metal conduit shall be made wrench-tight so that good bonding is made.

(C) **Threadless Couplings and Connectors.** Threadless couplings and connectors are available for rigid metal conduit. They shall be made up tight to ensure electrical continuity as required for bonding for this section. Standard locknut or bushing shouldn't be used for the bonding required.

(D) **Bonding Jumpers.** Where concentric or eccentric knockouts are used in enclosures that have been punched or the bonding security has been otherwise impaired, the electrical connection

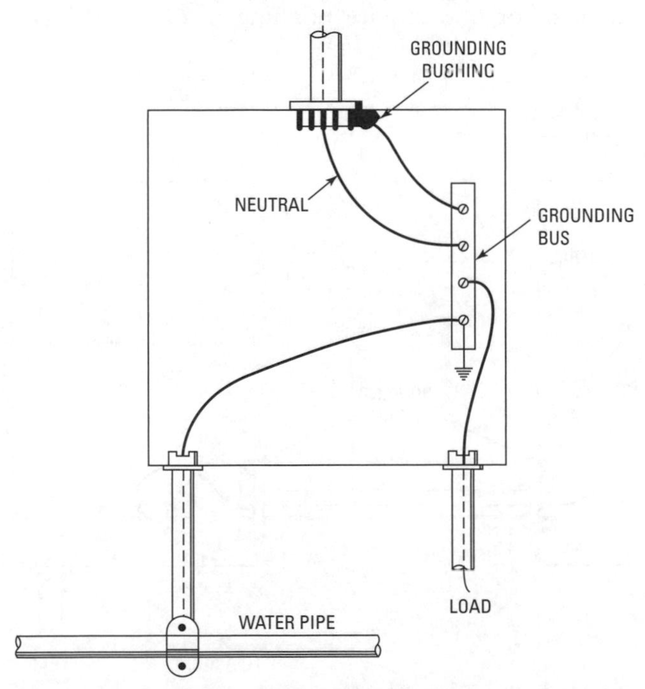

Figure 250-7 Grounding and bonding a service entrance.

to ground bonding jumpers meeting other requirements of this article shall be used in such places—for instance, bonding type bushings, or locknuts, or otherwise bonded around.

This requires that, in services and entrance equipment, as well as other points where there are eccentric or concentric knockouts, bonding jumpers be used around concentric or eccentric knockouts. It is also necessary to use bonding around any flexible conduit that might be used in conjunction with service entrances (Figure 250-3).

(E) Other Devices. Lockout and bushings of the conventional type are not approved for continuity at service entrances. Devices shall be approved for the purpose, that is, bonding-type bushings, wedge nuts, and so on.

250.96: Bonding Other Enclosures

All types of metal raceways, armor cable, cable sheath, enclosures, frames, and other metal noncurrent-carrying parts that serve as the equipment-grounding conductors, whether using a supplementary grounding conductor or not, require bonding to ensure the safe

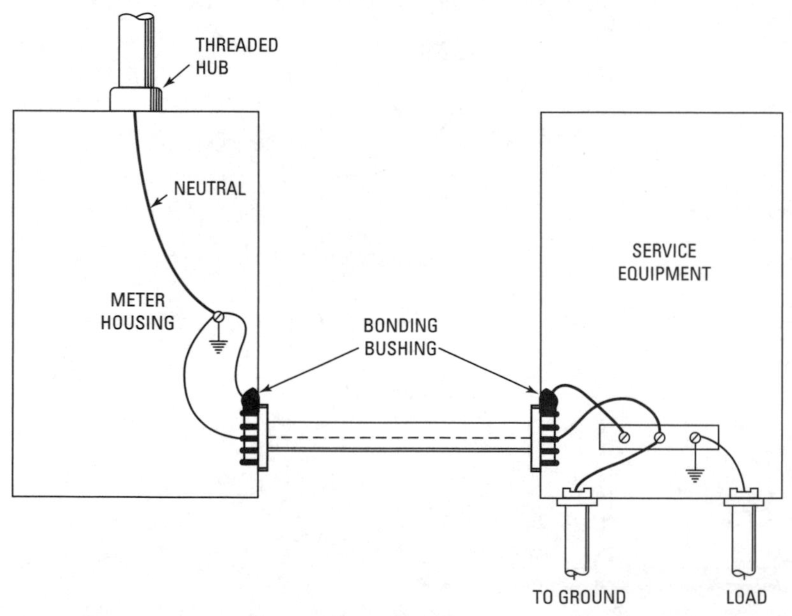

Figure 250-8 Grounding and bonding a typical service entrance.

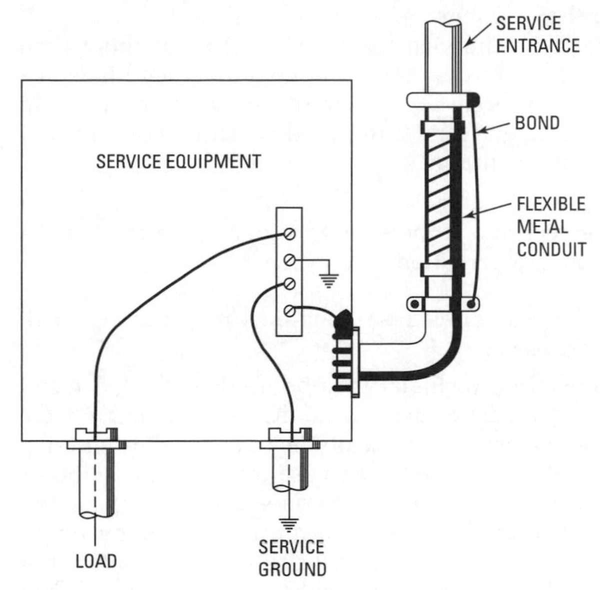

Figure 250-9 Bonding flexible metal conduit in a service entrance.

handling of any fault currents that may be imposed and to ensure electrical continuity. Sometimes you will run into nonconductive paint or similar coatings. These must be removed at the threads or other contact points, and these clean surfaces shall be connected by means of fittings or bonding jumpers designed to make such removal unnecessary for service. On large buildings there are usually expansion joints in the buildings to take care of temperature changes and movements where these joints are crossed by a metal raceway. There are listed fittings to permit the movement of the building by bonding across the fitting externally so as not to interrupt equipment-grounding continuity.

The phrase "shall be effectively bonded where necessary" is the key to this section. It appears to become a design and an inspection problem to pass judgment as to where to bond and where not to bond.

Exception

When it is deemed necessary for the reduction of electrical noise(electromagnetic interference) in a grounding circuit, a listed nonmetallic raceway fitting may be installed between the enclosure and metal raceways. In such instances, a separate equipment-grounding conductor must be used to ground the enclosure properly.

250.97: Bonding for Over 250 Volts

Metal raceways and conduits enclosing conductors at more than 250 volts to ground (other than service conductors) shall have the electrical continuity ensured by one of the methods outlined in Section 250.94(B) through (E). Recall that this section covered continuity at service equipment.

Exception

Where oversized eccentric or concentric knockouts are not used, the following explanations are permitted:

- **(A) Threadless Fittings.** Threadless couplings may be used with metal sheath cables.
- **(B) Two Locknuts.** Two locknuts may be used with rigid metal conduit or intermediate metal conduit, one locknut on the outside of the enclosure and one on the outside when entering boxes or cabinets. You still have to use the appropriate bushings inside the enclosure. On some you will use the metal and on others, as described later, you must use insulated bushings.

  Grounding connections and circuit conductors that are not made tight cause many electrical faults and some fires. When

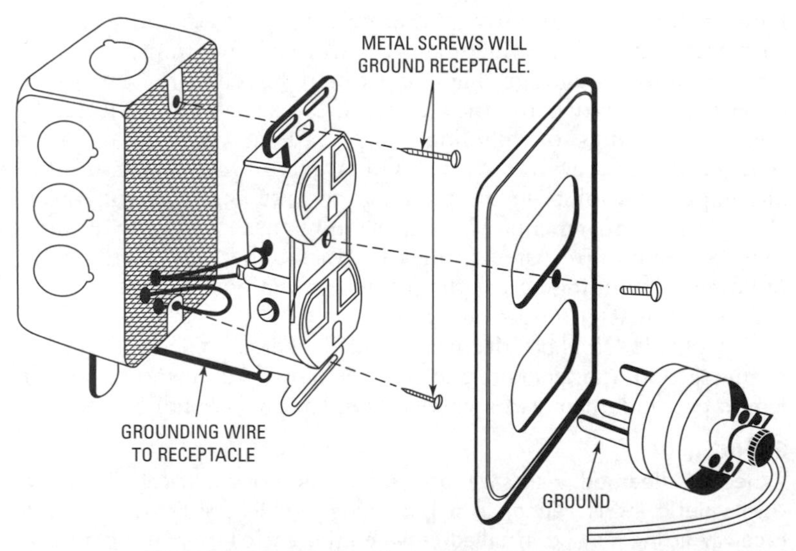

Figure 250-10 A grounding jumper shall be used from grounded boxes to the grounding terminal of the receptacle.

tightening screws that show inch-pounds, pressure should be used, with 12-inch-pound pressure as the minimum tightening of device screws. These locknuts are also to be made up tight. Elsewhere in the *Code* it is permissible to use one locknut and one bushing where it is not practical to use two locknuts and a bushing. Where the voltage exceeds 250 volts to ground, there are to be two locknuts as outlined above.

(C) **One locknut.** One locknut on the inside of the boxes or cabinets will suffice where the fitting shoulders seat firmly on the box or cabinet. These include electric metallic tubing connectors, cable connectors, and flexible metal conduit connectors.

250.98: Bonding Loosely Jointed Metal Raceways

This was discussed earlier for covering conduit crossing expansion joints or telescoping sections of raceways and using on bonding jumpers.

Metal trough raceways used in connection with sound recording and reproducing, made up in sections, shall contain a grounding conductor to which each section is bonded. In recording and reproducing, the bonding wire will reduce the conditions that affect the quality of recording and reproduction.

250.100: Bonding in Hazardous (Classified) Locations

Regardless of the voltage, the electrical continuity of raceways and boxes shall be ensured by one of the methods in Section 250.94(B) through (E). However, this does not cover the continuity in hazardous locations in entirety; the requirement for the specific type of hazardous condition involved should be looked up. These are covered in Sections 500 through 517. In reality, this section is rather broad and does not cover all hazardous locations.

250.104: Bonding of Piping Systems and Exposed Structural Steel

(A) **Metal Water Piping.** Regardless of whether the water piping is supplied by nonmetallic pipe to the building or structure, you are required to bond the interior metal water piping to the service-equipment enclosure. This bonding jumper is sized using Table 250.66. It may be run to the service-entrance equipment if the service-grounding conductor is of sufficient size, or it may be run to two or more grounding electrodes, if used. The jumper must be installed in accordance with Section 250.64(A) and (B).

The attachment points of all these bonding conductors must be accessible. For separately derived systems that use a grounding electrode according to Section 250.30(C)(3), the grounded conductor of the derived system must be bonded to the nearest metal water pipe.

Exception

In multiple occupancies, if the interior piping is metal and the piping systems of all occupancies are tied together, it is required that metal piping be isolated from all other occupancies by the use of nonmetallic pipe. Each individual metal water-pipe system is required to be bonded separately to the panelboard or switchboard enclosure, but not to the service-entrance equipment. Bonding jumper points of attachment shall be accessible and sized according to Table 250.122. This ensures that the metal piping system in each occupancy is at ground potential.

(B) **Other Metal Piping.** All interior metal piping that may be subjected to being energized shall be bonded to the service-equipment ground at the service enclosure to the common grounding conductor, if it is of sufficient size, to one or more of the grounding electrodes, if it is of sufficient size, or to one or more of the grounding electrodes, and this bonding jumper shall be sized according to *NEC* Table 250.122 (see Figure 250-11).

With circuit conductors that could cause energizing of piping, you are permitted to use the equipment ground run with the circuit conductors.

(C) **Structural Steel.** Exposed building steel frames that are not intentionally grounded must be bonded to either the service enclosure, the neutral bus, the grounding electrode conductor, or to a grounding electrode. The bonding conductor must have accessible connections and must be sized according to Table 250.66 and installed according to Section 250.64(A) and (B).

Note

It is a good idea, for additional safety, to bond all metal air ducts and piping on the premises.

250.106: Spacing from Lightning Rods

Metal electrical equipment parts must either be kept more than 6 feet (1.83 m) from lightning protection conductors, or else bonded to the conductors.

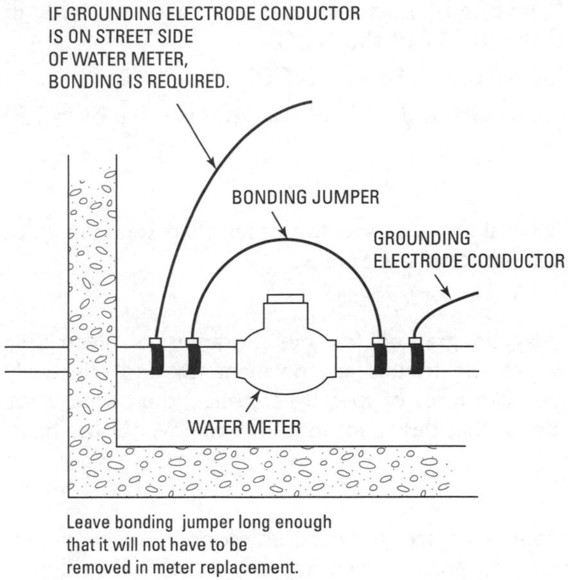

Figure 250-11 Proper bonding of a water meter.

VII. Equipment Grounding and Equipment-Grounding Conductors

250.110: Equipment Fastened in Place or Connected by Permanent Wiring Methods (Fixed)

The grounding of noncurrent-carrying metal parts of fixed equipment is always subject to becoming energized due to the results of abnormal conditions, and so this section requires the grounding of these noncurrent-carrying parts [see (A) through (F) below]. This shouldn't be taken lightly, and all precautions must be taken to keep these parts, should they become energized, at as close to zero potential with surrounding areas as possible.

- **(A)** **Vertical and Horizontal Distances.** Fixed equipment that is under 8 feet vertically or 4 feet horizontally of ground or grounded metal objects that may be contacted by persons. Concrete is an excellent ground and this should be included.
- **(B)** **Wet or Damp Locations.** Equipment in damp or wet locations that are not isolated.
- **(C)** **Electrical Contact.** See the *NEC*.

(D) **Hazardous (Classified) Locations.** These are covered in Articles 500 through 517 of the *NEC.*

(E) **Metallic Wiring Methods.** See the *NEC.*

(F) **Over 150 Volts to Ground.** Where the voltages are over 150 volts to ground.

Exception No. 1
If switches or circuit breakers are used for other than service equipment and only qualified persons have access.

Exception No. 2
Special permission may be granted for metal frames of electrically heated appliances where the frames are permanently and effectively insulated from ground. Take note of *special permission*; the purpose for this exception is no doubt that this is not intended to cover house heating in general.

Exception No. 3
If transformers and capacitors are mounted on wooden poles and are over 8 feet from grade or ground level and are used as distribution apparatus. Here again, care must be exercised to see that this equipment is high enough to avoid persons touching it while standing on grounded objects, and so on.

Exception No. 4
Equipment that is double-insulated and effectively marked as such, listed as processing, and listed as office equipment are not required to be grounded.

250.112: Fastened in Place or Connected by Permanent Wiring Methods (Fixed)—Specific

Equipment described in (A) through (K), irrespective of the voltage, must be grounded. This covers exposed noncurrent-carrying metal parts. For (A) through (K), see the *NEC.*

250.114: Equipment Connected by Cord and Plug

Exposed noncurrent-carrying metal parts of cord- and plug-connected equipment that may become energized must be grounded, as follows in (A) through (D):

(A) **In Hazardous (Classified) Locations.** These are covered in Articles 500 through 517.

(B) **Over 150 Volts to Ground.** If they are operated at over 150 volts to ground, guarded motors are exempt. Metal frames of heating appliances that are effectively insulated from ground may be exempted by special permission. Also see Exception No. 4 of Section 250.110.

(C) In Residential Occupancies. In residential occupancies:

(1) Refrigerators, freezers, and air conditioners.

(2) Clothes-washers, clothes-dryers, and dishwashers; sump pumps and electrical aquariums.

(3) Hand-held motor-operated tools.

(4) Motor-operated appliances such as hedge clippers, lawn mowers, snowblowers, and wet scrubbers.

(5) Portable hand-held lamps.

The use of ground-fault circuit interrupters will also be a great assistance to personnel safety.

Exception

Hand-held appliances that are doubly insulated or have equivalent protection shouldn't be required to be grounded. Double-insulated tools and the like shall be effectively marked as such.

A portable GFCI would give additional protection and assist in personnel safety.

(D) In Other Than Residential Occupancies.

(1) Refrigerators, freezers, and air conditioners.

(2) Clothes-washers, clothes-dryers, and dishwashers.

(3) Electronic computer plant and data-processing equipment.

(4) Sump pumps and electrical aquarium equipment.

(5) Motor-operated hand tools such as hedge clippers, lawn mowers, snowblowers, and wet scrubbers.

(6) Cord- and plug-connected appliances used in damp or wet locations, used by persons that may be standing on concrete ground or metal floors. This could include working inside metal tanks or boilers or concrete metal tanks.

(7) If likely to be used in damp or wet locations or any other conductive location, hand tools are included.

(8) Portable hand lamps.

(9) Stationary or fixed motorized tools.

Portable GFCIs are available and add a great deal of protection.

Exception No. 1

If the source of supply is an isolated transformer with an ungrounded secondary not over 50 volts, tools and hand lamps may be used in wet and conductive places and are not required to be grounded.

Exception No. 2

Double-insulated or distinctively marked portable tools and appliances, if listed, may be approved and shouldn't require grounding.

250.116: Nonelectric Equipment

This section covers grounding of metal parts of nonelectric equipment. Basically, any piece of nonelectric equipment that might accidentally become electrically energized must be properly grounded.

250.118: Types of Equipment-Grounding Conductors

Equipment-grounding conductors must be of the following types:

(1) A copper, aluminum, or copper-clad aluminum conductor. These may be solid or stranded, insulated or bare, or a bus bar of any shape.

(2) Rigid metal conduit.

(3) Intermediate metal conduit.

(4) Electrical metallic tubing.

(5) Flexible metal conduit that is listed for grounding, with fittings also listed for grounding.

2005

(6) Flexible metal conduit that is not listed for grounding, provided that: the fittings are listed for grounding, the circuit conductors are protected at 20 amperes or less, the length of the run is no more than 6 feet (1.8 m), and the conduit is not used for the sake of flexibility. If used for flexibility, a separate equipment grounding conductor must be used.

(7) Liquid-tight flexible metal conduit, provided that the fittings are listed for grounding, the circuit conductors are protected at 20 amperes or less for 3/8 inch or 1/2 inch conduit, the circuit conductors are protected at 60 amperes or less for 3/4 inch or 1 1/4 inch conduit, the length of the run is no more than 6 feet (1.8 m), and the conduit is not used for the sake of flexibility. If used for flexibility, a separate equipment grounding conductor must be used.

(8) Flexible metal tubing, provided that connectors listed for grounding are used, the circuit conductors are protected at no more than 20 amperes and the length of the run is no more than 6 feet (1.8 m).

(9) The armor of Type AC cable.

(10) Copper sheaths of Type MI cables.

(11) Type MC cables that are listed for grounding.

(12) Metal cable trays. See Section 392.7(B) for details.

(13) Cable bus framework (must be bonded properly).

(14) Other listed metal raceways and gutters (must be electrically continuous).

250.119: Equipment Fastened in Place or Connected by Permanent Wiring Methods (Fixed) Grounding

This covers the noncurrent-carrying metal parts of equipment. For other enclosures and raceways that are required to be grounded, (A) or (B) that follow shows two methods of this type of grounding:

Exception

As permitted by Sections 250.32, 250.140, and 250.142, equipment, raceways, and enclosures where grounded to the grounded circuit conductor.

(A) Equipment-Grounding Conductor Types. Section 250.54(B) covers the equipment-grounding conductors.

(B) With Circuit Conductors. When equipment-grounding conductors are contained in the same raceway cable or cord, or run with the circuit conductors, they may be bare, covered, or insulated. Equipment-grounding conductors covered or insulated must have insulation that is continuously of a green color or green with a yellow stripe.

It shall be a part of the cable or cord, or run in the same raceway with the circuit conductors to keep the impedance to the lowest value. When using NM, NMC, or UF cables, these cables are required to contain the equipment-grounding conductor as part of the cable. See *NEC* Section 300.3(B).

Exception No. 1

Should the insulated or covered conductor be larger than No. 6 copper or aluminum, this conductor shall be permanently marked at the time of installation, at each end or any other accessible point. It shall be identified by any one of the following means.

See the *NEC* for (A) through (C).

Exception No. 2

On DC circuits the equipment-grounding conductor may be run separately from the circuit conductors. The current is steady in DC, while in AC the current fluctuates 120 times per second, causing induction between the conductors. This is allowed because impedance does not enter into DC circuits.

Exception No. 3

Where maintenance and supervision is done only by qualified persons, one or more insulated conductors in a multiconductor cable may at the time of installation be permanently marked to indicate them as equipment-grounding conductors. This must be done at both ends and at any point where there is access to the equipment-grounding conductor.

- **(a)** By removing the insulation along the entire length of the exposed conductor.
- **(b)** By a painted or otherwise permanent color such as green tape where the equipment-grounding conductor is exposed.
- **(c)** Or marking by permanent tags or labels.

Note

Section 250.28 covers bonding jumper and the requirements thereof, and Section 400.7 covers the use of cords for fixed equipment.

250.122: Size of Equipment Grounding Conductor

Refer to the *NEC* and Table 250.122. Equipment-grounding conductors are sized in accordance with the rating or setting of an automatic overcurrent device in circuit ahead of equipment, conduit, etc.

If conductors are paralleled in several raceways, which you will find permitted in Section 310.4, and an equipment-grounding conductor has to be used (for instance, if nonmetallic conduit was used instead of metal), the equipment-grounding conductors shall be run in parallel—one in each raceway if more than one raceway is used. Each equipment-grounding conductor in parallel circuits must meet the requirements of Table 250.122.

Where current-carrying conductors have to be larger to compensate for voltage drop, the equipment-grounding conductors must be adjusted proportionately to the circular metal area.

The intent of this seems to be that the equipment-grounding conductor in each paralleled raceway circuit be sized to the ampere rating of the overcurrent device protecting these circuits at the service-entrance equipment. Table 250.122 covers the sizing of these equipment-grounding conductors.

Where a single equipment-grounding conductor is run with multiple circuits in the same raceway, it shall be sized for the largest overcurrent device protecting the conductors in the raceway or cable.

If the overcurrent device is a motor short-circuit protector or an instantaneous trip breaker (see Section 430.52), the size of the equipment-grounding conductor must be based on the rating of the protective device.

Exception No. 1

An equipment-grounding conductor as small as No. 18 AWG that is part of a lighting fixture may be used. It may not, however, be smaller than the circuit conductors in the lighting fixture.

Exception No. 2

The equipment-grounding conductor shouldn't be larger than the current-carrying conductors supplying any equipment.

Exception No. 3

See Sections 250.119(A) and 250.54(B) for raceway or cable armor or sheath used as the equipment-grounding conductor. Also refer to Section 250.2 regarding raceway or cable providing an effective grounding path.

Exception No. 4

If only qualified people will oversee the entire system, the grounding conductor for a high-impedance grounded-neutral system can be sized according to the available ground-fault current. (See Sections 250.36 and 250.186.)

250.124: Equipment-Grounding Conductor Continuity

(A) **Separable Connections.** Where separable connections are made to draw out equipment, there are attachment plugs from receptacles to main equipment connectors. The equipment-grounding conductor is longer so that it will be the first connection made when plugging.

The main point here is that the grounding terminal on plugs makes contact before the current-carrying conductors do.

Exception

Interlocking plugs, receptacles, equipment, and connectors do provide for grounding continuity first.

(B) **Switches.** You should never install a switch or automatic cutout in the equipment-grounding conductor where a premise wiring system is involved.

Exception
Where all conductors are disconnected at once.

The main point of all this is that the equipment-grounding conductor is all important for safety, and all assurances must be made to see that the equipment grounding is intact before energizing circuits.

250.126: Identification of Wiring Device Terminals
All terminals for equipment-grounding conductors must be identified. In general, green hexagonal (six-sided) screws or nuts are acceptable, as are entrance holes marked "G," marked "GR," or marked with the standard grounding symbol as shown in Figure 250-12.

Figure 250-12 Grounding Symbol.

VIII. Methods of Equipment Grounding

250.130: Equipment-Grounding Conductor Connections
The equipment-grounding conductor shall be connected on the supply side of the service-entrance disconnecting means and shall be made as required in (A) and (B) below. The equipment-grounding conductor from a separately derived system shall be connected at the source as required in Section 250.30.

(A) **For Grounded System.** The connection of the equipment-grounding conductor shall be made by bonding it to the grounded service conductor or neutral, and also to the grounding electrode. This can all usually take place in the service disconnecting means. See Figures 250-13 and 250-14.

(B) **For Ungrounded System.** You are required to have a grounding electrode, even though no conductor from the supply is grounded, and the equipment-grounding conductor (be it the metallic raceways or a conductor) shall be connected to the grounding electrode conductor.

Exception for (A) and (B) above
Nongrounded-type receptacles are to be used only for replacement on ungrounded systems and for branch circuit extensions only where the installations already exist and don't have equipment-grounding conductors in the branch circuits. When adding a grounding-type receptacle to

an ungrounded equipment wiring system, you may ground the receptacle-grounding terminal to a metallic water pipe. This is covered in the exception to this section and bonding is covered in Section 250.104(A).

Author's Note
Be sure the water pipe is continuous metal pipe and that there are no insulated couplings or unions separating the piping from ground.

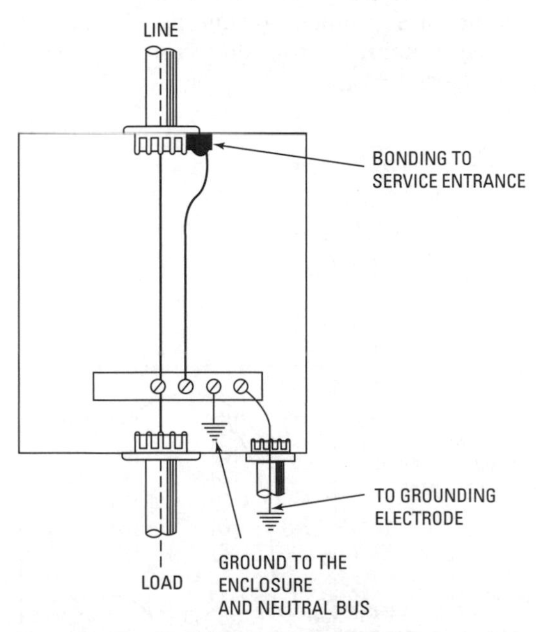

Figure 250-13 The grounding conductor that grounds the neutral must also be used for all other grounding.

Note
The exception in Section 210.22(D) permits the use of a GFCI-type of receptacle on systems that were installed without equipment-grounding conductors.

250.132: Short Sections of Raceway
When isolated sections of cable armor or raceway must be grounded, the grounding must be done in accordance with Section 250.119.

250.136: Equipment Considered Effectively Grounded

The noncurrent-carrying metal parts of equipment, as specified under (A) and (B) below, shall be considered as being effectively grounded.

(A) Equipment Secure to Grounded Metal Supports. Even if electrical equipment is secured on a good electrical contact with metal racks or supports, it shall be grounded by one of the means indicated in Section 250.119. Structural steel shouldn't be used as a conductor for equipment grounding. Recall that it was required that the equipment-grounding conductor be in with the circuit conductors.

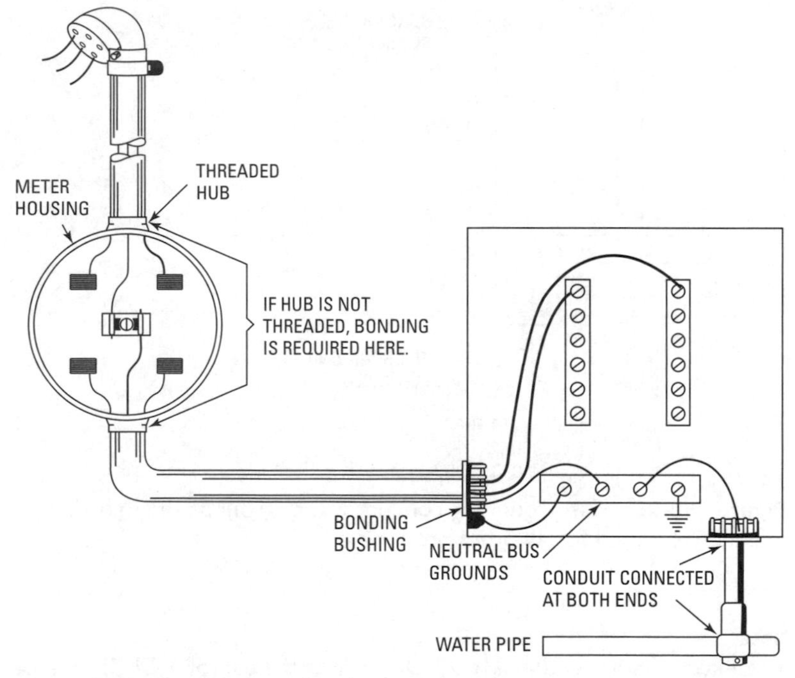

Figure 250-14 An example of proper bonding.

(B) Metal Car Frames. The metal car frames of elevators and similar devices are considered adequately grounded if they are attached to a metal cable running over or attached to a metal drum that is well grounded according to the grounding requirements of the *Code*. See Section 250.119.

250.138: Cord- and Plug-Connected Equipment

On equipment that has metal noncurrent-carrying parts that are to be grounded, the grounding may be accomplished by the following methods (A), (B), or (C):

(A) **By Means of the Metal Enclosures.** By means of the metal enclosure of the conductors feeding such equipment, provided an approved grounding-type attachment plug is used, one fixed contacting member being for the purpose of grounding the metal enclosure, and provided, further, that the metal enclosure of the conductors is attached to the attachment plug and to the equipment by approved connectors.

The preceding means that the outlet shall be of the grounded receptacle-type, and that the attachment plug shall be of the type approved for grounding from a grounding-type receptacle, and that the cord shall carry a grounding conductor, one end of which is attached to the grounding terminal of the attachment plug and the other end to the frame of the portable equipment. Since it was found that it is not always possible to attach a portable appliance or equipment to a grounding-type receptacle because the grounding terminal of the attachment plug was often broken off or cut off to fit an old type of receptacle, the following exception has therefore been added:

Exception

The grounding contact member of grounding-type attachment plugs on the power supply cord of portable hand-held, hand-guided, or hand-supported tools or appliances may be of the movable self-restoring type. In other words, there are approved attachment plugs that have a hinged grounding prong with a spring so that it can be folded out of the way and yet will restore itself to the normal position for use on grounding-type receptacles.

(B) **By Means of a Grounding Conductor.** The grounding conductor may be run in the cable or flexible-cord assembly, provided that it terminates in an approved grounding-type attachment plug having a fixed grounding-type contact member. This grounding conductor in the cable or cord assembly may be bare, but if insulated, it shall be green or green with a yellow stripe.

You will notice that *NEC* says "fixed grounding contact member"—again, there is an exception. This exception is exactly as the preceding exception for (A), that is, the contact member may be a restoring type.

(C) Separate Flexible Wire or Strap. The equipment can be grounded by using a flexible wire strap. The strap should be protected from damage as well as is practical.

250.140: Frames of Ranges and Clothes-Dryers

Frames of electric ranges and electric clothes-dryers shall be grounded as provided for in Sections 250.119 and 250.138, that is, by metal raceways or a grounding conductor. Outlet and/or junction boxes shall also be grounded. There is another alternative on 120/240-volt, three-wire circuits, or 120/208-volt circuits derived from a three-phase, four-wire supply. This type of circuit may be grounded to the neutral conductor, provided that the neutral or grounded circuit conductor is no smaller than No. 10 copper. This applies to wall-mounted ovens and countertop cooking units as well. This is the one instance where the neutral may be used also as the grounding conductor, providing all other conditions are met.

Where Type SE cable with a bare neutral is used for ranges or dryer branch circuits, it shall originate only for service equipment. If it were permitted to originate from feeder panels, we would have an isolated neutral bus with a bare conductor attached. With a bare conductor, it could easily touch the enclosure and defeat the protection that the isolated neutral affords. This does not apply to mobile homes or recreational vehicles.

If grounding-type receptacles are furnished with the equipment, they shall be bonded to the equipment.

250.142: Use of Grounded Circuit Conductor for Grounding Equipment

(A) Supply-Side Equipment. A neutral conductor is allowed to ground metal equipment enclosures or raceways in the following circumstances:

- **(1)** On the line side of a service disconnect.
- **(2)** On the line side of the main disconnecting means for separate buildings, as specified in Section 250.32.
- **(3)** On the line side of the disconnect or overcurrent protection device of a separately derived system.

(B) Load-Side Equipment. The grounded neutral shouldn't be used for grounding noncurrent-carrying metal parts of equipment, but is supplied from the load side of a service disconnecting means or on the load side of the separately derived system disconnecting means or the overcurrent devices.

Exception No. 1
The metal frames of appliances, as specified in Section 250.140.

Exception No. 2
As permitted in Section 250.32 for separate buildings.

Exception No. 3
Meter enclosures may be grounded to the grounded circuit conductor (neutral) if the grounding is on the supply side of the service disconnecting means, and: (a) Where no ground-fault protection has been installed on the service, (b) the meter enclosures in the above are located near the service disconnecting means, and (c) where the grounded circuit conductor is no smaller than the size required by Section 250.122 for equipment-grounding conductors.

Exception No. 4
Refer to Sections 710.72(E)(1) and 710.74 of the *NEC*.

Exception No. 5
DC systems are allowed to be grounded on the load side of the disconnect, as per Sections 250.119 and 250.138.

250.144: Multiple Circuit Connections
There will be cases where the equipment will be required to be grounded, that is, it is supplied by one or more circuits of grounded wiring systems on the premises; in this case, separate equipment-grounding conductors shall be supplied from each system as specified in Sections 250.119 and 250.138.

250.146: Connecting Receptacle-Grounding Terminal to Box
Grounding continuity between a grounded outlet box and the grounding circuit of the receptacle shall be established by means of a bonding jumper between the outlet box and the receptacle-grounding terminal (Figure 250-5).

Exception No. 1
When a box, such as a handy box, is surface-mounted, direct metal contact may be made metal-to-metal with the box and the yokes of the receptacle. This shall be accepted as a ground connection to the receptacle. The above won't apply to mounted receptacles unless you have ensured the positive ground connection between the receptacle yoke and box.

Exception No. 2
There are contact devices and yokes specifically listed for proper grounding when used in conjunction with supporting screws to ensure

proper grounding between the yoke and flush-type mounted boxes or receptacles.

Exception No. 3

If floor-mounted boxes are listed and designed specifically for making good electrical contact when this receptacle is installed, they may be used.

Exception No. 4

When electrical noise (electromagnetic interference) occurs in the grounding circuit, a receptacle on the grounding circuit in which the grounding terminal or receptacle has purposely been insulated from the mounting means or yoke should be permitted. In this case, an insulated equipment-grounding conductor may run to the isolated terminal with the circuit conductors. This insulated grounding conductor may pass through one or more panelboards in the same building without being connected thereto as covered in Section 384.27, so it terminates at the equipment-grounding terminal at that derived system or service.

See the *NEC* for the fine-print note.

Grounding similar to that described in Exception No. 4 may also be found to be advantageous for grounding computers or solid-state counters on machines. However, if metal conduit is used as the raceway, it should be insulated from the machine in question and then the insulated grounding conductor run back from the equipment to the source as explained in Exception No. 4.

The results of the above established that a grounding jumper shall be used from grounded boxes to the grounding terminal of the receptacle in all cases, including boxes on conduit circuits and EMT circuits, unless the box is surface-mounted so that the mounting screws may be tightened to make a secure grounding connection between the device yoke and the box. Of course, if a device is approved for making the proper grounding connection, the bonding won't be necessary (Figure 250-10).

250.148: Continuity and Attachment of Branch-Circuit Equipment-Grounding Conductors to Boxes

When more than one equipment grounding conductor enters a box, they shall be spliced or connected to any device provided for connecting them to part of the box. Soldering is not permitted. Splices shall be made to meet the requirements of Section 110.14(B), but no insulation is required. If there is a receptacle connected in that box, a separate pigtail shall be connected where the wires are spliced to serve that receptacle. This is done so that after that receptacle is removed, the grounding continuity won't be interrupted.

This has caused considerable concern in the field. The grounding conductor is to be attached to a screw used for no other purpose or by approved means. Boxes are available with two or more 10-32 tapped holes for this purpose. The clamp screws on loom-wire boxes are not intended to be used as grounding screws. All grounding conductors entering the box should be made electrically and mechanically secure, and the pigtail made from them should serve as the grounding connection to the device being installed. Thus, if the device is removed, the continuity won't be disrupted. Soldering is not permitted. Pressure connectors should be used—twisting the wires together is not satisfactory because poor contact may result. Most inspectors require that all of this be done at the time of the rough-in inspection so that they don't have to open all of the outlets to see that they have been properly made up (see Figure 250-5).

Where nonmetallic boxes are used, the equipment-grounding conductors do not have to be attached to them. The other requirements listed above are to be complied with. [See the *NEC* parts (A) and (B) of this section.]

Exception

An isolated grounding conductor, installed according to Section 250.146, does not have to be connected to the other grounding conductors in the box.

IX. Direct Current Systems

250.160: General

All direct current systems are required to comply with Part VIII and other portions of Article 250 that are not intended for AC systems.

250.162: Direct-Current Circuits and Systems to Be Grounded

(A) **Two-Wire Direct Current Systems.** Systems supplying two-wire DC must be grounded.

There are cases in which this could cause electrolysis on underground pipes and the like, such as in the past when trolley cars had the ground at positive potential one day and negative potential the next day to counteract damage by electrolysis.

Exception No. 1

A system that supplies limited areas of industrial equipment.

Exception No. 2

DC systems operating at 50 volts or less.

Exception No. 3
DC systems operating at 300 volts or more.

Exception No. 4
Section 250.5 covers DC systems obtained from a rectifier supply by AC current where superfluous.

Exception No. 5
Part III of Article 700 exempts signaling devices that have a current not over 0.030 amperes.

(B) Three-Wire, Direct-Current Systems. The neutral conductor supplying premises with three-wire DC supply must be grounded.

250.164: Point of Connection for Direct-Current Systems
On direct-current systems that are to be grounded, the grounding shall be done at one or more supply stations, but not at the individual services. This refers to the grounding of the supply conductors and not to the service equipment itself. The equipment should be grounded at the service, but the neutral should be isolated from the equipment at the service-entrance equipment. These requirements are greatly different from those for alternating-current systems because any passage of current between the ground at the supply station and at the service-entrance equipment may cause objectionable electrolysis, which must be avoided.

250.166: Size of Direct-Current System Grounding Conductor
The size of a grounding conductor for direct current systems must conform to the following:

(1) It can't be smaller than the neutral (in three-wire, DC systems).

(2) For systems other than three-wire systems, the grounding conductor can't be smaller than the system's largest conductor.

(3) It can't be smaller than No. 8 copper or No. 6 aluminum. See the *NEC* for exceptions.

X. Instruments, Meters, and Relays

250.170: Instrument Transformer Circuits
Current transformers (CT) and potential transformers (PT) shall have the secondary windings grounded when the primary windings are connected to 300 volts or more to ground, regardless of the

primary voltage. The exception to this is when the primary windings are connected to less than 1000 volts with no live parts or wiring exposed or accessible to other than qualified persons.

250.172: Grounding

Where accessible only to qualified persons, grounding is required for instrument transformers, cases, and frames.

Exception

On current transformers where the primary is not over 150 volts to ground and they are used exclusively for current meters.

250.174: Cases of Instruments, Meters, and Relays—Operating at Less Than 1000 Volts

Any instruments, relays, or meters operating with winding or working parts at less than 1000 volts shall be grounded as in (A), (B), or (C) following:

- **(A) Not on Switchboards.** All cases and other exposed metal parts of instruments, meters, and relays that are not on switchboards shall be grounded when the windings or working parts operate at 300 volts or more to ground and when they are accessible to other than qualified persons.
- **(B) On Dead-Front Switchboards.** All instruments, meters, and relays that are noted on a switchboard that has no live parts on the front of the panel shall be grounded, whether direct-connected or supplied by current and potential transformers.
- **(C) On Live-Front Switchboards.** All instruments, meters, and relays that are mounted on a switchboard that has live parts on the front of panels shouldn't have their cases grounded, but there shall be mats of insulating rubber or other suitable floor insulation wherever the voltage to ground exceeds 150 volts.

250.176: Cases of Instruments, Meters, and Relays—Operating Voltage 1 kV and Over

Any of these listed in the title that are 1 kV or over to ground have to be isolated by elevation or copper guarding, and the cases are to be left ungrounded.

Exception

Where electrostatic ground detectors are used, the internal ground segments of the instrument are connected to the instrument case and grounded. Here, the ground detector shall be isolated by elevation.

250.178: Instrument-Grounding Conductor

The minimum wire size for secondary instrument transformer-grounding conductors or instrument case-grounding conductors is No. 12 copper or No. 10 aluminum. The cases of meters and other control instruments are considered grounded when they are mounted directly on a grounded enclosure; no separate grounding conductor is required.

XI. Grounding of Systems and Circuits of 1 kV and Over (High Voltage)

250.180: General

High-voltage systems that are grounded shall comply with preceding sections of this article, as modified and supplemented by following articles or sections.

250.182: Derived Neutral Systems

Where a neutral is derived from a grounding transformer, permission is granted to use it when grounding high-voltage systems.

250.184: Solidly Grounded Neutral Systems

- **(A)** Neutral Conductor. The neutral must be insulated for at least 600 volts.

Exception No. 1

Bare copper wires may be used for the neutral of services and directly buried feeders.

Exception No. 2

Bare conductors can be used for the neutral of overhead circuits.

- **(B)** Multiple Grounding. A solidly grounded neutral can be grounded at more than one place for services, the parts of directly buried feeders having a bare neutral, or overhead portions (outdoors only).
- **(C)** Neutral Grounding Conductor. When protected from damage and insulated from phase wires, a neutral grounding conductor may be bare.

250.186: Impedance Grounded Neutral System

All systems using an impedance ground must comply with the following:

(A) Location. The impedance must be inserted into the grounding circuit between the grounding electrode and the neutral point of the transformer or generator.

(B) Identified and Insulated. The neutral of this type of system must be marked and insulated with the same quality of insulation as the phase conductors.

(C) System Neutral Connection. The system can only be connected to neutral through the impedance.

(D) Equipment-Grounding Conductors. In this kind of system, equipment-grounding conductors can be connected to the ground bus and grounding electrode conductor, and brought to the system ground. It may be bare.

250.188: Grounding of Systems Supplying Portable or Mobile Equipment

Any systems that supply movable high-voltage equipment (temporary substations are excepted) must comply with the following:

(A) Portable or Mobile Equipment. Movable high-voltage equipment must be supplied by a system that has an impedance-grounded neutral. Delta-connected systems must have a derived neutral.

(B) Exposed Noncurrent-Carrying Metal Parts. Such parts must be connected to the neutral impedance grounding point by an equipment-grounding conductor.

(C) Ground-Fault Current. The maximum voltage that can be found between the equipment frame and ground, due to the flow of fault current, is 100 volts.

(D) Ground-Fault Detection and Relaying. Ground-fault protection and relaying must be used to shut down automatically any component of the system that develops a ground fault. The continuity of the grounding conductors must also be monitored, and the current shut off if the continuity fails at any time.

(E) Isolation. The grounding electrode to which this system is connected may not be within 20 feet (6.1 m) of any other grounding electrode of any other system, nor can there be any type of connection to any other system's grounding electrode.

(F) Trailing Cable and Couplers. Any high-voltage trailing cables or couplers must meet the requirements of Part III of Article 400 and Section 710.45.

250.190: Grounding of Equipment

All metal parts of movable equipment (including fences, housings, supports, etc.) must be grounded.

Exception No. 1

When the equipment is isolated from ground such that no person can contact grounded metal parts when the equipment is energized.

Exception No. 2

Distribution equipment that is mounted on poles (See Section 250.110).

Grounding conductors that are not part of an assembly of cables can't be smaller than No. 6 copper or No. 4 aluminum.

Article 280—Surge Arresters

I. General

280.1: Scope

The general requirements are incorporated in this article governing the installation and connection requirements of surge arresters (also called *lightning arresters*). Don't confuse them with *surge capacitors*, which are covered in other places in the *Code*, especially in Section 502.3.

280.2: Definition

Surge arresters are devices to protect equipment from surge line voltages. They absorb some of the surges, but are also capable of stopping the flow of the surge current by absorbing it, and they maintain their capability of repeating such functions.

These surge arresters break down at voltages higher than the supply voltage, allowing the higher voltage and accompanying currents to flow to ground, thus protecting the equipment on the system. After the surge passes to ground, the arrester heals itself, shutting off flow current from the supply system.

A case in point: There was a 7½-horsepower irrigation-pump motor out in a field under a large tree. Because of lightning, motor rewinds were required on an average of once every two years. Special permission was granted to install surge arresters and surge capacitors to the service. Approximately 30 years later, there has not been a motor burnout due to lightning.

280.3: Number Required

When using surge arresters, they shall be connected to each ungrounded circuit conductor.

Surge arresters are available in single units to connect to only one ungrounded circuit conductor; here one would be required for each ungrounded conductor. They also are available with three units in one enclosure, which would thus take care of a three-phase supply. If there are other supply conductors, such as supplied from a farm service pole, as illustrated in Figure 225-1, one set of surge arresters would be sufficient. But in Figure 225-1 there are a number of service drops, and so surge arresters should be installed at the load-end of each service drop.

280.4: Surge Arrester Selection

(A) **On Circuits of Less Than 1000 Volts.** On circuits of less than 1000 volts, it is required that the voltage rating of the surge arresters be equal to or greater than the maximum voltage of the phase-to-ground voltage available. Thus a service of 480 volts to ground (RMS) would be 0.707 of the maximum voltage to ground; so maximum voltage would be 480 divided by 0.707, or 679 volts. By the same token, a service of 20 volts to ground (RMS) would be 170 volts maximum.

(B) **On Circuits of 1 kV and Over.** This requires that the surge arrester have a rating of not less than 125 percent of the maximum phase-to-ground voltage. Again, don't confuse RMS voltage and maximum voltage. Use the examples in (A) above.

Note

See ANSI *Standard C62.2*for further information. Arresters are usually metal oxide. See the *NEC* for this fine-print note.

II. Installation

280.11: Location

Surge protection of surge arresters may be located in the system either indoors or outdoors, but shouldn't be accessible to unqualified persons.

In deciding whether to place surge arresters indoors or outdoors, take into consideration that some have exploded. Although this is not a *Code* requirement, I would suggest that if they are placed indoors, they should be kept away from combustible items.

Exception

There are some surge arresters that are listed to be located in accessible places.

280.12: Routing of Surge Arrester Connections

The connections from the supply system to the surge arresters should be as short as possible. Also, there should be as few bends in the leads and grounding as possible. Lightning takes a direct path to the ground.

III. Connecting Surge Arresters

280.21: Installed at Services of Less Than 1000 Volts

Conductors used to connect surge arrestors can't be smaller than No. 14 copper or No. 12 aluminum. The conductor that is used to ground the arrestor must be connected to the grounded service conductor, the grounding electrode conductor, the grounding electrode, or the equipment-grounding terminal of the service panel.

280.22: Installation on the Load Side of Services of Less Than 1000 Volts

This section tells us that line and grounding conductors shouldn't be smaller than No. 14 copper or No. 12 aluminum. Some judgment must be used here, and, if possible, check with the authority that has jurisdiction as to the size of conductors to use.

Surge arresters may be connected to any two ungrounded conductors, grounded conductors, or grounding conductors. The grounded conductors and the grounding conductors should be interconnected only during normal operation of the surge arrester when a surge occurs.

The shortest method of getting the surge to ground is always the best method. Avoid bends as much as possible.

280.23: Circuits of 1 kV and Over—Surge Arrester Conductors

Here we are limited to No. 6 copper or aluminum for conductors connecting the surge arrester to both the ungrounded conductors and the ground.

280.24: Circuits of 1 kV and Over—Interconnection

Where circuits are supplied by 1 kV and over, the grounding conductor from surge arresters that protect a transformer supplying a secondary distribution system must be interconnected as follows.

(A) **Metallic Interconnection.** An interconnection to the secondary neutral may be made if the direct grounding is made to the surge arrester, provided that both of the following conditions are met:

(1) If the secondary has the grounded conductor connected elsewhere to a continuous metal underground water-piping system. If in urban areas there is a minimum of four water-pipe grounding connections in a distance of one mile, the direct ground from the surge arrester may be eliminated and the secondary neutral used as the grounding for the surge arrester.

(2) In many instances the primary is four-wire wye, with the neutral grounded periodically. In these cases the secondary neutral is usually interconnected with the primary neutral. If the primary neutral is grounded in a minimum of four places in each mile, plus the secondary service ground, the surge arrester ground may be interconnected with the primary and secondary grounds in addition to the surge arrester grounding electrode.

(B) Through Spark Gap. If the surge arrestor is not connected in accordance with part (A) of this section, or if it is not grounded as in part (A), but is grounded according to Section 250.83, a connection may be made through a spark-gap device, but only according to the following rules:

(1) In ungrounded or ungrounded systems, the spark-gap arrestor must have a breakdown voltage of more than two times the circuit's normal operating voltage, but not necessarily more than 10,000 volts. There must also be at least two grounding electrodes, placed no more than 20 feet (6.1 m) apart.

(2) In multi-grounded neutral primary systems, the spark-gap arrestor must have a breakdown voltage of no more than 3000 volts, and there must be at least two grounding electrodes, placed no more than 20 feet (6.01 m) apart.

(C) **By Special Permission.** The authority that has jurisdiction may grant special permission for an interconnection of the surge arrester ground and the secondary neutral other than permitted in (A) and (B) of this section.

280.25: Grounding

As is usually the case in grounding, the grounding of surge arresters shall comply with the requirements in Article 250, except as otherwise indicated in this article.

Chapter 3

Wiring Methods and Materials

Article 300—Wiring Methods

I. General Requirements

300.1: Scope

(A) **All Wiring Installations.** This article is very broad in its coverage; it discusses the fundamentals that apply to all wiring installations.

Exception No. 1
Class 1, Class 2, and Class 3 circuits, as provided for in Article 725.

Exception No. 2
Fire protective signaling circuits, as provided for in Article 760.

Exception No. 3
Article 770, covering optical fiber cables.

Exception No. 4
Communication systems, as provided for in Article 800.

Exception No. 5
Community antenna television and radio distribution systems are referenced in Article 8.

(B) **Integral Parts of Equipment.** Wiring that is an integral part of motor controllers, motors, motor control centers, or other factory-assembled control equipment is not covered by this article.

300.2: Limitations

(A) **Voltage.** Unless there are specific limitations in a section of Chapter 3, the wiring methods in Chapter 3 shall apply to voltage of 600 volts, nominal, or less. Elsewhere in the *NEC* you will find that some voltages over 600 volts, nominal, are specifically permitted under Chapter 3.

(B) **Temperature.** Section 310.10 covers limitations as far as temperature is concerned for conductors.

300.3: Conductors

(A) Single Conductors. The single conductors that are specified in Table 310.13 are only part of a recognized wiring method if permitted in Chapter 3.

(B) Conductors of the Same Circuit. All the conductors of a circuit (including any grounded and grounding conductors) must be run together in any raceway, gutter, tray trench, cable, or cord. Following are the four exceptions to this rule:

(1) Paralleled Installations. Conductors may be run in parallel under the provisions of Section 310.4. Under these provisions, each parallel group of conductors must be run in the same raceway, gutter, tray trench, cable, or cord. Grounding conductors must be sized according to Section 250.122, and runs in tray must comply with Section 392.8(D). One exception is made: Conductors run underground in nonmetallic raceways may be arranged in isolated phase arrangements. In such cases, the raceways must be near each other, and the conductors must comply with Section 300.20(B).

(2) Grounding Conductors. Equipment grounding conductors may be run outside a raceway or cable when the installation conforms to the requirements of Section 250.130(C) (for existing installations) or Section 250.134(B), Exception 2 (for DC circuits). Bonding conductors may be installed outside raceways per Section 250.102(E).

(3) Nonferrous Wiring Methods. The provisions of Section 300.20(B) may be followed where nonmetallic or nonmagnetic raceways, gutters, trays, cables, or cords are used. Single-conductor Type MI cables with nonmetallic sheaths must comply with Section 330.16.

(4) Enclosures. In the case where an auxiliary gutter is run between a pull box and a column-width panelboard and there are neutral terminations in the pull box, the neutral conductors supplied from the panelboard may originate in the pull box.

(C) Conductors of Different Systems.

(1) 600 Volts, Nominal, or Less. If more than one circuit is involved and each of these circuits is 600 volts, nominal, or less, regardless of whether AC or DC, they may be enclosed together in the same enclosure, whether a cable or raceway. By necessity the voltage rating of the insulation of the conductors must be equal to the maximum voltage involved, or higher. The equipment-grounding conductor may, of course, be bare or insulated, and identified. When circuits of different voltages occupy the same raceway or enclosure, some identification by color of insulation or other appropriate means should be used to identify the difference of voltage of the circuits involved.

Exception

In the case of solar photovoltaic systems, refer to Section 690.4(B) of the *NEC*.

Note

See Section 725.38(A)(2) for Class 2 and Class 3 circuit conductors in the *NEC*.

(2) Not Over 600 Volts, Nominal. Conductors over 600 volts may not occupy the same raceway, cable, or enclosure as conductors 600 volts or less, except as allowed under (a) through (e) below.

(a) Electric-discharge lamps with a secondary voltage of 1000 volts or less may have their conductors in the same fixture enclosure as the branch-circuit conductors, providing that the secondary conductors are insulated for the voltage involved. This takes care of the higher voltages that are encountered in discharge types of lighting fixtures.

(b) Leads of electric discharge lamp ballasts are permitted if the primary leads of the ballast are insulated for the primary voltage to the ballast, and may occupy the same enclosure of the fixture of the branch circuit conductors. An example is a 227-volt ballast that might have primary leads rated at 300 volts. These primary leads may be in the same enclosure as the branch-circuit conductors having 600-volt insulation.

(c) Conductors such as excitation, control, relay, and ammeter conductors used in connection with any individual motor or starter may occupy the same enclosure as the motor circuit conductors.

(d) Conductors of different voltages may be used in motors, switchboards, and similar equipment.

(e) Conductors of different voltages may be used in manholes, provided the conductors of different systems are separated and securely and appropriately fastened.

Conductors with nonshielded insulation and operating at different voltages may not occupy the same cable, enclosure, or raceway.

300.4: Protection Against Physical Damage
This title is self-explanatory. Additional and corrective action may be required.

(A) Cables and Raceways Through Wood Members.

(1) Bored Holes. When installing exposed or concealed conductors in insulating tubes, or installing a cable or raceway-type wiring method in holes bored through studs, joists, rafters, or wood members, the holes shall be drilled as near to the center of the wood member as possible but not less than 1¼ inches (31.8 mm) from the nearest edge. In drilling, notching, and so on, the structure shouldn't be weakened. It is not within the scope of the *Code* to cover this fully, but all building codes should be adhered to.

If the holes can't be bored more than 1 inch from the nearest edge, the cable or raceway shall be protected by a plate or bushing of at least 1⁄16 inch (1.59 mm) in thickness, of sufficient size to protect from penetration of nails and the like, and of appropriate length and width to cover the area of wiring.

(2) Notches in Wood. If it won't weaken the wood, notching for cables or raceways of all types is allowed by the *Code*, provided that the notch is covered by a steel plate at least 1⁄16 inch (1.59 mm) thick to protect the cable from nails. [For the intent of (A) and (B), see Figure 300-1.]

See the *NEC* for raceways in Articles 342, 344, 352, and 358, which are exceptions.

(B) Cable and Electrical Nonmetallic Tubing Through Metal Framing Members.

(1) This part can be extremely important, as metal framing members can cut into nonmetallic-sheathed cables where they pass through any hole in metal framing members, whether factory or field punched or made by any other means. Great care shall be taken so that the cable is thoroughly protected by proper bushings or grommets that won't readily deteriorate and are fastened securely in the metal opening before passing any cable through them.

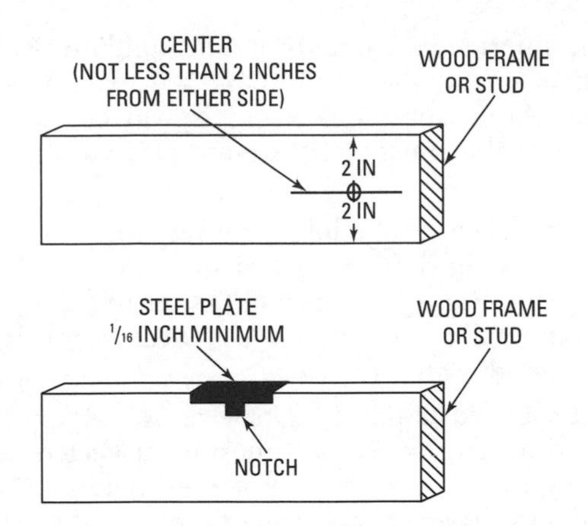

Figure 300-1 Proper drilling and notching procedures.

(2) If there is any chance of the building contractor penetrating nonmetallic-sheathed cable by nails or screws, you must install a steel sleeve for the cable to pass through or cover the building studs, and so on, by means of a steel plate or steel clip. This also applies to electrical nonmetallic tubing. This steel shall be at least 1⁄16 inch in thickness to keep the nails or screw from penetrating the cable or tubing.

(C) Cables Through Spaces Behind Panels Designed to Allow Access. The cables or raceways must be supported according to their applicable articles.

Exception No. 1
IMC, EMT, rigid metal conduit, and rigid nonmetallic conduit are exempt from these requirements.

Exception No. 2
Mobile homes and recreational vehicles are also exempt from these requirements.

(D) Cables and Raceways Parallel to Framing Members. When run parallel to framing members, cables and raceways are to be set back 1¼ inches from the surface of the framing member. (The surface is the part that is susceptible to penetration by a nail or screw.) If this distance can't be obtained due to the materials used, a 1⁄16-inch steel plate or equal is necessary for protection.

Rigid metal conduit, intermediate metal conduit, rigid nonmetallic conduit, and electrical metallic tubing are exempt from this requirement. Also, this requirement is waived in cases where a raceway or cable must be fished into place or in mobile homes or RVs.

(E) **Cables and Raceways Installed in Shallow Grooves.** In these cases, the cable or raceway must be protected by a 1⁄16-inch steel plate or the equivalent. Rigid metal conduit, intermediate metal conduit, rigid nonmetallic conduit, and electrical metallic tubing are exempt from this requirement.

(F) **Insulated Fittings.** Ungrounded conductors of size No. 4 or larger that enter an enclosure from a raceway must be protected by an insulating bushing or an equivalent insulating material. The insulating material must have a temperature rating equal to or greater than that of the conductors it protects. Integral threaded hubs are considered acceptable.

300.5: Underground Installations

(A) **Minimum Cover Requirements.** Table 300.5 in the *NEC* gives the depths at which buried cables or conduits or other raceways are installed. These are minimum requirements.

See the *NEC* at this point for the definition of "cover."

On buried cables and raceways, it is the author's contention that added protection should be required. For instance, a buried cable shouldn't be buried directly against the rocky bottom of a trench. There should be a layer of fine sand laid on the rocks, and after the cable is laid another layer of fine sand should be added so that the rocks in the backfill don't cut. Personal experience has shown that direct-buried raceways, unless covered by a manufacturer-produced coating of plastic, are very vulnerable to electrolysis. Referring back to the table in this book covering soil resistance, you can see that different soils have different electrolysis characteristics. In soil in the author's area, a tar coating over the raceway does not stand up long. In some cases, therefore, it might be a good idea to consider encasing the raceway in concrete. I suggest 2000-pound concrete with 3⁄8-aggregate maximum and a 7-inch slump, which allows it to form readily around the raceway.

See the *NEC* for Table 300.5, "Minimum Cover Requirements, 0 to 600 Volts, Nominal."

Exception No. 1

If a 2-inch (50.8 mm) thick pad or equivalent in physical protection is placed in the trench over the underground installation, the minimum cover requirements may be reduced by 6 inches (152 mm). This does not refer to rigid metal conduit or intermediate metal conduit.

(B) Grounding. Grounding and bonding of underground installations shall be by methods that are described in Article 250 on grounding.

(C) Underground Cables Under Buildings. Section 230.49 explains the protection of service conductors run underground. Some inspectors will require a sand bed and a sand covering for protecting underground conductors, especially in rocky areas and where freezing temperatures are encountered. They may also require a mechanical protection over the conductors. The mechanical protection required is the inspector's prerogative. Some may require redwood or treated boards; others may require a plastic strip that has the word caution printed along the entire length.

UF cable under concrete floors, patios, and so on shall be installed in raceway that extends to points away from or up through the concrete (check for transition and continuity).

A short time ago, UF cable could be buried below a cement floor of a building, and conduit used to sleeve it run through the concrete. The new *Code* ruling is a very welcome change; most inspectors never considered it good workmanship to pour a floor over a cable because it was not readily replaceable.

(D) Protection from Damage. Direct-buried cables and conductors where they emerge from the ground shall be protected by proper raceways, as covered in Section 300.5(A) and Table 300.5, from a point below grade to at least 8 feet above grade. In no case shall this requirement of encasing exceed 18 inches.

When conductors are brought from underground to above ground, several factors definitely should be taken into consideration. When direct burial conductors are in a trench, they tend to settle, especially if the trench is not properly compacted. Thus, should these be service lateral conductors entering a conduit up to a meter enclosure and should any settling result from the trench, the conductors could be pulled out of the meter enclosure and possibly result in a fire.

To avoid this, a loop should be formed at the bottom of the trench to provide slack.

Also, in areas where hard freezing occurs, the conductor insulation is often cut where the conductors enter the conduit underground. This may be prevented by using an insulated bushing on the conduit underground and by not making a 90-degree bend where the conductors enter the conduit underground. Make a bend of more than 90 degrees so that any water that accumulates will flow down into the ground, thus aiding in eliminating damage to the insulation of the buried conductors where they enter the conduit.

You must protect conductors that enter a building at the point of entrance.

If the raceway entering a building is subject to physical damage, rigid metal conduit or intermediate metal conduit or Schedule 80 rigid nonmetallic conduit or the equivalent is required.

When using rigid metal conduit, consider that it must be protected from corrosion in the earth, as specified in Section 344.1.

When service laterals are not encased in concrete and are buried 18 inches or more below grade, a warning ribbon must be placed at least 12 inches above them in the trench.

(E) Splices and Taps. Boxes are not required to make splices or taps in underground cables, but the splices or taps have to be made by methods and with identified material that have been approved by the electrical inspector.

(F) Backfill. Materials removed from the trench excavation may often contain large rocks, paving material, cinders, large or sharply angular substances, or corrosive materials. This shouldn't be used for backfill, as it may cause damage to cable, ducts, or other raceways. As stated earlier, it is often a good idea to use sand bedding and a layer of sand-fill to prevent damage from frost heave.

If necessary to prevent physical damage to the raceway or cable, granular selected material may be used for covering, or suitable running boards such as redwood or any other suitable means to protect raceway or cable from damage may be used.

(G) Raceway Seals. Sealing will be required at one or both ends of conduits or raceways where moisture may contact energized parts. Earlier in the book a case was cited in which moisture caused a short out of the service-entrance equipment.

Note

Where a raceway enters a building underground, there is always a chance of hazardous gases or other vapors entering into the building through the conduit. Therefore, you must seal raceways entering buildings.

(H) Bushing. Conduits terminating underground for direct-burial wiring methods shall have bushings on the ends. The author would recommend the insulating type of bushings. Also, the end of the conduit in the earth should be sloped down to drain, as freezing has often caused faults in the conductors. A seal that provides physical protection from damage may be used instead of a bushing.

(I) Single Conductors. All conductors of a circuit shall be installed in the same raceway, including the neutral and equipment-grounding conductor. Single conductors in a trench shall be installed in close proximity. The purpose of this is to keep the impedance of the circuit as low as possible.

Exception No. 1

Conductors in parallel may be installed in the same raceway if they contain all the conductors of the circuit, including the grounding conductors.

Exception No. 2

As permitted in Section 310.4, and under the conditions of Section 300.20 if met, isolated phase installations are allowed in nonmetallic raceways enclosed to the point where the conductors are paralleled, as described in the two aforementioned sections.

(J) Ground Movement. Over time, the backfill in trenches or in other places may tend to settle, thereby putting substantial stress on conductors. In addition, frost can also be a hazard. The installers of cables in below-ground locations must take this into account and must avoid such problems whenever possible. Leaving loops of extra cable in the trench generally works well to avoid this situation and is recognized as acceptable by a fine-print note to this section.

300.6: Protection Against Corrosion

Metal raceways, armor cable, cable sheathing, cabinets, elbows, couplings, fittings, boxes, and all other metal hardware necessary to the installation must be made of material that will be suitable for the environment encountered.

(A) General. Ferrous raceways, cable armor, boxes, cable sheathing, cabinets, and all other fittings of metal—including supports but not including the threads at joints—shall be protected by a coating of material that will resist corrosion. Use zinc, cadmium, enamel, or some other (corrosion-resistant) material for coating. If the protective coating is enamel only, installation shouldn't be out of doors or in wet locations. This will be further discussed in item (C) of this list. If the boxes or cabinets have approved coatings and are marked or listed for outdoor use, they may be used outside.

Exception

Electrically conducting material may be used on threads at joints.

(B) **In Concrete or in Direct Contact with the Earth.** Ferrous and nonferrous metal raceways, cable armor, boxes, and all other fittings involved in the installation may be installed in concrete. They may be in direct contact with the earth. If the surrounding has severe corrosive influences, then they may be in direct contact with the earth if they are judged to be suitably provided with corrosion-resistant material approved for the purpose. Refer to Table 250-1 of this book, "Resistivities of Different Soils."

The above is rather broad in scope. In one area, galvanized rigid conduit may last only 1 year. Asphalt painting may not be adequate in another area. So the corrosion-resistant material must be approved by the inspection authority, who is familiar with the earth's ohm/centimeter resistivity in the area.

(C) **Indoor Wet Locations.** Such locations might include dairies, laundries, canneries, and the like. These may also include places where walls and floors have to be frequently washed down, because that moisture may be absorbed. Therefore the entire wiring system, including boxes, fittings, and so forth, must be mounted so that there is at least a ¼-inch air space between them and the wall or other supporting surface.

Meat-packing plants, tanneries, hide cellars, casing rooms, glue houses, fertilizer rooms, salt storage facilities, some chemical works, metal refineries, pulp mills, sugar mills, round houses, some stables, and similar locations are judged to be occupancies where severe corrosive conditions are likely to be present.

It can be seen from the preceding discussion that precautions have been taken for protection from corrosive elements, including excess moisture. There may be conditions in which the best raceway might be the nonmetallic type. If this type is used, the article covering the installation of it must be adhered to. Plastic-coated metal raceway conduits are also available.

Note

Acids and alkali chemicals that may cause severe corrosion may be stored or handled in some of these indoor wet locations, such as meat-packing plants, tanneries, glue factories, certain stables, areas adjacent to seashores, swimming pools, places where chemical deicers are used, fertilizer storage places, damp cellars, salt or chemical storage places, and so on.

300.7: Raceways Exposed to Different Temperatures

(A) **Sealing.** Where portions of interior raceways may be subject to widely different temperatures, such as conduit going from a normal room to a freezer storage room, circulation of the air can cause the warmer air to rise and flow into the cooler section. At such places the raceway shall be sealed to prevent the flow of air from one temperature to the other, thereby stopping condensation.

While not specifically covered, conditions will be found in which the wirer runs a service mast from the outside of the house down an inside wall of a house to the service-entrance equipment. The temperature in the house may vary from the outside temperature by as much as 100 degrees. This will cause a chimney effect. The warm air will rise and be cooled, thus causing moisture to run down inside the conduit into the panel.

(B) **Expansion Joints.** These were briefly covered earlier in this book. In many places in large buildings, expansion joints in raceways must compensate for the expansion or contraction due to thermal conditions.

Deciding where these joints are to be used requires common sense. When using expansion joints it is necessary to provide bonding jumpers, as required under Section 250.77, so as to make good electrical contact around the expansion joints.

See *NEC* Index for *Code* sections that cover expansion joints.

300.8: Installation of Conductors with Other Systems

You may not enclose electrical conductors in raceways, cable trays, and so forth, if they contain pipes or tubes or the equivalent to be used with steam, water, gas, or any service other than the electrical service.

300.10: Electrical Continuity of Metal Raceways and Enclosures

All raceways, armored cable, and other metal enclosures shall be connected to any boxes or other fittings and cabinets so that effective electrical continuity of grounding is acquired.

Locknuts, bushings, and connectors supply this continuity, except where other measures must be taken, such as bonding at services, expansion joints, eccentric and concentric knockouts, and in hazardous locations. Where nonmetallic boxes are used, as permitted in Section 370.3, they don't have to make a connection with the metal raceway, but the box must include an integral bonding means. Short sections of metal enclosure that are used only to provide protection or support don't have to be grounded.

300.11: Securing and Supporting

(A) **Secured in Place.** All raceways, boxes, cabinets, and fittings must be securely fastened in place. Suspended ceiling support wires can be used only to support equipment if they are not part of a fire-rated floor or roof assembly. Furthermore, they can be used only to support branch-circuit wiring associated with equipment that is somehow attached to the suspended ceiling.

Above a fire-rated floor or ceiling assembly, the required independent support wires must be identified by color, tagging, or other suitable means. The support wires of appropriately tested support systems need not be identified.

(B) **Raceways Used As Means of Support.** Raceways may not be used to support other raceways, cables, or equipment, except as follows:

(1) If the raceway or support is identified for such use.

(2) Raceways containing Class 2 circuit conductors or cables used strictly for the purpose of connection to the equipment control circuits may be supported from the general circuit raceways.

(3) If the raceway is used to support boxes or conduit bodies per Section 370.23 or fixtures per Section 410.16(F).

300.12: Mechanical Continuity—Raceways and Cables

There shall be mechanical continuity between boxes, cabinets, or other fittings and enclosures, outlets or raceways, and cables, metal or nonmetallic raceways, armored cable, or cable sheaths.

Exception

Short pieces of raceway used only for support or protection (see Section 250.33).

300.13: Mechanical and Electrical Continuity—Conductors

(A) General. Conductors are to be continuous between outlet devices and shall never be spliced in a raceway, except as permitted for auxiliary gutters in Section 366.8, for wire-ways in Section 376.6, for boxes and fittings in Section 300.15(A), and for surface metal raceways in Section 386.7. In other words, conductors in raceways shall be in one piece and without splice except as used above. (See Figure 300-2.)

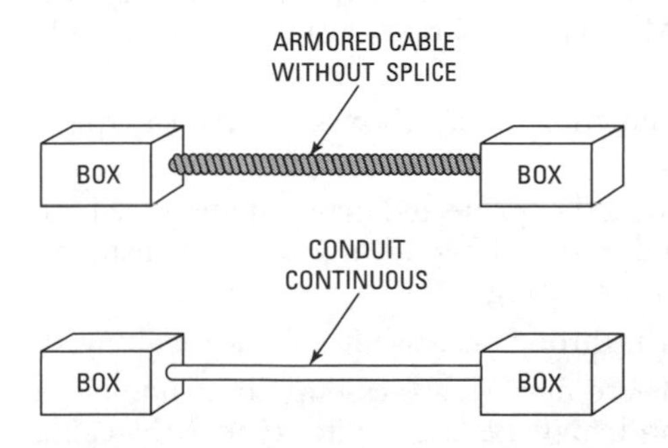

Figure 300-2 Approved method of connecting conduit and armored cable between boxes.

(B) Device Removal. On multiwire circuits, you may not remove a device such as a lampholder or receptacle and leave the grounded conductor open. On multiwire circuits, the grounded connector or neutral shall always be installed so that if a device is removed, the grounded conductor will remain continuous throughout the entire multiwire circuit.

300.14: Length of Free Conductors at Outlets and Switch Points

A minimum of 6 inches (152 mm) of free conductor length shall be left at each outlet and switch point, except if a conductor loops through without connection to any device and is without a splice. This purpose of this ruling is to ensure that ample conductor length for making connections to the outlet or switchis provided, and that some spare length is also provided in the event that a piece should break off in the makeup. Where the box is less than 8 inches in any dimension, every conductor must be long enough to extend at least three inches past the opening.

Exception

If the conductors are not terminated or spliced at a box or junction point, the above paragraph won't apply.

300.15: Boxes, Conduit Bodies, or Fittings—Where Required

- **(A) Wiring Methods With Interior Access.** Except as allowed by the following list (B) through (M), a box or conduit body must be installed at every splice point, outlet, switch point, junction, or pulling point where the wiring method is conduit, EMT, AC cable, MC cable, MI cable, MN cable, or other cables.

  A box must be installed at every outlet and switch point of knob-and-tube wiring.

  In all cases, all connections must be made with fittings and connectors that are listed for such use. In other words, you may not connect EMT with a fitting designed for flexible metal conduit.
- **(B) Equipment.** Integral junction boxes (boxes built into equipment) are acceptable.
- **(C) Protection.** Cables must be protected from damage where they enter or exit conduit or tubing. Fitting must be installed on the ends of conduit or tubing.
- **(D) Type MI Cable.** Straight-through accessible splices are allowed.
- **(E) Integral Enclosure.** These devices are essentially a one-piece switch/box combo and may be used with type MN cable instead of a box or conduit body.
- **(F) Fitting.** Fittings that are identified for the use and don't contain splices or terminations may be used instead of a box or conduit body.
- **(G) Direct-Buried Conductors.** As allowed by Section 300.5(E).
- **(H) Insulated Devices.** As allowed by Section 334.21.
- **(I) Enclosures.** As allowed by Section 314.8 for switches and Section 430.10(A) for motor controllers.
- **(J) Luminaires (Fixtures).** As per Section 410.31.
- **(K) Embedded.** As per Sections 424.40, 424.41(D), 426.22(B), 426.24(A), and 427.19(A).

2005

(L) Manholes and Handhole Enclosures. In manholes and handhole enclosures accessible only to qualified personnel. See Section 314.30.

(M) Closed Loop. With devices listed for installation without a box on these systems.

300.16: Raceway or Cable to Open or Concealed Wiring

(A) Box or Fitting. When changing from conduit, electrical metallic tubing, nonmetallic-sheathed cable, Type AC cable, Type MC cable, or mineral-insulated, metal-sheathed cable and surface raceway wiring to open wiring or concealed knob-and-tube wiring, a box or fitting with bushed openings or holes for each conductor may be used. The box or fitting shall contain no splices, nor shall it be used at fixture outlets (Figure 300-3).

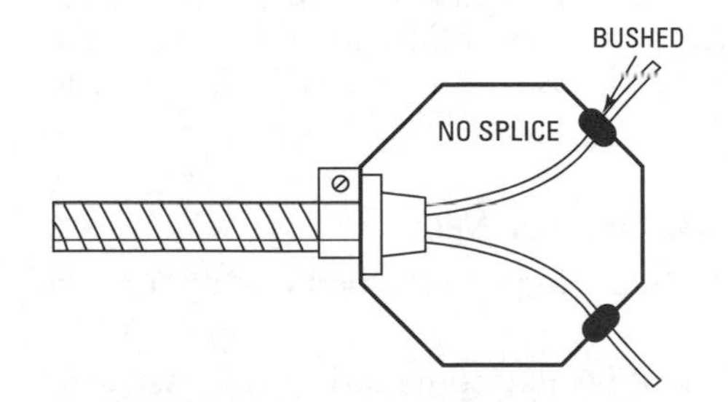

Figure 300-3 Method of transferring from cable to concealed knob-and-tube wiring.

(B) Bushing. At the end of a terminal fitting, conduit, or electrical metallic tubing where the raceway ends behind an open (unenclosed) switchboard or other uncontrolled or similar equipment, a bushing may be used in lieu of a box or terminal fitting at the end. Insulating bushings must be used, except for lead-covered conductors.

300.17: Number and Size of Conductor in Raceway

The most commonly used raceways are conduit and EMT. See *NEC* tables in Chapter 9. Table 1 gives us the percent of cross-section of conduit and tubing for conductors. Table 2 gives us fill for fixture wires. Tables 3A, 3B, and 3C apply to complete conduit or tubing systems and don't apply to sections of conduit used for physical protection only. Nipples not exceeding 24 inches may be filled to 60 percent.

Note

The fine-print note explains the NEC sections applicable, so refer to your NEC for this (FPN).

300.18: Raceway Installations

(A) **Complete Runs.** Aside from busways or exposed raceways with hinged covers, raceways must be complete between outlets or junctions prior to the installation of any conductors. Where necessary for the installation of equipment, raceways may be installed without the terminating connection and may be completed once the equipment is in place. Prewired raceway systems are permitted only where specifically permitted in the *Code.*

(B) **Welding.** Welding is not an acceptable method of supporting, connecting, or terminating metal raceways. Welding is allowed only for systems specifically designed for it or where specifically allowed by the *Code.*

300.19: Supporting Conductors in Vertical Raceways

(A) **Spacing Intervals—Maximum.** Conductors in vertical raceways shall be supported at the top of the run or as close to the top as practical. There shall also be supports as specified in Table 300.19 (A) of the *NEC.*

Exception No. 1

An exception to Table 300.19(A). See your *NEC.*

(B) **Support Methods.** The following covers methods of support used:

(1) On clamping devices of insulating material with insulated wedges inserted in the ends of the conduit, if the clamping does not adequately support the cable conductor, the wedges should be approved so as not to damage the insulation by wedging it to the conduit. Varnished cambric and thermoplastic insulation is inclined to slip on the conductor. Therefore, wedges might not hold this type of conductor and insulation effectively.

(2) Boxes may be inserted at the required intervals. These boxes are to have insulated supports installed and secured adequately to stand the weight of the conductors to which they are to be attached. The boxes shall be provided with covers.

(3) In addition to the boxes protecting the conductors at not less than 90 degrees, carrying them horizontally to a distance equal to not less than twice the conductor diameter, two or more supports shall be used for carrying these conductors or cables. The support intervals when this method is used shall be no greater than 20 percent of the distances mentioned in the preceding table. The necessity of supporting at not more than one-fifth of the distance in the table would seem to make this method less practical, because additional expense would be involved.

(4) Equally effective methods may be used, but approval should be secured from the authority that has jurisdiction.

300.20: Induced Currents in Metal Enclosures or Metal Raceways

To accomplish the minimizing of heating effects, the phase conductors, neutral, where used, and equipment-grounding conductor, where used, shall all be in the same enclosure. See the *NEC* for further clarification.

300.21: Spread of Fire or Products of Combustion

Although we don't often think of electrical installations being responsible for fire spread, it is a definite problem when wiring is installed in hollow spaces, vertical shafts, or ventilation of air-handling equipment ducts, or even up through a ceiling to another floor. All of these installations shall be made so as to greatly limit the spread of fire penetration through walls, partitions, floors, or ceilings that are fire rated for fire resistance as much as possible, and only approved methods for installing fire-stops at these points shall be used to maintain the fire rating of the building or structure.

Even slight openings will increase the risk of possible fire spread. This section is very important. Many new products have been approved to close these spaces.

300.22: Wiring in Ducts, Plenums, and Other Air-Handling Spaces

This section applies to air-handling spaces and applies to the use and installation of wiring and equipment in ducts, plenums, or any air-handling spaces.

Note

Article 424, Part VI covers electric duct heaters.

(A) **Ducts or Dust, Loose Stock, or Vapor Removal.** No wiring whatsoever shall be installed in ducts handling flammable vapors or loose stock. No wiring shall be installed in ducts or duct shafts containing only such ducts that are used for vapor removal or for ventilation of commercial-type cooking.

(B) **Ducts or Plenums Used for Environmental Air.** MI cable, intermediate metal conduit, EMT, rigid metal conduit, metal-sheathed cable, type MC cable using a smooth or corrugated impervious metal sheath without an overall nonmetallic covering, and flexible metal tubing are all used for environmental air. When equipment is permitted in these types of ducts or plenums, flexible metal conduit not to exceed 4 feet (1.22 m) in length may be used to connect physically adjustable equipment and devices.

When flexible metal conduit is used, the fittings shall effectively close any openings or connections only if necessary where there is direct action upon or sensing of the contained air, should any equipment or device be installed in such ducts or chambers. If illumination must facilitate repair and maintenance, only gasketed-type fixtures may be used.

Note

What has been discussed here is only applicable if ducts or plenums are used exclusively for the environmental air.

(C) Other Space Used for Environmental Air. Cables of the following types are permitted in space for environmental air: MI cable, metal-sheathed cable, type MC cable without an overall nonmetallic covering, type AC cable, and other factory-assembled multiconductor control or power cable where specifically listed for the use. Any of the other types of cables or conductors shall be installed in EMT, intermediate metal conduit, wireway with metal covers, metal solid-bottom cable tray with solid metal covers, flexible metal conduit, surface metal raceways, or flexible metallic tubing. Totally enclosed, nonventilated busways may be installed in such spaces. Plug-in busways may not.

Unless prohibited elsewhere in the *Code*, electrical equipment in either metal or nonmetal enclosures may be used, provided the nonmetal enclosure is listed for the use, if it is adequately fire-resistant, and produces very little smoke. Along with the wiring methods that are suitable for this installation, other electrical equipment may be permitted in this space when provided with a metal enclosure or with listed nonmetallic enclosures of a type that has adequate fire-resistant and low smoke-producing characteristics where all associated wiring materials are suitable for ambient temperatures that may be involved.

Note

Paragraph (C) includes other spaces, such as the space above hung ceilings, that are being used for environmental air-handling purposes.

Exception No. 1
If liquid-tight flexible metal conduit is not over 6 feet long, it is permitted in these spaces.

Exception No. 2
Fans that are specifically approved for such use are permitted.

Exception No. 3
Rooms that are habitable or areas of the building that are not used primarily for air-handling are not included in this section.

Exception No. 4
Cable assemblies of metallic manufactured wiring system, if they don't have nonmetallic sheath and are listed for this use, may be permitted.

Exception No. 5
Joist or stud spaces in dwelling units are not included in this section if wiring passes through such places in such a way that it is perpendicular to the long dimension of such spaces.

Exception No. 6
Underfloor areas are covered in Section 645.2(C)(3) and are not covered by this section.

(D) Data Processing Systems. Article 645 covers spaces under raised floors where there is electrical wiring for air-handling areas of data processing systems.

300.23: Panels Designed to Allow Access

When cables, raceways, or equipment are installed behind access panels (including suspended ceiling panels), they must be arranged so that the panels can be removed.

II. Requirements for Over 600 Volts, Nominal

300.31: Covers Required

Boxes, fittings, and similar enclosures must have covers suitable for preventing accidental contact with energized parts or physical damage to parts or insulation.

300.32: Conductors of Different Systems

High-voltage and low-voltage systems have conductors that may not occupy the same enclosure or pull box and junction box.

Exception No. 1
Both high- and low-voltage conductors may be installed in the same enclosure as motors, switchgears, or control assemblies.

Exception No. 2

When high-voltage and low-voltage conductors may be installed in a manhole if they are well separated.

300.34: Conductor Bending Radius

There is a difference in the bending radius of shielded and nonshielded conductors. The radius of a bend for nonshielded conductors shouldn't be less than eight times the diameter of the conductor. For shielded or lead-covered conductors, the radius of a bend shouldn't be less than 12 times the diameter of the cable during or after installation.

Care should be exercised when pulling conductors off a reel to ensure that the reels are placed so that the natural bends of the cable as it leaves the reels enter the raceways without reversing the natural bends. This is highly important, especially with shield cables, as damage may occur to the shielding if these precautions are not taken.

300.35: Protection Against Induction Heating

Care must be taken when arranging conductors in metal ducts to avoid heating by induction. In other words, all conductors of the same circuit shall be installed in the same raceway. Also, if nonmetallic ducts are used and placed in a bank and poured in concrete, you must secure them so that the spacing is maintained and flotation won't occur. To accomplish these things, it is often necessary to secure them by typing them together or down in the trench. No metallic wires or the like should encircle the duct bank. All ties shall be made to eliminate a closed loop around the duct bank.

300.36: Grounding

Article 250 on grounding applies to the wiring of equipment installations.

300.37: Aboveground Wiring Methods

Conductors above 600 volts run above ground must be in rigid metal conduit, IMC, EMT, rigid nonmetallic conduit, cable trays, cable bus, other identified raceways, or as open runs of metal-clad cable. Open runs of MV cable and bare conductors or bus bars are permitted where only qualified personnel will have access.

300.39: Braid-Covered Insulated Conductors—Open Installation

If the braid on these conductors is not flame-retardant, a flame-retardant saturant must be applied. See the *NEC* for more detail.

300.40: Insulation Shielding

Metal and semiconducting shielding must be stripped back and stress reduction provided. See the *NEC* for details.

300.42: Moisture or Mechanical Protection for Metal-Sheathed Cables

Where conductors emerge from a metal sheath, and where conditions warrant, moisture and/or damage protection shall be provided.

300.50: Underground Installations

(A) **General.** Underground conductors must be suitable for the location where they are installed and must meet the requirements of Section 310.7 and Table 300.50. Also see Section 250.2(D) for the grounding of type MC cables. Nonshielded cables must be installed in rigid or intermediate conduit or rigid nonmetallic conduit encased in 3 inches of concrete.

(B) **Protection from Damage.** Where conductors emerge from the ground, they must be enclosed in a listed raceway such as rigid metal conduit, IMC, or Schedule 80 PVC from the depth requirement of Table 300.50 to 8 feet above grade. Conductors entering buildings must be protected from the depth requirement to the point of entrance. All installations must be suitable to the ground conditions.

(C) **Splices.** Direct-burial cables must be spliced using suitable materials. They must also be watertight and be protected from damage. See the *NEC* for shielding issues.

(D) **Backfill.** Rocks, paving material, or the like shouldn't be used as backfill where it could damage the electrical installation. Protection must be provided where required.

(E) **Raceway Seal.** When a raceway enters a building from underground, a seal must be provided to keep gasses out of the building. The seal must be made with an identified compound.

Article 310—Conductors for General Wiring

310.1: Scope

The purpose of this article is to ensure that conductors for general wiring have mechanical strength, have adequate current-carrying capacity (ampacity), and are adequate for the particular usage to which they are to be put (such as wet or dry locations), for the temperature to which they will be subjected, and so on. A large part of this article is in the form of tables, which will be explained as they are covered.

Conductors that are an integral part of a motor, motor controller, or other equipment, or which are provided for elsewhere in the *Code*, are not covered in this article.

Note
Flexible cords and cables are covered in the NEC in Article 400, and fixture wire in Article 402.

310.2: Conductors

(A) **Insulated.** Conductors shall be insulated, except when covered or bare conductors are specifically permitted elsewhere in the *Code*.

Note
The NEC in Section 250.152 covers insulation for neutrals on high-voltage systems.

(B) **Conductor Material.** Conductors in this article may be copper, copper-clad aluminum, or aluminum, and unless the material is specified, they shall be copper.

310.3: Stranded Conductors

All conductors of size No. 8 or larger installed in raceways shall be stranded.

Exception
As permitted in other sections of the *NEC* [see Section 680.22(B)].

310.4: Conductors in Parallel

This section applies to aluminum, copper-clad aluminum, or copper conductors connected in parallel so as to in reality form a common conductor. This part is commonly abused. It is recommended that the following conditions for paralleling be followed closely.

When necessary to run equipment-grounding conductors with paralleled conductors, these equipment-grounding conductors are to be treated the same as the other conductors and sized according to Table 250.95. Size 1/0 or larger, comprising each phase or neutral, may be paralleled, that is, both ends of the paralleled conductors electrically connected together so as to form a single conductor.

Exception No. 1
You will find an exception in Section 620.12(A)(1).

Exception No. 2
Conductors smaller than 1/0 AWG may be run in parallel with supply control power to indicating instruments, contactors, relays, solenoids, and similar control devices if the following conditions are met: (a) They

are in the same raceway or cable; (b) the ampacity of each conductor is sufficient to carry the entire load shared by the parallel conductors; and (c) if one parallel conductor should happen to become disconnected, the ampacity of each conductor is not exceeded.

Exception No. 3

Conductor sizes smaller than 1/0 may be run in parallel at frequencies of 360 Hertz and higher—if all of the conditions in Exception No. 2(a), (b), and (c) are met.

The parallel phase conductors or neutral or parallel conductors shall conform to the following requirements:

(1) All parallel conductors shall be the same length.

(2) Parallel conductors shall be made of the same conductive material.

(3) Parallel conductors shall all be the same circular mil area.

(4) The insulation of all the conductors shall be of the same material.

Note

The question was asked in a question-and-answer session, why the insulation of all the parallel conductors has to be of the same type. If you study insulation-resistance properties and the effect of different temperatures on the insulating properties of different conductor insulation, the reason will become apparent. A case in point: On an industrial building, space called for RHH-USE conductors to be run in conduits above a building exposed to sunlight and temperature variations. The first test of insulation was run early one cool morning, some time before energizing, and the resistance of insulation met the specification requirements. When it came time to energize the insulation, resistance was again measured, and it dropped to a very low figure. Energizing was not permitted at that time, and the inspector began studying insulation resistance. He found that THW had been used instead of the specified-type insulation, and when the sunlight raised the ambient temperature, the insulation resistance dropped drastically. The point of this example is that if different insulation types are used, ambient temperature conditions may have a great deal to do with leakage through the insulation, thus upsetting the impedance of the conductors in parallel.

(5) Terminating all parallel conductors in the same manner is essential. If paralleled conductors are run in separate raceways, the conductors in each raceway must have the same physical characteristics, so that conductors in each paralleled raceway will pull equal amounts of current.

See (FPN) in the *NEC*.

Equipment-grounding conductors used in parallel circuits shall meet the requirement of this article, and also be sized as shown in the table in Section 250.95.

Consideration must be given to spacing in enclosures that have parallel circuits run to them. This is for bending radius and space considerations. For further information, refer to Articles 312 and 314.

Note 8 to Tables 310-16 through 310-31 shall be complied with when installing parallel conductors.

The above applies to parallel conductors in the same raceway or cable and also to single conductors or multiconductor cables that are stacked or bundled together and are not installed in raceways. Unless more than three current-carrying conductors are installed in one raceway, and more than one set of such a raceway and conductors are paralleled together, no derating will be required.

Exception No. 4

For existing installations only: Grounded neutral conductors No. 2 AWG or larger can be run in parallel. This may be done only under engineering supervision.

310.5: Minimum Size of Conductors

Conductors shouldn't be smaller than No. 14 copper or No. 12 aluminum or copper-clad aluminum, whether they are solid or stranded.

See Table 310.5 in the *NEC,* which gives minimum sizes of conductors for a number of different voltages, and see Table 310-64, "Definitions."

Following the above, there are eight exceptions that won't be repeated here. Refer to these exceptions in the *NEC.*

2005

310.6: Shielding

The insulation shall be ozone-resistant for solid dielectric insulated conductors if they are being operated over 2400 volts and are in permanent installations. They shall also be shielded. At high voltages, ozone is created, and ozone has deteriorating effects on many types of insulation, which is the reason ozone-resistance is required. When cable is shielded metallically, as by a ribbon of tin-coated copper or bare copper conductor, these shields shall be grounded at both ends and bonded across any splices made in the cables. The requirement of Section 250.51 shall be followed. Shielding contains voltage stresses to the insulation. Stress-cones are used at terminations and splices, and it is a good idea to use the manufacturer's instructions for the cable being used.

Exception

If nonshielded insulated conductors are listed by a testing laboratory, they may be used when voltages don't exceed 8000 volts and under the following conditions:

- **(a)** High-voltage insulation shall resist electric discharge or surface tracking. If this is not the case, then the insulated conductors must be covered by a material not affected by ozone, electric discharge, or surface tracking.
- **(b)** In wet locations the insulated conductors shall have a nonmetallic jacket over all the insulation where there shall be a continuous metallic sheath.
- **(c)** See the *NEC*.
- **(d)** Table 310.63 covers the thickness of insulation and jacket.

310.7: Direct-Burial Conductors

As was stated earlier, cables operating above 2000 volts must be shielded. Any cables used for direct burial shall be of a type that has been identified and listed to be used thus.

Exception

Multiconductor, nonshielded cables from 2001 to 5000 volts are permitted, but the cable itself must have an overall metal sheath or armor.

The requirements of Section 250.51 shall be met for metallic shield, sheath, or armor grounded effectively through the grounding path.

Note

You are referred by the *NEC* to Sections 300.5 and 710.3(B).

The authority enforcing the *Code* may require supplemental mechanical protection, such as a covering board, concrete pad, or raceway. In rocky soil, and more especially where frost is prevalent, the inspection authority will usually require in addition that a bed of fine sand be provided both under and over the conductors and the protection board. Rocks subject to frost heave will cause damage to the insulation.

310.8: Locations

(A) Insulated Conductors. The following types of cable may be used in wet locations: Lead-covered, MTW, RHW, RHW-2, TW, THW, THW-2, THHW, THHW-2, THWN, THWN-2, XHHW, XHHW-2, ZW, or a type listed for such use.

Conductors in damp and dry locations may be FEP, FEPB, MTW, PFA, RH, RHH, RHW, RHW-2, SA, THHN, TW, XHH, THW, THW-2, THHW, THHW-2, THWN, THWN-2, XHHW, XHHW-2, Z, or ZW.

Conductors installed in direct sunlight must be so listed.

Author's Note

In going through these identifications of conductor insulation, you will find that the Hs may be used at a higher temperature, the Ws for wet locations. The Ns are nylon covered, and the double Hs are for higher temperatures than one H.

(B) Cables. Only types of cable containing one or more conductors shall be used in wet locations unless listed for use in such locations.

Direct-burial conductors must be listed for direct burial.

310.9: Corrosive Condition

Where conductors are exposed to (1) oils, (2) greases, (3) vapors, (4) gases, (5) fumes, (6) liquids, or other substances that might cause deterioration of the conductor insulation, a type approved for the condition is required. A case in point is a gasoline-dispensing island where lead-covered cable or cable with an approved nylon jacket is required. You should check the UL listings to be certain whether or not they are approved for the location.

310.10: Temperature Limitation of Conductor

Table 310.13 lists temperature ratings of various insulations. These are the temperature ratings under load-carrying conditions, not ambient temperature. For instance, it is incorrect to use TW wire for wiring boiler controls—a high-temperature wire is required. Also, TW wire is not generally applicable for running to fluorescent lighting fixtures because of the heat of the ballasts.

No conductor is to be used in any manner that will cause the insulation to reach a temperature greater than the temperature for which the insulation was designed.

Note

See the (FPN) at this location in the *NEC*.

The principal problems of operating temperature are:

(1) Ambient temperature refers to the temperature in the area in which the conductors are run. It may affect certain parts of a run because of a higher room temperature in certain locations, and it may change from time to time: winter to summer, daylight to dark, and so on.

(2) Current flow in a conductor will always cause some heat because of the resistance of the conductor.

(3) The amount of heat dissipated by the conductors will depend upon whether the ambient temperature is high or low. When conductors are surrounded with a thermal insulation, it will also affect the rate of heat distribution.

(4) Conductors adjacent to each other will affect the raising of the ambient temperatures and slow up the heat dissipation.

310.11: Marking

(A) Required Information. All conductors must be marked, showing:

- **(1)** Maximum rated voltage.
- **(2)** The type of wire or cable, as designated by a letter code, such as THHN or XHHW.
- **(3)** Some type of name or marking that identifies the manufacturer of the wire.
- **(4)** The size of the wire.

(B) Methods of Marking:

- **(1) Surface Marking.** All common types of conductors must be marked on their surface, showing the wire size at least every 24 inches, and all other markings at least every 40 inches.
- **(2) Marker Tape.** Continuous marking tapes are required inside metal-covered multiconductor cables. Type AC and MI cables are accepted.
- **(3) Tag Marking.** Certain types of cables must be marked by tags attached to the coil of cable. These types are MI cable, switchboard wire, metal-covered single conductor cables, AC cable, and asbestos-covered conductors.
- **(4) Optional Marking of Wire Size.** Types MC cable, TC cable, irrigation cable, power-limited tray cable, and power-limited fire protective signaling cable are allowed to have their marking on the surface of the individual conductors.

(C) Suffixes to Designate Number of Conductors. Suffixes to the letter designations of wires may be used to identify to different types of cables. The suffixes are:

D—for two parallel conductors with a nonmetallic covering.
M—for two or more spirally-twisted conductors with a nonmetallic covering.

(D) Optional Marking. Cable types listed in Table 310.3 are allowed to have additional markings that identify special characteristics of the cable.

310.12: Conductor Identification

(A) Grounded Conductors. Grounded conductors of size No. 6 or smaller shall be identified as being grounded conductors or neutral by being white or natural gray color. Multiconductor flat cables No. 4 or larger shall have the grounded or neutral conductor identified by a ridge of the conductor, or by three white stripes 120 degrees apart along the conductor's entire length.

As an inspector, the author found attempts made to use colors other than white, natural gray, or with three continuous white stripes as the grounded or neutral conductor. It is very dangerous to deviate from what is called for by the *Code*.

Exception No. 1
Multiconductor cables insulated with cambric or varnished cloth.

Exception No. 2
Fixture wires as covered in Article 402.

Exception No. 3
This exception covers mineral-insulated cable and metal-sheathed cable. Conductors of MI cable are bare, so they must be sleeved with sleeves of proper colors, such as white or natural gray for the grounded conductors, and appropriate colors for phase conductors.

Exception No. 4
Section 210.5(A) covers branch circuits. A grounded conductor of different systems must be identified per this section.

Exception No. 5
This exception allows grounded conductors in multiconductor cable to be identified at their terminations at the time of their installation by distinctive white markings, which are permanent, or by other effective means. The main conditions involved here are that maintenance and supervision of such will be done by only qualified persons.

If it is aerial-mounted, cable shall be identified as described in the preceding item, (A), or by a ridge so located on the outside of the cable as to clearly identify it as a grounded cable.

If conductors have a white or natural gray covering, with a colored tracer thread in the braid that identifies the source or manufacturer, they shall meet the requirements of this section.

Tracer threads are sometimes advantageous; for example, when several grounded conductors are in the same location, the threads aid in identifying the source of the grounded conductors.

Section 200.6 covers identification requirements for conductors larger than No. 6.

(B) Grounding Conductors. When using insulated equipment-grounding conductors, the insulation shall be continuous green or green with one or more yellow stripes. The conductor may, of course, be bare.

Exception No. 1

Insulated or covered conductors larger than No. 6 may be black or another color, but when installing them, the conductor shall be identified at every point where the conductor is accessible by one of the following means:

- **(a)** Stripping the insulation or covering from the entire exposed length.
- **(b)** Coloring the exposed insulation or covering green.
- **(c)** Marking the exposed insulation or covering with green colored tape or green colored adhesive labels.

Exception No. 2

At the time of installation, a multiconductor cable may be permanently identified as being the equipment-grounding conductor at each end, and at every point where the conductor needs to be accessible. This applies only where maintenance supervision and qualified persons service the installation and the marks permanently identify it as the grounding conductor. Whenever an end starts or appears as accessible equipment, follow the *NEC* requirements in this exception.

(C) Ungrounded Conductor. Ungrounded conductors shall be clearly distinguishable from grounded or grounding conductors. This applies whether they are single conductors or in cable form. They shall never be white, natural gray, green, or green with yellow stripes. Ungrounded conductors may be solid colors or colors with stripes in a regularly spaced series of identical marks; such stripes shouldn't be white, gray, or green, and shouldn't conflict with markings as required in Section 310.11(B)(1).

Exception

As permitted by Section 200.7.

310.13: Conductor Constructions and Applications

Insulated conductors must meet the requirements of one or more of the following tables: 310.13, 310.61, 310.62, 310.63, 310.64, 310.65, 310.66, and 310.67.

If specified in the respective tables, all of these conductors shall meet wiring methods that are covered in Chapter 3.

Note

Thermoplastic insulation is very susceptible to temperature variations. If temperatures are cooler than minus 10°C (or plus 14°F) during installation, care is required when installing above the temperatures. At normal temperatures, thermoplastic insulation may also be deformed by pressure. This requires extra care during the installation and at any point where the conductor is supported.

Insulation resistance will vary a great deal with temperature. The resistance is higher at lower temperatures and steadily decreases as the temperature rises. A case in point: On one job I was testing insulation on 500 kcmil THW insulation. The first test was run early in the morning. A portion of the conduit run was in the open on a roof and this insulation test was very high. Later, just before energizing the feeder circuit, another insulation-resistance test was run, and the resistance was found to be very much lower than the first test. One question was whether the insulation had been deformed. After analyzing the problem, I realized that the sun's heat was raising the temperature on the outdoor conduit and thus giving a much lower insulation-resistance reading.

310.14: Aluminum Conductor Material

When aluminum conductors came into use, many problems arose. Now when solid aluminum conductors Nos. 8, 10, and 12 AWG are installed, they must be of an aluminum alloy conductor material. The balance of this is thoroughly covered in the *Code* at this point. Please refer to your *NEC*.

310.15: Ampacity

The term *electrical ducts* as used in this particular article recognizes any electrical conduit as covered in Chapter 3 that is suitable for use underground. Other raceways, round in cross-section, are listed to be embedded in concrete underground.

Conductors rated 0 through 2000 volts may have their ampacities rated by one of two specified methods. Type V conductors for 2001 through 5000 volts are rated the same as the conductor types for 0 through 2000 volts. The two methods are as follows:

(A) General. The general method of determining ampacity is by referring to Tables 310.16 through 310.20 and their notes for 0- through 2000-volt cables, and Tables 310.67 through 310.86 and their notes for 2001- through 35,000-volt cables.

(B) Engineering Supervision. When done under engineering supervision, ampacities can be determined by the "Neher-McGrath" formula.

Note

In these tables the ampacity involves temperature alone. Voltage drops will have to be calculated.

Note

If the voltage drop does not exceed 5 percent, this will allow for reasonable efficiency.

(C) Selection of Ampacity. If more than one calculated ampacity is found to be acceptable for a given circuit length, you must use the lowest calculated value.

A very controversial item is whether the neutral conductor of a four-wire wye system has a bearing on derating. It is the author's opinion that this question can't be answered in a broad sense, but depends upon the types of load and whether or not the neutral is carrying any appreciable amount of current.

Exception

When two different ampacity ratings apply to adjacent parts of the same circuit (such as a circuit that goes from free-air into a conduit), the higher rating may be used for 10 feet or 10 percent of the conductor's high-ampacity length past the point of transition, whichever is less.

Article 312—Cabinets and Cutout Boxes

312.1: Scope

This article covers the installation of cabinets, cutout boxes, meter socket enclosures, and their construction.

I. Installation

312.2: Damp or Wet Locations

Cabinets and cutout boxes used in wet or damp locations shall be suitable for the location and installed in such a manner that moisture is not likely to enter or accumulate in the enclosure. The boxes for the cabinets must be mounted ¼ inch between the enclosure and the wall and any other type of members used for support. Only weatherproof cabinets and cutout boxes shall be used.

A recommendation that would be good to follow is that boxes of nonconductive material be used with nonmetallic-sheathed cable when such cable is used in locations where moisture is likely to be present.

An interesting case involving this sort of installation concerned a 2-inch service mast that ran straight down to a service-entrance cabinet with branch-circuit breakers included. The location was a turkey brooder house that had high humidity and was warm, but the outside temperature was below zero. The mast acted as a chimney drawing the warm moist air up and out of the interior. The moisture in the air condensed on the interior of the cold mast and continually dripped into the service-entrance equipment, causing a short that could have possibly developed into a fire. An inspector found the condition and required a seal-off next to the service-entrance equipment. This is mentioned because such an area might be overlooked as a damp location.

Note

You are referred to Section 300.6 of the NEC for protection against corrosion.

In hazardous (classified) locations, Articles 500 through 517 shall be conformed to.

312.3: Position in Wall

The requirements here are the same as for outlet boxes, namely, that in concrete, tile, or other noncombustible walls, the boxes or cabinets may be set back not to exceed ¼ inch from the finished surface, and on walls of combustible materials, they shall be flush with or project from the finished surface.

312.4: Unused Openings

Openings in cabinets and cutout boxes that are not used must be closed with a cap or plug that provides protection equal to the wall of the box. Metal plugs used in nonmetallic boxes must be recessed ¼ inch.

312.5: Conductors Entering Cabinets or Cutout Boxes

All conductors entering cabinets or cutout boxes shall be protected from abrasion and shall meet the following requirements:

(A) Openings to Be Closed. You must properly close enterings through which conductors pass. In other words, where conduit is used, the hole in the cabinet or box shall be the proper size for the trade size of conduit used.

(B) **Metal Cabinets and Cutout Boxes.** When metal cutout boxes or cabinets are used with open wiring or knob-and-tube wiring, the conductors entering these boxes shall enter the metal by means of an approved insulating bushing, or you may use flexible insulating tubing that extends from the last support of the wiring and is firmly secured to the cabinet or cutout box.

(C) **Cables.** Each cable must be adequately secured to cabinets or cutout boxes.

Exception

Cables with nonmetallic sheaths can enter cabinets through raceways not over 18 inches long if they are supported within 12 inches of entering the box and if a fitting at the end of the conduit protects the cables from abrasion. Conduits containing cables entering through the top must be plugged to prevent debris from entering the box.

312.6: Deflection of Conductors

This covers the deflection of conductors in cabinets and cutout boxes at terminals, or conductors entering or leaving cabinets and cutout boxes.

(A) **Width of Wiring Gutters.** Table 312.6(A) gives us the criteria by which conductors may be deflected in a cabinet or cutout box, which is the gutter width that is provided in these locations. The purpose is not to injure the conductors or insulation in installation. The number of conductors in parallel in accordance with Section 310.4 should be used to judge parallel conductors in cabinets or cutout boxes. See the *NEC* for Table 312.6(A), "Minimum Wire Bending Space at Terminals and Minimum Width of Wiring Gutters in Inches," and the note therewith.

(B) **Wire Bending Space at Terminals.** The space for bending wiring to terminals is covered in (1) or (2) below:

(1) When the conductors don't enter or leave the enclosure through a wall that is opposite the terminal, Table 312.6(A) shall apply.

Exception No. 1

If conductors enter or leave an enclosure that is located opposite its terminal, the *Code* permits said conductors to enter or leave the enclosure if the gutter is joined to a gutter adjacent thereto and if it has a width that conforms to Table 312.6(A). The gutter or enclosure or cutouts that we are covering here refer to the space that does not involve terminals or switches.

Exception No. 2

If a conductor is 300 kcmil or less, it may enter or leave the enclosure that only contains a meter socket(s) through the wall opposite its terminal, and only if the terminal is the lay-in type and either of the following conditions is met:

- **(a)** The terminal faces the opposite enclosure wall, but the offset is not greater than 50 percent of the bending space required, as specified in Table 312.6(A), or
- **(b)** The terminal is turned in the enclosure, and the terminal is within a 45-degree angle facing directly to the enclosure wall.

Note

This explains something about the offset. It is the distance on the enclosure wall from the centerline of the terminal to a line that passes through the center of the opening in the enclosure.

(2) When a conductor enters or leaves the enclosure through a wall that is opposite this terminal, Table 312.6(B) shall be used.

For removable (single-barrel) compression-type (crimp-type) lugs, bending space may be reduced as follows:
300kcmil-750 kcmil: 3 inches
AWG-259 kcmil: 2 inches

But reduction can be applied only to the extent that minimum wire bending space is reduced to 6 inches. In no case may the bending space be reduced to a value less than that in Table 312-6(A).

The above may seem confusing, but read it carefully and follow the instructions, because the results are highly important.

(C) Insulated Fittings. Where No. 4 or larger conductors are ungrounded and enter a raceway in a cabinet, pull box, junction box wireway, or auxiliary gutter, protection for the conductors shall be provided by a fitting with a smooth and rounded insulating surface. This is usually in the form of an insulated bushing screwed on the raceway end. If the conductors are separated by an insulating material that is substantially and securely fastened in place, a bushing may be used instead.

Exception
Threaded hubs may be mounted as part of the enclosure, and the above fitting and insulation may be omitted, provided that the hub itself is smoothly rounded or flared where the conductors enter the enclosure.

None of the above permits conduit bushings made entirely of insulating material to secure the raceway. The temperature rating of the insulated conductors shouldn't be higher than the temperature rating of the insulating material used in the fitting or other material.

Insulated bushings and inserts are made in various forms. A metal bushing with a fiber insert that will lock into place may be used; an insulating bushing made entirely of insulating material may be used but a locknut installed ahead of it is a must; or bushings that are metal and have insulation incorporated as a part of the bushing may be used (Figure 312-1).

312.7: Space In Enclosures
Enclosures and cabinets must be designed to provide sufficient room for all the conductors inside of them.

312.8: Enclosures for Switches or Overcurrent Devices
This section has caused much discussion in the field and left much on the shoulders of the authority that has jurisdiction to make the judgment as to adequate spaces within enclosures. The enclosures for overcurrent devices and switches may not be used as junction boxes, auxiliary gutters, or raceways for conductors tapping off or feeding through to other switches or overcurrent devices. See Figures 312-2, 312-3, and 312-4.

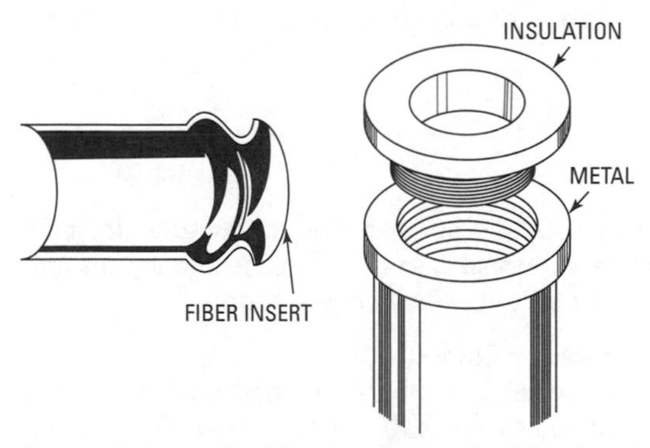

Figure 312-1 Bushing insulation for No. 4 or larger conductors.

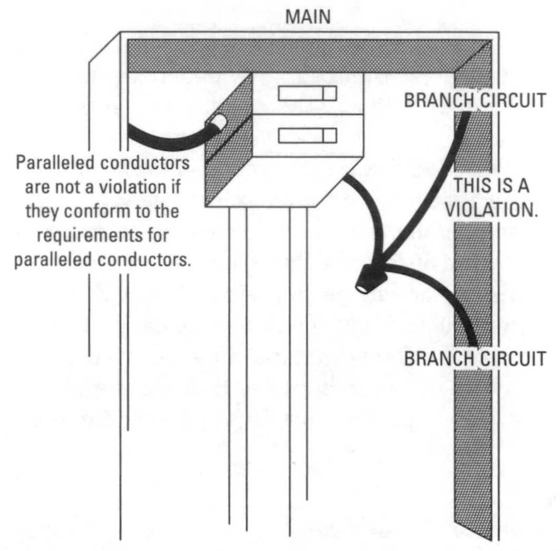

Figure 312-2 Cabinets and cutout boxes shouldn't be used as junction boxes.

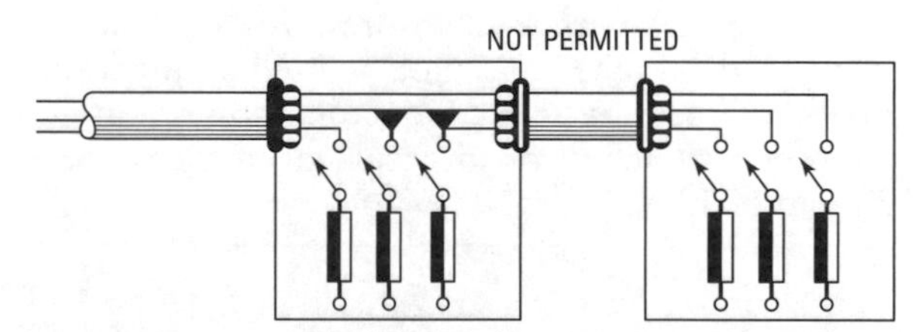

Figure 312-3 Improper method of connecting more than one switch enclosure.

Exception

Where adequate, spacing is supplied so that the conductors don't fill over 40 percent of the cross-sectional area of the space and splices and taps do not fill more than 75 percent of any cross-section.

312.9: Side of Back Wiring Spaces or Gutters

Metal cabinets and cutout boxes must be furnished with adequate wiring space, auxiliary gutters, or special compartments for wiring, as required by Section 373.11(C) and (D).

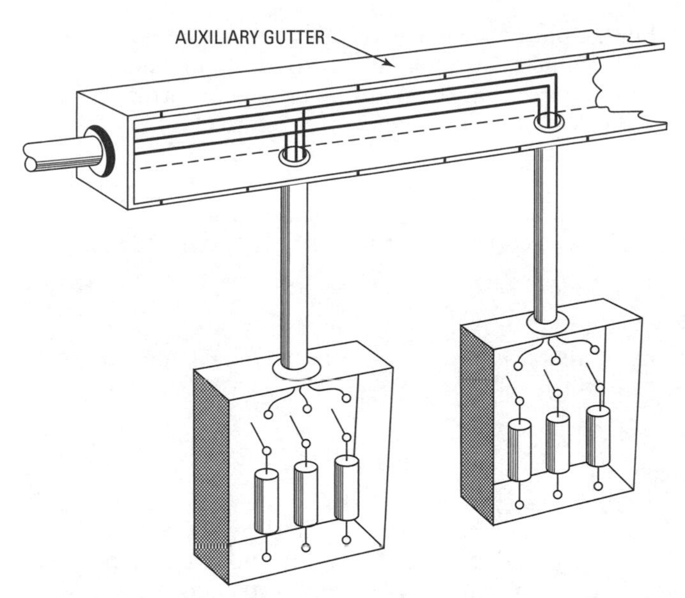

Figure 312-4 Proper manner of connecting more than one switch enclosure.

II. Construction Specifications

312.10: Material

This section covers cabinets and cutout boxes:

(A) Metal Cabinets and Cutout Boxes. Cabinets and cutout boxes shall be coated inside and outside.

Note

For corrosion protection the *NEC* refers you to Section 300.6.

(B) Strength. Ample strength and rigidity must be incorporated in the design and construction of cabinets and cutout boxes. When these are made of sheet metal or metal without the coating, they shall have a thickness of not less than 0.053 inch.

(C) Nonmetallic Cabinets. If nonmetallic cabinets are to be used, they must be approved or listed before they are installed.

312.11: Spacing

The mere fact that a switch enclosure is rated at X number of amperes does not mean that the space inside is adequate for the job. For instance, the enclosure might be designed for single-conductor installation but would be overcrowded if conductors were paralleled. Thus, the engineer, installer, and inspector must analyze the intended use and ensure that the bending space conforms to Table 312.6(A). See the *NEC* for the balance of Section 312.11.

2005

Article 314—Outlet, Device, Pull, and Junction Boxes, Conduit Bodies, Fittings, and Handhole Enclosures

This is one of the most important articles in Chapter 3, and should be carefully examined.

I. General

314.1: Scope

This article covers the installation of all boxes, conduit bodies, fittings, and handhole enclosures required in Section 300.15. Those that are covered in Section 300.15 and may be used as pull boxes, junction boxes, or outlet boxes shall conform to the provisions of this article, depending upon the purpose for which you are using them. These boxes and the like may be made of cast material, sheet metal, or nonmetallic material. Other boxes, such as FS, FD, or larger, are not considered to be conduit bodies. Elbow fittings that are capped, including service-entrance elbows, shouldn't be classified as conduit bodies. When we speak of conduit bodies, we generally speak of condulets.

Note

Part IV of this article covers systems of over 600 volts.

314.2: Round Boxes

Round boxes create a problem with the use of locknuts, bushings, and connectors. There are usually knockouts in the bottom of the boxes, so where these types of connections are used, they shouldn't be connected to the sides of the box. Round boxes are usually found on existing jobs. Boxes that are now manufactured are in accordance with the *NEC*.

314.3: Nonmetallic Boxes

Nonmetallic boxes may be used only with Type NM or NMC cable, nonmetallic raceways, concealed knob-and-tube wiring, or open wiring on insulators.

Exception No. 1
Nonmetallic boxes can be used with metal raceways or metal-jacketed cables as long as internal bonding means and provisions for installing grounding jumpers are provided.

Exception No. 2
An integral bonding means between all threaded entries for metal raceways or metal-sheathed cables is required when using nonmetallic boxes for these cables and raceways.

314.4: Metal Boxes

All metal boxes shall be properly installed so as to carry the continuity of the equipment grounding system, which is covered in Article 250.

314.5: Short-Radius Conduit Bodies

No splices, taps, or devices are permitted in short-radius conduit bodies such as service elbows and capped elbows.

II. Installation

314.15: Damp or Wet Locations

When used in damp or wet locations, boxes and fittings shall be installed in such a way as to prevent the accumulation of moisture or to prevent water from entering the boxes or fittings. All boxes and fittings that are installed in wet locations shall be weatherproof.

In hazardous (classified) locations, all boxes—in fact, the entire wiring system—shall conform to Articles 500 through 517.

Where boxes are mounted in floors, they shall be listed for use in wet locations. The reason for this is that, in scrubbing floors, for example, water may enter these boxes and be a source of trouble. Section 314.27(B) covers boxes in floors. For protection from corrosion, see Section 300.6.

Conduit bodies are also included in this section.

314.16: Number of Conductors in Outlet, Device, and Junction Boxes, and Conduit Bodies

The interior of the boxes, and so on, shall be large enough that there is plenty of free space for the conductors installed in them.

This is a basic and broad statement that, in itself, is quite sufficient; however, a complete analysis of this will be made here as it is in the *Code*. The main point is that in the installation of conductors in boxes, it should be unnecessary to force the conductors into the box, as this is a potential source of trouble. The *Code* spells out what is good practice, as well as the minimum requirements. The installer should, however, always bear in mind the purpose for these requirements and, if necessary to do a good job, go even further than the minimum requirements.

The limitations imposed by Section 314.6(A) and (B) are not intended to apply to terminal housings supplied with motors. See Section 430.12.

Conductors No. 4 or larger shall comply with the provisions of Section 314.18 when used in boxes and conduit bodies.

Section 314.18 provides for space to make proper bends without injury to the conductors.

(A) Box Volume Calculations. The volume of a wiring enclosure (box) shall be the total volume of the assembled sections, and where used, the volume of the space provided by plaster rings, domed covers, extension rings, and so on. These are marked with their volumes in cubic inches, and shall be added together to calculate the total box fill volume.

(1) Standard Boxes. Table 370.6(A) gives us the maximum number of conductors that are permitted in standard boxes. Section 314.18 covers boxes or conduit bodies when used as junction or pull boxes.

(2) Other Boxes. Other boxes that are less than 100 cubic inches (1650 cm^3) and are not covered in Table 314.6(A), such as conduit bodies having more than two conduit entries and nonmetallic boxes, must be durably marked with their cubic-inch interior capacity when manufactured. To find the maximum conductor allowed in these boxes, use Table 314.6(B); you are allowed to take the deductions provided for in Section 314.6(A)(1). These deductions for volume are based on the largest conductor entering the box. If boxes have a larger cubic-inch capacity than that which is designated in Table 314.6(A), their cubic-inch capacity shall be marked as is indicated in this section. The maximum allowable number of conductors to enter such a box may be calculated using the volume allowed for each conductor, as found in Table 314.6(B). Please refer to Tables 314.6(A) and 314.6(B) in your *NEC*.

(B) Box Fill Calculations. Table 370.6(A) covers the maximum number of conductors that will be permitted in outlet and junction boxes. There is no allowance in this table for fittings or devices—such as studs, cable clamps, hickeys, switches, or receptacles—that are contained in the box. These must be taken into consideration, and deductions must be made for them. A deduction of one conductor shall be made for each of the following: one or more fixture studs, cable clamps, and hickeys. One conductor shall be deducted for one or more grounding conductors that enter the box.

There shall be a further deduction of two conductors for one or more devices mounted on the same yoke or strap.

2005

A conductor that runs through a box is counted as only one conductor, except if it is more than twice the minimum length required by Section 300.14 (6 inches), then it is to be counted twice. Therefore, a loop of wire in a box that is more than 12 inches long counts as two conductors.

A conductor originating outside of the box and terminating in the box is counted as one conductor. Conductors entirely contained within the box are not counted. When different-sized conductors are in the box, the conductor deducted for a fixture stud, and so on, must be the largest size in the box; the conductors deducted for the device strap must be the largest size connected to the device strap.

Boxes are often ganged together with more than one device per strap mounted in these ganged boxes. In these cases, the same limitations will apply as if they were individual boxes.

Please refer to Figures 314-1 through 314-4 for illustrations concerning this section. Caution must be taken to use boxes large enough to accommodate the counting of these grounding conductors. This has been interpreted in various ways in the past, but it is now very clear. Table 314.6(A) now has the cubic capacity listed for the popular sizes of boxes, and has also been expanded.

The volume of a wiring enclosure (box) shall be the total volume of the assembled sections, and where used, the volume of the space provided by plaster rings, domed covers, extension rings, and so on. These are marked with their volumes in cubic inches, and shall be added together to calculate the total box fill volume.

Exception

A grounding conductor and/or up to four No. 14 fixture wires that enter a box under a fixture canopy can be omitted from these calculations. Additionally, branch-circuit conductors that go directly into a fixture canopy don't have to be counted as leaving the box.

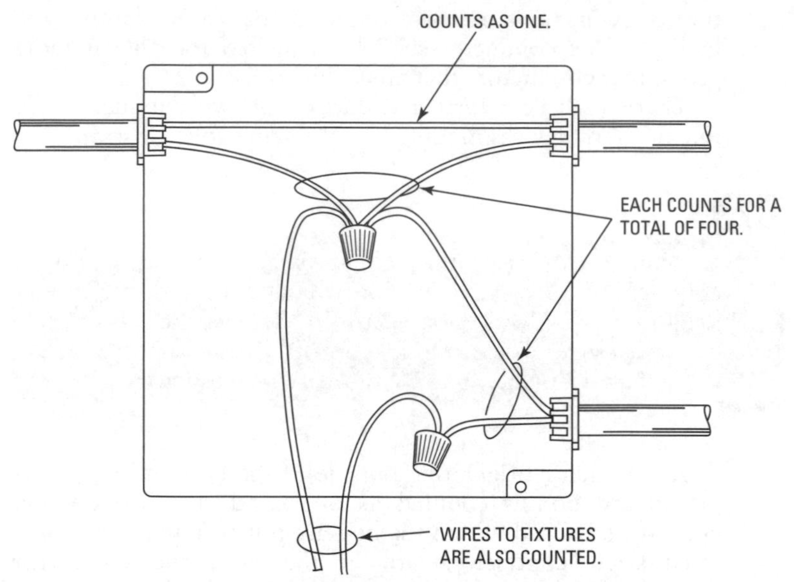

Figure 314-1 Which conductors to count in junction box.

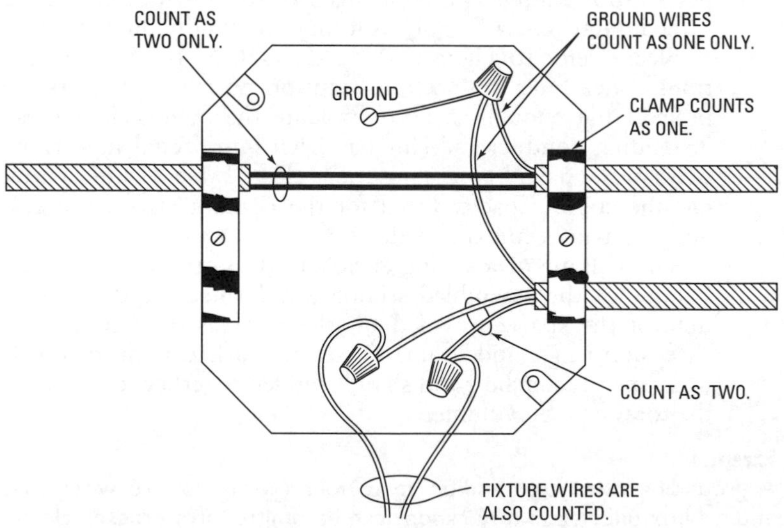

Figure 314-2 Grounding wire does not count.

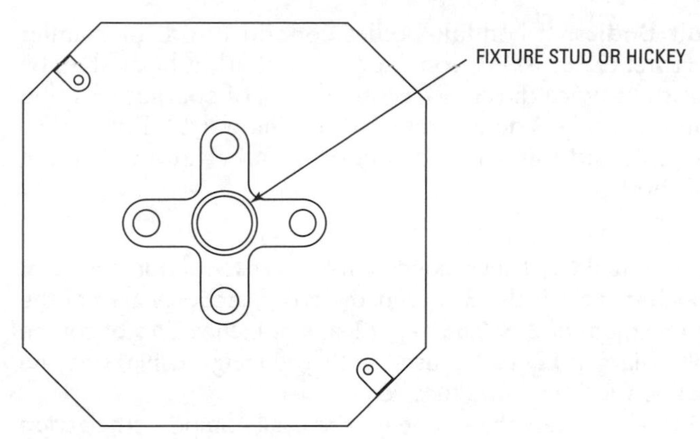

Figure 314-3 Fixture stud or hickey counts as one conductor.

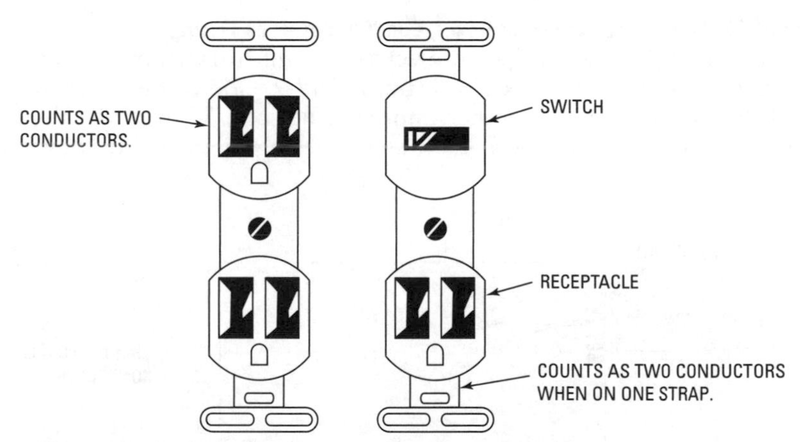

Figure 314-4 How to count devices when figuring fill.

There will be numerous occasions where Table 314.6(A) won't be applicable. In such cases, refer to Table 314.6(B), from which the number of conductors that will be permitted in the box can be calculated. The various wire sizes and the space that be allowed per conductor are given. It is necessary to calculate only the cubic space in the box, and multiply by the free space within the box for each conductor. If the calculation exceeds the cubic size of the box, the next larger box will have to be used or an extension added.

(C) Conduit Bodies. If conduit bodies contain No. 6 or smaller conductors, the cross-sectional area of a conduit body shall be not less than twice the cross-sectional area of conduit entering the conduit body. You are referred to Chapter 9, Table 1 for conduit fill, and this will indicate the number allowed in the conduit body.

Explanation

From Chapter 9, Table 1, 1-inch conduit has a cross-sectional area of 0.86 square inches. If a 1-inch LB is used, the cross-sectional area of the LB shall be a minimum of $2 \times 0.86 = 1.72$ square inches. This of course would permit a larger LB to be used with reducing bushings at the threaded hubs to fulfill the two times requirement.

Conduit bodies with less than three entries shall comply with Section 370.5; otherwise, no taps or splices are permitted and the conduit body shall be supported in a secure manner.

314.17: Conductors Entering Boxes, Conduit Boxes, or Fittings

Care shall be exercised to protect the conductors from abrasion where they enter the boxes or fittings. With conduit, this is accomplished by bushings or other approved devices. With NM cable, as shown in Figure 314-5, the outer covering of the cable should

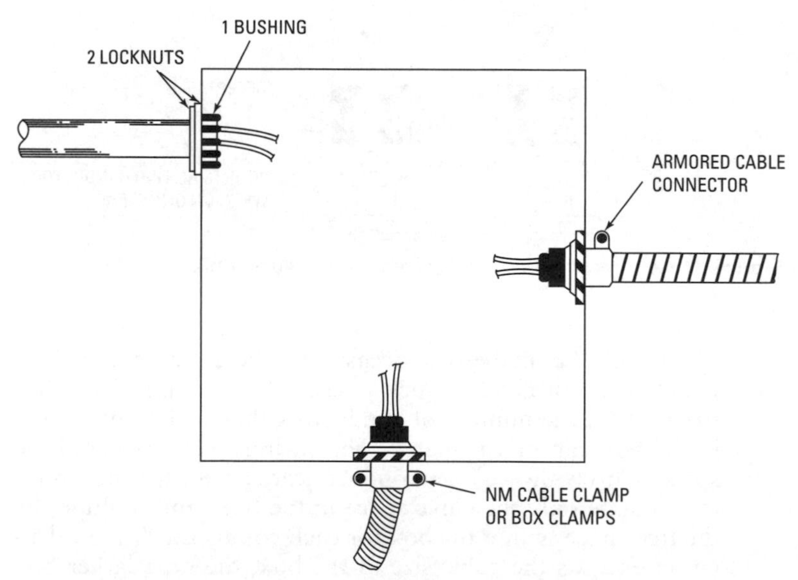

Figure 314-5 Connection of cables and conduit to boxes.

protrude from the clamp to provide this protection. With armored cable, fiber bushings should be inserted between the conductors and the armor to prevent any abrasion. The following shall be complied with:

(A) Opening Must Be Closed. Where conductors enter any openings, they shall be properly closed. Where single conductors enter the boxes, loom covering is to be provided; with cable, cable clamps shall be used or the boxes provided with built-in cable clamps; with conduit, the locknuts and bushings will adequately close the openings.

(B) Metal Boxes, Conduit Bodies, and Fittings. When metal boxes, conduit bodies, or fittings are used with open wiring, proper bushings shall be used. In dry places, flexible tubing may be used and extended from the last conductor support into the box and secured (Figure 314-6).

Where a raceway or cable enters the box or fitting, the raceway or cable should be properly secured to the box or fitting. With conduit, two locknuts and a bushing should be used. With armored cable, approved connectors shall be used. With NM cable, a connector or built-in clamps shall be used.

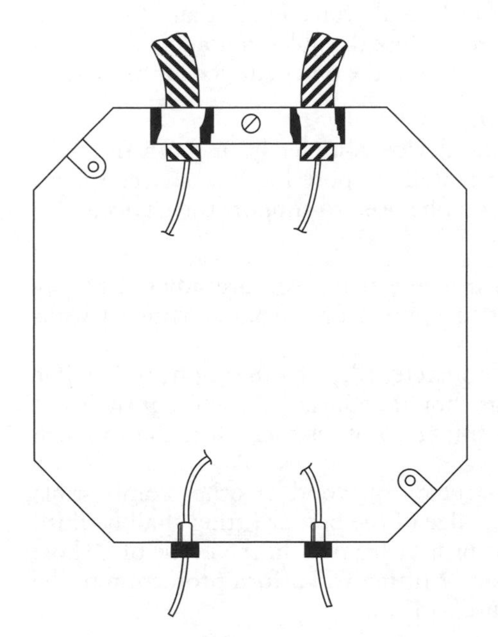

Figure 314-6 Open wiring into boxes.

(C) Nonmetallic Boxes. Where nonmetallic boxes are used with either concealed knob-and-tube work or open wiring, the conductors shall pass through individual holes in the box. If flexible tubing is used over the conductor, it shall extend from the last conductor support into the hole in the box at least ¼ inch.

Where nonmetallic cable is used, it shall extend through the opening in the box at least ¼ inch. Individual conductors or cables do not have to be clamped if the individual conductors or cables are supported within 8 inches of the box. When nonmetallic conduit is used with nonmetallic boxes, the conduit shall be connected to the box by approved means.

Nonmetallic boxes shall be rated at the lowest temperature rating of the conductor that enters the nonmetallic box.

In all instances, all permitted wiring methods shall be secured to the boxes.

(D) Conductors No. 4 AWG or Larger. Section 312.6(C) covers the installation of these.

314.18: Unused Openings

Any unused openings of boxes, conduit bodies, or fittings, where the knockout has been removed, shall be effectively closed by an approved means that will afford equal protection to that of the original. Metal plugs or plates used to close the holes in nonmetallic boxes shall be recessed at least ¼ inch from the outer surface of the box.

314.19: Boxes Enclosing Flush Devices

Such boxes must enclose the device completely on both the back and sides and provide substantial support for the device. Screws that hold the box in place can't be used to support the device.

314.20: In Wall or Ceiling

This section is much abused in the field. You are advised to pay close attention to this part to prevent fires from starting in walls and ceilings.

In walls and ceilings of concrete, tile, or other noncombustible materials, boxes and fittings should be installed such that the front or outer edge of the box or fitting is not set back more than ¼ inch from the finished surface.

In walls or ceilings constructed of wood or other combustible materials, the outer or front edge of the box or fitting shall be flush with the finished surface or project from it. In the event of a short circuit or any arcing, the box or fitting will afford protection to the combustible materials (Figure 314-7).

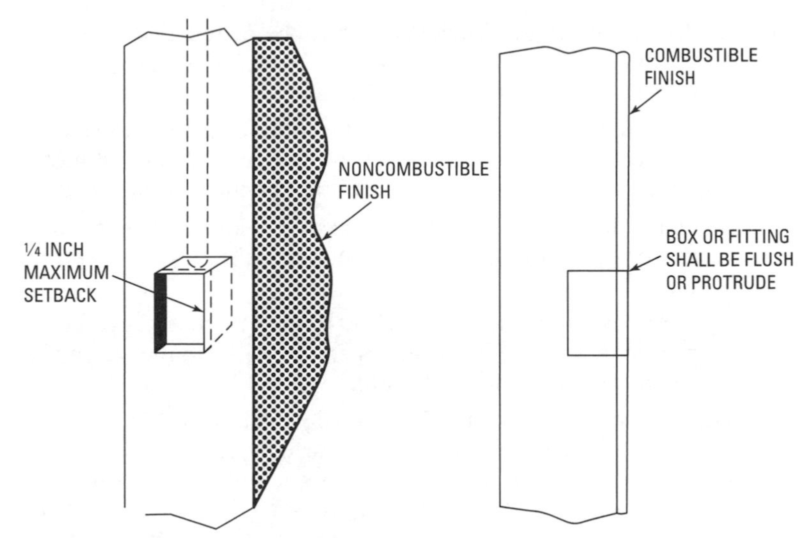

Figure 314-7 Setback of boxes in walls and ceilings.

314.21: Repairing Plaster and Drywall or Plasterboard

Any of these surfaces that are damaged or incomplete must be repaired so that there is no gap between the box and the plaster, or the like, greater than ⅛ inch.

314.22: Exposed Surface Extensions

Exposed extensions from boxes or fittings are very often desirable or necessary. In making these extensions, there are a number of approved wiring methods that may be used, but a box extension ring or blank cover shall be used for attaching to the concealed wiring and shall be electrically and mechanically secured to the original box and extensions from it, and shall be in accordance with the regulations provided in other articles of Chapter 3.

Exception

An extension may be made from a cover mounted on a concealed box. The wiring method must be flexible and arranged with a grounding means that is independent from the connection between the box and the cover. The cover must have a means for preventing the cover from being removed if the screws become loose.

314.23: Supports

Enclosures covered by this article must be securely mounted in accordance with the following:

(A) Surface Mounting. Enclosures must be fastened to the surface, unless the surface is not sufficient, in which case the installation must conform with (B).

(B) Structural Mounting. They must be supported from the structural member either directly or by use of a wood or metal brace. Support wires are not permitted as the sole support.

- **(1)** Nails may be used, and may pass through the box, as long as they are located within ¼ inch from the back or end of the box.
- **(2)** Metal braces must be at least .020 inch thick and protected from corrosion. Wood braces shall be no smaller than the common 1 × 2 (nominal) size.

(C) Nonstructural Mounting. You may install a flush box in existing covered surfaces, provided there is adequate support by clamps, anchors, or fittings. Framing members of suspended ceilings may be used as supports, provided they are fastened to each other and the building structure sufficiently. Fastening to such members must be by bolt, screw, rivet, or by an approved clamp.

(D) Raceway-Supported Enclosure(s), without Devices or Fixtures. Threaded enclosures of 100 cubic inches or smaller that contain no devices and don't support fixtures may be supported by two or more conduits terminated in the threaded hubs, and if the conduits are supported within 3 feet of the box or enclosure. Conduit bodies of a size no larger than the conduit that supports them may be supported by EMT or conduit.

(E) Raceway-Supported Enclosures, with Devices or Fixtures. The requirements of (D) apply here, except the conduits must be supported within 18 inches of the box or enclosure.

(F) Enclosure(s) in Concrete or Masonry. Enclosures may be supported by being embedded.

(G) Pendant Boxes. These boxes must be supported by a multiconductor cord or cable. The conductors must be protected from strain.

314.24: Depth of Outlet Boxes

In no case shall a box have a depth of less than ½ inch, and if flush devices are to be mounted in it, the depth shall be no less than 15⁄16 inch.

314.25: Covers and Canopies

In completed installations each outlet box shall be provided with a cover unless a fixture canopy is used.

(A) Nonmetallic or Metal Covers and Plates. Either nonmetallic or metallic plates and covers may be used with nonmetallic boxes, but when metallic plates or covers are used, the grounding provisions of Section 250.110 apply—it will be necessary to see that they are properly grounded.

Note

The *NEC* refers you to Sections 410.18(A) and 410.56(C) for metal faceplates.

As always, the *Code* is concerned with the practical side of things. Any metal part of a wiring system is subject to becoming energized, and should therefore be grounded whenever there is a possibility of anyone touching the device and a grounded surface at the same time. Concrete floors and walls are, as a rule, considered grounded surfaces, even if covered with block tile. See Section 250.42 in your *NEC*.

(B) Exposed Combustible Wall or Ceiling Finish. This paragraph is often overlooked. When mounted on combustible walls or ceilings, the fixture canopy or pan used, or the ceiling finish exposed between the edge of the canopy or pan and the outlet box, shall be covered with a noncombustible material (Figure 314-8).

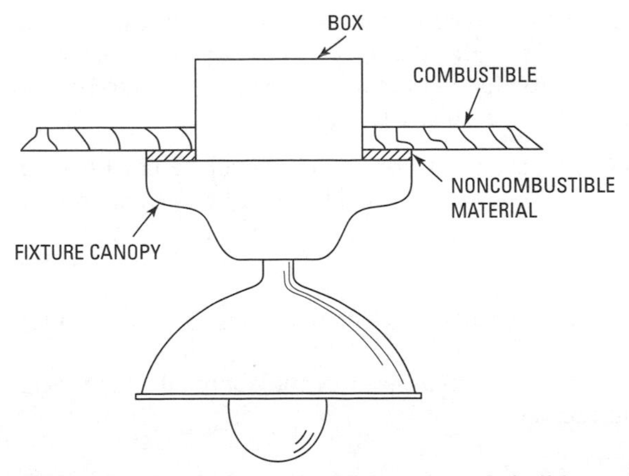

Figure 314-8 Noncombustible material shall be installed between a canopy and a combustible ceiling.

(C) Flexible Cord Pendants. When outlet boxes or conduit fittings have covers through which flexible cords or pendants are to run, these holes shall be smooth, shall have bushings designed for the cord to pass through, and shall have well-rounded surfaces. Hard-rubbercord or rubber-type bushings shouldn't be used, as they are subject to deterioration. Where the cord comes through the bushed hole in the cover, a suitable "electrician's" knot or some other suitable knot shall be provided above the plate to take the strain off the tap to the supply conductors.

314.27: Outlet Boxes

(A) Boxes at Lighting-Fixture Outlets. Any boxes used for lighting-fixture outlets should be designed for that purpose and all outlets used exclusively for lighting fixtures shall have the boxes designed or installed so that the lighting fixture may be attached. Attention should be given to the weight of the fixture. The provisions of Article 410 pertaining to the hanging of fixtures shall be complied with. The maximum size wall-mounted fixture that can be mounted to a standard outlet box is one that is no more than 16 inches in any dimension and weighs no more than 6 lb.

(B) Floor Boxes. Floor boxes especially approved for the purpose shall be used for receptacles located in the floor.

Exception

Standard listed types of flush receptacle boxes are permitted where receptacles are located in elevated floors of show windows or other locations when the authority that has jurisdiction judges them to be free from physical damage, moisture, and dirt.

(C) Boxes at Fan Outlets. Outlet boxes shall never be used for the support of ceiling-panel-type fans. They shall be attached by a secure means into a structural member of the ceiling.

Exception

If the boxes are listed for supporting ceiling fans, they may be used for the support of the fan.

This part should be adhered to as well as the approval of standard receptacle boxes and the like.

314.28: Pull and Junction Boxes

The following items (A) through (D) should be adhered to when boxes and conduit bodies are used for pull or junction boxes:

(A) Minimum Size. Where raceways contain conductors of size No. 4 or larger, or for cables that contain conductors of size No. 4 or larger, the minimum dimensions of pull or junction boxes to be used with these raceways or cables are as follows:

(1) Straight Pulls. The width of a box for a straight pull is governed by the size of conduits used and the space required for the locknuts and bushings. The length, however, shouldn't be less than eight times the trade size of the largest raceway. In Figure 314-9, for example, there is a 4-inch conduit, a 2-inch conduit, and a 1-inch conduit. Therefore, the length will be 4 × 8, or a minimum of 32 inches in length, and the width will be approximately 12 inches to accommodate the locknuts and bushings without crowding.

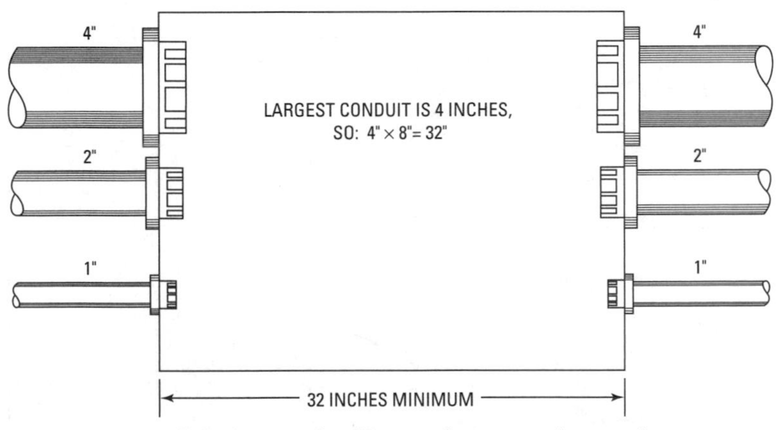

Figure 314-9 Calculation of pull boxes for use without splices or taps.

(2) Angle or U Pulls. When angle or U pulls are made, the distance between the point where the raceway enters the box and the opposite wall shall be not less than six times the trade-size diameter of the largest raceway. If more than one entry is made into the box, the distance just mentioned shall be increased by the sum of the diameters of all the other raceways that enter the same box in one row and on the same wall as the box. The row that provides the maximum distance shall be used.

Exception

If the entry into the box of the row of conduits is on a wall opposite the cover of the box, and if the distance from that wall to the cover meets the requirements of the column for one wire per terminal, as covered in Table 312.6(A).

When entries of raceways contain the same conductor, they shall be not less than six times the trade-size diameter of the largest raceway.

In Figure 314-10, the dimensions shown are the minimums. It is very possible that the 30-inch figure won't agree with the diagonal figures; this will depend on the actual locations of the conduits. Nevertheless, the figures are minimum-size and, if necessary, a larger box may be required.

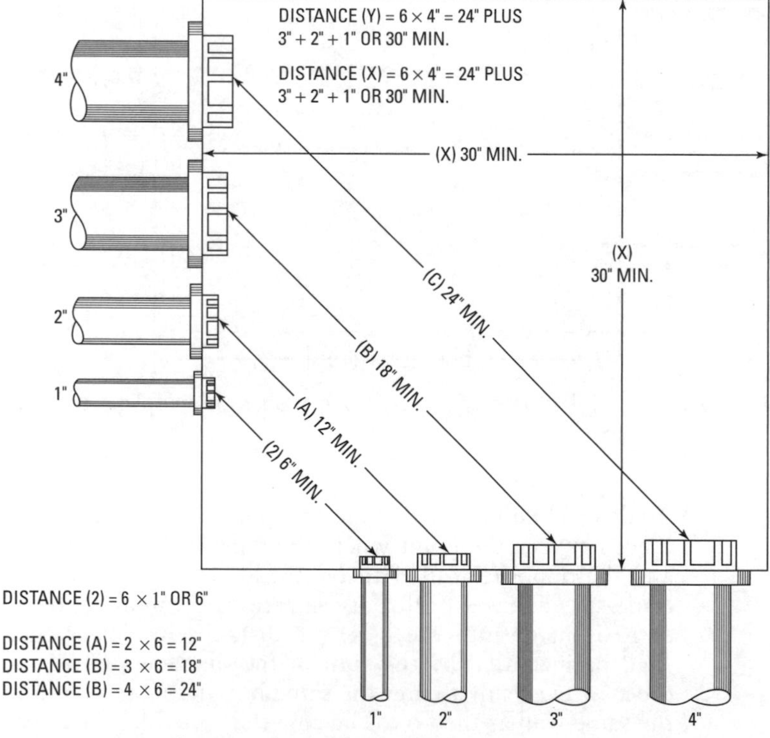

Figure 314-10 Junction box calculations.

If cable size is transformed into raceway size in preceding items (1) and (2), you shall use what would be the minimum-size raceway required to contain the size and number of the cable.

(3) Boxes of lesser dimensions than those required in the preceding subsections, (1) and (2), may be used for installations of combinations of conductors that are less than the maximum conduit-fill (of conduits being used) permitted in Table 1, Chapter 9, provided the box has been approved for and is permanently marked with the maximum number of conductors and the maximum AWG size permitted.

Exception

On motors, the terminal size must comply with Section 430.12.

(B) Conductors in Pull or Junction Boxes. In pull or junction boxes that have any dimension over 6 feet (1.83 m), it is necessary to either rack or cable the conductors. This is required not only to maintain some sort of support for the conductors, but it will also tend to keep the conductors of the same circuit together and, in this way, to keep magnetic induction to a minimum. This is especially important in circuits that carry heavy currents where magnetic induction may affect the voltage balance between phase (see Figure 314-11).

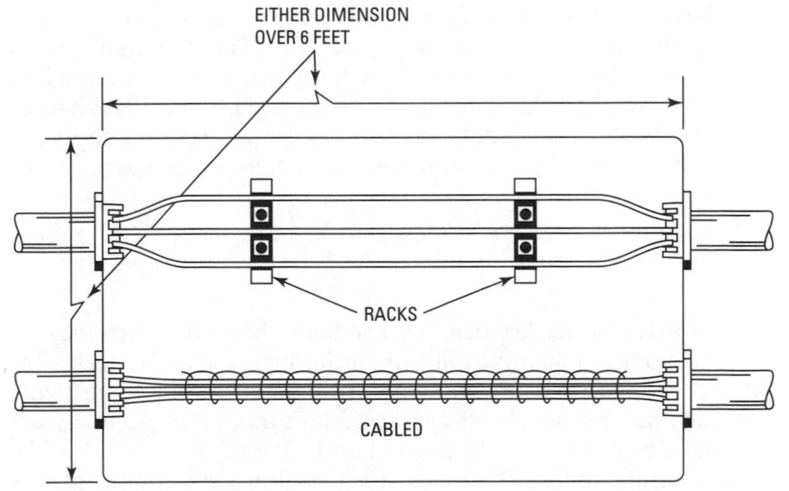

Figure 314-11 Cabling or racking conductors in large boxes.

Reference is made to Section 312.6(C), which requires insulated bushings or an equivalent insulation at terminations where No. 4 or larger conductors are used (see Figure 314-12). These bushings may be fiber or plastic, a combination metal and plastic (where a grounding-type bushing is required), or fiber inserts.

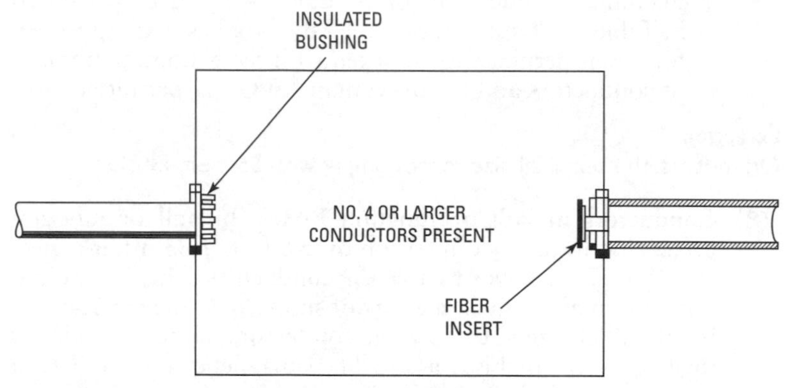

Figure 314-12 Use of insulation at bushings.

(C) Covers. Covers shall be compatible with the box with which they are used. This covers pull boxes, conduit boxes, junction boxes, and fittings construction, and they shall all be suitably applicable for the condition of use. The requirements of Section 250.110 must be met where metal covers are used; in other words, where metal covers are used, the covers shall be effectively grounded. Covering is essential, in case a short circuit or ground fault should occur, to contain the sparks in the box or fitting.

(D) Permanent Barriers. When barriers are required in a box, each section made by the barrier is treated as a separate box.

314.29: Conduit Bodies, Junction, Pull, and Outlet Boxes to Be Accessible

Conduit bodies, junction, pull, or outlet boxes shall be accessible without the necessity of removing any part of the building structure, paving, or sidewalk. It can be certain that an inspector won't consider a box buried in the ground and covered over as accessible. Junction, pull, and outlet boxes are there for a specific purpose—either for pulling in conductors, for junctioning conductors, or for the connection of a device. Should it ever be necessary to rework the system, these boxes must be accessible. Boxes in accessible attics or crawl spaces are considered as accessible, as are boxes in drop ceilings with removable panels.

Exception

Listed boxes, where covered by gravel or light aggregate, or noncohesive granulated soil, are permitted if their location is permanently marked and identified, and if they are located so that they may be easily excavated. Here again, protection against electrolysis in gravel and aggregate must be at least the minimum of any other type of covering, but added protection would not be amiss to ensure that the boxes are watertight against ground space.

III. Construction Specifications

Wirer are not too concerned in most cases with the construction specifications of boxes. They purchase approved boxes that will, of course, meet specifications. However, it sometimes becomes necessary to build or assemble a box to meet a certain purpose. Therefore, it is a good idea to be aware of what these specifications are.

314.40: Metal Boxes, Conduit Bodies, and Fittings

Metal boxes, conduit bodies, and fittings shall conform to the following:

(A) **Corrosion-Resistant.** The covering of metal boxes, conduit bodies, and fittings, of course, shall be suitable for the conditions that prevail in the area in which they are used. They shall be protected by one of the following means (the one to be chosen shall be applicable for the conditions): (1) corrosion-resistant metal, (2) well galvanized, (3) enameled, or (4) otherwise properly coated.

Reference is made to Section 300.6, which prohibits the use of enamel in certain places. Also, coating with a conductive material is recommended, such as cadmium, tin, or zinc, since these will secure a better electrical connection.

(B) **Thickness of Metal.** Boxes not over 100 cubic inches in volume must be made of metal not less than .0625-inch thick. The walls of cast or malleable iron boxes must be at least 3⁄32 inch thick. Other special metal boxes may have walls of 1⁄8 inch.

(C) **Metal Boxes Over 100 Cubic Inches.** These boxes must be of adequate strength. If of sheet metal, the metal must be at least .053-inch thick (uncoated).

(D) Grounding Provisions. Any box designed to be used with nonmetallic raceways or cables must have a means for the connection of a grounding conductor. Usually, this is a 10/24 threaded hole in the back of the box for the screw of a grounding pigtail.

314.41: Covers

Covers must meet the same construction specifications as the boxes they are installed with, or must be lined with at least 1/32 inch of insulation. Covers made of insulating materials can be used if they provide sufficient strength and protection.

314.42: Bushings

Holes through which pendants pass must have approved bushings or smooth, rounded surfaces. When conductors not in cords pass through a cover, and the like, each conductor must pass through its own bushed hole. Such holes shall be connected by a slot. See Section 300.20.

314.43: Nonmetallic Boxes

For nonmetallic boxes, supports must be either outside the box, or the box must be designed to prevent contact between the conductors and the supporting screws.

314.44: Marking

The manufacturer's name or trademark shall be durably and plainly marked on all boxes, conduit bodies, covers, extension rings, plaster rings, and so on.

IV. Manholes and Other Electric Enclosures Intended for Personnel Entry

314.50: General

Manholes and similar enclosures must be large enough to provide a safe workspace, especially around exposed live parts. With water and gas seepage, manholes can be difficult and sometimes hazardous locations, and it is advisable to oversize and maintain them well.

314.51: Strength

Manholes and similar enclosures must be designed under engineering supervision and must be able to bear whatever loads are placed upon them. Again, it is best to oversize. See the *National Electrical Safety Code* for information on loading.

314.52: Cabling Work Space

A clear space 3 feet wide must be provided when cables are located on both sides and 2½ feet when cables are located on one side only. Headroom may not be less than 6 feet. See the *NEC* for exceptions.

314.53: Equipment Work Space

When live parts are present, the requirements of Sections 110.26 (600 volts or less) or 110.34 (over 600 volts) must be followed.

314.54: Bending Space for Conductors

See Section 314.28(A) for 600 volts or less and Section 314.71 for over 600 volts.

314.55: Access to Manholes

See the *NEC* for specifics on these manufacturing requirements.

314.56: Access to Vaults and Tunnels

- **(A)** **Location.** Openings may not be located directly over electrical equipment or conductors.
- **(B)** **Locks.** In addition to the requirements of Section 110.34(C), the locking mechanisms of vaults and tunnels should, if at all possible, be arranged so that people can't be locked inside.

314.57: Ventilation

Ventilation should be provided where practical.

314.58: Guarding

Where objects could enter the manhole through some type of grating, protection of equipment and conductors must be provided.

314.59: Fixed Ladders

Fixed ladders must be corrosion-resistant.

V. Pull and Junction Boxes for Use on Systems Over 600 Volts, Nominal

314.70: General

Not only do the general requirements in Article 314 apply, but the rules in Part V shall also specifically apply.

314.71: Size of Pull and Junction Boxes

The following will apply to the installation of conductors in pull and junction boxes, with adequate dimensions for their installation:

Note

The following is very essential in the installation of high-voltage cables. Most of them are shielded cables and great care must be taken so as not to disturb the shielding to any great extent during installation. Also, there will be splices; since high-voltage splicing is a great deal different from low-voltage splicing, a great deal more space is appropriate.

(A) **For Straight Pulls.** The length of the box shall be not less than 48 times the diameter of the conductor and insulation or cable entering the box. Remember, this figure of 48 times is a minimum, as in many instances, one may wish to increase this length for ease in handling of the cables and for splicing.

(B) **For Angle or U Pulls.** Between each conductor or cable that enters a box and the opposite wall of the box the distance shall be not less than 36 times the diameter of the outside of the largest cable or conductor. Where there are additional entries, the above distance shall be increased to the sum of the outside dimensions over sheath of the cables or conductor entries that enter through the same wall of the box.

Exception No. 1

Where the cables or conductors exist in the box, the 36-times-the-diameter requirement on the entries may be reduced to 24 times the outside diameter of the cable and cable sheath or conductors.

Exception No. 2

If the provisions of Section 300.34 are followed, conductor or cable entries that enter the wall of the box opposite a removable cover are permitted.

Exception No. 3

Terminal housings on motors must conform to Section 430.12.

(C) **Removable Sides.** This requires that one or more sides of a pull box be removable.

314.72: Construction and Installation Requirements

(A) **Corrosion Protection.** Some means must be used to make both the inside and outside of boxes corrosion-resistant. This may be accomplished by making them of material that is already corrosion-resistant or is suitable for such protection as enameling, galvanizing, plating, or other means. Of course, the authority that has jurisdiction is the judge of whether the means of corrosion protection is adequate.

(B) **Passing Through Partitions.** Where cables or conductors pass through partitions, suitable means must be provided to prevent damage to the cable or conductors. This may be suitable bushings, shields, or fittings that have smooth or rounded edges.

(C) **Complete Enclosure.** Provision shall be made to enclose conductors contained in boxes completely.

(D) **Wiring Is Accessible.** In accordance with Section 110.34, the boxes must be installed so that wiring is accessible without the necessity of removing any part of a building to get to the box. Adequate workspace shall also be provided for working with conductors or cables after opening the box.

(E) **Suitable Covers.** Suitable covers must be installed to cover these boxes, and a means of securely fastening the cover in place shall be provided. If underground box covers weigh over 100 pounds, the weight of the cover shall be considered equivalent to fastening the cover in place. It is essential that a permanent marking be made on the outside of the box cover that says "WARNING—HIGH VOLTAGE—KEEP OUT." Half-inch-high block-type letters shall be used, and they shall be readily visible.

(F) **Suitable for Expected Handlings.** Both the box and cover must be made so that they will withstand any handling to which they may be subjected.

Article 320—Armored Cable: Type AC

I. General

320.2: Definition

Type AC cable is a flexible metallic enclosure with circuit conductors installed at the time of manufacturing. The *NEC* refers you to Section 320.100.

II. Installation

320.10: Uses Permitted

Unless otherwise specified in the *NEC* and if not subject to physical damage, Type AC cable is permitted in both exposed and concealed work for both branch circuits and feeders. Type AC cable may be embedded in plaster that is applied to brick or masonry. It is allowed in cable trays, where approved for such use. It may be run or fished in the air voids of masonry block or tile walls; where such walls are exposed or subject to excessive moisture or dampness or below grade line, Type ACL cable shall be used.

320.12: Uses Not Permitted

The *Code* specifies other locations where AC cable can't be used, including the following locations:

(1) In theaters, except as permitted in Article 518.

(2) In motion picture studios.

(3) In hazardous locations, except as allowed by Section 501.4(B) and 504.20.

(4) In areas where it might be exposed to corrosive fumes or vapors.

(5) In storage battery rooms.

(6) On elevators or in hoistways, except as allowed by Section 620.21.

(7) In commercial garages, as prohibited by Article 511.

320.15: Exposed Work

The cable on exposed work shall follow the surface of the building finish or of running boards. Please remember the limitations on bends and don't try to form the cable to a right angle, as this would damage the cable.

320.17: Through or Parallel to Framing Members

The same protection must be taken as outlined in Section 300.4 to protect the cable from being punctured by nails, and so on.

320.23: In Accessible Attics

Where run across the top of floor joists or within 7 feet of the floor or floor joists across the face of rafters or studding in the attics and roof spaces that are accessible, the cables are to be protected by substantial guard strips that are as high as the thickness of the cable. Where such spaces are only accessible by scuttle holes or the equivalent, protection is required within 6 feet of the scuttle hole. Where cable is run along the sides of rafters, studs, or floor joists, guard strips and running boards aren't be required.

This may sound like a lot of extra work for space that will probably not be used. Remember, though, that plans change. The attic may become a storage space or room, in which case damage to the cable may result and the work will have to be redone.

320.24: Bending Radius

For Type AC cable, the bend radius is taken from the inner edge of the bend and shouldn't be less than five times the diameter of the cable.

320.30: Securing and Supporting

AC-type cable shall be secured at intervals not to exceed 4½ feet and within 12 inches of a box or fitting, except where fished and except that the 12 inches may be extended to not over 24 inches at terminals where there is a necessity for flexibility. See the *NEC* for other exceptions.

320.40: Boxes and Fittings

(A) Approved fittings suitable for Type AC cable shall be used at all terminals where Type AC cable is used.

(B) For Type AC cable, approved fittings to prevent abrasion of the conductors and their insulation shall be used. In addition to this, an approved insulating bushing must be inserted at the end between the conductors and the outer metallic covering. The connection to the fitting or box must be designed in such a way that the insulated bushing will be visible for inspection without removing the fitting. This is an excellent requirement, because it seems that in the haste of installation, this bushing is so often overlooked, and it is a common spot for a breakdown in the insulation. Any splices or connections to other insulated bushing are not required where a lead covering is used on the conductors, such as ACL type. Any splices or connections to other raceways must be in approved junction boxes.

III. Construction

320.100: Construction

AC cable is approved. It has an acceptable extra metal covering, and insulated conductors inside the metal covering.

AC cable should be used for branch circuits and feeders. The armor must be flexible metal tape. AC cable shall have a bonding strip that touches the outer flexible metal covering for the full length. Where the bonding strip is required in regular AC cable, it shall make good contact with the outer sheath for the full length.

320.104: Conductors

The types of conductor permitted in AC cable are listed in Table 310.13. If not listed there, they must be identified for use in AC cable. Around all the conductors in the cable a moisture-resistant and fire-retardant covering should be applied. This is a fibrous-type covering. The ampacity of the conductors is determined in Section

310.15. If the cable is Type ACT, moisture-resistant fibrous covering shall be wrapped around each conductor.

Exception

When Type AC cable is installed in thermal insulation, the insulation of the individual wires must be rated for 90°C, but the ampacity for those same wires must be rated as if the wires were only rated for 60°C.

320.108: Equipment Grounding

AC cable must provide an adequate grounding path. See Section 250.4.

320.120: Marking

Marking shall be required, as described in Section 310.11. The maker of the AC cable must be distinctively marked on the cable sheath throughout its entire length.

Article 322—Flat Cable Assemblies: Type FC

This product was new with the 1971 *NEC*. As an introduction, it might be a good idea to give some of the submitter's supporting comment as it appeared in the 1971 preprint:

This new article is basically a busway system. In an approved busway system, all of the basic components are factory assembled. In this proposed wiring system, the basic components are intended to be field assembled.

The conductors are formed into a flat cable assembly of three or four conductors and are completely encased in an insulating material, properly spaced. Special spacing insulation is extruded integrally with the cable assembly to facilitate the location of the cable within the metal raceway. Refer to Section [322.2] of the NEC.

Article [322] (Wireways, Chapter 3) more or less completes the requirements for a class of wiring systems intended for use and field assembly of standard components. Article [368] (Busways) more or less begins the requirements for wiring systems consisting of completely factory-wired assemblies.

This proposed Article [322] is a transition from the field assembly of standard components to the field of completely wired factory installations.

322.2: Definition

FC cable is an assembly of parallel conductors imbedded in insulating material, specifically designed for installation in surface metal raceways.

II. Installation

322.10: Uses Permitted

Type FC cable is for use only with branch circuits to supply tap devices that are suitable for lighting, small appliances, or small power loads, and is to be used only in installation for exposed work, where not subject to severe physical damage.

FC cable shall be installed in the field only in metal surface raceways that are identified for this purpose. The surface metal raceway must be installed as a complete system before any of the FC cable is pulled into the raceway.

When installed within 8 feet (2.44 m) from the floor, FC cable must be protected with a metal cover.

322.12: Uses Not Permitted

It shouldn't be installed where there are corrosive vapors unless it is suitable for the location. It shouldn't be installed outdoors or in wet or damp locations unless identified. It can't be used in hazardous (classified) locations or hoistways.

Branch-circuit rating may not be over 30 amperes.

322.30: Securing and Supporting

FC cable's special design features are to be used for mounting the FC cable assemblies within the surface metal raceways. Instructions for supporting the surface metal raceways will differ, and only the supporting means for a specific raceway shall be used.

322.40: Boxes and Fittings

(A) **Dead Ends.** Dead ends shall be covered with only approved end caps. The metal-raceway dead ends shall be closed only by identified fittings.

(B) **Fixture Hangers.** Fixture hangers from flat cable assemblies shall be installed only with identified fittings.

(C) **Fittings.** All fittings used with FC cable shall be designed and the insulation made to prevent physical damage to the cable assemblies.

(D) **Extensions.** All extensions from flat cable assemblies shall be made from the terminal blocks enclosed within the junction boxes, installed at either end of the flat cable assembly runs.

322.56: Splices and Taps

(A) **Splices.** Splices shall be made only in junction boxes only by approved wiring methods for making the splices.

(B) Taps. Taps between phase and neutral or other phases may be made by the use of fittings and devices that are approved for the purpose, and tap devices shall be rated not less than 15 amperes and not more than 300 volts. Color-coding is required, and is covered in Section 363.20.

III. Construction

322.104: Conductors

FC cable may be in assemblies of two, three, or four conductors. This comes in only one size, namely No. 10 AWG special-stranded copper wires.

322.112: Conductor Insulation

Materials of which the cable is to be made and that form the insulation for all the conductors shall be material found in Table 310.13 for general-purpose branch-circuit wiring.

322.120: Marking

In addition to the requirements for the marking of conductors in Section 310.11, FC cable must be marked at least every 24 inches, showing its temperature rating.

The grounded conductor (neutral) must be identified with a white or gray color.

Terminal blocks used with FC systems must be marked with a color or word coding. The common coding is white for neutral, black for A phase, red for B phase, and blue for C phase.

Article 324—Flat Conductor Cable: Type FCC

I. General

324.1: Scope

This article covers cable and the necessary associated accessories, which are defined in this article. This cable was designed for use under carpet squares.

324.2: Definitions

Type FCC Cable. Take note of the fact that the conductors are flat and are to be only of copper. Three or more flat copper conductors that are placed edge to edge and entirely enclosed with an assembly of insulation make up FCC cable.

FCC System. FCC systems are designed for complete branch circuits, which, as we stated, are designed to be installed under carpet squares. FCC systems should be complete with all the necessary fittings, including associated shielding, the connector terminals, adapters, and receptacles that are designed only for the Type FCC cable. Standard fittings, and the like should not be used. All of the items must be properly identified for the purpose for which they are used.

Cable Connector. FCC cable is designed for use where installations join connectors without using junction boxes.

Insulated End. The ends of FCC cable shall have an insulator for the ends designed for the purpose.

Top Shield. The circuit components of an FCC system must have a grounded metal shield that covers the flat under-carpet conductors. This is for protection against physical damage.

Bottom Shield. Protective layer installation must be made between the flat conductor cable and the floor. This may or may not be an integral part of the cable; it is required for protection from physical damage.

Transition Assembly. There shall be approved assembly for the connection of the FCC to other approved types of wiring. This may be an approved means of making the interconnection, or it may be a box or covering that will provide electrical safety and protect all conductors from physical damage.

Metal Shield Connections. When metal shields are connected to other metal shields, transition assemblies, receptacle housings, or self-contained devices, the connections must be mechanically and electrically sound.

324.3: Other Articles

Other articles that may be applicable include some of the provisions of Articles 210, 220, 240, 250, and 300.

324.4: Uses Permitted

(A) **Branch Circuits.** Type FCC installations may be used with individual branch circuits for general purpose and appliance branch circuits.

(B) **Floors.** FCC cable may be used on floors that are of sound construction and smooth. The continuous floor surface may be made of many different materials, such as ceramic, concrete, wood, or composition flooring, or other similar materials.

(C) **Walls.** If metal surface raceways are used on walls, Type FCC systems are permitted.

(D) **Damp Locations.** Since rugs are sometimes cleaned with a liquid cleaner, Type FCC systems must be approved for damp locations.

(E) **Heated Floors.** If floors are heated in excess of 30°C (86°F), the material used for the heated floors must be identified as being suitable for use at these temperatures.

324.5: Uses Not Permitted

FCC systems shouldn't be used in the following places: They are not to be installed out of doors, where they will be in wet locations; wherever subject to corrosive vapors; in hazardous (classified) locations; or in residences, hospitals, or schools.

324.6: Branch-Circuit Ratings

(A) **Voltage.** Ungrounded conductors used in FCC installations shouldn't be permitted to be over 300 volts between the ungrounded conductors, and not over 150 volts between the ungrounded conductor and the grounded conductor.

(B) **Current.** Individual branch circuits may be rated up to 30 amperes for appliance and general-purpose circuits. FCC installations shouldn't exceed 20 amperes.

II. Installation

324.10: Coverings

The FCC-type insulation, including all parts on the floor, shall be covered by carpet squares that are not larger than 3 feet square. These squares shall be held down by adhesive, and this shall be a release-type of adhesive.

324.11: Cable Connections and Insulating Ends

All FCC cable shall use connectors that are identified for the use. The installation shall have electrical-continuity insulation and dampness sealing from spilled liquids; it is up to the installer to see that all of these are provided. Also, the cable ends must be thoroughly sealed against dampness and spilled liquids. Only listed insulation ends should be used.

324.12: Shields

(A) **Top Shield.** See the definition at Section 342.2 of this book. Refer to Section 328.3 and you will find that Article 250 is mentioned; this article covers grounding.

(B) Bottom Shield. See the definition in Section 324.3 of this book.

324.13: Enclosure and Shield Connections

The equipment-grounding conductor of the supply branch circuit shall be electrically connected to all shields, boxes, receptacle housings, and self-contained devices. Identified connectors only shall be used for making this electrical continuity. In this electrical equipment grounding that is required, the electrical resistivity of the sealed system shouldn't be more than the electrical resistivity of one of the regular conductors used in the installation of Type FCC cable. Here, again, grounding of the shield is mentioned, as in Section 324.12. To properly check the resistivity of the shielding and compare it with the resistivity of one conductor requires a very-low-reading ohmmeter, or Ducter.

324.14: Receptacles

All receptacles, including the receptacle housing and self-contained devices that are used in the installation of any FCC system, shall be positively identified for the purpose you are using them for. A good connection shall be made to the Type FCC cable and metal shields. At each receptacle, a connection to the grounding conductor of Type FCC cable shall be made to the grounded shielding at each receptacle.

324.15: Connection to Other Systems

The transition system for connecting the FCC system to other systems where they originate shall be the only system identified for this purpose.

324.16: Anchoring

For anchoring the FCC system to the floor, adhesive or mechanical anchoring may be used, but they shall be identified for this purpose. The floors shall be in such shape that the FCC system will be firmly anchored to the floor until the carpet squares are properly in place.

324.17: Crossings

Only two Type FCC cables are allowed to cross at any point. It is also allowable for a flat communications cable or signal cable to cross over or under a Type FCC cable. In all instances, a grounded metal shield must separate the two cables.

324.18: System Height

Tapering or feathering at the edges of FCC systems is required if the height above the floor level exceeds 0.090 inches (2.29 mm).

324.19: FCC Systems Alterations

Alterations in FCC systems may be made. Added cable connectors may be used at new connection points if alterations are needed. It is permissible to leave unused portions of the FCC-type system in place and they may still be energized, but all cable ends shall be covered with insulation ends.

324.20: Polarized Connections

All connections and receptacles must be installed so that the polarization of the system is maintained.

III. Construction

324.30: Type FCC Cable

Only approved Type FCC cables shall be used when they are approved for use with FCC systems. They may consist of three, four, or five flat copper conductors. One of these shall be the equipment-grounding conductor. Only moisture-resistant and flame-retardant insulation shall be used.

For Sections 324.31, "Markings," 324.32, "Conductor Identification," 324.33, "Corrosion-Resistance," 324.24, "Insulation," 324.25, "Shields," 324.36, "Receptacles and Housings," and 324.37, "Transition Assemblies," refer to the *NEC*.

Article 326—Integrated Gas Spacer Cable: Type IGS

IGS cable is a factory assembly of individually insulated conductors enclosed in a nonmetallic conduit that is filled with pressurized gas. See the *NEC* for details.

Article 328—Medium Voltage Cable: Type MV

328.1: Definition

Type MV cable stands for medium voltage cable. It may be multiconductor or a single conductor that is a solid dielectric cable. The insulation rating of MV cable is 2001 volts or higher.

Later in this article you will find this covers Type MV cable rated up to 35,000 volts.

328.2: Other Articles

To conform to other *Code* requirements, this article on Type MV cable must also meet other requirements in the *Code*, especially Articles 300, 305, 310, 392, 501, and 710.

328.3: Uses Permitted

Type MV cable is permitted for many purposes: It can be used for up to 35,000 volts, nominal, for power systems. It may also be used in wet or dry locations, or in raceways or cable trays; this part is covered in Section 392.2(B). It may be used for direct burial if installed as per Section 710.3(B), and it may also be used in messenger-supported systems.

328.4: Uses Not Permitted

This type of cable shouldn't be used when exposed to direct sunlight, unless specifically identified for that purpose. Unless specified for this purpose in Section 392.2(B), it shouldn't be used in cable trays.

328.5: Construction

Copper, aluminum, or copper-clad aluminum conductors may be used in MV cable, and the construction shall conform to the requirements of Article 310.

328.6: Ampacity

Section 310.15 gives the allowable ampacity when Type MV cable is used.

Exception

In Section 392.13, when Type MV cable is installed in cable trays.

328.7: Marking

As with many other cables, the marking on Type MV shall conform to Section 310.311.

Article 330—Metal-Clad Cable: Type MC

I. General

330.2: Definition

MC cable is factory-produced and is composed of one or more conductors, each of which is individually insulated. These insulators are enclosed within a metal enclosure or interlocking tape, or a smooth or corrugated tube.

II. Installation

330.10: Uses Permitted

Unless otherwise specified in the *Code*, and if not subject to physical damage, the following uses for Type MC cable are permitted:

(1) Branch circuits, feeders, and services.

(2) Signal circuits, control circuits, and power and lighting.
(3) Indoors or outdoors.
(4) When concealed or exposed.
(5) If identified for the use, may be used for direct burial.
(6) Installed in cable trays.
(7) Installed in any approved raceway.
(8) Open runs of cable.
(9) On messenger cable as an aerial conductor.
(10) As provided in Articles 501, 502, 503, and 504 in hazardous (classified) locations.
(11) In dry locations.
(12) If any of the following conditions are met, it may be installed in wet locations:
 (a) The metallic covering won't be damaged by moisture.
 (b) A water-resistant lead covering is applied under the metallic covering.
 (c) The insulation of the conductors is approved for wet locations.

Note
Section 300.6 covers corrosion protection.

Note
The conditions that must be met in (12) above must also apply to (5) above.

2005

330.12: Uses Not Permitted
MC cable shouldn't be used where corrosive conditions exist. One of the corrosive conditions is direct burial in earth, causing electrolysis of the metal sheath or cinder-fills, which will consume the metal sheath rather rapidly. Some of the corrosive chemicals are hydrochloric acid, chlorine vapor, strong chlorides, and caustic alkalis. In addition, Type MC cable may not be used where exposed to physical damage.

Exception

If the metal sheath is proved capable of withstanding any of the mentioned corrosive substances, it may be made of material suitable for the conditions.

It leaves much up to the inspection authority to judge if the material is suitable for the conditions.

330.17: Through or Parallel to Framing Members

MC cables passing through or parallel to framing members must comply with Section 300.4.

330.23: In Accessible Attics

MC cables that are installed in attics or roof spaces must fulfill the requirements of Section 320.23, which lists the rules for AC cable.

330.24: Bending Radius

Bends in MC cable must be made so that the cable is not damaged. The radius of the curve of the inside bend shouldn't be less than is shown below:

(A) Smooth Shcath.

- **(1)** MC cable not more than ¾ inch in diameter shall have a bending radius of a minimum of 10 times the diameter of the cable.
- **(2)** For MC cable more than ¾ inch in diameter but not more than 1½ inches in diameter, the bending radius shall be a minimum of 12 times the diameter of the cable.
- **(3)** For MC cable more than 1½ inches in diameter, the bending radius shall be a minimum of 15 times the diameter of the cable.

(B) **Interlocked-Type Armor or Corrugated Sheath.** The bending radius shouldn't be less than seven times the diameter of the cable, measured for the radius from the inside of the cable.

(C) **Shielded Conductors.** Bending radius may be 12 times the diameter of one of the individual conductors, or the bending radius seven times the diameter of the multiconductor cable, including the sheath. Whichever one of the two gives the larger bending radius shall be used.

330.30: Securing and Supporting

The intervals of support of MC cable shouldn't exceed 6 feet, except where fished in finished walls. In addition, MC cables that are used for residential branch circuits must be supported within 12 inches of a box.

330.31: Single Conductors

Where single conductors are used, their installation must conform to Section 300.22.

330.40: Boxes and Fittings

All fittings shall be identified for use in connecting Type MC cable to boxes, cabinets, and so on. If single conductor cables enter ferrous metal boxes or cabinets, Section 300.20 shall be complied with to prevent inductive heating.

330.80: Ampacity

The ampacity of Type MC cable is controlled by Sections 310.15 or 310.60.

Exception No. 1

Type MC cable ampacities installed in cable trays must be in conformance with Sections 392.11 and 392.13.

Exception No. 2

No. 16 or No. 18 conductors in MC cables must have their ampacity determined according to Table 402.5.

Note

Temperature limitations of the ampacity of the conductors will be found in Section 310.10.

III. Construction Specifications

Refer to the *NEC*, Sections 330.104 through 330.116.

Article 332—Mineral-Insulated, Metal-Sheathed Cable: Type MI

I. General

332.1: Definition

Mineral-insulated, metal-sheathed Type MI cable is composed of one or more conductors insulated by a highly compressed refractory mineral insulation and enclosed in a gastight metal-tube sheath.

The highly compressed refractory mineral insulation is magnesium-oxide powder. The conductors are normally copper. MI cable is also made up into a heating cable in which the outside or sheath is a seamless phosphorized copper tubing. As a word of explanation, magnesium oxide is the material used in many range burners as the insulation in the tube elements to enclose and insulate the nichrome elements from the outer sheath of the units (Figure 332-1). MI cable may be used only with approved fittings for terminating and connecting boxes, outlets, and so on (Figure 332-2).

332.2: Other Articles

As with other wiring materials, Type MI cable must comply not only with this article, but also with the provisions of Article 300 in the *Code*.

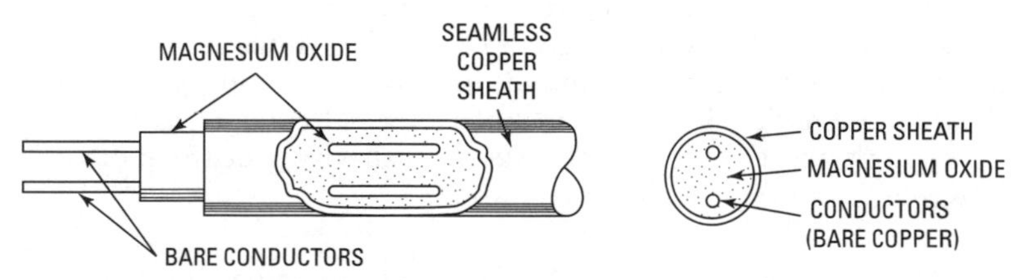

Figure 332-1 Construction of MI cable.

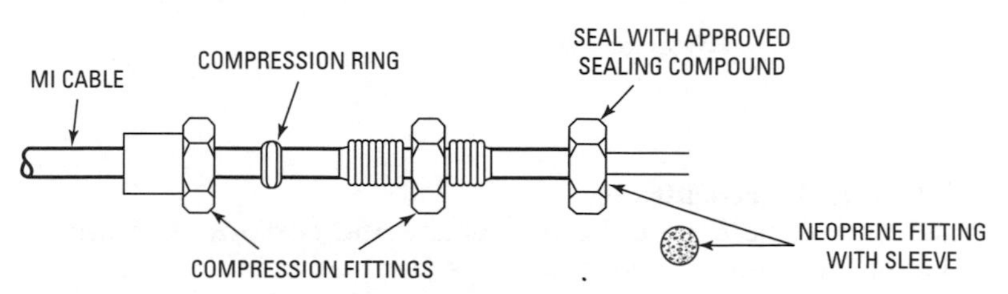

Figure 332-2 Fittings for MI cable.

II. Installation

332.10: Uses Permitted

Mineral-insulated metal-sheathed MI cable may be used for (1) services, (2) feeders, (3) branch circuits, (4) open wiring (exposed), (5) concealed wiring, and (6) wet locations.

MI cable may be used in practically all locations except that, if used in a highly corrosive location (such as cinder-fill), it shall be protected. It may be used in all of the following situations: (1) all hazardous areas (Class I, Class II, and Class III locations); (2) under plaster extensions; (3) embedded in plaster finish; (4) on brick or masonry; (5) when exposed to weather or continuous moisture; (6) embedded in concrete or fill; (7) buildings under construction; (8) when exposed to oils and gasoline; and (9) any other location that won't have a deteriorating effect on the copper sheath.

The following is not a part of the *Code*, but it is felt that it should be mentioned here as a precaution in the installation of MI cable: Magnesium oxide will draw some moisture when exposed to the air. Therefore, when the cable is cut, the end should be taped. When using a piece of MI cable, about 6 inches should be cut off from the end, or start back from the cut end a couple of feet and heat it with a torch, working toward the cut end, thus driving any moisture out of the cable (Figure 332-3). The temperature to which the cable may be heated is limited only by the melting point of the sheath. In Article 310 it is mentioned that MI cable is 90°C cable. This won't prevent the use of a torch in driving out the moisture, however.

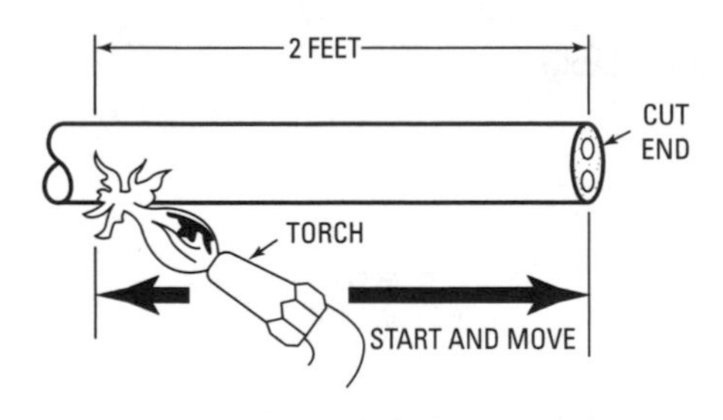

Figure 332-3 One method of driving out moisture.

332.12: Uses Not Permitted

Where it will be exposed to destructive and corrosive conditions, MI cable may not be used.

Exception

If the MI cable is made of material suitable for the condition.

Exception
Fished-in MI cable.

332.17: Through or Parallel to Framing Members

Reference is made to Section 300.4, which permits notching, providing a steel plate of not less than 1⁄16 inch is provided so as to give protection from nails.

332.24: Bending Radius

All bends shall be so made that the cable won't be damaged, and the radius of the curve of the inner edge of any bend shall be not less than five times the diameter of cable 3⁄4 inch or less in diameter (see Figure 332-4). Ten times cable diameter is permitted for cables of greater than 3⁄4-inch diameter.

332.30: Securing and Supporting

Mineral-insulated metal-sheathed cable shall be securely supported by approved staples, straps, hangers, or similar fittings, and so designed and installed as not to injure the cable. Cable shall be secured at intervals not exceeding 6 feet (1.83 m), except where cable is fished or in cable trays. When MI cable is used for its fire rating, it must be supported every 3 feet in horizontal runs, and every 6 feet in vertical runs. Please note the term "approved." Consideration must always be given to electrolysis, which is ever present between dissimilar metals.

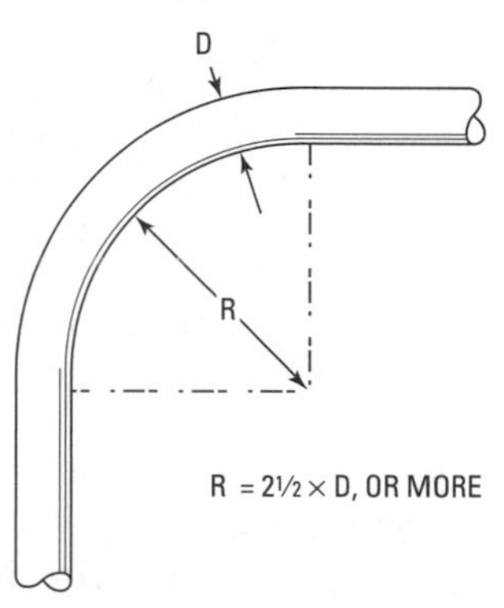

Figure 332-4 Radius of bends.

332.31: Single Conductors

When single conductor MI cables are used, the voltages induced on their sheaths must be minimized. This is done by grouping all cables of circuits together, including neutral conductors of these circuits. Where these conductors enter metal enclosures, induction heating must be prevented (see Section 300-20).

332.40: Boxes and Fittings

(A) Fittings. Identified fittings shall be used with MI cable for the conditions of service. Type MI cable of single conductor must be used when entering metal boxes, as outlined in Section 300.20, to prevent induced currents from causing heating.

(B) **Terminal Seals.** Where MI cable is terminated, approved seals shall be used. Refer to Figure 330-2. These seals have a compression ring and fittings similar to those used on copper tubing. In addition, there is a neoprene bushing that slips into these fittings and is sealed with an epoxy resin, after which a neoprene bushing with holes for the conductors and the sleeves is inserted. This sealing should be done immediately after stripping or heating so as to prevent the entrance of any moisture. The bare conductors are, of course, insulated with sleeving, the color of which must be correct for the particular use of the conductor.

MI cable of the heating type is often used for ice and snow melting. When used for these purposes, an insulation (Megger®) test should be made to check the insulation resistance of the conductors to the metal shield to see that proper insulation values are maintained. The same test should be run after the cable is embedded in the concrete slab. A case in point found small holes in the metal sheath that permitted moisture to enter the cable. The concrete had to be removed and new cable installed because the installation was worthless.

332.80: Ampacity.

The ampacity for MI cable is determined by referencing Section 310.15.

(1) MI cables in cable trays have their ampacities determined by Section 392.11

(2) Single MI cables can be Free-Air-rated per Table 310.17 if they are grouped together per Section 332.31, installed on messenger wires, and have air spaces of at least 2.15 times the outer diameter of the largest conductor.

III. Construction Specifications

Note

The *Code* does not specify the sizes that are available in MI cable, and so the following information in Table 332-1 is included for the reader's convenience:

Table 332-1 Sizes Available in MI Cable

Cable	*Size Available*
Single-conductor	No. 16-4/0
Two-conductor	No. 16-No. 4
Three-conductor	No. 16-No. 4
Four-conductor	No. 16-No. 6
Seven-conductor	No. 16-No. 10

Remember that No. 14 wire is the smallest allowable under the *Code* for general wiring; smaller sizes may be used for low-voltage thermostat wiring, as permitted. There is also MI low-energy cable rated at 300 volts available in two-, three-, four-, and seven-conductor cable, but a check should be made to be certain that this low-energy cable has been tested by UL.

Mention was also made of heating cable. This is not considered a wiring cable, and so it would have to be tested by UL for the purpose for which it is to be used.

332.104: Conductors

Only copper conductors are to be used in MI cable, and they shall be of standard AWG sizes.

332.112: Insulation

The insulation between the conductors and the MI cable must be highly compressed, of a refractory mineral, and provided with proper spacing of the cables.

332.116: Outer Sheath

Copper is used for the outer sheath and shall be of continuous construction, provide mechanical protection, provide moisture seal, and the sheath be used for grounding purposes, but an adequate continuous path must be maintained. MI cable that has a stainless steel outer sheath must have an equipment-grounding conductor sized in accordance with Table 250.95 and Section 250.57(B).

Article 334—Nonmetallic-Sheathed Cable: Types NM and NMC

I. General

334.2: Definition

Nonmetallic-sheathed cable is an assembly of two or more conductors that has an outer sheath of moisture-resistant, flame-resistant, nonmetallic material. This is commonly known as Romex.

Nonmetallic-sheathed cable shall be of an approved type. This type of cable is available in sizes from No. 14 to and including No. 2. This cable may have an uninsulated conductor or a green insulated conductor for equipment-grounding purposes in addition to the current-carrying conductors. Table 250.95 in the *NEC* lists the size of this grounding conductor in reference to the current-carrying conductors.

Type NM cable is the type most commonly used, especially in residential occupancies, but there is an NMC type that has an overall covering that is not only flame-retardant and moisture-resistant, but also fungus-retardant and corrosion-resistant.

There must be a distinctive marking on the exterior of the cable for its entire length that specifies cable type and the name of the manufacturing company. The box in which the cable is packaged will have the UL listing, if any, on it. NM cable is also made in No. 12 through No. 2 sizes for aluminum conductors.

A new type of cable, NMS, has been added to the 1996 edition of the *NEC*. NMS is a composite type of cable that contains both line voltage and low-voltage conductors under the same cable sheath. It is designed specifically for intelligent home systems and for closed-loop systems in particular (see Article 780).

II. Installation

334.10: Uses Permitted

Unless not permitted in Section 334.12, Types NM, NMC, and NMS cable may be used in one- and two-family dwellings and other structures. When installed in cable trays, NM cable must be listed for such use.

Note

Section 310.10 covers temperature limitations for the conductors in NM and NMC cable.

Nonmetallic cable may be used for the following:

- **(A)** **Type NM.** Type NM cable may be used for concealed or exposed installation, installed or fished in the hollow voids of masonry or tile walls that are not exposed to excessive moisture or dampness, or installed in voids in masonry block or tile walls where such walls are not exposed or subject to excess moisture or dampness.
- **(B)** **Type NMC.** Moisture- and corrosion-resistant Type NMC cable may be used for both exposed and concealed work in dry, moist, damp, or corrosive locations, and in outside or inside walls of masonry block or tile. When NMC cable is

installed in a shallow chase of concrete, masonry, or adobe, it must be protected by a 1/16-inch steel plate.

(C) **Type NMS.** This type of cable is permitted exposed or concealed in dry locations. It can be fished into masonry voids unless the locations have excessive moisture.

334.12: Uses Not Permitted

(A) **Types NM, NMC, and NMS.** The uses not permitted include service-entrance cables, in commercial garages, in theaters and assembly halls (except as provided in Article 518), in motion-picture studios, in storage-battery rooms, in hoistways, in hazardous locations, or embedded in poured cement, concrete, or aggregate. The first floor of a building is the floor that has 50 percent or more of its outside wall surface level with or above finished grade. In addition to the three levels above grade specified in this subsection, one additional level for vehicle storage, and so on (not for habitation), is permitted.

(B) **Types NM and NMS.** Types NM and NMS cable may not be used where exposed to corrosive vapors or fumes; embedded in masonry, concrete, fill, or plaster; run in shallow chase in masonry or concrete and covered with plaster or similar finish; for direct burial; or in adobe or similar finish.

334.15: Exposed Work

In exposed work, the cable is to be installed as follows: It shall follow the surface of the building finish or of running boards. It shall be protected from physical damage where necessary by conduit, intermediate metal conduit, pipe, guard strips, or other satisfactory means, and when passing through a floor, it shall be protected by conduit or pipe to a minimum height of 6 inches above the floor.

334.17: Through or Parallel to Framing Members

See Section 300.4.

334.23: In Accessible Attics

Where run across the top of floor joists, or within 7 feet of the floor or floor joists across the face of rafters or studding in attics and roof spaces that are accessible, the cables should be protected by substantial guard strips that are as high as the thickness of the cable. Where such spaces are only accessible by scuttle holes, or the equivalent, protection is required for any cable within 6 feet of the scuttle hole. Where the cable is run along the sides of rafters, studs, or floor joists, guard strips or running boards are not required.

This may sound like a lot of extra work for space that will probably not be used. Remember, however, that plans change. The attic may become a storage space or room, in which case damage to the cable may result and the work will have to be redone.

334.24: Bending Radius

All bends should be made so as not to damage the cable or its protective covering, but no bend shall have a radius of less than five times the diameter of the cable.

334.30: Securing and Supporting

Approved staples, straps, or other fittings shall be used to support nonmetallic-sheathed cable and shall be so designed and installed so as to not cause damage to the cable. Too often it is the last blow of the hammer that should never be given. Inspectors find much insulation damaged in this fashion. It is also easy to use a nail bent over to secure a cable. This should never be tolerated.

Cables shall be secured at intervals not exceeding 4½ feet and within 12 inches of the box or fitting, except that in concealed work in finished buildings or finished panels of prefabricated buildings, it may be fished into the walls. Two conductor cables may not be stapled on edge. Passing through a joist, rafter, and so on, is considered to be a support. Insulated staples must be used for cables with conductors smaller than No. 8. Cable ties are acceptable as a supporting means.

Exception No. 1

The cable may be fished between accessible points where installed as concealed, worked in finished panels, or when the buildings are prefabricated. You can see that fastening would be impractical.

Exception No. 2

Some fitting devices are identified for installation of NM or NMC cable without outlet boxes. They are equipped with the cable clamp. The cable must be secured in place at intervals not exceeding 4½ feet and within 12 inches of this device. There shall also be an unbroken loop of cable 6 inches or more that will become available on the interior side of a finished wall to provide extra cable if replacements are made or some of the conductors are broken at the device.

If the cable is run at an angle to joists in unfinished basements, assemblies not smaller than two No. 6 or three No. 8 conductors may be fastened directly to the lower edges of the joists. Small assemblies shall be either run through bored holes or on running boards. Cables of any size that are run parallel to the joists may be attached to the sides of the joists or to the face.

334.40: Boxes and Fittings

Nonmetallic outlet boxes are approved for nonmetallic cable and provision for their use is made in Section 314.3.

334.80: Ampacity

The ampacity of NM cables is determined by the temperature rating of the cables and by Section 310.15.

III. Construction Specifications

334.100: Construction

For most of this section, see the *NEC*. In copper of small sizes No. 14 AWG and for copper-clad aluminum of small sizes No. 12 AWG, there shall be an equipment-grounding conductor in these cables. It may be bare or insulated.

- **(A)** Types NM and NMS. These cables must be flame-retardant and moisture-resistant.
- **(B)** Type NMC. With NMC cable, the overall covering shall be flame-retardant, moisture-resistant, fungus-resistant, and corrosion-resistant.

334.104: Conductors

The conductors used in these cables are to be of types listed in Table 310.13, where they are suitable for branch-circuit wiring. Other types are permitted if identified for use in the cables.

Although the conductors in the cable shall have a rating of 90°C (194°F), the ampacity of NM and NMC cables shall be rated only at 60°C (140°F), and Section 310.15 shall be complied with.

Article 336—Power and Control Tray Cable: Type TC

336.2: Definition

TC cable is a factory-made cable assembly most commonly used in cable trays, but also supported by messenger wires or in raceways.

II. Installation

336.10: Uses Permitted

Type TC cable may be used where lighting and power, control and signal, and communication circuits may be used in cable trays and raceways outdoors, where supported by messenger wire. In some cases it may be used as stated above in hazardous (classified) areas; these cases are covered in Articles 392 and 501, and they will be for industrial establishments in which maintenance, servicing, and installation are done by only qualified persons. It may be used in Class I locations, which is covered in Article 725.

Note

The temperature limitations of TC cable come under Section 310.10.

336.12: Uses Not Permitted

Cable tray for TC cable shouldn't be installed where it will be subject to physical damage. It can't be mounted on cleats or brackets in an attempt to use it for open wiring, which is prohibited. It must be listed for exposure to direct sunlight if used in such a location. Direct burial is not permitted unless the TC cable is specifically listed for that use.

336.24: Bending Radius

Whatever bends are made in TC cables must be made in such a way that the cable is not damaged. Bending radii vary from four-times cable diameter for cables of one inch or less, to five-times diameter for cables between 1 and 2 inches, and six times cable diameter for cables over 2 inches.

336.80: Ampacity

If the TC tray cable is smaller than No. 14 AWG, its ampacity is covered in Sections 400.5 and 392.11.

III. Construction Specifications

336.100: Construction

The following covers Type TC tray cable:

(1) Insulated conductors shall be sizes 18 AWG through 1000 kcmil copper; sizes 12 AWG through 1000 kcmil aluminum or copper-clad aluminum.

(2) Insulation: Insulated conductors (copper) in sizes 14 AWG and larger and (aluminum) sizes 12 AWG and larger, including copper-clad aluminum, shall have insulation of one of the types listed in Table 310.13. The insulation must be suitable for branch-circuit and feeder conductors. Insulated conductors of size 18 and 16 AWG copper shall be in accordance with Article 726.16.

(3) The outer sheath shall be a flame-retardant, nonmetallic material. Where installed in wet locations, Type TC cable shall be resistant to moisture and corrosive agents.

Article 338—Service-Entrance Cable: Types SE and USE

There has been considerable misuse of service-entrance cables. It will be attempted here to demonstrate some of these abuses so that you may secure a better understanding of the uses for which service-entrance cable is intended.

338.2: Definition

Service-entrance cable may be a single conductor or a multiconductor assembly. It is usually supplied with an outer covering, but it may be supplied without the covering. Its primary use is for services. The following are the different types:

(A) **Type SE.** A flame-retardant and moisture-resistant covering is required in Type SE cable. This may be one of two types—one with a bare neutral conductor and the other with all conductors insulated and with or without a grounding conductor for the purpose of grounding equipment. This is mentioned because it is very important, as explained in the following information.

Type SE cable has no inherent mechanical protection against abuse. Therefore, it may be necessary to provide mechanical protection should the *Code*-enforcing authority require it.

(B) **Type USE.** This is a direct burial cable and is recognized for use underground. The insulation of the USE shall be moisture-resistant, but not flame-retardant. It is intended that USE cable be a conductor assembly provided with a suitable overall covering. This brings up the point of single conductors that might be marked "RHW-USE." Most certainly, if the *Code*-enforcing authority approves these conductors for direct burial, they will demand that they be buried side-by-side and possibly require other provisions in their installation, such as mechanical protection. If the conductors are marked USE, they are identified for use underground for direct burial.

Single-conductor-type USE cable, which has been recognized for underground use, may have bare copper conductor cabled with it in the assembly. If USE is single-conductor paralleled or cabled conductors recognized for underground burial, a bare copper conductor may be applied. These constructions are not required to have an outer overall cover.

(C) **One Uninsulated Conductor.** When USE cable or SE cable consists of two or more insulated conductors, they may have a bare conductor. This is wrapped around the insulated conductors and serves as a neutral conductor.

338.10: Uses Permitted As Service-Entrance Conductors

When SE cable is used for service-entrance conductors, the requirements of Article 230 shall be adhered to. (See Figures 338-1 through 338-6.)

USE conductors are permitted as service laterals above-ground outside of buildings. [See Section 300.5(D) regarding protection of these conductors.]

338.12: Uses Permitted as Branch Circuits or Feeders

(A) Grounded Conductor Insulated. SE cables may be used for branch circuits and feeder on interior wiring, where all of the circuit conductors of the cable are of rubber-covered or thermoplastic type. Where an equipment-grounding conductor is required, it may be a bare conductor in the cable.

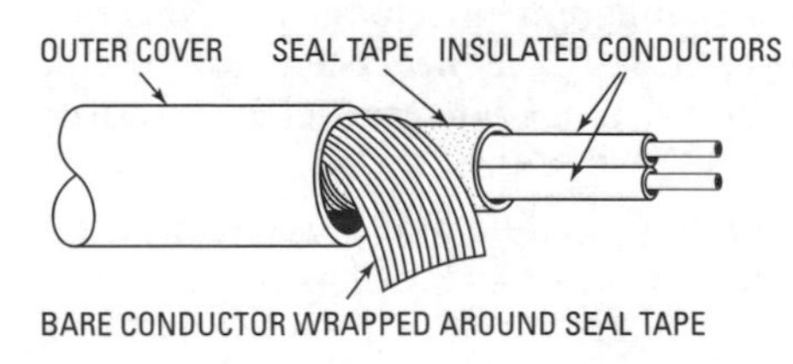

Figure 338-1 SE cable with bare neutral.

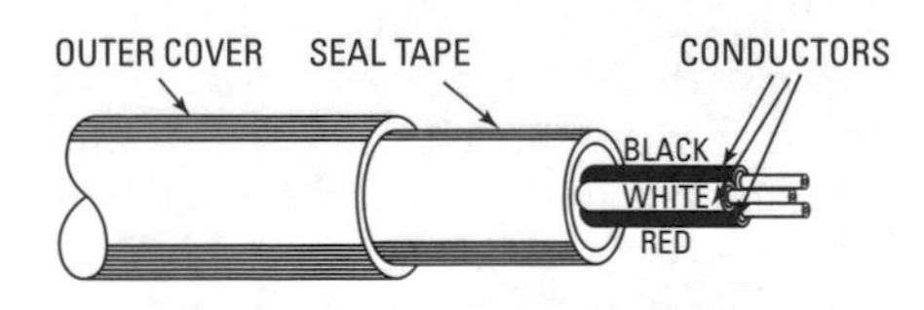

Figure 338-2 SE cable with all insulated conductors.

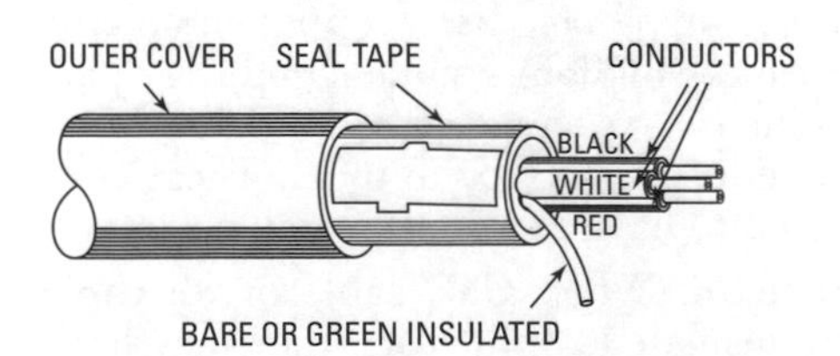

Figure 338-3 SE cable with insulated conductors and equipment-grounding conductor.

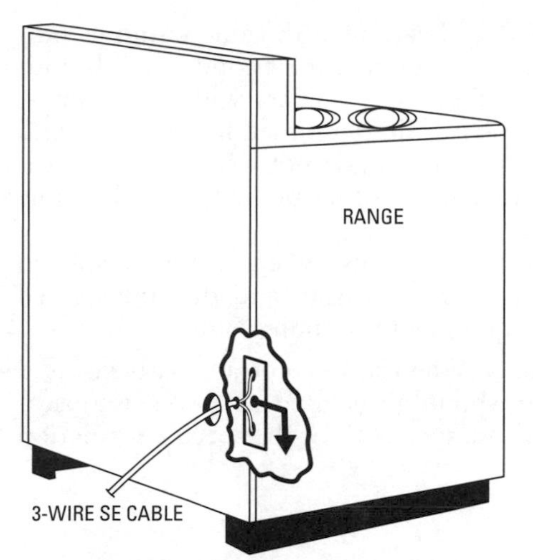

Figure 338-4 SE cable for ranges and dryers.

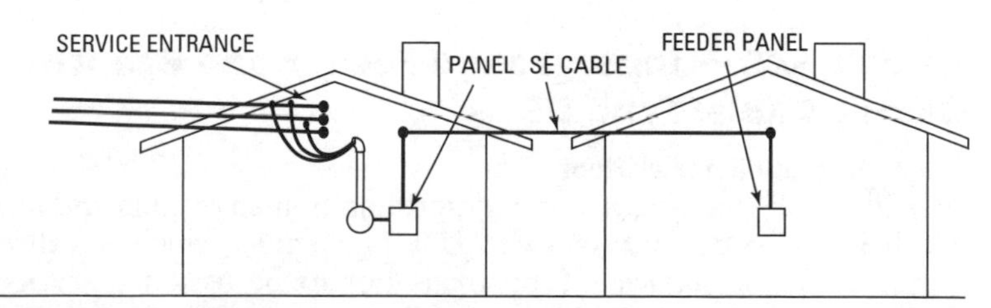

Figure 338-5 SE cable for feeder or branch circuit to other buildings on the same premises.

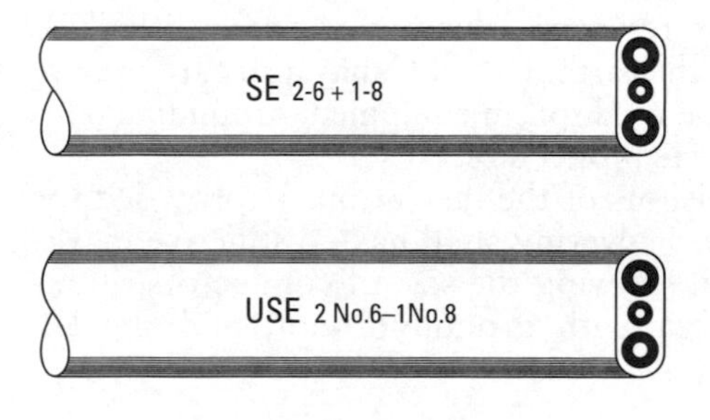

Figure 338-6 Marking of SE cable.

(B) Grounded Conductor Not Insulated. SE cable without individual insulation on the grounded circuit conductor shouldn't be used for branch circuits or feeders within buildings, except that a cable that has a final nonmetallic covering and is supplied by alternating current at not over 150 volts-to-ground may serve as a feeder to other buildings on the same site.

SE cable is allowed for interior use when the fully insulated conductors are used for circuit wiring and the uninsulated conductor is used for equipment grounding only.

(C) Temperature Limitations. When service-entrance cable is used to supply appliances, it shouldn't be used where the temperatures are higher than those specified for the specific insulation in the SE cable.

338.24: Bends

As with most cables, SE and USE cables must be installed (and handled) so that they are bent on a radius of no less than five times cable diameter.

Article 340—Underground Feeder and Branch-Circuit Cable: Type UF

340.2: Description and Marking

This is a cable that is used for underground branch circuits and feeders. It is not to be confused with USE-type cable, which is a direct burial cable for services. Type UF is not to be used for services. Approved UF cable comes in conductor sizes from No. 14 copper or No. 12 aluminum through No. 4/0. The conductors within the outer sheath may be TW, RHW, and so on, as long as they are identified for this use. Over the conductors and their insulation is placed an outer covering that shall be flame-retardant, moisture-resistant, fungus-resistant, and corrosive-resistant—this outer covering shall be suitable for direct burial in the earth. UF cable may also carry a bare or insulated conductor to be used for an equipment-grounding conductor, providing that it is the proper size.

In addition to the provisions of the insulation, as provided for in Section 310.11, the outer covering shall have distinctive markings along its entire length showing the size of conductors, number of conductors, whether with ground or not, and the UF marking.

II. Installation

340.10: Uses Permitted

(1) UF cable may be used as a direct burial cable for branch circuits and feeders, if installed properly and the conductor sizes are protected by proper overcurrent devices and if all other conditions as outlined are met. It would be a good idea to refer to Section 300.5, which states that when this cable is installed under a building, it must be encased in a raceway, and this raceway shall extend through the floor and outside the building. Your attention should be called to the fact that any raceway that is installed or buried in the earth will be subject to electrolysis, unless this is taken into consideration.

(2) UF single-conductor cables may be installed for feeders, subfeeders, and branch circuits, including the neutral. However, they shall be run together in the same trench or raceway. In mentioning the neutral, be assured that in no case will a bare neutral be permitted in direct burial. Also, in running these circuits, an equipment ground will have to be used and must be run in the same trench or raceway as the other conductors. It must also be remembered that bare equipment-grounding conductors are not permitted for direct burial.

Exception

Refer to the *NEC*, Section 690.31 for solar photovoltaic systems.

(3) The *Code* stipulates a minimum of 24-inch burial unless the conductors or cable are protected by a covering board, concrete pad, raceway, and so on. It will also be found that many *Code*-enforcing authorities require, in addition to the 24-inch burial (and especially in rocky ground), the protection of a sand bed and covering for the cable. This is easy to understand, for in frost areas, rocks may damage the insulation of the cable. The enforcing authority has the right to require this, as all work shall be done in a proper and workmanlike manner. See Section 300.5 for UF cable buried underground.

(4) The use of UF cable for interior wiring is not prohibited provided that it meets the requirements of the *Code*. If used as nonmetallic-sheathed cable, Article 334 applies. It is very common practice to use UF cable in concrete-block walls (especially basement walls) instead of NMC cable, as it will stand any dampness that is present. When used for interior wiring, UF cable shall consist of multiple conductors, unless it is to be used for nonheating leads of heat cable, as allowed in Section 424.43.

Exception
Section 424.43 permits to the use of a single-conductor UF cable as the nonheating leads at the ends of heat cable. See Section 690.31 when used with solar photovoltaic systems.

When cable trays are used to support UF cable, the cable shall be of a multiconductor type.

Note
Section 310.10 covers temperature limitations for the conductors in UF cable.

Author's Note
Type UF cable can be very difficult to strip, especially in cold weather. Care must be taken that the removal of the outer cable jacket does not result in the nicking or ripping of the insulation of the conductors.

340.12: Uses Not Permitted

There are a number of places where UF cable may not be used, such as service-entrance cable, theaters, commercial garages, motion picture studios, rooms having storage batteries, in hazardous (classified) locations, and in hoistways, and it shall never be embedded in poured concrete or aggregate (it may, however, be used as a nonheating lead, as covered in Article 424, where it may be embedded in plaster); and it shouldn't be used in sunlight, unless specifically listed for such use.

Exception
See the *NEC*, Section 501.4(B), Exception. This section covers Class I hazardous areas.

340.24: Bending Radius

UF cables must be bent so that the cable remains undamaged. In no case may the radius be less than five times the cable diameter.

Article 342—Intermediate Metal Conduit

I. General

342.2: Definition

Intermediate metal conduit is a metal conduit—similar to rigid metal conduit—that may be threaded and fitted with approved fittings.

II. Installation

342.10: Uses Permitted

(A) **All Atmospheric Conditions and Occupancies.** Use of intermediate metal conduit is permitted: (1) under all atmospheric conditions; (2) in all occupancies; (3) providing, where practical, dissimilar metals are not in contact anywhere in the system, to eliminate the possibility of galvanic action; or (4) to be used as an equipment-grounding conductor. See Section 250.91.

Exception

The *Code* permits aluminum fittings and enclosures to be used with steel intermediate metal conduit. The author prefers to use steel alone, because, in the electrolysis series, steel is more noble than aluminum, and therefore, if any electrolysis takes place, the steel will affect the aluminum.

(B) **Corrosion Protection.** Intermediate metal conduit may not be used unless made of material judged suitable for the condition, or unless corrosion protection is provided. Intermediate metal conduit, elbows, couplings, and fittings may be installed in concrete in direct contact with the earth, or in areas subject to severe corrosive influence, but the conduit, and so on, shall be protected from corrosion if the protection is judged suitable for prevailing conditions.

This places considerable responsibility on the inspector to use discretionary powers in its enforcement. It is recommended that if in doubt, the inspection authority be contacted for a decision before installation.

Note

The *Code* refers you to Section 300.6 for protection against corrosion.

(C) **Cinder-Fill.** Cinder-fill causes considerable corrosion; therefore, unless intermediate metal conduit is of corrosion-resistant material that will withstand this corrosive condition, it shouldn't be buried under or in cinder-fill unless protected by a noncinder concrete covering no less than 2 inches thick, or unless it is buried a minimum of 18 inches below the fill. See Section 300.6, and Figures 344-1 and 344-2 in the following article of this book.

(D) **Wet Locations.** All means of support shall be made of corrosion-resistant materials. Referral is made to Section 300.6, which requires ¼-inch air space between the wiring system, including boxes and fittings in damp or wet locations.

Also note that the above requires that screws and bolts used in the installation shall be corrosion-resistant or plated by corrosion-resistant materials. This prohibits the use of common steel nails, screws, and bolts in these locations. Galvanized ¼-inch spacers may be purchased for the purpose of giving the ¼-inch air space required. Common steel washers are not permitted, as they are not coated with corrosion-resistant material.

342.20: Size

The minimum size of intermediate metal conduit that may be used is ½ inch, and the maximum size is 4 inches.

342.22: Number of Conductors in Conduit

Table 1, Chapter 9 may be used for the percentage fill for conductors. Conduit dimensions appear in Table 4, Chapter 9.

342.24: Bends—How Made

This is the same as Section 344.24.

342.26: Bends—Number in One Run

No more than four 90-degree bends, or their equivalent (360 degrees of total bends) are allowed.

342.28: Reaming and Threading

Cut ends of intermediate conduit shall have the inside smoothed by reaming or other satisfactory means, removing all the rough edges to prevent abrasion of conductor insulation. With threaded intermediate metal conduit, taper dies shall be used. See your *NEC* for additional information.

342.30: Securing and Supporting

As in Section 300.18, intermediate metal conduit is to be installed and fastened securely in place as a complete system before pulling in conductors. The first fastening from an outlet or junction box shall not be over 3 feet away; other support shall be not less than 10 feet apart. In cases where structural members don't allow for IMC to be supported within 3 feet of a box, it may be supported at up to 5 feet from the box. Passing through a structural member is considered to be a support.

Exception No. 1

If threaded couplings are used in straight runs of intermediate conduit they are to be used only at stresses at terminations where the conduit is deflected between supports.

Exception No. 2

If the conduit is made up of threaded couplings and firmly supported at the top and bottom of the vertical run and if no other visible means of supporting the rigid intermediate conduit is available, you may supply supports for the conduit that are not to exceed 20 feet in a vertical run.

Exception No. 3
By special permission, IMC may have fewer than the specified supports when used as a service, branch circuit, or feeder mast. Support from framing members is considered the same as other supports.

342.42: Couplings and Connectors
This is the same as Section 344.42.

342.46: Bushings
This is the same as Section 344.46. See Section 314.6 for the protection of conductors at bushings.

342.56: Splices and Taps
Splices shall be made only in junction outlet boxes and conduit bodies; there shall be none in the conduit. Article 314 supplements this.

III. Construction Specifications

Intermediate metal conduit shall comply with the following:

- **(A)** **Standard Lengths.** Standard length in intermediate metal conduit is 10 feet. If specific conditions require it, lengths longer than 10 feet may be shipped with or without couplings.
- **(B)** **Corrosion-Resistant Material.** Intermediate metal conduit that is made of noncorrosive materials or corrosive-resistant materials shall be suitably marked.
- **(C)** **Marking.** This conduit shall be plainly marked at 2⅜-feet intervals with the letters “IMC.” It shall also be marked as covered in the first sentence in Section 110.21.

The above construction specifications are quite important to assist the inspection authority in identifying the conduit and to provide assurance that it has approved labeling.

Article 344—Rigid Metal Conduit

Rigid conduit is the old standby in wiring methods. It may be used in all atmospheric conditions and occupancies, but there are some provisions that cover its use. Ferrous conduit and fittings that have enamel protection from corrosion can only be used indoors and even then shouldn’t be subject to severe corrosive influences. Where practical, ferrous conduit shall be used with ferrous fittings, and nonferrous conduit shall be used with fittings of similar material. This is to avoid galvanic action between the dissimilar metals.

Exception

Steel rigid metal conduit may be used with aluminum fittings and enclosures. Also, aluminum rigid metal conduit may be used with steel fittings and enclosures.

Unless made of a material judged suitable for the condition, or unless corrosion protection approved for the condition is provided, ferrous or nonferrous metallic conduit, elbows, couplings, and fittings shouldn't be installed in concrete or in direct contact with the earth, or in areas subject to severe corrosive influences. This places considerable responsibility upon the inspector to use discretionary powers in its enforcement. It is recommended that, if in doubt, the inspection authority be contacted for a decision before installation. The following item was brought to my attention: A 6-inch galvanized rigid metal conduit was installed in direct contact with the earth in a temporary wiring situation during construction on an industrial building. Less than one year later this conduit was removed, and the portion in contact with the earth was all but eaten away. This can be prevented; rigid steel conduit is now available with a heavy nonmetallic coating that makes it practically noncorrosive.

Note

See Section 300.6, where rigid metal conduit is solely protected by enamel. Conduits, fittings, and so on (solely protected by enamel), are not to be used outdoors or in locations judged corrosive or damp.

Please also refer to Section 344.1 in the *Code*.

II. Installation

344.10: Uses Permitted

(A) **All Conditions and Occupancies.** In general, rigid metal conduit is permitted in all locations, except in very corrosive areas.

(B) **Cinder-Fill.** Cinder-fill causes considerable corrosion. Therefore, unless conduit is of a corrosion-resistant material that will withstand this corrosive condition, it shouldn't be buried under or in cinder-fill unless protected by a noncinder concrete covering of a minimum of 2 inches thickness, or unless it is buried a minimum of 18 inches below the fill. See Figures 344-1 and 344-2.

(C) **Wet Locations.** All means of support of rigid metal conduit, including bolts, straps, or screws, and so on, shall either be protected from corrosion or be of a noncorrosive material.

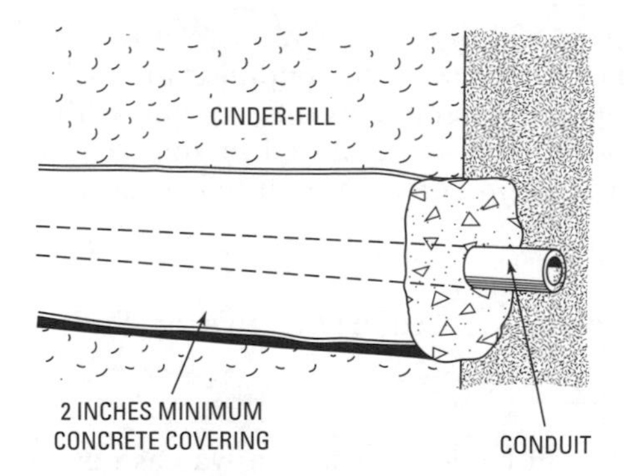

Figure 344-1 Concrete covering for conduit under cinder-fill.

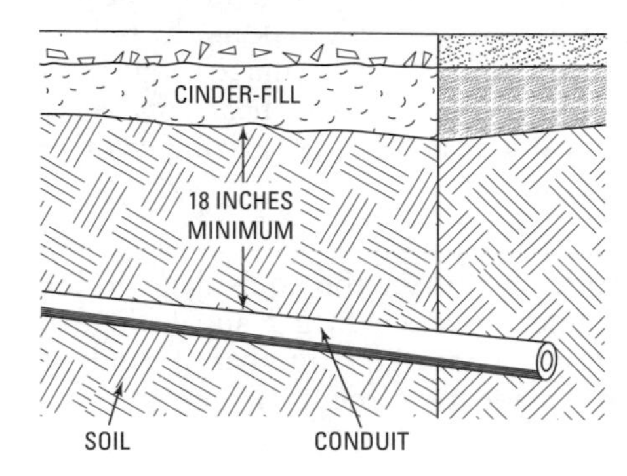

Figure 344-2 No concrete is required if buried a minimum of 18 inches below cinder-fill.

Refer to Section 300.6, which requires a 1⁄4-inch air space between the wiring system, including boxes and fittings in damp or wet locations. Also note that the above requires that screws and bolts used in the installation shall be corrosion-resistant or plated by a corrosion-resistant material. This prohibits the use of common steel nails, screws, and bolts in these locations. Galvanized 1⁄4-inch spacers may be purchased for the purpose of giving the required 1⁄4-inch air space—common steel washers are not permitted, as they are not coated with corrosion-resistant material.

344.14: Dissimilar Metals

Wherever possible contact between dissimilar metals is to be avoided. Aluminum may be used with RMC, but copper is to be avoided.

344.20: Size

For practical purposes, ½-inch trade-size rigid conduit is the smallest allowable. However, for underplaster extensions, conduit with a minimum inside diameter of 5⁄16 inch is permitted in some cases, and ⅜-inch conduit is permitted in some cases listed in Section 430.145(B).

344.22: Number of Conductors

Conduit fill of conductors shouldn't exceed the fills given in Table 1 of Chapter 9.

344.26: Bends—How Made

Conduit may be bent, but it shall be done in such a way as not to damage the conduit or reduce its internal diameter. Conduit is often kinked while being bent. Installation of kinked conduit is not permitted, as the internal diameter will be reduced, making the pulling of conductors increasingly difficult and making damage to the insulation more probable. Torches have been used to heat conduit to facilitate the bending, but this will damage the galvanizing or other protective coating and is not permitted.

344.28: Bends-Number in One Run

There is a limit of four 90-degree bends, or the equivalent of 360 degrees, between outlet and outlet, between fitting and fitting, or between outlet and fitting. This is to control the number of bends so that pulling of conductors is not made too difficult (Figure 344-3).

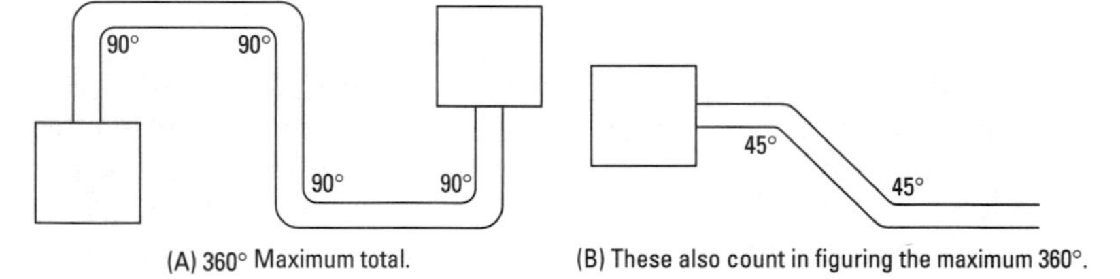

Figure 344-3 A maximum of 360 degrees allowed between outlets and/or fittings.

There are times with large conductors that a total of 360 degrees of bends will be too much. These may be calculated. In these cases damage may be caused to the insulation of the conductors. This is an engineering problem and also up to the authority that has jurisdiction.

344.28: Reaming and Threading

(A) **Reamed.** All ends of this conduit shall be carefully reamed to remove rough edges that might damage the conductors.

(B) **Threaded.** The die you should use to thread rigid metal conduit must have a ¾-inch taper per foot.

Where conduit is threaded in the field, it is assumed that a standard conduit cutting die providing ¾-inch taper per foot will be employed. It must be kept in mind that conduit couplings are different from water-pipe couplings—conduit couplings have no taper in the threads inside the coupling, whereas pipe couplings do.

344.30: Securing and Supporting

Rigid metal conduit shall be installed as a complete system, as provided in Section 300.18, and shall be securely fastened in place. Conduit shall be firmly fastened within 3 feet of each outlet box, junction box, cabinet, or fitting. Conduit shall be supported at least every 10 feet except that straight runs of rigid conduit made up with approved threaded couplings may be secured in accordance with Table 344.30(B)(2), provided such fastening prevents transmission of stresses to terminus when conduit is deflected between supports. In cases where structural members don't allow for RMC to be supported within 3 feet of a box, it may be supported at up to 5 feet from the box. Passing through a structural member is considered to be a support.

On vertical risers from machine tools and the like, the distance between supports may be increased to 20 feet. There are restrictions: (1) The conduit must be provided with threaded couplings; (2) the conduit is firmly attached at the top and the bottom; and (3) there is no other means of intermediate support readily available. By special permission, rigid conduit may have fewer than the specified supports when used as a service, branch circuit, or feeder mast. Note also that support from framing members is considered the same as other supports.

344.42: Couplings and Connectors

(A) **Threadless.** Threadless couplings and connectors shall be made tight to ensure electrical continuity. Concrete-type fittings only shall be used when rigid metal conduit is buried in concrete, and if it is in wet locations, the fittings must be of the rain-tight type. They, or the box they are contained in, should contain the listed marking of the testing laboratory.

(B) Running Threads. It was common years ago to use running threads in place of conduit-type unions. For those not familiar with running thread, it is made by threading a piece of the conduit for a considerable length so that a coupling can be screwed down on the running thread, and then screwed back to the conduit you are trying to connect to. This is not done now. There is no taper to running threads—all of the thread is the same size throughout its length, tending to make a loose fit. In addition, there is no corrosion-resistant covering on the threaded portion.

344.46: Bushings

Bushings are required on conduits wherever they enter boxes or fittings to protect the wire from abrasion, unless the design of the box or fitting is such that it provides an equivalent protection. Section 314.6(C) requires insulated bushings where No. 4 or larger conductors are used. This may be an insulated bushing, a fiber insert for a metal bushing (provided that it is an approved insert), or it may be a grounding-type bushing, which is a combination of metal and insulation, where grounding bushings are required, or the approved fiber insert may be used with a grounding-type of bushing.

344.56: Splices and Taps

All splices and taps must be made according to Section 300.15.

III. Construction Specifications

The requirements of this section are very plain. The requirements of this section are covered in (A) through (C) below:

(A) Standard Lengths. Rigid metal conduit is shipped in 10-foot lengths. One coupling is furnished with each length, and it shall come with threads on each end and be properly reamed. If longer lengths are required, the conduit may be ordered in lengths longer than 10 feet with or without couplings or threads. Steel conduit shall have an inferior coating of a character and appearance so as to readily distinguish it from ordinary pipe commonly used for other than electrical purposes. Great care must be exercised when applying the interior coating on conduit in order that no abrasion to the insulation results.

(B) Corrosion-Resistant Material. If the conduit is of nonferrous corrosion-resistant material, it shall be plainly marked to indicate this.

(C) Durably Identified. Each length of rigid metal conduit shall be identified every 10 feet. This also refers to Section 110.12.

This is highly important so that it will be easy to identify what type of rigid conduit is being used.

Article 348—Flexible Metal Conduit

This article covers the use of flexible metal conduit, commonly called "Greenfield."

348.2: Definition

Flexible metal conduit is a raceway made of a helically wound interlocking metal strip.

II. Installation

348.10: Uses Permitted

Flexible metal conduit may be used in wet locations only when the conductors are of types approved for the specific conditions, or if they are lead-covered conductors and there is very little likelihood of water entering under raceways or enclosures that might be connected to the flexible metal conduit.

348.12: Uses Not Permitted

Flexible metal conduit shouldn't be used in storage-battery rooms or hoistways, nor shall it be used in hazardous (classified) locations, except as permitted in Sections 501.4(B) and 504.20. If it is likely to be exposed to oil, gasoline, or other deteriorating materials affecting rubber, then rubber-covered conductors shouldn't be used. Flexible metal conduit shouldn't be used underground for direct burial, nor shall it be used embedded in poured concrete or aggregate. It can't be used where subject to physical damage.

Flexible metal conduit offers high impedance, having three to four times the impedance of AC cable. Therefore, many inspectors will require bonding or the use of an equipment-grounding conductor when using flexible metal conduit.

348.20: Size

For practical purposes, the minimum-size metal conduit is ½-inch electrical trade-size. There are a few exceptions that will permit ⅜-inch electrical trade-size to be used. There is sometimes a tendency to use ⅜-inch conduit in the wrong places and for the wrong purposes.

Exception No. 1

As permitted for motors by Section 430.145(B).

Exception No. 2

If the lengths are not over 6 feet, ⅜-inch, nominal, trade-size is permitted by the *Code*. For approved assemblies, or by installation in the field, and as required in Section 410.67(C), flexible metal conduit is permitted for tap connections to lighting fixtures, or for lighting fixtures.

Exception No. 3
Section 604.6(A) permits 3⁄8-inch trade-size flexible metal conduit to be used in manufactured wiring systems.

Exception No. 4
The *NEC* refers you to Section 620.21.

Exception No. 5
3⁄8-inch flexible metal conduit can be used to connect wired fixtures, when part of a listed assembly. See Section 410.77(C). See Table 348.22 in the *NEC*.

348.22: Number of Conductors
The number of allowed conductors must be as specified in Table 1, Chapter 9 (for ½″ through 4″ sizes), or in Table 348.22 (for the 3⁄8″ size).

348.26: Bends
A maximum of 360 degrees of bend is allowed between pulling points. In concealed installations, angle connectors shouldn't be used.

348.28: Trimming
Except where screw-in fittings are used, all cut ends of flexible metal conduits must be trimmed to remove sharp edges.

348.30: Securing and Supporting
Flexible metal conduit is a wiring method and may be used to wire buildings, and so on, unless prohibited in certain locations, such as in most hazardous locations. It shall be supported by approved means at least every 4½ feet and within 12 inches of each side of every outlet box or fitting. There are, however, a few exceptions to this:

Exception No. 1
Flexible metal conduit may be fished. Here, of course, it would be impossible to support it within walls and the like.

Exception No. 2
If flexibility is necessary at terminals, the length shouldn't be over 3 feet.

Exception No. 3
Tap connections are required for fixtures, as covered in Section 410.67(C). Lengths of not more than 6 feet from a fixture terminal are permitted. I am certain that almost any inspection authority would

require that the flexible metal conduit not be draped in such a manner as to lay on drop-in ceilings and the like.

348.56: Splices and Taps

Splices and taps must be made according to Section 300.15. See Article 314 for rules on the use of boxes and conduit bodies.

348.60: Grounding

Flexible metal conduit may be used as a grounding means where both the conduit and the fittings are listed for grounding.

Exception No. 1

Flexible metal conduit may be used as an equipment-grounding means if the total length of the conduit (and thus the ground return path) is 6 feet or less. Only fittings listed for grounding for this type of conduit are permitted, and the overcurrent devices protecting the conductors are to be rated not over 20 amperes.

Exception No. 2

The grounding shall be installed where flexible metal conduit is connected to equipment that requires flexibility.

Basically, flexible metal conduit is not to be used as an equipment-grounding conductor. This exception was put in the *Code* since there is no way of knowing the problems resulting from the resistance of the conduit in these lengths.

Article 350—Liquid-Tight Flexible Metal Conduit and Article 356—Liquid-Tight Flexible Metal Conduit, cover two types of flexible conduit. One is liquid-tight flexible nonmetallic, the other is liquid-tight flexible metal conduit.

Liquid-tight flexible metal conduit is not intended as a cure-all, but has a very definite purpose. However, when it is used, care must be taken that only approved terminal fittings are employed. When conventional fittings are used, the grounding that is normally provided by the conduit is often destroyed.

350.2: Definition

This is a raceway of circular cross-section with the outside covering being a nonmetallic, sun-resistant, liquid-tight jacket over the metal flexible core. Only listed couplings, connectors, and other fittings shall be used.

II. Installation

350.10: Use Permitted

When listed and marked, liquid-tight flexible metal conduit may be used for direct burial in the earth, and for both concealed and exposed installations:

(1) It may be used in installations where maintenance conditions require not only flexibility, but also protection from liquids, solids, or vapors.

One very practical use for this material might be where there is a service pole located close to a pumphouse. The movement of the pole might prohibit the use of rigid metal conduit as the wiring method. The installation is exposed to the elements, so wiring with liquid-tight flexible metal conduit would be very practical, provided that all of the requirements of proper grounding continuity are met.

(2) Sections 501.4(D), 502.4, and 503.3 are referred to. It may be used as per these sections—if it is listed for approval. It may also be used as covered in Section 553.7(B) for floating buildings.

The aforementioned sections don't give a complete release to use liquid-tight flexible metal conduit in all hazardous locations, but there are some places where it is permitted.

350.12: Uses Not Permitted

Where not permitted:

(1) Any place that is subject to physical damage.

(2) If ambient temperatures and/or conductor temperature for liquid-tight flexible metal conduit are in excess of that for which the material or covering was approved.

350.20: Size

The sizes of liquid-tight flexible metal conduit shall be a minimum of ½-inch to a maximum of 4-inch electrical trade-size.

Exception

Section 350.10 permits the use of ⅜-inch size.

350.22: Number of Conductors

(A) **Single Conduit.** Table 1, Chapter 9 of the *NEC* gives the percentage of conductor fill permitted in a single conduit of ½ inch through 4 inches.

(B) **⅜-Inch Liquid-Tight Flexible Metal Conduit.** See Table 348.22 of the *NEC* for the maximum number of conductors allowed in ⅜-inch liquid-tight flexible metal conduit.

350.24: Bends

When this type of conduit extends from outlet to outlet, fitting to fitting, or outlet to fitting, the total bends of the conduit shouldn't exceed 360 degrees. This includes any offsets or bends that occur immediately at the outlet or fitting.

Concealed raceway installations shouldn't contain angle connectors.

350.30: Securing and Supporting

Where liquid-tight flexible metal conduit is used as a fixed raceway, it shall be secured (1) by approved methods, (2) at intervals not exceeding 4½ feet (1.37 m), and (3) within 12 inches (305 mm) of outlet boxes and fittings.

Exception No. 1

Where liquid-tight flexible metal conduit is acceptable for fishing—but, of course, it can't be fastened at supportive locations.

Exception No. 2

Where flexibility is required, the length shouldn't exceed 3 feet.

Exception No. 3

Section 410.67(C) requires that there be not more than 6 feet from the fixture to the terminal connection.

350.42: Coupling and Connectors

Only listed terminal fittings shall be used with liquid-tight flexible metal conduit.

This wiring method has connectors and the like that are listed for liquid-tight flexible metal conduit to give electrical and mechanical continuity; substitutions shouldn't be permitted.

350.60: Grounding

If listed for grounding purposes, the liquid-tight flexible metal conduit is permitted for use as the equipment-grounding conductor, and the fittings shall also be listed. Wherever bonding jumpers must maintain equipment-grounding continuity, Section 250.79 shall be followed.

If used where flexibility is required and where connected to equipment, an equipment-grounding conductor shall be installed.

Exception

Liquid-tight flexible metal conduit of sizes 1¼-inch and smaller may be used as a grounding conductor if the total length does not exceed 6 feet, but the conduit must be terminated in fittings that are listed for grounding. If the overcurrent devices protecting the circuits are rated at 20 amperes or less, they shall be for ⅜- and ½-inch trade-sizes, and if 60 amperes or less, they shall be for ¾-inch through 1¼-inch trade-sizes.

Note

The *NEC* refers you to Sections 501.16(B), 502.16(B), and 503.16(B).

Article 352—Rigid Nonmetallic Conduit

Rigid nonmetallic conduit is any system of nonmetallic conduit and fittings that is resistant to moisture and chemicals. For above-ground use, it must also be sunlight resistant, crush resistant, heat resistant, and flame retardant. Certain types of rigid nonmetallic conduit are listed to be installed underground in continuous lengths from a reel.

Note

Nonmetallic conduit may be made of a number of different materials (such as PVC, asbestos cement, fiberglass epoxy, fiber, soapstone, or high-density polyethylene). Note that Polyvinyl Chloride (PVC) is the most commonly used type.

II. Installation

352.10: Uses Permitted

The use of rigid nonmetallic conduit and fittings follows, but be sure that the type you are using is listed for the following conditions. Some of the types listed in the fine-print note in the section above are permitted only for certain conditions.

Note

Some nonmetallic conduit becomes very brittle in cold, which would cause it to be more easily damaged.

- **(A)** **Concealed.** Concealed wiring in walls, floors, and ceilings of buildings or structures is permitted.
- **(B)** **Corrosive Influences.** Section 300.6 covers locations that are severely corrosive. Therefore, use the type of rigid nonmetallic conduit that is approved for the specific chemicals in the specific corrosive atmosphere.
- **(C)** **Cinders.** Cinders have corrosive influences; nonmetallic conduit may be used under cinders where specifically approved.
- **(D)** **Wet Locations.** Dairies, laundries, canneries, and so on, are wet locations, requiring frequent washing. For such places, the entire system, including conduits, boxes, and fittings, shall be installed so as to prevent water from entering the conduit system. Straps, bolts, screws, and so on, shall be made only of materials that are proved resistant to corrosion.
- **(E)** **Dry and Damp Locations.** Section 352.10 permits installation in dry or damp locations.

(F) Exposed. If it won't be subject to physical damage, it may be used for exposed wiring installation when it is approved for exposure.

(G) Underground Installations. Referral is made to Sections 300.5 and 710.3(B) where installation of rigid nonmetallic conduit is made underground.

(H) Support of Conduit Bodies. Rigid nonmetallic conduit may support nonmetallic conduit bodies that don't contain devices or support fixtures. The trade-size of the conduit body may not be larger than the trade-size of the conduit.

352.12: Uses Not Permitted

Rigid nonmetallic conduit may not be used in the following:

(A) Hazardous (Classified) Locations. Rigid nonmetallic conduit may not be used in any hazardous (classified) locations, except those covered in Sections 514.8 and 515.5. There is an exception to this in Class 1, Division 2 locations as found in the exception to Section 501.4(B). Article 514 covers gasoline dispensing and service stations. Section 514.8 explains where rigid metal conduit or, where buried not less than 2 feet in the earth, nonmetallic rigid conduit may be installed. When nonmetallic rigid conduit is installed, there shall be an equipment-grounding conductor installed, conforming in size to Table 250.95, for the purpose of grounding the metal noncurrent-carrying parts of any equipment. The use of rigid nonmetallic conduit in this type of location won't change the use of seal-offs, as covered elsewhere in Article 514. Also see Sections 503.3(A) and 515.5, and Figure 352-1.

(B) Support of Fixtures. Rigid nonmetallic conduit shouldn't be used for supporting fixtures or any other equipment.

(C) Physical Damage. It shouldn't be used unless identified for locations where subject to physical damage.

(D) Ambient Temperatures. It shouldn't be used where subject to ambient temperatures higher than 50°C (122°F), unless listed for such use.

(E) Insulation Temperature Limitations. If the conductor limitations are higher than the temperature limitations for which the conduit is approved, it shouldn't be used.

(F) Theaters and Similar Locations. Rigid nonmetallic conduit can't be used in theaters and similar locations, except as allowed by Articles 518 and 520, and Section 352.20.

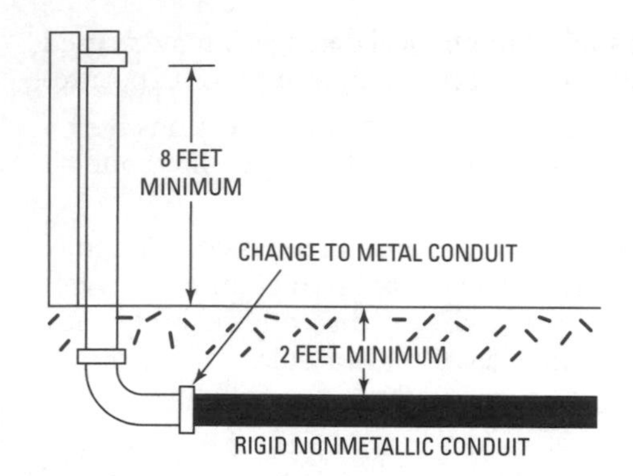

Figure 352-1 Running nonmetallic conduit above ground. PVC may be run above ground if approved for the purpose and not subject to physical damage.

Rigid nonmetallic conduit is made in trade-sizes from ½ inch through 6 inch.

352.22: Number of Conductors

The fill for both new work and rewire are governed by the same rules as are applicable to rigid metal conduit. This won't be repeated here but refer back to Section 344.22 and the tables of *NEC* Chapter 9, along with the applicable notes to these tables.

352.24: Bends—How Made

Field bends are to be made only by the use of a bending machine designed for the purpose, and Table 344.24 governs the radius of the bends. Only identified methods of bending shall be used. Never use flame torches for bending.

In one case the wirer failed to use the bending machine and used a torch to heat the points to be bent, with the result that the conduit did not bend evenly. There were numerous kinks in it, and it all had to be replaced.

352.26: Bends—Number in One Run

A maximum of 360 degrees of bends are allowed in any single run.

352.28: Trimming

Trimming and smoothing the inside and outside of ends to remove rough edges is required. This is for the same reason that rigid metal conduit has to be reamed, namely to prevent damage to the conductor insulation.

352.30: Securing and Supporting

Rigid nonmetallic conduit shall be supported within 3 feet (914 mm) of each box, cabinet, or other conduit termination, and elsewhere according to Table 354.30(B) in the *NEC*.

352.44: Expansion Fittings

Where expansion and contraction due to temperature differences may exceed 1/4 inch between rigidly mounted boxes or enclosures, approved rigid nonmetallic conduit expansion joints shall be used. Practically all substances expand and contract with temperature changes, which will put undue strain on the conduit. Expansion joints absorb this expansion and contraction, thus avoiding damage to the conduit and fittings.

352.46: Bushings

This ruling is practically the same as for rigid metal conduit, except that it says "bushing" or "adapter." As with metal conduit [Section 314.6(C)], an insulated bushing is required where No. 4 or larger conductors are used.

352.48: Joints

In the makeup of rigid nonmetallic conduit, various methods are used, depending on the material. The instructions for the particular conduit shall be followed when making up lengths and between conduit and couplings, fittings, and boxes. Use of a solvent or similar substance is to be applied to the joints and fittings before putting them together.

352.56: Splices and Taps

Splices and taps are never to be made in the conduit, only in junction boxes, outlet boxes, or conduit bodies. This is also covered in Article 314.

Article 354—Nonmetallic Underground Conduit with Conductors

I. General

354.2: Description

Nonmetallic underground conduit with conductors is a factory-made plastic conduit with conductors already installed in it. It is delivered to a job-site on a large reel and is generally chosen because of its labor-saving characteristics. (It requires no labor for the installation of conductors.)

II. Installation

354.10: Uses Permitted

This wiring system is allowed in the following circumstances:

- **(1)** Directly buried. (Must meet the cover requirements of Section 300.5.)

(2) In concrete.

(3) In cinder-fill.

(4) In underground corrosive areas. See 300.6.

354.12: Uses Not Permitted

This wiring may not be used exposed, inside buildings, or in hazardous locations. There are two exceptions to this:

(1) Conductors that are removed from the plastic outer conduit may be run in areas suitable to their ratings.

(2) There are a few areas where this wiring is allowed in hazardous areas. See Sections 503.3(A), 504.20, 514.8, and 515.5.

354.20: Size

The smallest size allowed is ½ inch trade size, and the largest allowed is 4 inch.

354.24: Bends—How Made

Bends must be made by hand. The conduit is flexible enough to do so.

354.26: Bends—Number in One Run

No more than four 90-degree bends (or the equivalent, 360 degrees) are permitted in one run.

354.28: Trimming

To terminate this wiring method, the last foot or more of conduit must be removed from each end. When this is done, the conduit must be trimmed to remove rough edges, both inside and outside.

354.46: Bushings

Where underground nonmetallic conduit with conductors enters a box, appropriate bushings must be used to protect the conductors. If the box has adequate protection, a separate bushing is not required.

354.48: Joints

All joints of any kind must be made by approved methods.

354.50: Conductor Terminations

All wire terminations must be made by approved methods.

354.56: Splices and Taps

Splices and taps must be made in approved junction boxes or enclosures.

354.60: Grounding

Where grounding is required, an assembly with a separate grounding conductor must be used.

III. Construction

See the *NEC* for manufacturing specifications of this wiring method.

Article 356—Liquid-Tight Flexible Nonmetallic Conduit

I. General

356.2: Definition

There are three basic types of liquid-tight flexible nonmetallic conduit. They are:

(1) A smooth inner core and cover that are bonded together, with at least one layer of reinforcement between the core and the cover.

(2) A smooth inner surface with reinforcement inside of the conduit wall.

(3) A corrugated inner and outer surface with no reinforcement in the conduit wall.

II. Installation

356.10: Uses Permitted

Liquid-tight flexible nonmetallic conduit may be installed in either exposed or concealed locations:

Note

If exposed to extreme cold temperatures, it is more likely to be damaged from physical contact.

(1) It may be used where flexibility is required in installation, maintenance, or operation.

(2) If protection from vapors, liquids, and solids is required for the circuit conductors installed therein, it may be used.

(3) It must be suitably marked and listed if used outdoors.

(4) For direct burial, when listed for the purpose.

(5) In lengths more than 6 feet when secured properly.

(6) As a listed, manufactured, pre-wired assembly.

356.12: Uses Not Permitted

(1) Where it can be subjected to physical damage.

(2) If, due to ambient and conductor temperatures, the conduit is subjected to higher temperatures than it has been approved for.

(3) Voltage of the circuit conductors used therein shall be 600 volts, nominal, or less. The one exception to this rule is for electric signs [see Section 600.31(A)].

(4) In lengths greater than 6 feet.

356.20: Size

Sizes of liquid-tight nonmetallic conduit include ½-inch through 4-inch electrical trade-size.

Exception

⅜-inch liquid-tight nonmetallic conduit can be used to enclose motor leads [see Section 430.145(B)], as whips to lighting fixture or equipment [see Section 410.67(C)], or for electric signs [see Section 600.31(A)].

356.22: Number of Conductors

The number of conductors in a single conduit is the same as for regular conduit, as covered for percent of fill in Table 1, Chapter 9.

356.24: Bends

Runs of liquid-tight nonmetallic conduit can't have more than 360 degrees of bend (four quarter-bends) between pull boxes, junction boxes, and so on.

356.30: Securing and Supporting

When liquid-tight nonmetallic conduit is used as a fixed raceway, it must be supported within 12 inches of every box, cabinet, or fitting. Thereafter, it must be supported every 4 feet.

Exception

It may be fished into place without support; where flexibility is required, it may be supported up to 3 feet from a box or fitting; and it may be used in lengths up to 6 feet for fixture whips.

356.42: Couplings and Connectors

Terminal fittings shall be identified for the use when used on liquid-tight nonmetallic conduit.

356.60: Grounding and Bonding

When an equipment-grounding conductor is required, which will be in practically all cases, the equipment-grounding conductor shall be run in the conduit with the circuit conductors as required by Article 250. See Sections 250.134(B) and 250.102.

Article 358—Electrical Metallic Tubing

Electrical metallic tubing is commonly known as EMT, thin wall conduit, or merely thin wall.

EMT may be used for exposed work and concealed work. If during or after installation EMT is subject to severe physical damage, it shouldn't be used. As with rigid metal conduit, it shouldn't be used if protected by enamel only. In cinder-fill or cinder concrete, if subjected to permanent moisture, it shouldn't be used unless it is encased in regular concrete and has at least a 2-inch casing around it. It may, however, be buried at least 18 inches beneath the surface under a cinder-fill. Care shall be taken to keep it from contacting dissimilar metals, to eliminate the possibility of electrolysis. See the table showing resistivity of soils. There is a great difference among soils in their reaction to EMT and conduits. EMT can't be used to support light fixtures or other equipment.

Exception

Aluminum fittings and enclosures may be used on steel EMT.

II. Installation

358.10: Uses Permitted

EMT may be used either exposed or concealed or in direct contact with concrete. In corrosive areas it must be suitably protected. See Section 300.6 for more information on corrosion protection.

> **Wet Locations.** When used in wet locations, all bolts, straps, and so on, that are used with the EMT must be corrosion-resistant.

358.12: Uses Not Permitted

EMT may not be installed in the following areas:

(1) Where subject to severe damage.

(2) Where protected from corrosion only by enamel.

(3) In cinder concrete or fill subject to permanent moisture, unless protected on all sides by 2 inches of noncinder concrete, or at least 18 inches below the cinder concrete or fill.

(4) In any hazardous location, except as permitted by Sections 502.4, 503.3, and 504.20.

358.20: Size

(A) Minimum. Minimum size of EMT shall be ½-inch trade-size. Smaller sizes may be used as under plaster extensions. In Section 430.145(B) you will find that it can sometimes be used to enclose motor leads.

(B) Maximum. The maximum trade-size permitted is 4 inches.

358.22: Number of Conductors

The requirements for conductors in EMT are exactly the same as for rigid metal conduit as provided in Section 352.22. See Table 1, Chapter 9, for percentage fill permitted in a single tubing.

The cross-sectional area of conductors is covered in the tables of Chapter 9 and the notes therewith.

358.28: Reaming and Threading

Basically, EMT is not meant to be connected to boxes or fittings or coupled together by means of threads in the wall of the tubing. An integral coupling is permitted where it is factory-threaded. All ends of EMT shall be reamed or otherwise finished so as not to leave any rough edges. EMT cuts easily and thus there is a tendency not to worry too much about reaming. Reaming is very important, however, to keep from damaging the insulation of the conductors.

358.30: Securing and Supporting

EMT shall be installed as a complete system, as was covered in Article 300. EMT shall be fastened securely within 3 feet of boxes, and so on; otherwise, supports shouldn't be placed over 10 feet apart. This ruling is a definite asset as there has always been much question as to what constituted properly secured EMT in an installation. As required in Article 300, raceways shall be installed completely before installing the conductors.

When structural members don't allow for EMT to be supported within 3 feet of every box, unbroken (no couplings) pieces are allowed to extend up to 5 feet from the box before they are supported. Also, it may be fished behind panels for concealed work in finished buildings.

358.42: Couplings and Connectors

As with all fittings, EMT fittings, couplings, and so on, shall be made up tight for electrical continuity. If EMT is buried in masonry or concrete, the fittings shall be concrete-tight and approved. In wet locations the fitting shall be identified as being rain-tight. There are a great variety of connectors and couplings available, but regardless of the type used, care should be taken to make up the connection tightly. It will be recalled that EMT will, in most cases, be serving as the equipment ground. Loose connections won't provide proper grounding and could be the cause of a fire or injury.

Where permitted in concrete, couplings and fittings might be damaged during the installation of the concrete, and once buried, repairs are very costly. Care should be taken to make sure that such damage does not occur.

358.24: Bends—How Made

Table 344.24 covers the radius of bends (minimum). In making bends in EMT, care shall be taken not to injure the EMT or decrease the interior dimensions. These are the same requirements as for rigid metal conduit. In making field bends, the proper equipment should always be used.

358.26: Bends—Number in One Run

These requirements are the same as for rigid metal conduit—a maximum of the equivalent of four 90-degree bends, or a total of 360 degrees—between outlet and outlet, between fitting and fitting, or between outlet and fitting.

358.56: Splices and Taps

Splices and taps shall never be made in the EMT itself, but only in junction boxes, outlet boxes, and conduit bodies. This also is covered in Article 314.

Conductors, including splices and taps, shouldn't fill a conduit body to more than 75 percent of its cross-sectional area at any point. All splices and taps shall be made by approved methods. Also see Article 314.

Conduit bodies are commonly known as LB condulets, T condulets, and so on.

Article 360—Flexible Metallic Tubing

Flexible metallic tubing is a raceway for electrical conductors. It is circular in cross-section, and must be flexible, liquid-tight, metallic, and without a nonmetallic covering.

II. Installation

360.10: Uses Permitted

Flexible metallic tubing may be used in dry locations and in accessible locations, provided it is protected from physical damage. It may be also used for concealed work above a suspended ceiling. It may be used for branch circuits, and also for voltages up to 1000 volts.

360.12: Uses Not Permitted

The following uses are not permitted for flexible metallic tubing: It may not be used underground as a direct-burial raceway, nor may it be embedded in poured concrete or aggregate. It shouldn't be permitted in hoistways or battery-storage rooms. It shouldn't be used in hazardous (classified) locations. It can be used in these locations only where it might be permitted in other articles of the *Code*. It may not be used in lengths over 6 feet or when subject to physical damage.

360.20: Size

(A) Minimum. ½-inch electrical trade-size flexible metallic tubing is the smallest size permitted.

Exception No. 1

In accordance with Sections 300.22(B) and (C), ⅜-inch trade-size shall be permitted. In referring to these two sections, you will find that they cover ducts or plenums used for environmental air and other spaces used for environmental air.

Exception No. 2

See Section 410.67(C). As part of an approved assembly, ⅜-inch trade-size may be used in lengths of not over 6 feet, and it may also be used in lighting fixtures.

(B) **Maximum.** The maximum allowable trade-size of flexible metallic tubing is ¾ inch.

360.22: Number of Conductors

(A) **½-inch and ¾-inch Flexible Metallic Tubing.** Table 1, Chapter 9, governs the maximum cross-sectional fill of conductors allowed in flexible metallic tubing.

(B) **⅜-inch Flexible Metallic Tubing.** Table 350.3, which describes fill of insulated conductors for ⅜-inch flexible metal conduit, also applies to fill for flexible metallic tubing.

Note

For conductor cross-sectional area, see Chapter 9 of the *NEC* and its notes.

360.24: Bends

Refer to the *NEC*; (A) covers bends where flexible metallic tubing is infrequently flexed in service after installation and Table 360.24(A) covers this bending.

Fixed Bends

Table 360.24(B) covers bends in flexible metallic tubing for fixed bends. Refer to Tables 360.24(A) and 360.24(B).

360.40: Fittings

The terminating fittings for flexible metallic tubing shall be listed for the purpose and shall satisfactorily close any openings in that connection.

Note

Sections 300.22(B) and (C) cover use in air ducts, plenums, and so on, that are for carrying environmental air.

360.60: Grounding

Section 250.118(8) applies to flexible metallic tubing.

Article 362—Electrical Nonmetallic Tubing

I. General

362.2: Definition

Electrical nonmetallic tubing is corrugated plastic tubing that can be easily bent by hand, which provides a smooth inner surface that facilitates the pulling of wires. It is flame-retardant and is resistant to chemical atmospheres and moisture.

II. Installation

362.10: Uses Permitted

The uses for electrical nonmetallic tubing are listed below:

(1) ENT may be used for exposed work where not exposed to damage, and where the building is no more than three stories.

(2) It may be used in walls, ceilings, and floors provided that there is a thermal barrier that is composed of material that will have a minimum of a 15-minute rating for finish and is identified in listings of fire-rated material.

(3) In places such as those covered in Section 300.6, where such places or locations may be subjected to severe corrosive influence. This is where the material of the electrical nonmetallic cable has been specifically approved.

(4) If not prohibited by Section 362.12, it may be used in concealed dry and damp locations.

(5) This type of conduit may be used above suspended ceilings—if this ceiling is provided with a thermal-barrier material with at least a 15-minute rating listing of fire-rated assemblies.

(6) If fittings are identified for embedding in concrete, flexible nonmetallic tubing may be installed in poured concrete.

(7) In indoor wet locations, or in concrete slabs on grade, if approved fittings are used.

Note

Flexible nonmetallic tubing is susceptible to extreme cold temperatures, and at these low temperatures it becomes brittle. For this reason it will become more vulnerable to physical damage, especially during installation.

362.12: Uses Not Permitted

ENT can't be installed for these uses:

(1) Hazardous locations, except as allowed by Sections 504.20 and 505.15(A)(1).

(2) As a means of supporting fixtures or other equipment.

(3) When subject to ambient temperatures higher than its rating.

(4) For conductors whose temperature ratings are higher than the conduit's rating.

(5) Direct burial.

(6) For voltages over 600 volts.

(7) Exposed, except as allowed by Section 362.10(1) and (5).

(8) In theaters, except as provided for by Articles 518 and 520.

(9) Exposed to sunlight unless identified for such use.

362.20: Size

(A) **Minimum.** Tubing shouldn't be smaller than ½-inch electrical trade-size.

(B) **Maximum.** Tubing shouldn't be larger than 2-inch electrical trade-size.

362.22: Number of Conductors

Table 1, Chapter 9 in the *NEC* governs the percentage of fill of conductors in nonmetallic flexible conduit.

362.24: Bends—How Made

The radius of the bends in the internal side of the bend shouldn't be less than covered in Table 344.24 for rigid metal conduit. Also, bends in nonmetallic electrical tubing shall be made so as not to reduce the internal diameter of the tubing. No special tools are required for bending; they are made by hand.

362.26: Bends—Number in One Run

The number of bends between outlet and outlet or outlet and fitting shall be the same as for rigid metal conduit, that is, maximum 360 degrees total, including bends located immediately at outlets or fittings.

362.28: Trimming

All cut edges, both inside and outside, shall be trimmed to eliminate rough edges.

362.30: Securing and Supporting

Nonmetallic electrical tubing installation shall be fully installed before pulling wires, and it shall be fastened within 3 feet of any box, junction box, and so on, where the ends terminate. Passing through a joist, rafter, or the like is considered a support. Elsewhere, the tubing shall be secured every 3 feet.

Exception

Lengths up to 6 feet can be used without support as a connecting whip to a lighting fixture.

362.46: Bushings

Where the tubing enters boxes and the like, bushings must protect the wiring from abrasion. If the design of the box or fitting is such that abrasion to the wiring bushings is prevented, it won't be required.

Note

It is suggested that you refer to Section 314.6(C) for the protection required with No. 4 AWG or larger.

362.48: Joints

Only approved fittings, couplings, and boxes shall be used by approved methods.

362.56: Splices and Taps

See Article 314. Splices shall never be made within the tubing, but only at junction boxes, outlet boxes, or conduit boxes.

362.60: Grounding

Separate grounding conductors must be used wherever grounding is required.

Article 366—Auxiliary Gutters

366.1: Use

For all appearances, auxiliary gutters could, in a sense, be termed "wireways" (Article 362) or "busways" (Article 364). The main difference is the purpose for which they are to be used and some of the installation requirements. Auxiliary gutters are a supplemental wiring method to be used at meter centers, distribution centers, switchboards, and similar points of wiring systems. They may enclose conductors or bus bars, but are never to be used to enclose switches, for overcurrent devices, for other appliances, or for other apparatus.

366.2: Extension Beyond Equipment

The only place where auxiliary gutters may be extended beyond a distance of 30 feet (9.14 m) is for elevator work (Section 620.35). Whenever they extend beyond this 30-foot distance, they fall in the category of wireways or busways and come under the provisions of wireways (Article 376) and busways (Article 368). See Figure 366-1.

366.3: Supports

Metal auxiliary gutters must be securely fastened at intervals not exceeding 5 feet. Nonmetallic gutters must be supported every 3 feet and at each end or joint.

366.4: Covers

The covers shall be securely fastened to the gutter—this is usually done by means of screws.

Auxiliary gutters are troughs with a removable lid, and may be purchased in almost any length. The lid is usually fastened to the trough with screws (see Figure 366-2): when auxiliary gutters are used outdoors in damp or wet locations, they shall be weatherproof or waterproof, whichever is applicable.

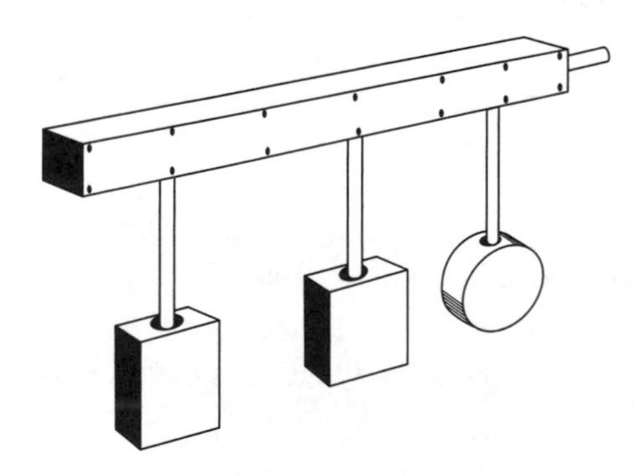

Figure 366-1 Purpose of auxiliary gutters.

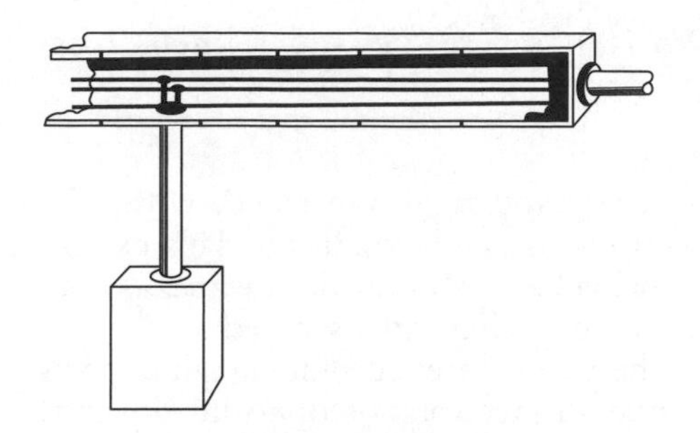

Figure 366-2 Interior and construction of auxiliary gutters.

366.5: Number of Conductors

The ruling is that sheet metal auxiliary gutters shall contain no more than 30 current-carrying conductors at any cross-section. However, there is no limit to the number of conductors used only for signal circuits or the number of controller conductors between a motor and its starter if these control wires are used only for starting duty. In addition, the number of conductors that may be installed in a gutter shouldn't exceed 20-percent fill, regardless of the use of the conductors.

Nonmetallic gutters may not be filled to more than 20 percent of their cross-sectional area.

Exception No. 1

Section 620.35 for use with elevators is an exception.

Exception No. 2

The conductors used for the following are not considered as current-carrying conductors: (a) signal circuits, and (b) control conductors between motor starters when used for starting use duty only.

Exception No. 3

Correction factors are covered in Note 8(a) or Tables 310.16 through 310.31. There shall be no limitation to the number of current-carrying conductors. Instead, the sum of the cross-sectional area of all conductors is not to exceed 20 percent of the cross-section of area of the auxiliary gutter.

366.6: Ampacity of Conductors

The ampacity of insulated aluminum and copper conductors is listed in Tables 310.16 and 310.31, respectively. Conductors in auxiliary gutters are not subject to *NEC* Section 310.15(2)(A) [30 or less] for derating.

When copper bars are used in gutters, their ampacity is limited to 1000 amperes per square-inch of the cross-section of the bar. Aluminum is limited to 700 amperes per square-inch cross-section.

366.7: Clearance of Bare Live Parts

Bare conductors shall be securely and rigidly fastened, with adequate allowance made for contraction and expansion. This expansion and contraction may be provided for by various means, as long as the bars or bare conductors are mechanically secured.

The minimum clearance between bare current-carrying parts that are mounted on the same surface, but of opposite polarity, shall be 2 inches. There shall also be a minimum of 1-inch clearance between bare metal conductors (current-carrying) and the metal surfaces of the gutters.

366.8: Splices and Taps

(A) Within Gutters. Splices and taps are permitted in gutters, but shall be made and installed by approved methods only. They must be accessible by means of removable covers or doors. Not more than 75 percent of the area of the gutter shall be taken up by the conductors, plus the taps and splices.

(B) Bare Conductors. Taps that are made from bare conductors (such as buses) shall leave the gutter opposite their point of terminal connection to the bus, and the taps shouldn't come into contact with any bare current-carrying parts with different potentials.

(C) Suitably Identified. All taps from gutters shall be suitably identified as to the circuit or the equipment that they supply. The *Code* does not spell out exactly how to do this, but the means of identifying the circuits or equipment supplied must meet the approval of the inspection authority. This could be done by tagging the leads in the gutter, but a very practical method is to mark the circuits served on the gutter lid. See Figure 366-3.

(D) Overcurrent Protection. Taps from conductors in a gutter are all subject to the provisions outlined in Section 240.21. In brief, Section 240.21 states that tap conductors not over 10 feet in length shall have an ampacity of not less than the ampacity of the one or more circuits or loads that they supply. Taps not over 25 feet in length shall have an ampacity of at least one-third of the ampacity of the conductors to which they are tapped, and they shall terminate in a single set of fuses or circuit breakers, which will limit the current to that of

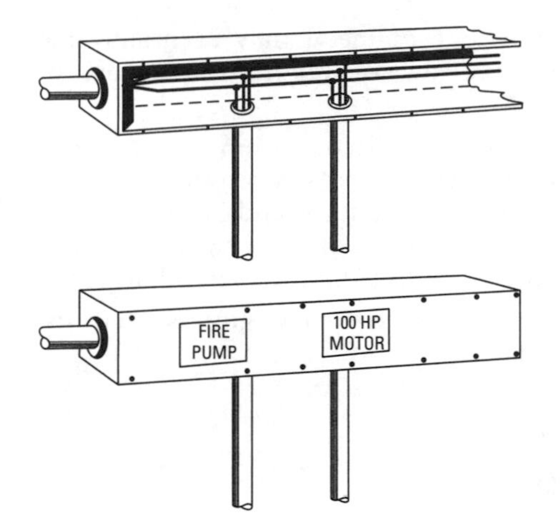

Figure 366-3 Identifying circuits or loads from gutters.

the tap conductors or less. There is considerably more to this "tap rule" in Article 240, but it won't be repeated here.

366.9: Construction and Installation

Auxiliary gutters shall be constructed in accordance with the following:

(A) Electrical and Mechanical Continuity. Gutters shall be such that their installation will maintain electrical and mechanical continuity for the complete system. Mechanical and electrical security are two important points that are ever present in electrical wiring and must never be overlooked.

(B) Substantial Construction. The construction of gutters shall be such that a complete enclosure is provided for the conductors. Suitable protection from corrosion shall be provided both inside and outside the gutters. If corner joints are used, they shall be made tight to prevent entry of moisture and the like. If rivets or bolts are used in this construction of gutters, 12-inch spacing between rivets and bolts is the maximum allowed.

(C) Smooth, Rounded Edges. Protective means shall be applied any place where conductors may be subject to abrasion. In other words, bushings, shields, or other fittings necessary to protect the conductors shall be used at all joints, around bends, between gutters and cabinets or junction boxes, or at any other possible location where abrasion might occur.

Insulation laying on the edge of metal may eventually cut through to the conductor because of weight, temperature, or vibration.

(D) **Deflected Insulated Conductor.** Section 366.6 and Table 366.6(A) will apply wherever insulated conductors are deflected within the auxiliary gutter, at the ends, or where conduit, fittings, or other raceways enter or leave the gutter, or at any point where the gutter is deflected more than 30 degrees.

(E) **Outdoor Use.** Only rain-tight auxiliary gutters shall be installed in wet locations.

Article 368—Busways

I. General Requirements

368.1: Scope

This article covers busways and associated fittings used as service-entrance feeders and branch circuits.

368.2: Definition

Busway, as covered in this article, means a grounded metal enclosure used to contain bare or insulated conductors, which may be either copper or aluminum, and which may be bars, rods, or tubes, and are factory-mounted.

Note

For cable bus, refer to Article 370.

368.4: Use

(A) **Use Permitted.** Busways must be installed in the open, and are to be visible.

Exception

Concealed busway installation is allowed under the following conditions:

(a) Only totally enclosed, nonventilating types of busway may be concealed.

(b) All joints and fittings must be accessible.

(c) It must be installed behind panels that allow access.

(d) It may not be installed in air-handling spaces. Or, if in air-handling spaces, it may not be plug-in-style.

(B) **Use Prohibited.** Busways shouldn't be installed under the following conditions: (1) where physical damage or corrosive conditions exist, (2) in hoistways, or (3) in hazardous (classified) locations [in Section 501.4(B) they are approved in these hazardous locations for some specific uses]. Unless specifically approved for such use, they may not be installed in wet or damp locations or outdoors.

Lighting or trolley busways may not be installed lower than 8 feet above the floor or working platform unless protected with a cover that is identified for such use.

368.5: Supports

Busways are to be supported at intervals not exceeding 5 feet unless specifically approved for support at greater distances.

368.6: Through Walls and Floors

Busways may be extended through walls transversely (horizontally), providing that they go through dry walls and are in unbroken lengths (see Figure 368-1).

Busways may be extended vertically through dry floors when totally enclosed, provided this total enclosure extends for a minimum distance of 6 feet above the floor through which they pass and are adequately protected from physical drainage with a 4-inch-high curb (see Figure 368-2).

368.7: Dead Ends

A dead end of a busway shall be properly closed.

368.8: Branches from Busways

Branches from busways shall be made with busways, rigid or flexible conduit (metal) electric metallic tubing, surface metal raceways, rigid metal conduit, metal-clad cable, listed bus drop cables,

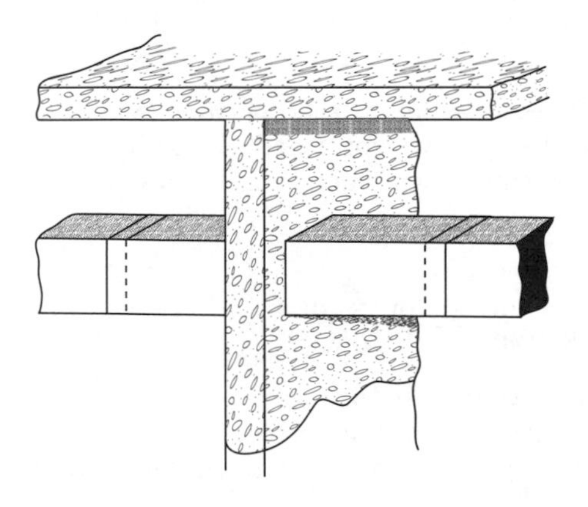

Figure 368-1 Busways may extend transversely through walls in unbroken lengths.

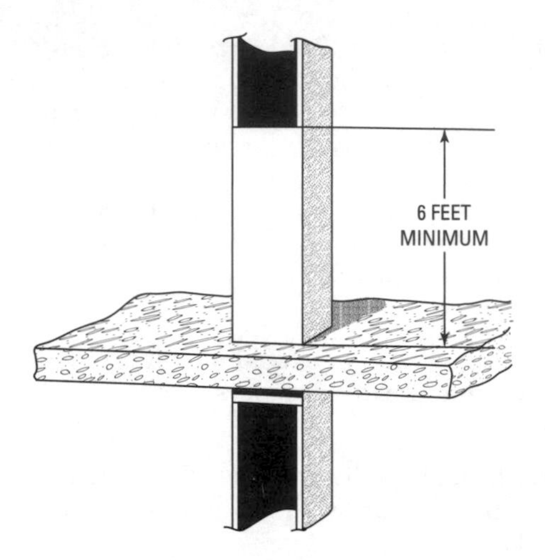

Figure 368-2 Busways may extend vertically through floors (dry), if not ventilated, for a minimum height of 6 feet.

liquid-tight flexible metal conduit, electrical nonmetallic tubing, or with suitable cord assemblies approved for hard usage with portable equipment or to facilitate the connection of stationary equipment to aid in interchanging said equipment.

There are provisions for use of flexible cord assemblies. These connections may be made directly to the end terminals of a busway that has a plug-in device installed. These cords must not extend more than 6 feet horizontally from the busway. There is no vertical limit if suitable means are used to take up the tension on the cord. Again we must bear in mind the continuity of equipment-grounding conductors; therefore, if nonmetallic raceways are used, connection to the equipment-grounding conductor shall meet the requirements of Sections 250.8 and 250.12.

Exception

Lengths of cord longer than 6 feet may be installed between a busway plug-in device and a tension take-up device provided the cord is supported every 8 feet and that the entire installation is under the supervision of qualified people.

368.9: Overcurrent Protection

It is necessary to provide overcurrent protection with busways in accordance with Section 368.10 through 368.13.

368.10: Rating of Overcurrent Protection—Feeders

If the overcurrent protection device does not correspond to the rating of the busway, the next larger size of overcurrent protection may be used. However, this only applies up to 800 amperes. Above 800 amps, the next larger size may not be used. Exceptions are made under the provisions of Sections 240.3 and 450.6(A)(3).

368.11: Reduction in Size of Busway

Busways in industrial establishments may be reduced in size without the use of overcurrent protection at the point of reduction providing that the following condition or conditions are met: (1) The smaller busway does not extend over 50 feet; (2) the smaller busway has a current rating of at least one-third that of the larger busway; (3) the smaller busway is protected by overcurrent capacity of not over three times the rating of the smaller busway; and (4) the smaller busway doesn't come in contact with combustible material.

368.12: Subfeeder or Branch Circuits

Where busways are to be used as a feeder, devices or plug-in connections for tapping off subfeeders or branch circuits shall contain the overcurrent devices to protect the subfeeder or branch circuit. These may be either externally operated circuit feeders or externally operated usable switches. If these overcurrent devices are not readily reachable, disconnecting means shall be provided for opening them. This means of opening or disconnecting the overcurrent device may be ropes, chains, or suitable sticks that may be operated from the floor.

Exception No. 1

For the overcurrent protection of taps, refer to Section 240.21, which allows placing the overcurrent protection at various distances from the busway, depending on conditions.

Exception No. 2

On cord-connected fixtures of the fixed or semifixed type, the overcurrent device may be a part of the fixture cord plug.

Exception No. 3

The overcurrent protection may be mounted on cordless fixtures that are plugged directly into the busway.

368.13: Rating of Overcurrent Protection—Branch Circuits

Busways may be used as a branch circuit of any type of branch circuit covered in Article 210. When a busway is used as a branch circuit, the overcurrent device protecting the busway usage determines the ampere rating of the branch circuit. Any requirements of Article 210 that apply to busway branch circuits shall be complied with.

368.15: Marking

Busways must be marked, showing the voltage and currents for which they were designed, and with the manufacturer's name or trademark.

II. Requirements for over 600 Volts, Nominal

368.21: Identification

Each run shall have a nameplate permanently marked with the voltage rating and the continuous current rating. If the busway is force-cooled, both the normal forced-cooled current rating and the self-cooled rating shall be marked; the rated frequency, the rated impulse withstanding voltage, the rating at 60 AC Hertz of the voltage it will withstand, and the momentary overcurrent it may withstand shall also be marked.

368.22: Grounding

Metal-enclosed busways that meet the requirements of Article 250 must be installed for equipment grounding continuity.

368.23: Adjacent and Supporting Structures

Busways that are metal-enclosed shall be installed in such a way that any temperature rises that may occur from induced currents circulating in any adjacent metallic parts will in no way be hazardous to people or in any manner constitute a fire hazard.

368.24: Neutral

The neutral bus shall be designed to carry any loads that might be placed on it, especially with discharge lighting. Harmonic currents will increase the current in the neutral. Any momentary short circuit requires that the neutral be capable of handling it. See Note 10 in Article 310.

368.25: Barriers and Seals

Any busways installed outside the building that enter the building shall incorporate a vapor seal to prevent the interchange of interior and outdoor air.

Where floors or ceilings are penetrated, or drop ceilings or walls are penetrated, fire barriers must be installed.

Exception

If the busway is forced-cooled, vapor seals won't be required.

368.26: Drain Facilities

For the removal of any condensed moisture at low points in the bus run, drain plugs, filter drains, or other similar methods shall be provided.

368.27: Ventilated Bus Enclosures

Article 710, Part IV covers ventilated bus enclosures, unless the design is such that foreign objects can't be inserted into the energized parts of the busway.

368.28: Terminations and Connections

If the busway enclosures terminate at machines that are cooled by flammable gases, proper seal-off bushings, baffles, or any other means that will keep the flammable gas from entering the bus enclosures are required.

Temperature changes will cause the buses to expand or contract; therefore, flexible or expansion connections will be required. If building vibration-insulation joints are used, means shall be provided to adjust for temperature changes.

All conductor terminations and any connecting hardware shall be readily accessible for installation, connection, and maintenance.

368.29: Switches

Switches must have the same momentary rating as the busway. Disconnecting links must be marked so as not to be removed when energized. Switches not designed to open under load must be interlocked to prevent this from happening. Disconnecting link enclosures must be interlocked, so that there can be no access to live parts.

368.30: Low-Voltage Wiring

If secondary control wiring is provided as a part in the bus that is metal-enclosed, it shall be insulated from all primary circuit elements by a flame-retardant barrier. This does not include short lengths of secondary conductors such as those running to instrument transformers.

Article 370—Cablebus

370.1: Definition

Cable bus is an assembly of conductors, fittings, and terminations in an enclosed and ventilated housing. It must be designed to carry fault currents and to withstand the resultant forces.

370.2: Use

(A) 600 Volts or Less. Cable bus can be used at any voltage or current for which the conductors are rated, and may be installed exposed only. It must be identified for the use when installed in wet, corrosive, hazardous, or damp locations. The framework of the cablebus may be used as an equipment ground when properly bonded.

(B) Over 600 Volts. Cable bus may be used over 600 volts.

370.3: Conductors

(A) Types of Conductors. Conductors in cable bus must have a temperature rating of at least 75°C and must be suitable for the locations, as per Articles 310 and 490.

(B) Ampacity of Conductors. Ampacity must comply with Tables 310.17 and 310.19 for installations under 600 volts or Tables 310.69 and 310.70 for installations over 600 volts.

(C) Size and Number of Conductors. The conductors must match the design specifications of the cablebus, but may not be smaller than 1/0.

(D) Conductor Supports. The insulated conductors must be supported by blocks or other means designed for the purpose. For horizontal runs, the conductors must be supported at least every 3 feet; and for vertical runs, at least 1½ feet. Spacing between the conductors can't be more than one conductor diameter.

370.5: Overcurrent Protection

If the allowed ampacity of the conductors does not match a standard rating, the conductors may be protected at the next higher standard rating, but only up to 800 amps. Above 800 amps, the next larger size may not be used.

Exception

For systems operating at over 600 volts, overcurrent protection according to Section 240.100 may be used (but is not required).

370.6: Support and Extension Through Walls

(A) Support. Cablebus must be supported at least every 12 feet. When longer spans are necessary, the product must be designed for such support spacing.

(B) Transversely Routed. Cablebus may be installed through walls (not firewalls), as long as the walls are continuous, protected from damage, and not ventilated.

(C) Through Dry Floors and Platforms. Cablebus can be extended vertically through floors and platforms (except where fire steps are needed) if the cablebus is totally enclosed where it goes through the floor, and for 6 feet above the floor or platform.

(D) Through Floors and Platforms in Wet Locations. The same requirements for dry locations apply for wet locations. In addition, water must be prevented from flowing through the floor opening by curbs or some other means.

370.7: Fittings

Cablebus systems must have approved fittings for horizontal or vertical changes in direction, dead ends, terminations to equipment or enclosures, and guards for extra protection (used only where required).

370.8: Conductor Terminations

An approved means of terminating must be used for connections to conductors in a cablebus.

370.9: Grounding

Cablebus sections must be bonded together and grounded in accordance with Article 250, but not according to Section 250.86.

370.10: Marking

Each section must be marked, showing the manufacturer's name, and the maximum number rating, voltage, and current of the conductors.

Article 372—Cellular Concrete Floor Raceways

This article is very similar to the following article, "Cellular Metal Floor Raceways." Because of this similarity, only those points that are treated differently will be covered in order to prevent repetition.

Figure 372-1 shows the construction of cellular concrete floor raceways. They are constructed of precast concrete with cells or openings provided for the wiring conductors. Since this type of raceway is made of concrete, it can't be made electrically continuous. Therefore, an equipment ground of the proper size must be used in the installation, and all header ducts, junction boxes, and inserts shall be electrically secured to this grounding conductor.

Refer to the *NEC* for coverage of this article.

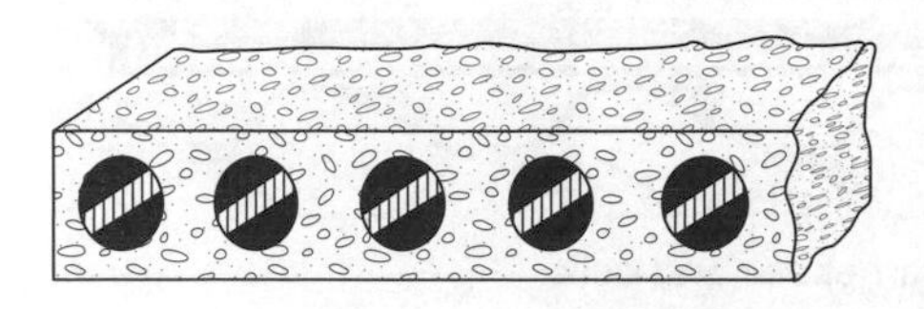

Figure 372-1 Cross-sectional view of a cellular concrete floor raceway.

Article 374—Cellular Metal Floor Raceways

This article is similar in many respects to the preceding article on underfloor raceways.

374.2: Definition

In order to gain a better understanding of cellular metal floor raceways, illustrations will be used. Figure 374-1 shows a cross-section of cellular metal floor raceway. Raceways are made not only for electrical conductors, but also for telephone lines, signal circuits, steam, and hot and cold water pipes. Figure 374-2 illustrates a cell. Notice that it is just one single enclosed tubular area. Figure 374-3 illustrates a header that is transversely connecting two cells. With this arrangement, conductors may be run at right angles to the cells, as well as in the cells of the raceway.

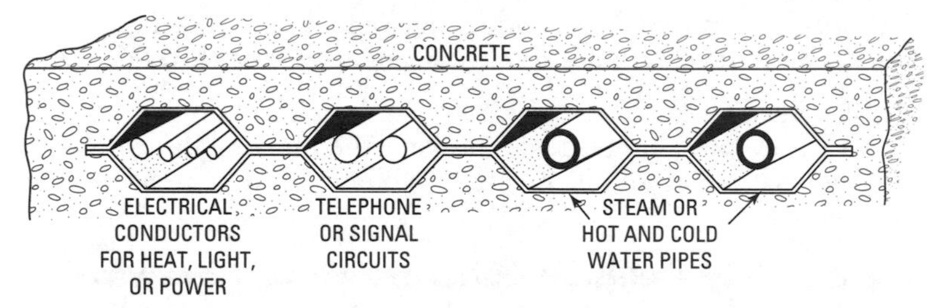

Figure 374-1 Cross-sectional view of cellular metal floor raceways used for installation of electrical and other systems.

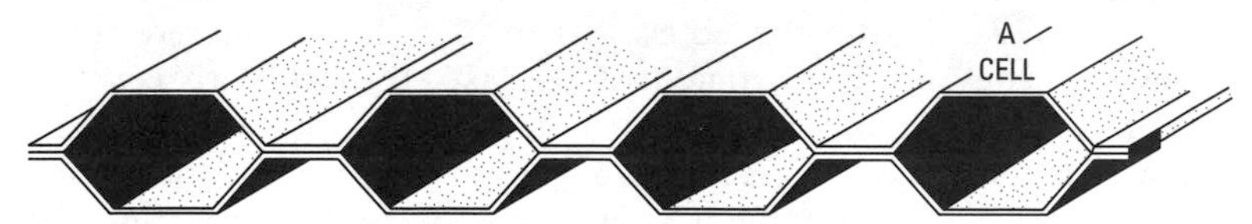

Figure 374-2 Illustration of a cell.

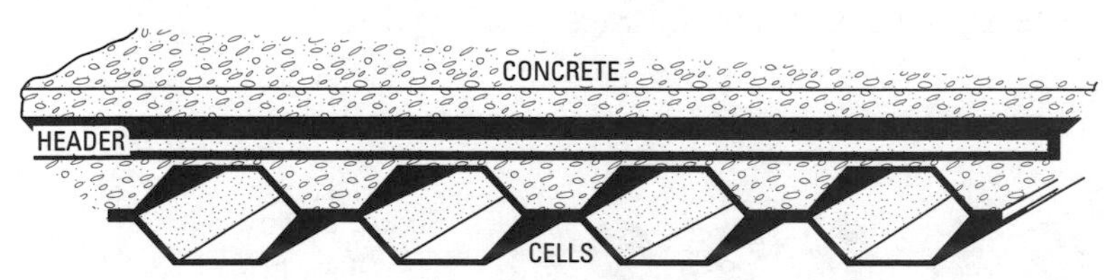

Figure 374-3 Illustration of a header and cells.

374.2: Use

Conductors are not to be installed in cellular metal floor raceways (1) where corrosive vapors are present; (2) in any hazardous (classified) locations, except, as indicated in the *NEC*, for Class I, Division 2 locations in some instances as permitted in the Exception to Section 501.4(B); or (3) in commercial garages (except where they supply outlets or extensions to the area below the floor but not above). See Section 300.8 in the *NEC*, and Figure 374-1.

II. Installation

374.4: Size of Conductors

Except by special permission, conductors larger than 1/0 shouldn't be installed. Notice the definition of "special permission" as it appears in Article 100—"the written consent of the authority enforcing this *Code*."

374.5: Maximum Number of Conductors in Raceway

The fill shall be no more than 40 percent of the cross-sectional area of the raceway. In other words, the total cross-sectional area of all the conductors shouldn't fill the raceway to over 40 percent of its capacity. Tables in Chapter 9 of the *NEC* give the cross-sectional area of various conductors and their insulations. These figures are to be used in the calculation of the total cross-sectional area of the conductors. The requirements just mentioned won't apply where Type AC metal-clad cable or nonmetallic-sheathed cables are used in the raceways. The area of the raceway will have to be calculated mathematically unless the specifications are available to the wirer.

374.6: Splices and Taps

Splices and taps are allowed only in junction boxes with underfloor raceways. With cellular metal floor raceways, they are also permitted in header access boxes. Refer to Figure 374-3, which shows a header.

374.7: Disconnected Outlets

Conductors that supply outlets that are being discontinued shall be removed from the raceway back to a junction box. They can't be merely taped up and left in place.

374.8: Markers

Markers are to be extended through the floor for the purpose of locating the cells and the wiring system in the future. There should be enough markers to properly assist in the location of the raceways. It is also recommended that the location of these markers be indicated on the final set of electrical plans that is to be given to the owner.

374.9: Junction Boxes

Junction boxes are to be installed level with the floor grade and are to be sealed against the entrance of water and dirt. They shall be made of metal and shall be made electrically continuous with the rest of the system. Although not mentioned at this point, the metal of which the junction boxes are made should be such as to not cause corrosion or electrolysis.

374.10: Inserts

Inserts (such as for outlets) shall be made of metal and made electrically continuous with the rest of the system. They shall be installed level with the floor grade and made watertight. When cutting the raceway for the installation of these inserts, no rough spots shall be left, and the dirt and chips removed, so as not to cause abrasion to the insulation of the conductors. When installing inserts, the tools used when cutting through the raceway shall be such as not to cause damage to the conductors that have been installed.

374.11: Connection to Cabinets and Extensions from Cells

Flexible metal conduit may be used for connections from raceways, distribution centers, and wall outlets if they are not concrete. If in concrete, rigid metal conduit, intermediate metal conduit, EMT, or approved fitting shall be used. If provisions have been made for equipment-grounding conductors, nonmetallic conduit or electrical nonmetallic tubing is permitted.

Article 376—Metal Wireways

376.2: Definition

Wireways are troughs made of sheet metal, with either hinged or removable covers. They are used for protecting wires or cables (Figure 376-1).

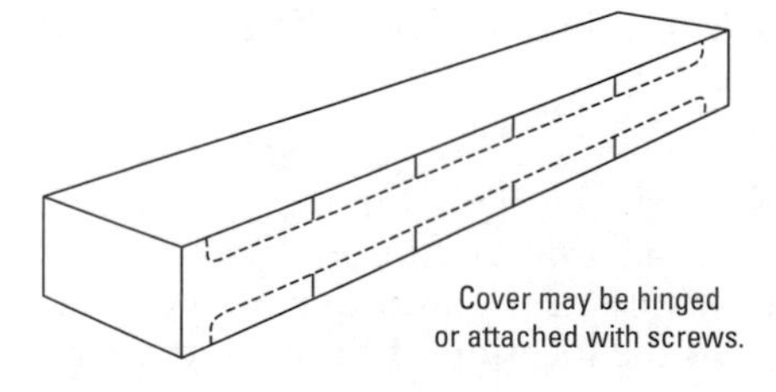

Figure 376-1 Illustration of wireway.

376.10: Uses Permitted

Wireways shall be used only for exposed work, and may be used out of doors if of an approved rain-tight construction.

Wireways shouldn't be installed where they can be physically damaged, or where there are corrosive vapors. They can't be installed in hazardous (classified) locations. There is an exception for Class I, Division 2 and Class II, Division 2 in these sections: 501.4(B) and 502.4(B).

When passing through walls, metallic wireways shall be in unbroken lengths, but the conductors must remain accessible.

376.21: Size of Conductors

The design of the wireway shall govern the maximum size conductor that will be permitted.

376.22: Number of Conductors

There is a limit of 30 current-carrying conductors in any wireway, at any cross-sectional area. This is the maximum number of conductors allowed.

For this purpose conductors for signal circuits, or controller circuits between starter and motor, shouldn't be considered current-carrying conductors.

No more than 20 percent of the cross-sectional area of a wireway may be used for conductors. This is based on the total cross-sectional area of the conductors against the cross-sectional area of the wireway.

There are derating factors specified in Section 310.15(B)(2)(a) that are not applicable when the 30 current-carrying conductors are at the 20-percent fill specified above.

In elevators and dumbwaiters, you may use up to 50 percent of the cross-sectional area of the wireway. The 30-conductor limit does not apply for theaters. See Section 520.5.

Note

Cross-sectional area of conductors is covered in the tables and the notes at the beginning of Chapter 9.

376.23: Insulated Conductors

(A) **Deflected Insulated Conductors.** When conductors are deflected (bent) in the wireway (for whatever reason) or if the wireway deflects more than 30 degrees, the wire-bending dimensions of Section 314.6 must be followed.

(B) **Metallic Wireways Used As Pullboxes.** When insulated conductors No. 4 or larger are pulled through a wireway, the

Code, in effect, considers the wireway to be a pullbox. Thus the rules of Section 314.28(A) apply to such installations.

376.30: Securing and Supporting

Wireways may be screwed or bolted to a wall or supported by hangers or any other suitable and acceptable means, but in any case they shall be supported at intervals not to exceed 5 feet (1.52 m). In no case shall the distance between supports exceed 10 feet (3.05 m). (See Figure 376-2.)

Exception

Vertical run supports for wireways shall not be over 15 feet apart. There shouldn't be more than one joint between supports. Any adjoining wireways shall be fastened to the other wireway so as to provide a rigid joint.

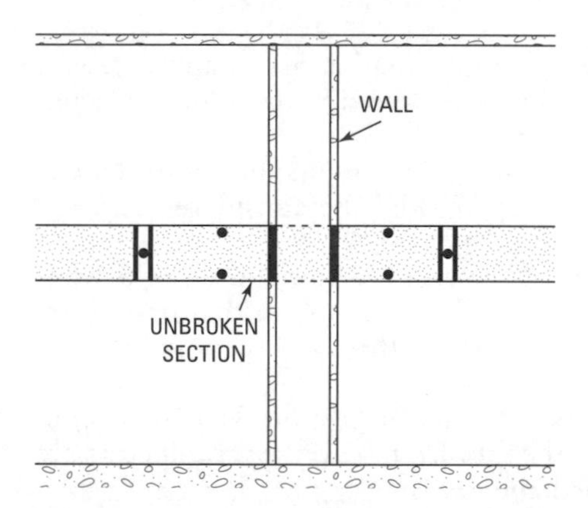

Figure 376-2 Wireways may extend transversely through a wall in unbroken lengths.

2005

376.56: Splices and Taps

This is one wiring method in which splices and taps are allowed, but with the following restrictions: (1) they must be accessible; (2) they must be insulated by approved means; and (3) they shouldn't fill the wireway to more than 75 percent of its area at that point. If the splices or taps are staggered slightly, more room can be obtained.

Power distribution blocks may not have exposed live parts inside the wireway after the installation is complete.

376.58: Dead Ends

Dead ends shall have caps to close the ends of the wireways.

376.60: Grounding

The provisions of Article 250 shall be adhered to.

376.70: Extension from Wireways

Extension from wireways may be made with many different types of raceways. Refer to the *NEC* for this section. However, we are always faced with equipment-grounding continuity, and if it is required by some of this type of raceway, continuity must be maintained throughout. See Sections 250.113 and 250.118.

Article 378—Nonmetallic Wireways

378.2: Definition

Nonmetallic wireways are troughs made of sheet metal, with either hinged or removable covers. They are used for protecting wires or cables.

378.10: Uses Permitted

Nonmetallic wireways shall be used only for exposed work, and may be used out doors if of an approved rain-tight construction.

Nonmetallic wireways shouldn't be installed where they can be physically damaged, or where there are corrosive vapors. They can't be installed in hazardous (classified) locations. There is an exception for Class I, Division 2, and Class II, Division 2, in these sections: 501.4(B) and 502.4(B).

When passing through walls, nonmetallic wireways shall be in unbroken lengths, but the conductors must remain accessible.

378.21: Size of Conductors

The design of the wireway shall govern the maximum size conductor that will be permitted.

378.22: Number of Conductors

There is a limit of 30 current-carrying conductors in any wireway, at any cross-sectional area. This is the maximum number of conductors allowed.

For this purpose, conductors for signal circuits, or controller circuits between starter and motor, shouldn't be considered current-carrying conductors.

No more than 20 percent of the cross-sectional area of a wireway may be used for conductors. This is based on the total cross-sectional area of the conductors against the cross-sectional area of the wireway.

There are derating factors specified in Section 310.15(B)(2)(a) that are not applicable when the 30 current-carrying conductors are at the 20-percent fill specified above.

In elevators and dumbwaiters, you may use up to 50 percent of the cross-sectional area of the wireway. The 30-conductor limit does not apply for theaters. See Section 520.5.

Note

Cross-sectional area of conductors is covered in the tables and the notes at the beginning of *NEC* Chapter 9.

378.23: Insulated Conductors

(A) Deflected Insulated Conductors. When conductors are deflected (bent) in the nonmetallic wireway (for whatever reason) or if the wireway deflects more than 30 degrees, the wire-bending dimensions of Section 314.6 must be followed.

(B) Metallic Wireways Used As Pullboxes. When insulated conductors No. 4 or larger are pulled through a wireway, the *Code*, in effect, considers the wireway to be a pullbox. Thus the rules of Section 314.28(A) apply to such installations.

378.30: Securing and Supporting

Wireways may be screwed or bolted to a wall or supported by hangers or any other suitable and acceptable means, but in any case they shall be supported at intervals not to exceed 5 feet (1.52 m). In no case shall the distance between supports exceed 10 feet (3.05 m).

Exception

Vertical run supports for wireways shall not be over 15 feet apart. There shouldn't be more than one joint between supports. Any adjoining wireways shall be fastened to the other wireway so as to provide a rigid joint.

2005

378.56: Splices and Taps

This is one wiring method in which splices and taps are allowed, but with the following restrictions: (1) They must be accessible; (2) they must be insulated by approved means; and (3) they shouldn't fill the wireway to more than 75 percent of its area at that point. If the splices or taps are staggered slightly, more room can be obtained.

Power distribution blocks may not have exposed live parts inside the wireway after the installation is complete.

378.60: Grounding

The provisions of Article 250 shall be adhered to.

378.58: Dead Ends

Dead ends shall have caps to close the ends of the wireways.

378.70: Extension from Wireways

Extension from wireways may be made with many different types of raceways. Refer to the *NEC* for this section. However, we are always faced with equipment-grounding continuity, and if it is required by some of this type of raceway, continuity must be maintained throughout. See Sections 250.113 and 250.118.

Article 380—Multioutlet Assembly

Multioutlet assemblies consist of either a flush or surface raceway that has been designed to hold receptacle outlets and has either been factory-assembled or assembled in the field. They are used especially where there are a number of outlets required in a relatively short space, such as in show rooms where the connections for appliances for demonstration, for outlets along a workbench, and so on.

Some multioutlet assemblies have receptacles that are movable. They make connections with some type of bus and may be slid along to obtain different spacings. Other assemblies are designed with outlets at fixed intervals so as to facilitate the connection of appliances, tools, and so on. There are many different makes available, so care should be taken to pick a multioutlet assembly that is approved and to install it properly.

II. Installation

380.2: Uses

Multioutlet assemblies are to be installed only in dry locations. They are not to be installed: (1) where concealed, except that the back and sides of metal multioutlet assemblies may be surrounded by the building finish, and nonmetallic multioutlet assemblies may be recessed in the baseboard (Figures 380-1 and 380-2); (2) where subject to severe physical damage; (3) where the voltage is 300 volts or more between conductors, unless the assembly is of metal that has a thickness of at least 0.040 inch; (4) where subject to corrosive vapors; (5) in hoistways; or (6) in any hazardous (classified) locations.

It may be used in Class I, Division 2, locations as permitted in Section 501.4(B).

380.3: Metal Multioutlet Assembly Through Dry Partitions

Metal multioutlet assemblies shouldn't be run in partitions, but may be run through dry partitions providing that the covers outside of the partition are arranged so that they may be removed, and providing that there are no outlets inside the partition (Figure 380-3).

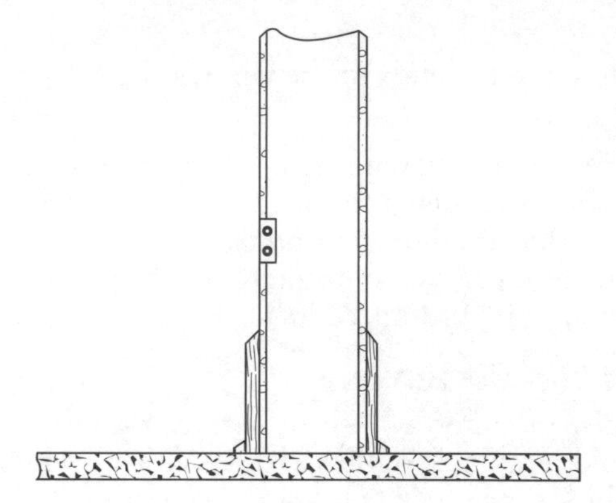

Figure 380-1 Metal multioutlet assemblies may be installed in building finishes.

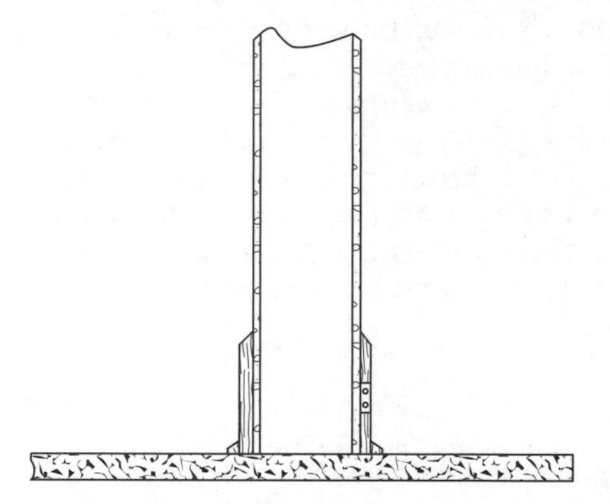

Figure 380-2 Nonmetallic assemblies may be installed in baseboards.

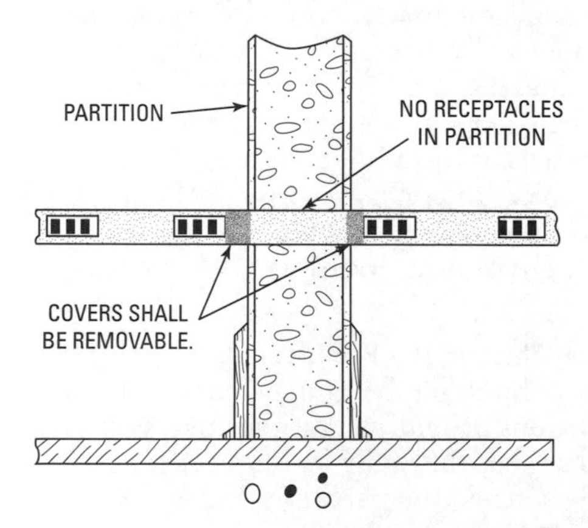

Figure 380-3 Multioutlet assemblies may pass through dry partitions, provided there are no outlets in the partition and the covers are removable.

Article 382—Nonmetallic Extensions

382.2: Definitions

Nonmetallic extension is a wiring method that is intended for specific purposes, mainly that of extensions either on the surface or in the form of an aerial-cable assembly. Reference to Figure 382-1 will give an idea of its construction and will facilitate its application. This illustration shows a cross-section and how it is mounted on the surface.

II. Installation

382.10: Uses Permitted

The use of nonmetallic extensions is permitted where the following conditions are met:

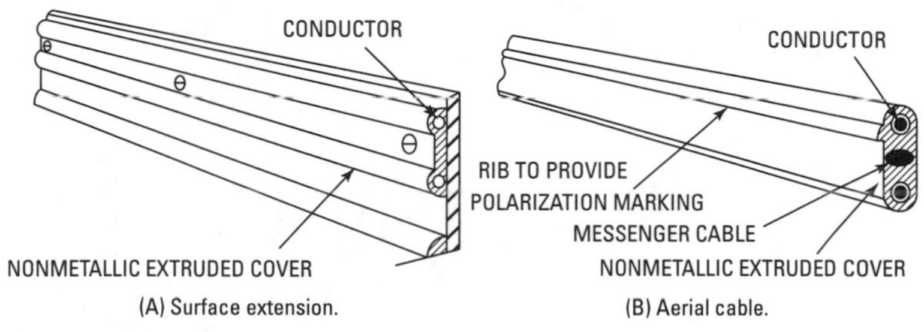

Figure 382-1 Cross-sectional view of nonmetallic surface-extension cable and nonmetallic aerial cable.

- **(A) From an Existing Current.** These extensions are only permitted from existing 15- to 20-ampere branch circuits that meet all of the requirements in Article 210.
- **(B) Exposed and in Dry Locations.** The extension is never to be run concealed and must always be used in dry locations.
- **(C) Nonmetallic Surface Extensions.** Nonmetallic surface extensions may be used in residential areas or in offices.
  - **(C1)** [**Alternative to** (C)] Aerial cable is to be used for industrial purposes where flexibility is required for connecting equipment. It is not intended for use in offices and residences.

Note

Temperature limitations for nonmetallic surface extensions are covered in Section 310.10.

Figure 382-2 illustrates the use of nonmetallic surface extensions. Figure 382-3 illustrates the use of aerial nonmetallic cable with fluorescent lighting.

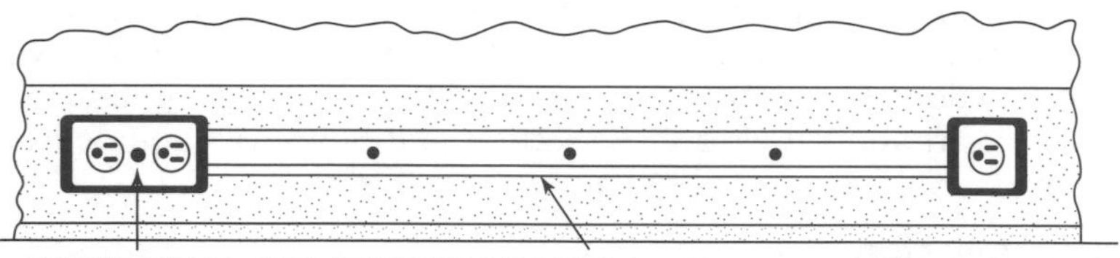

Figure 382-2 Purpose of nonmetallic surface extensions.

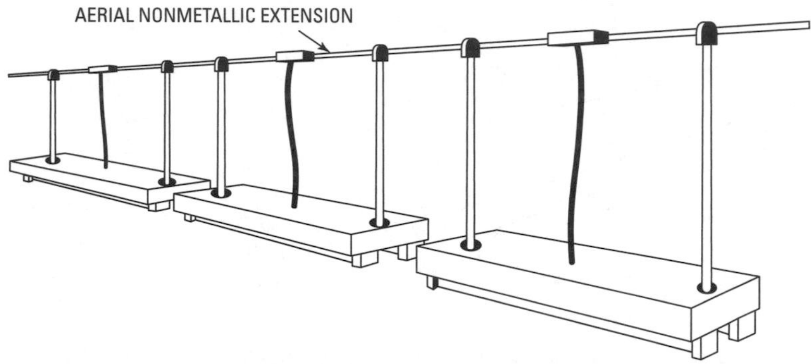

Figure 382-3 Aerial nonmetallic extension.

Nonmetallic extensions shall be installed in conformity with the following requirements:

(1) One, or more than one, extension may be run in any direction from the existing outlet from which it is supplied. This won't apply to extensions run on the floor or within 2 inches of the floor. (See Figure 382-4.)

(2) Nonmetallic surface extensions shall be secured in place by approved means at intervals not exceeding 8 inches, except that where connection to the supplying outlet is made by means of an attachment plug, the first fastening may be placed 12 inches or less from the plug. There shall be at least one fastening between each two adjacent outlets supplied. An extension shall be attached only to woodwork or plaster finish, and shouldn't be in contact with any metal work or other conductive material except with metal plates on receptacles (see Figure 382-5).

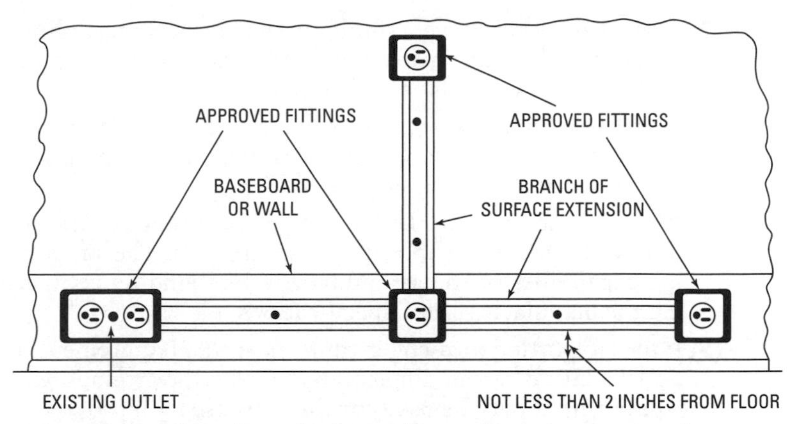

Figure 382-4 Branches of nonmetallic surface extensions and clearance.

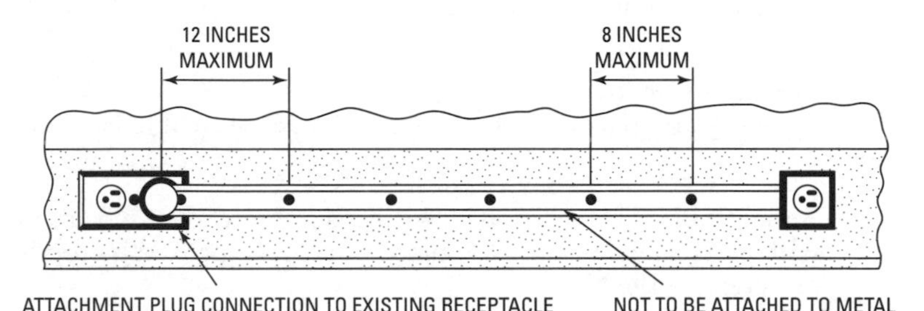

Figure 382-5 Supporting distances for nonmetallic surface extensions.

(3) A cap shall be furnished to protect the assembly from physical damage when bends reduce the normal spacing of conductors.

(D) Aerial Cable.

(1) Aerial cable shall be supported by the messenger cable in the assembly, not by the conductors or insulation. The messenger cable is to be fastened securely at both ends by means of approved cable clamps and turnbuckles for taking up any sag in the assembly. If the span is over 20 feet, the assembly shall be supported by approved hangers at intervals not to exceed 20 feet. The cable shall be so suspended as to eliminate excessive sag and shall clear any metal by a minimum of 2 inches. The assembly is not to contact any metal and the messenger cable is to take the strain of supporting the assembly with added supports as required.

(2) Aerial cable shall have a minimum height of 10 feet above areas of pedestrian traffic only and a minimum height of 14 feet above vehicular traffic (see Figure 382-7).

(3) Cable assemblies that are over workbenches, but not over vehicular or pedestrian traffic areas, may have a minimum height of 8 feet (see Figure 382-8).

(4) Where the strength of the messenger cable is not exceeded, it may also serve to support lighting fixtures. The supporting capabilities of the assembly may be found by contacting the manufacturer for specifications.

(5) If the supporting messenger cable meets the requirement of Article 250 for grounding conductors for the grounding of equipment, it may be used for this purpose if all provisions covering the same in Article 250 are met, but under no condition is to be used as a grounded or hot conductor of a branch circuit.

382.12: Uses Not Permitted

The following are places where nonmetallic surface cable can't be used:

(A) **Aerial Cable.** This method of wiring is not intended to be used with aerial cable to take the place of other approved methods provided by the *Code*.

(B) **Unfinished Areas.** They are never to be used in unfinished basements, attics, or roof spaces, but are intended for use only in finished places and must be exposed.

(C) **Voltage Between Conductors.** The maximum voltage shouldn't exceed 150 volts between conductors for nonmetallic surface extensions and shouldn't exceed 300 volts between conductors for aerial cables.

(D) **Corrosive Vapors.** They shouldn't be used where subject to corrosive vapors.

(E) **Through a Floor or Partition.** They are to be installed only within the room in which they originate and are not to be run through walls, floors, or partitions.

382.24: Bends

Bends that reduce the spacing between conductors must be covered with a cap. (See Figure 382-6.)

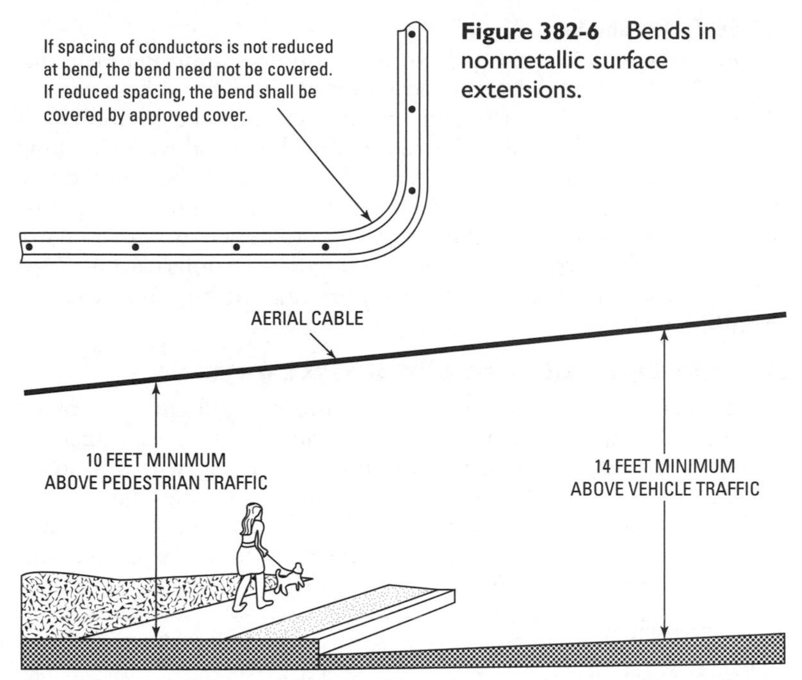

Figure 382-6 Bends in nonmetallic surface extensions.

Figure 382-7 Clearances above floors for aerial nonmetallic surface extensions.

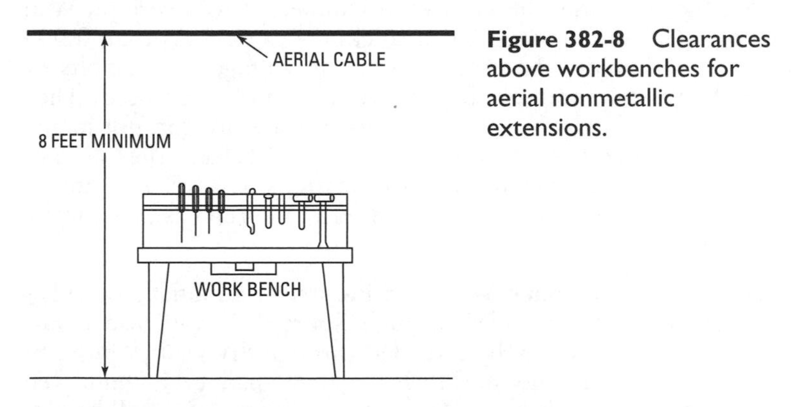

Figure 382-8 Clearances above workbenches for aerial nonmetallic extensions.

382.40: Boxes and Fittings

The fittings and devices used with this method of wiring are to be of a type identified for the use and each run shall be terminated so that the end of the assembly is covered and no bare conductors exposed.

382.56: Splices and Taps

Nonmetallic extensions shall be continuous, unbroken lengths. There shall be no splices and no exposed conductors between fittings. If approved coverings are used, taps may be permitted. If used as aerial cable, and approved material is installed for making taps, provision must be made for polarization of the conductors. Taps of the receptacle-type shall be provided with locking-type devices, rather than the ordinary receptacle device. Refer to Figure 382-1, which shows a rib on the aerial cable for polarization. Taps other than by means of devices approved for this purpose are prohibited.

Article 386—Surface Metal Raceways

Metal surface raceways provide a wiring method that has many advantages. This wiring method is not intended for new construction, but is quite valuable in additions to existing wiring systems that must be expanded without cutting into the existing building to add conduit and other components. Metal surface raceways have been used for many years and have been found quite satisfactory for the purpose for which they are intended.

II. Installation

386.10: Uses Permitted

Surface metal raceways may be used in dry locations. They shouldn't be used as follows: Unless otherwise approved and listed, they shouldn't be used where subjected to severe physical damage. Unless the metal has a thickness of not less than 0.040 inch, they shouldn't be used. Where the voltage is 300 volts or more between circuit conductors, they shouldn't be used. They won't be approved where corrosive vapors are present, nor in any hoistways. Except as permitted in Section 501.4(H), they are not permitted in hazardous (classified) locations, Class I, Division 2. Nor are they permitted where concealed, except as specifically allowed.

(A) **Extension Through Walls and Floors.** Multioutlet assemblies are not to be extended through floors and walls. Metal surface raceways may be extended through dry walls, dry partitions, and dry floors. However, they shall be in unbroken lengths where they pass through, so that no joint will be hidden. Also, the conductors must remain accessible.

Note

The *NEC* refers you to Article 380 for multioutlet assemblies.

386.21: Size of Conductors

The size of the raceway determines the size of the conductors that it may be used for. The design of the size of the metal surface raceway is a governing factor. This will make it necessary to know what size conductor the raceway was designed for.

386.22: Number of Conductors in Raceways

The design of the raceway governs the number of conductors that may be installed. Also, the number will depend upon the size of the conductors. The manufacturer of the raceway can be contacted for this information.

If all of the following conditions are met, the derating factors in *NEC* Note 8(a) accompanying Tables 310.16 through 310.31 shouldn't apply: The cross-sectional area of the raceway exceeds 4 square-inches; not more than 30 current-carrying conductors are installed; the sum of the cross-sectional area of all conductors installed in metal surface raceways does not exceed a maximum of 20 percent of the interior cross-sectional area of the raceway.

Note

The cross-sectional areas of different sizes of conductors and insulation can be found in Tables and Notes at the beginning of Chapter 9.

386.56: Splices and Taps

Splices and taps are permitted if the metal surface raceways have removable covers that are readily accessible when the installation is completed. Seventy-five percent of the raceway area is the limitation placed on splices and taps in the area. If there are no removable covers, splices and taps shall be made only in junction boxes. Of course, all splices and taps are to be made by approved methods.

386.60: Grounding

Surface metal raceways that provide a transition from other wiring methods must have a means of connecting to an equipment-grounding conductor. (See Figures 386-1 and 386-2.)

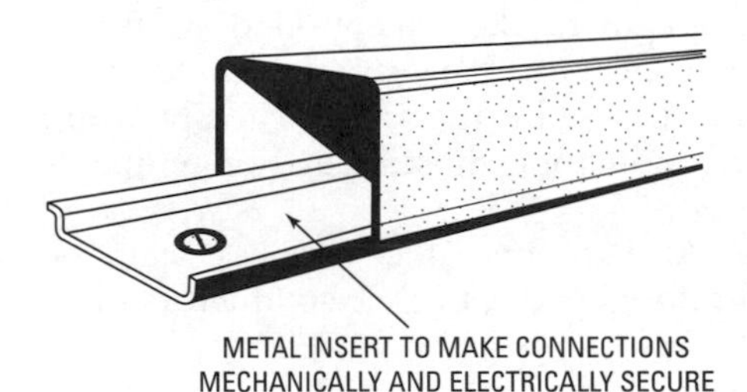

Figure 386-1 Connectors shall be electrically and mechanically secure.

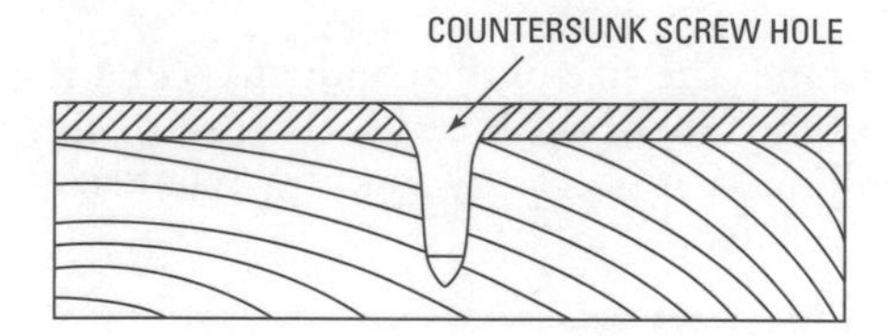

Figure 386-2 Holes for screws shall be countersunk to protect wire from damage.

386.70: Combination Raceways

Take special note of this section, as it is quite different from conduit systems. Sometimes combinations for signal, lighting, and power circuits may be necessary to run in the same metal surface raceway. Where this is required, two sections shall be made in the raceway by a metal partition, and each separate compartment in the raceway shall be identified by a different color. These colors shall be maintained throughout the premises.

Article 388—Surface Nonmetallic Raceways

388.2: Description

This section covers a raceway made of nonmetallic material. It is resistant to moisture and chemical atmospheres. Heat resistance is important, as well as resistance to impact and distortion. The temperature under which it is likely to be used shouldn't cause distortion, and it shall also be resistant to low temperature effects.

II. Installation

388.10: Uses Permitted

It should be used in dry locations. It should not be used in concealed locations, where subject to physical damage. Voltages between conductors shall be 300 volts or less. It is not to be used in hazardous (classified) locations, with the exception of Class 1, Division 2 where permitted in Section 501.4(B). It shouldn't be used in hoistways. It shouldn't be used if the ambient temperature exceeds 50°C (122°F), or for conductors whose insulation temperature exceeds 75°C (167°F).

Unbroken lengths of surface nonmetallic raceway can be installed through drywall and floor spaces, but the conductors must remain accessible.

2005

388.12: Uses Not Permitted

Surface nonmetallic raceways are not to be used: where they are concealed (except when run through walls as indicated above); where subject to damage; where the voltage is greater than 300 volts between conductors (unless specifically listed for the use); in hoistways, excessively hot locations; or in hazardous locations [see Section 501.4(B)(3) for an exception]. Conductors whose temperature limits are higher than the temperature rating of the raceway may not be used. They also may not be used in damp or wet locations.

388.21: Size of Conductors

The design of the nonmetallic surface raceways governs the conductor sizes that may be used.

388.22: Number of Conductors in Raceways

The design of the nonmetallic surface raceways and the size of the conductors governs the number of conductors.

388.56: Splices and Taps

Splices and taps are allowed in surface nonmetallic raceways, but may not fill the raceway to more than 75 percent of its cross-sectional area.

388.70: Combination Raceways

When nonmetallic surface raceways are used for both signaling and power circuits, they must be run in separate compartments, which are identified (by color-coding or other means). The position of each compartment must remain the same through the entire run.

III. Construction Specifications

388.100: Construction

These raceway systems must be distinguishable from other raceways and be designed so that the wires won't be abraded when installed.

Article 390—Underfloor Raceways

Underfloor raceways are extensively used in larger buildings, especially those with concrete floors. They have a number of advantages that make them convenient to use where a number of floor outlets are required (as in offices), and where more may be needed at some future date.

390.2: Use

Underfloor raceways may be installed beneath the surface of concrete or other flooring material (Figure 390-1), or in office occupancies, where laid flush with the concrete floor and covered with linoleum or equivalent floor coverings (Figure 390-2).

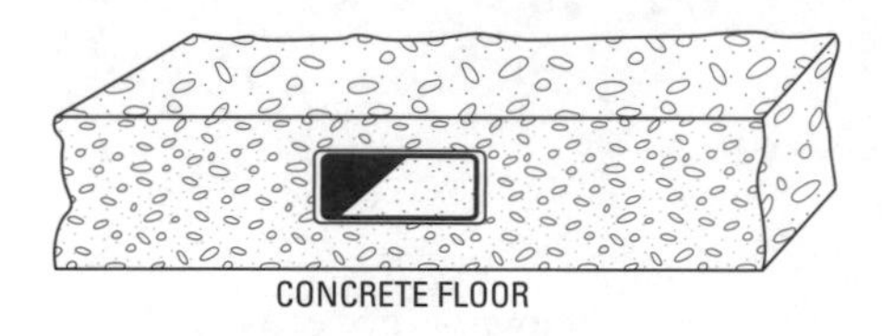

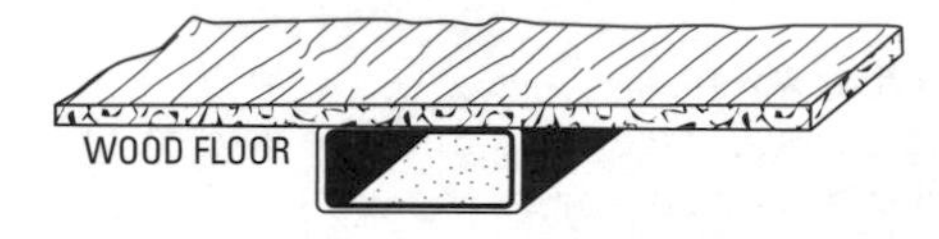

Figure 390-1 Underfloor raceways may be installed in concrete floors or under other floors.

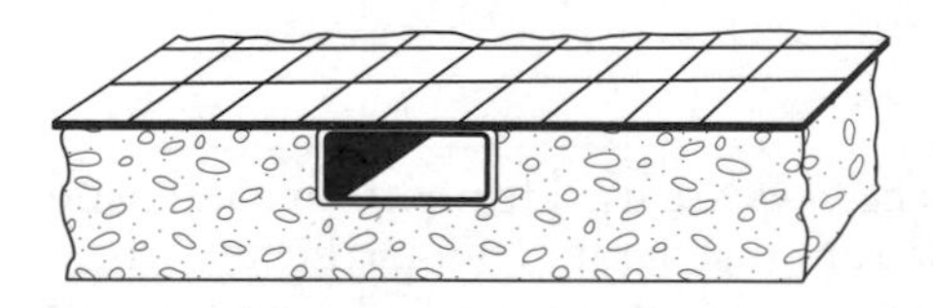

Figure 390-2 Underfloor raceways may be laid flush with the surface of the concrete in office occupancies if covered by linoleum or the equivalent.

Underfloor raceways shouldn't be installed (1) where subject to corrosive vapors, or (2) in any hazardous (classified) location unless made of material judged suitable for the condition. It may be used in Class I, Division 2 locations as permitted in Section 501.4(B).

Unless corrosion protection approved for the conditions is provided, ferrous or nonferrous metallic underfloor raceways, junction boxes, and fittings shouldn't be installed in concrete or in direct contact with the earth or in areas subject to severe corrosive influences. Formerly, open-bottom-type raceways were permitted where they were installed in a concrete fill. This has now been deleted

(Figure 390-3). These raceways shall be corrosion-protected and approved for the purpose where installed in concrete or in contact with the earth. This is the same requirement as for rigid metallic conduit.

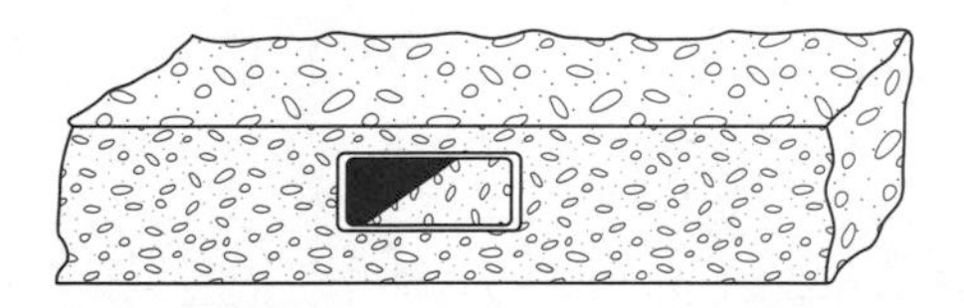

Figure 390-3 Open-bottom raceways are no longer permitted.

390.3: Covering

Raceway coverings shall comply with the following items (A) through (D):

(A) Raceways Not Over 4 Inches Wide. Raceways that are half-round or flat-top raceways that are not over 4 inches in width shall be covered with ¾ inch of concrete or wood (Figure 390-4). There is an exception to this in item (C) of this list for flat-top raceways.

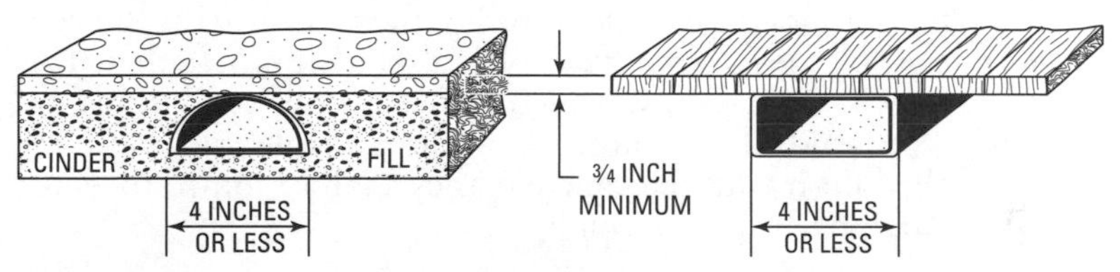

Figure 390-4 Minimum covering over 4-inch raceways.

(B) Raceways Over 4 Inches Wide but Not Over 8 Inches Wide. If underfloor raceways exceed 4 inches in width but are not over 8 inches in width, they are to be covered with concrete of not less than 1-inch thickness. If raceways are spaced less than 1 inch, a concrete covering over the raceway shall be 1½ inches (see Figure 390-5).

(C) Trench-Type Raceways Flush with Concrete. Trench-type underfloor raceways may be flush with the concrete floor or other surface. They shall be approved raceways with cover plates that are rigid and designed to give mechanical protection to the covers.

(D) Other Raceways Flush with Concrete. In office occupancies, if flat-top raceways are approved and are not over 4 inches in

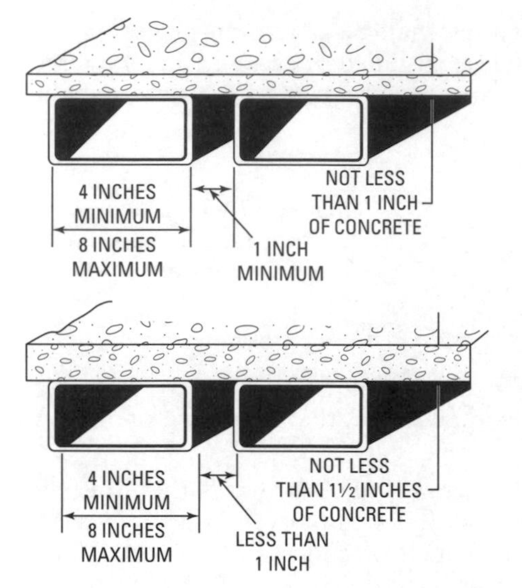

Figure 390-5 Minimum covering over 4- to 8-inch raceways.

width, permission is granted to lay them flush with the concrete surface if covered by linoleum that is not less than 1¹⁄₁₆ inch thick. Other equivalent coverings may be used instead of linoleum. When no more than three single raceways are installed flush with the concrete, they shall be joined together, thus making a rigid assembly.

390.4: Size of Conductors

The design of the raceway shall govern the maximum conductor size.

390.5: Maximum Number of Conductors in Raceway

To figure the number of conductors for underfloor raceways, the total cross-sectional area of all the conductors is calculated. This total area shouldn't exceed 40 percent of the cross-sectional area of the interior of the raceway. In arriving at the cross-sectional area of the conductors, use the appropriate tables in Chapter 9.

390.6: Splices and Taps

Splices and taps are not to be made in the raceway itself but only in junction boxes. For purposes of this section, loop wiring is not a splice or tap (see Figure 390-6). Your attention is also called to the next section.

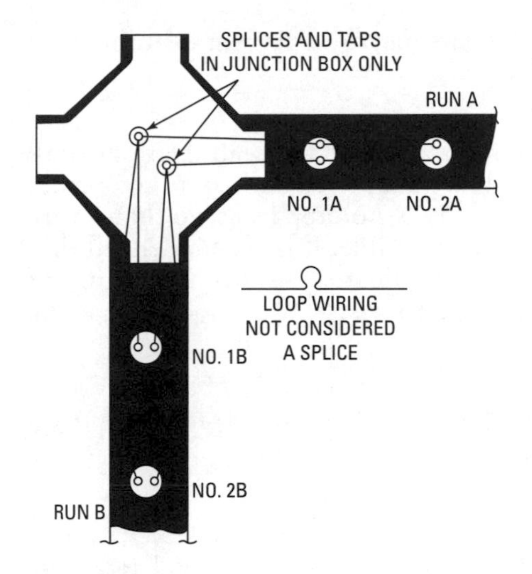

Figure 390-6 Splices and taps are to be made in junction boxes only.

Exception

In trench-type flush raceways that have removable covers, splices and taps are permitted if they are accessible after the installation is complete. When such taps or splices are made, they shouldn't fill the cross-sectional area of the raceway over 75 percent.

390.7: Discontinued Outlets

When an outlet is removed from underfloor raceways, the conductors supplying that outlet shall also be removed from the raceway. No splices or taps are allowed when abandoning new outlets, as would be the case on loop wiring.

To illustrate the essential parts of this ruling, refer to Figure 390-6. Should outlet No. 2B in Run B of the raceway be removed, the wires that served it shall be removed back to the junction box. Should outlet No. 1A also be removed, the conductors to No. 2A would also have to be removed and new conductors run to No. 2A from the junction box.

390.8: Laid in Straight Lines

It is necessary to lay underfloor raceways in a straight line from one junction box to the next, with the centerlines of the two junction boxes coinciding with the centerline of the raceway. This is essential so that the raceways may be located in the event that additional outlets are required later and so that the end markers covered by the next section have some meaning. The raceways must also be held firmly in place by appropriate means to prevent disturbing the alignment during construction.

390.9: Markers at Ends

A suitable number of markers that extend through the floor shall be installed at the end of line raceways, and at other places where the location of the raceway is not apparent, so that future location of the raceway is made possible. It is recommended that these identification markers be indicated on any blueprints of the building for future reference to assist in the location of the raceways.

390.10: Dead Ends

All dead ends of raceways shall be closed by suitable means that are approved by the inspection authority.

390.13: Junction Boxes

Junction boxes shall be leveled to the floor grade and sealed against the entrance of water. Junction boxes used with metal raceways shall be metal and shall be electrically continuous with the raceways. Metal underfloor raceways, junction boxes, and so on are to be electrically continuous, the same as metallic conduit and electrical metallic tubing. This is one of the most important parts of the installation.

390.14: Inserts

Inserts shall be leveled to the floor grade and sealed against the entrance of water. Inserts used with metal raceways shall be metal and shall be electrically continuous with the raceway. Inserts set in or on fiber raceways before the floor is laid shall be mechanically secured to the raceway. Inserts set in fiber raceways after the floor is laid shall be screwed into the raceway. In cutting through the raceway wall and setting inserts, chips and other dirt shouldn't be allowed to fall into the raceway, and tools shall be used that are designed so as to prevent the tool from entering the raceway and injuring conductors that may be in place. All of this merely means good workmanship. It might be a good idea to mention here that if fiber raceways are used, it will be necessary to install a separate conductor to be used as a grounding conductor, the same as required with nonmetallic rigid conduit.

390.15: Connections to Cabinets and Wall Outlets

Connection between distribution centers and wall outlets shall be made with flexible metal conduit or flexible nonmetallic conduit when not installed in concrete. Rigid metal conduit, intermediate metal conduit, EMT, or approved fittings may be used for connections between distribution centers and wall outlets.

Under certain conditions nonmetallic raceways are permitted. Also, see this section in the *NEC*.

This article concerns only underfloor raceways—this means that it is not intended to be run up walls to outlets and cabinets. Therefore, conduit or other approved means must be used for this purpose.

Article 392—Cable Trays

392.1: Scope

This article covering cable-tray systems includes a number of items. Among these are solid-bottom trays, troughs, channels, ladders, and so on.

392.2: Definitions

A *cable-tray system* is an assembly of rigid sections and fittings used to support cables.

392.3: Uses Permitted

(A) Wiring Methods. Continuous rigid cable supports may be used for the following, but the conditions covering the installation of each should be followed as per the articles referenced:

- **(1)** Mineral-insulated, metal-sheathed cable (Article 332).
- **(2)** Armored cable (Article 320).
- **(3)** Metal-clad cable (Article 330).
- **(4)** Power-limited tray cable (Section 725.40).
- **(5)** Nonmetallic-sheathed cable (Article 334).
- **(6)** Multiconductor service-entrance cable (Article 338).
- **(7)** Multiconductor underground-feeder and branch-circuit cable (Article 340).
- **(8)** Power and control-tray cable (Article 336).
- **(9)** Other factory-assembled, multiconductor control, signal, or power cables that are specifically approved for installation in cable trays; or any approved conduit or raceway with its contained conductors.

(B) In Industrial Establishments. In these establishments, provided that only qualified persons will service the cable-tray systems, items (1) and (2) of the following list are allowed in ladder, ventilated, and ventilated-channel trays.

(1) Single Conductor. Single conductor cables 1/0 though 4/0 may be installed in ladder-type trays with a minimum ladder-spacing of 9 inches, or in a ventilated tray. Single conductor cables larger than 4/0 are permitted in any type of listed cable tray.

Exception No. 1
Article 630, Part IV tells us where welding cables are permitted.

Exception No. 2
If single conductors are used for equipment grounds, they may be bare No. 4 AWG or larger.

(2) Multiconductor. Type MV multiconductors, as covered in Article 328, if exposed to sun, shall be identified for this purpose.

(C) Equipment-Grounding Conductors. Table 392.7(B)(2) gives minimum cross-sectional area of metal in square-inches for different ampere ratings or settings of largest automatic overcurrent device protecting any circuit in the cable-tray system. In industrial and commercial establishments only, where continuous maintenance and supervision ensure that only competent persons will service the installed cable-tray system, the cable trays may be used as equipment-grounding conductors.

(D) Hazardous (Classified) Locations. Cable-tray systems may be used in hazardous locations where the contained cables are specifically approved for such use. See Sections 501.4, 502.4, and 503.3.

Note
Cable trays are permitted in other than industrial use.

(E) Nonmetallic Cable Tray. These are permitted with corrosive areas and in areas requiring isolation of different voltages.

392.4: Uses Not Permitted

Cable trays are not permitted where subject to severe physical damage, for example in hoistways. Cable trays may not be used in environmental air-handling spaces except as a support for the wiring methods approved in these areas [see Section 300.22(C)].

392.5: Construction Specifications

See the *NEC*.

392.6: Installation

(A) Complete System. Cable trays shall be installed as a complete system from the point of origin to the point of termination, including bends and the like. The electrical continuity of the cable-tray system must be maintained, no matter what types of bends or modifications are made.

When individual conductors pass from one tray to another, or from trays to raceways or equipment, the conductors must be supported at least every 6 feet and must be protected from physical damage. Bonding jumpers per Section 250.102 must connect two sections of cable tray or the tray and raceway. See Section 250.96. (See Figures 392-1 and 392-2.)

(B) Completed Before Installation. Conductors can't be installed in cable trays until the run of the cable tray involved is entirely completed.

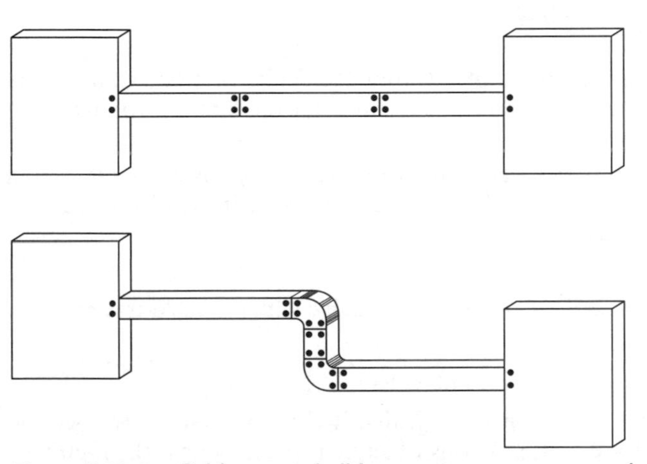

Figure 392-1 Cable trays shall be continuous as a complete system from the point of origin to the point of termination.

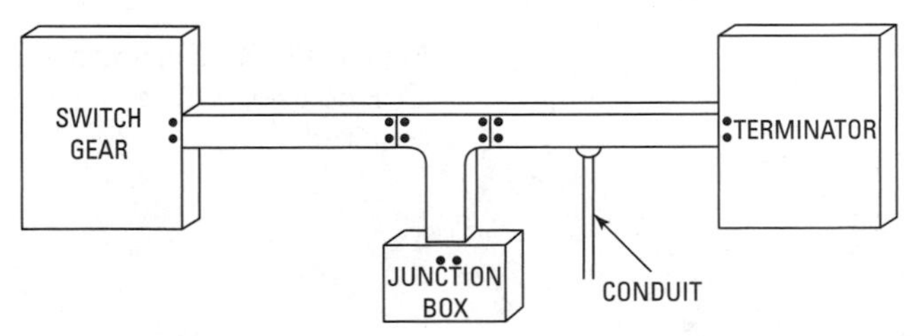

Figure 392-2 Cable trays shall be mechanically connected to all enclosures or raceways.

(C) Supports. Wherever cable for cable trays enters raceways or other enclosures, supports shall be provided.

(D) Covers. Wherever runs of cable trays might require additional protection, covers may be installed of material that is compatible with the material of which the cable tray is made (Figure 392-3).

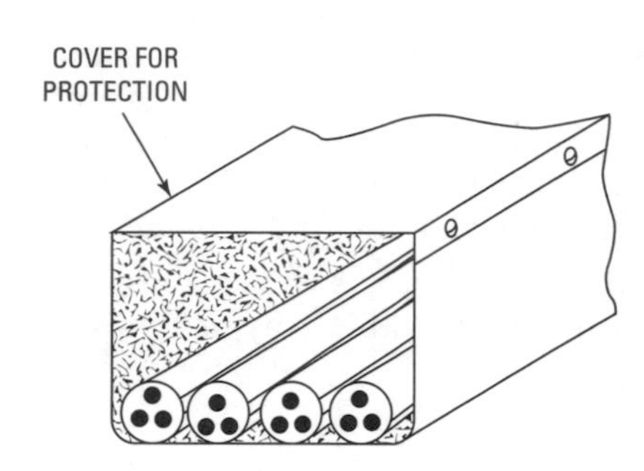

Figure 392-3 Covers shall be used where necessary for protection.

(E) Multiconductor Cables Rated 600 Volts or Less. Conductors may occupy the same cable tray regardless of the voltage, provided that it is 600 or less.

(F) Cables Rated Over 600 Volts. You may install cables rated over 600 volts in the same cable tray with cables rated 600 volts or less.

Exception No. 1
A barrier in the cable tray that is compatible with the cable-tray material.

Exception No. 2
If MC cable is used for over 600 volts.

(G) Through Partitions and Walls. If the requirements of Section 300.21 are met, portions of cable trays may pass through partitions or vertically through platforms or wet or dry locations. See Figures 392-4, 392-5, and 392-6.

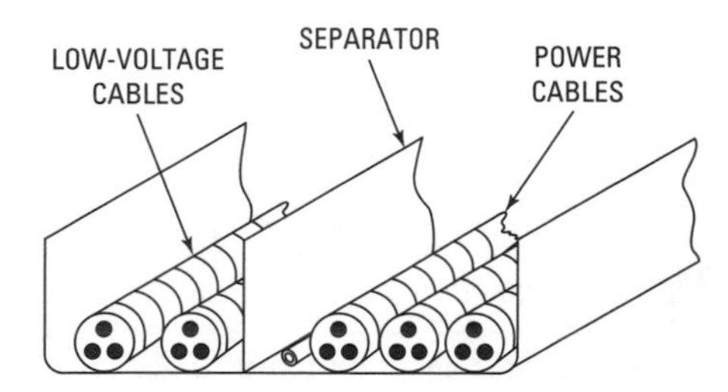

Figure 392-4 Separators shall be used where required.

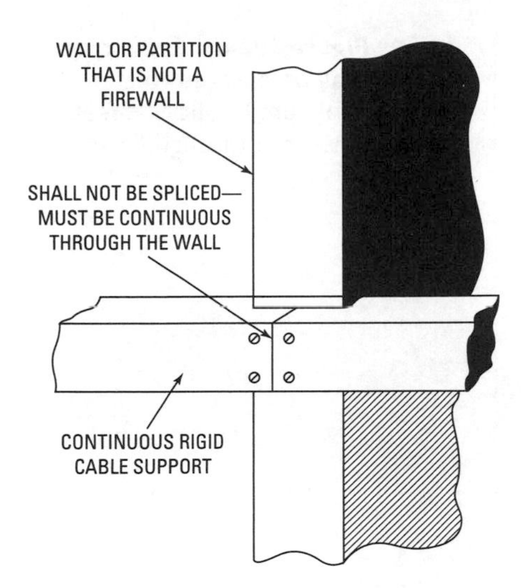

Figure 392-5 Cable trays shall be continuous through walls or partitions.

(H) **Exposed and Accessible.** Except as permitted by Section 392.6(G), cable trays must be accessible and exposed.

(I) **Adequate Access.** Cable trays shall be installed so that there is sufficient space provided to permit access for the installation and maintenance of the cables. This section does not give dimensions, but Figures 392-7 through 392-9 might give some idea of what clearances might be adequate.

(J) **Incidental Support from Cable Tray.** Approved cable trays may be used to support rigid nonmetallic conduit EMT and outlet boxes. See Article 314 for box support information.

Note
See the *NEC* Section 310.10, which covers temperature limitations of conductors.

392.7: Grounding

(A) **Metallic Cable-Tray Systems.** The grounding of metallic cable trays that support electrical conductors shall meet the same grounding requirements as for other conductor enclosures covered in Article 250.

(B) **Steel or Aluminum Cable-Tray Systems.** If the following requirements are met, cable trays made of steel or aluminum may be used as equipment grounding.

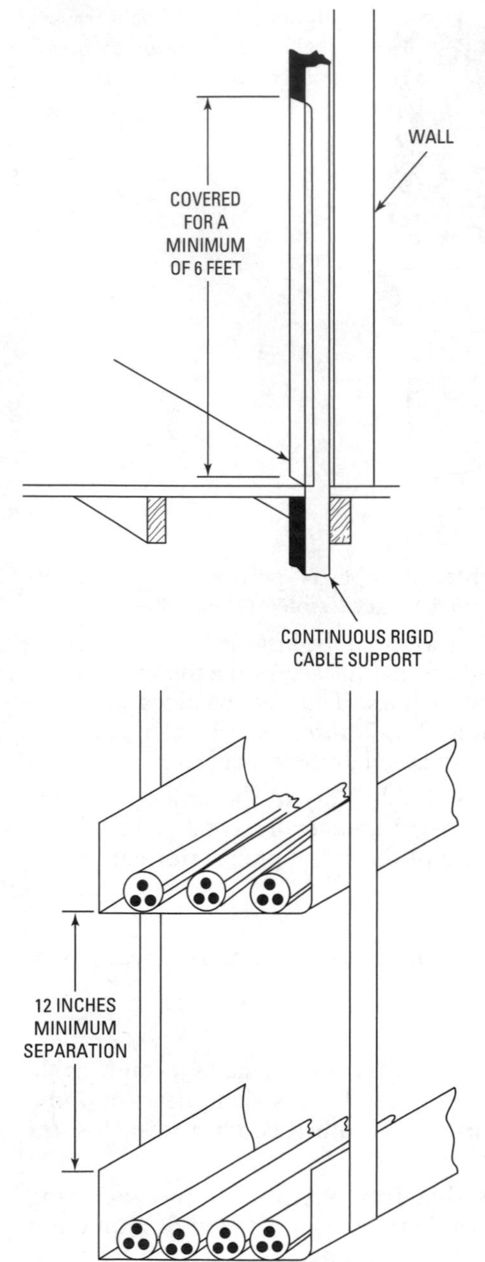

Figure 392-6 Cable trays shall be covered to a minimum height of 6 feet when run through floors.

Figure 392-7 A 12-inch minimum separation shall be maintained between continuous rigid cable supports in tiers.

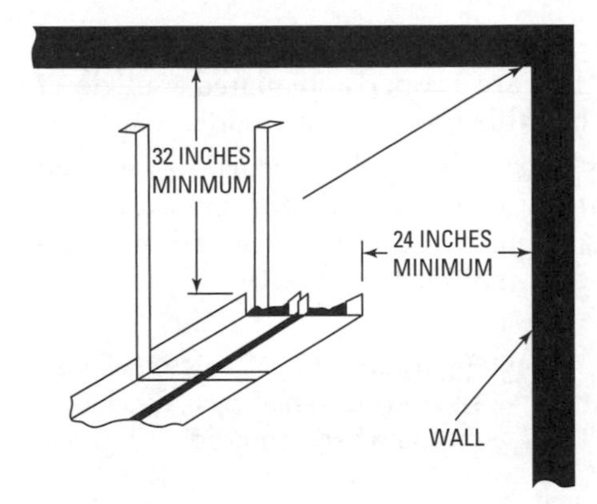

Figure 392-8 A 24-inch minimum horizontal and a 32-inch minimum vertical working space should be maintained.

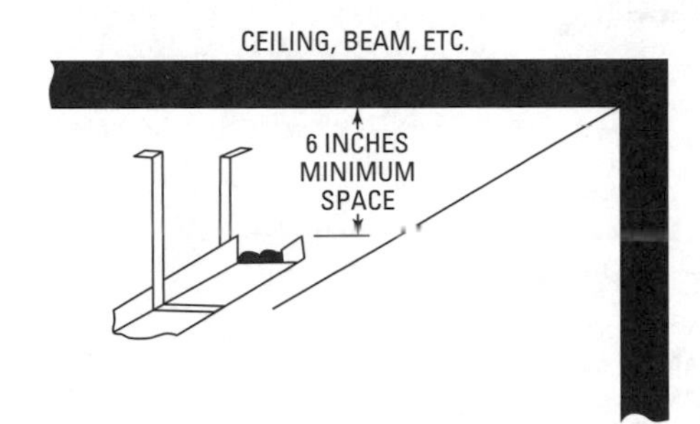

Figure 392-9 A minimum 6-inch clearance should be provided from the top of the cable support to the ceiling, beams, and so on.

(1) All the cable trays and fittings must be identified or listed as equipment-grounding conductors.

(2) They shall conform to the requirements in Table 392.7(B)(2) to provide the minimum cross-section area of cables that are used as equipment-grounding conductors. See Table 392.7(B)(2).

(3) Cable tray and fittings must be legibly and durably marked to show the area of the metal and all types of cable trays or ladders.

(4) Bondings of cable trays and fittings shall be bonded as covered in Section 250.96. This may be bolted mechanically or by bonding jumpers that are to be installed as per Section 250.102.

See the *NEC*, Table 392.7(B)(2), "Metal Area Requirements for Cable Trays Used As Equipment Grounding Conductors."

392.8: Cable Installation

(A) Cable Splices. Splices that are properly insulated and don't project above the sides of cable trays may be made.

(B) Fastened Securely. They must be securely fastened or fastened by other approved means of conductors in horizontal runs of cable trays. This keeps them in their proper location, especially if fault currents occur (see Figure 392-10).

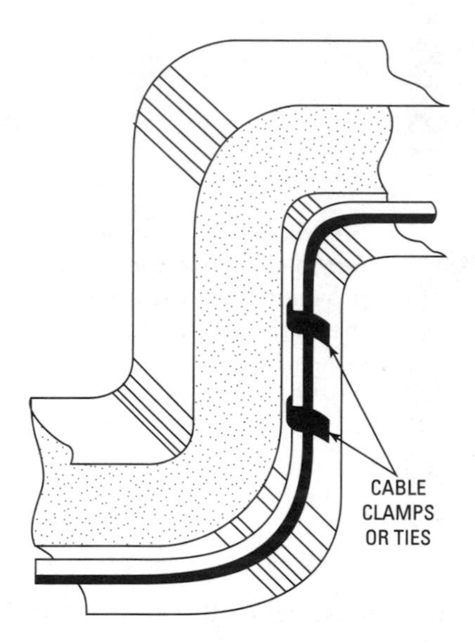

Figure 392-10 Cables shall be securely fastened by a suitable means when required.

(C) Bushed Conduit. If conduit has cables with conductors installed in it, junction boxes are not required, but the conduit must be properly bushed and supported from physical damage.

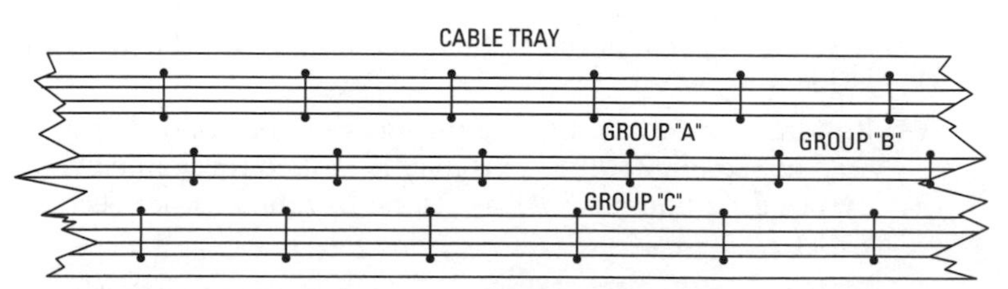

Figure 392-11 Single-conductor phase groups shall be bound in circuit groups.

2005

(D) **Connected in Parallel.** Where parallel conductors are installed in cable trays, the phase conductor and neutral of each parallel set is permitted as covered in Section 310.4. One of each of the phase conductors in the neutral shall be bound together to prevent unbalanced currents in the parallel conductors due to inductive reactance. Note that this appliesonly to alternating current systems. DC conductors are excluded from this type of installation.

Thus, by grouping and using such tie-wraps to keep three different phase conductors and the neutral together, the currents cancel each other; if this was not done, there would be an unbalance of phase currents at times, and cancellation would not take place.

Also single conductors must be bound in circuit groups to contain them to prevent excessive movement in the event of fault-current magnetic forces (see Figure 392-11).

Exception

Triplex of twisted conductors are an exception.

392.9: Number of Multiconductor Cables, Rated 2000 Volts, Nominal, or Less, in Cable Trays

This section covers the number of cables of 2000 volts or less that may be installed in a single cable tray. The conductor sizes that are covered in this section apply to both aluminum and copper conductors.

(A) **Any Mixture of Cables.** For ventilated trough cable trays that contain multiconductors or lighting cables and any mixture thereof of multiconductor power and lighting control and signal conductors, see the following for the maximum number of conductors that conform to such installations.

2005

((1) If all of the conductor cables are 4/0 AWG or larger, the sum of the cable diameters may not exceed the width of the cable tray, nor are cables permitted to be installed in multilayers—a single layer only is allowed!

Tables giving the diameters of conductors, with various types of insulation, are given in Chapter 9 of the *NEC*.

(2) Table 392.9 gives the allowable cable fill-area for multiple-conductor cables in ladder, ventilated-trough, or solid-bottom cable trays. Where all of the cables are smaller than 4/0 AWG, Table 392.9, Column 1, gives the allowable fill-area for the appropriate cable-tray width.

(3) Part of this is similar to (2), so it won't be repeated here. However, the sum of all cables smaller than 4/0 AWG shouldn't exceed the maximum allowable fill-area. Column 4 of Table 392.9 covers the computation for the proper width of a cable tray. Only a single layer of 4/0 AWG and larger cables are permitted, and no smaller cables are allowed in this layer.

(B) Multiconductor Control and/or Signal Cables Only. Ladder or ventilated trough cable trays that are used for multiple-conductor control and/or signal cables *ONLY* and which have a depth of 6 inches (152 mm) or less may have a conductor fill of not to exceed 50 percent of any cross-sectional area of the cable tray. Cable trays with a usable depth of more than 6 inches (152 mm) shall use 6 inches (152 mm) as the depth for fill, and not more than 6 inches (152 mm).

(C) Solid Bottom Cable Trays Containing Any Mixture. In solid-bottom cable trays for mounted conductor, power, lighting cables, or with control insertion cables, you shall comply with items (1), (2), and (3) of the following list:

(1) If all the cables installed are 4/0 AWG or larger, you are allowed only up to 90-percent fill in a single layer (Figure 392-12).

(2) Column 3 of Table 392.9, where the appropriate cable-tray widths are given, is to be used where all of the cables are smaller than 4/0 AWG. The sum of the cross-section areas of all cables involved shall comply with the maximum allowable in the above-stated column and section.

(3) Column 4 of Table 392.9 and the appropriate cable-tray widths shall be adhered to when 4/0 AWG or larger cables are installed. With cables smaller than 4/0 AWG in the same cable tray, the sum of the cross-sectional area of all cables smaller than 4/0 AWG shouldn't exceed the allowable fill-area when using the aforementioned column and table for the appropriate tray width, and the cables of 4/0 AWG and larger are to be installed in a single layer. No additional cables are to be placed on them.

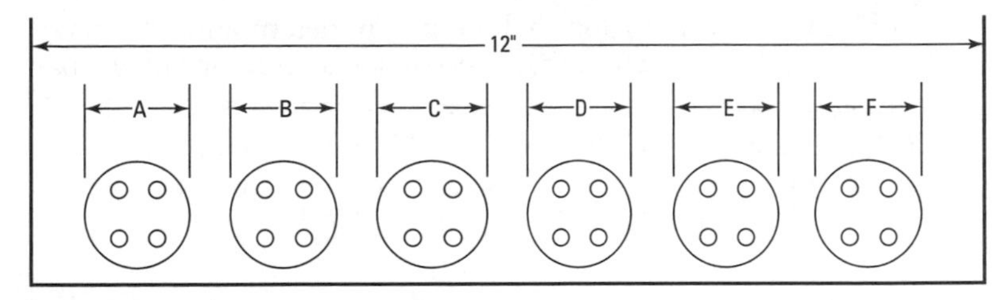

Figure 392-12 Cables (multiple conductor) diameter A + B + C + D + E + F shouldn't equal more than 10.8 inches when installed in a 12-inch cable tray. Single layer only.

Using Column 4 of Table 392.9 gives the square-inches allowable for various sizes of cable trays, for conductors (multiple) smaller than 4/0 AWG. Note the 4/0 AWG and larger conductors are to be in a single layer only.

(D) Solid-Bottom Cable-Tray Multiconductor Control and/or Signal Cables Only. A solid-bottom cable tray that has a usable inside depth of 6 inches or less and contains multiple-conductor control and/or signal cables only, the cross-sectional areas shouldn't be more than 40 percent of the cross-sectional area of the cable tray. If the cable-tray depth is over 6 inches, the usable cross-sectional area shall be figured as if the cable tray were 6 inches deep, even though it is deeper.

(E) Ventilated-Channel Cable Trays. Refer to the *NEC*.

Exception

Refer to the *NEC* and Table 392.9.

392.10: Number of Single Conductor Cables Rated 2000 Volts or Less in Cable Trays

When conductors are rated 2000 volts or less, the number of single conductor cables in a single cable-tray section must abide by the requirements of this section. Single conductors or cable assemblies shall be distributed evenly across the bottom of the cable tray in which they are being installed. This will make it necessary to secure them in place to hold them in the spacing required, and the conductors may be copper or aluminum conductors.

(A) Ladder or Ventilated-Trough Cable Trays. Cable trays as indicated in the title contain single conductor cables. The following additions to this portion shall be met in determining the maximum single conductors that the particular tray or ladder may hold:

(1) The sum of the diameter of all single conductors 1000 kcmil or larger shouldn't exceed the cable-tray width. Refer to Table 392.10. See Figure 392-13.

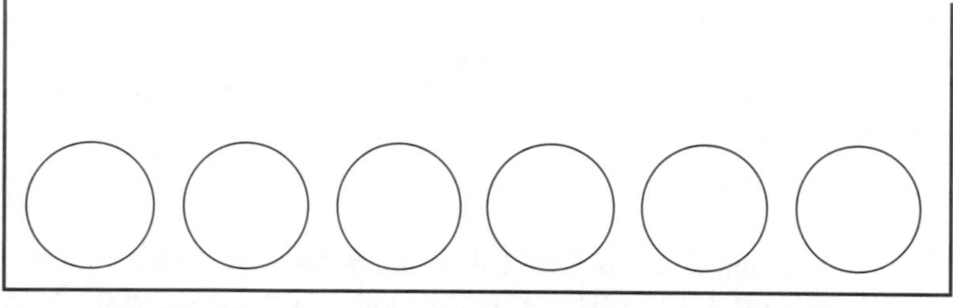

Figure 392-13 The sum of the diameter of a single-conductor 1000 kcmil or larger shouldn't exceed the width of the cable tray.

(2) Where all of the cables are smaller than 1000 kcmil, the cable-tray fill is governed by Column 1 of Table 392.10, and the sum of the cross-sectional areas of the cables shouldn't exceed the square-inches allowable in Column 1 for the appropriate cable-tray width.

(3) If 1000 kcmil or larger single-conductor cables are installed in a cable tray with single-conductor cables smaller than 1000 kcmil, the total cross-sectional area of the cables smaller than 1000 kcmil is controlled by the computation in Column 2 of Table 392.10 for the appropriate cable-tray widths.

2005

If any single-conductor cables are Nos. 1/0 through 4/0, the sum of all such cable diameters may not exceed the inside width of the channel.

Only a single layer of cables may be installed, although this does not prevent the installation of a group of conductors being wrapped together and installed in a cable tray.

(B) Ventilated Channel-Type Cable Trays. The sum of the diameter of all single conductors in 3-inch-, 4-inch-, or 6-inch-wide ventilated channel-type cable trays shouldn't exceed the inside width of the channel.

See the *NEC* for Table 392.10, "Allowable Cable Fill Area for Single Conductor Cable in Ladder or Ventilated Trough Cable Trays for Cable Rated 2000 Volts or Less."

392.11: Ampacity of Cables Rated 2000 Volts or Less in Cable Trays

The derating factor with Tables 310.16 through 310-31 covers Note 8(a) and applies only to multiconductor cables that have more than three current-carrying conductors. Derating is based upon the number of conductors in the cable and not on the number of conductors in the cable tray.

(A) **Multiconductor Cables.** Tables 310.18 and 310.22 cover the allowable ampacities and will apply to multiconductor cables nominally rated 2000 volts or less, provided that they are installed according to the requirements of Section 392.9.

Exception No. 1

Ampacities must be cut to 95 percent of the ampacity in Tables 310.22 and 310.31 where cable trays are continuously covered for more than 6 feet with a solid cover that is not ventilated.

Exception No. 2

When multiconductors in uncovered trays are installed in a single layer and maintain a spacing of one conductor diameter between cables, their ampacity must be determined according to Section 310.15(B) (ambient corrected ampacities of multiconductor cables rated 0–2000 volts). Refer also to Table B310-3 in the Appendix.

(B) **Single-Conductor Cables.** The following will apply to the ampacity of single-conductor cables or single-conductor cables that are twisted or bound together, such as triplex to quadraplex, and so on. This applies to nominally rated 2000 volts or less:

(1) If 600 kcmil or larger single-conductor cables are installed as per Section 392.10, uncovered cable trays shouldn't exceed 75 percent of the ampacities as shown in Tables 310.17 and 310.19. Where cable trays are continuously covered for more than 4 feet and unventilated covers are used, ampacities for 600 kcmil and larger cables shouldn't exceed the ampacity of 70 percent of the level ampacities shown in Tables 310.17 and 310.19.

Referring to Tables 310.17 and 310.19 in the *NEC,* you will notice that these tables are for single conductors in free air for copper and aluminum, respectively.

(2) When cables are installed according to Section 318.9, single-conductor cables from 250 kcmil through 500 kcmil installed in uncovered cable trays shall have their ampacities calculated as 65 percent of the allowable ampacities in Tables 310.17 and 310.19. If the cable trays are continuously covered for more than 6 feet with solid unventilated covers, then for 250 kcmil through 500 kcmil single-conductor cables, the ampacities shouldn't exceed 60 percent of the allowable ampacities as covered in Tables 310.17 and 310.19.

(3) Tables 310-17 and 310-19 give the ampacities of cables that shouldn't be exceeded under the following: single cables installed in single and uncovered cable trays, if spacing is maintained of not less than the diameter of one cable between each conductor; this covers cables of 25 kcmil and larger.

(4) The ampacity allowable in Table 310.23 applies to single cables arranged in a triangular form and in uncovered cable trays. The cables shall be spaced not less than 2.15 times the diameter of one cable, between the circuit conductors in a triangular configuration, and the ampacities are for No. 1/0 and larger cables.

Note

Remember that we have Correction Tables for Ambient Temperatures over 30°C (86°F). These appear following the tables in *NEC* Article 310.

392.12: Number of Type MV and Type MC Cables (2001 Volts or Over) in Cable Trays

The number of cables as covered in the title and rated nominal 2001 volts or over in a single tray shouldn't exceed the requirements of this section.

The sum of the diameters of the load conductor cables for single conductors that are installed in cable trays shall be installed in single layers, and the sum of the diameters shouldn't exceed the cable-tray width at which they are installed. Where single conductors or triplexed or quadraplexed cables are installed, they are to be in a single layer only. This can also apply to bundled conductors if they are all of a single circuit group. The cable-tray width shouldn't be exceeded by the sum of the diameter of the covered conductors. These shall be installed in single-layer arrangement.

392.13: Ampacity of Type MV and Type MC Cable (2001 Volts or Over) in Cable Trays

The requirements of Section 392.12 shall apply to this type of cable.

(A) Multiconductor Cables (2001 Volts or Over). The ampacity of multiconductor cables shouldn't exceed the ampacities allowable in Tables 310.75 and 310.76.

Exception No. 1

If the cable trays are covered with unventilated covers exceeding 6 feet with solid unventilated covers, the ampacities as shown in Tables 310.75 and 310.76 shall be rated to 95 percent of the ampacities shown in these tables.

Exception No. 2

The allowable ampacity in Tables 310.71 and 310.72 will be allowed for multiconductor cables provided that they are installed in a single layer and the cable trays are uncovered. Spacing of not less than one cable diameter shall be allowed between the multiconductor cables.

(B) Single-Conductor Cables (2001 Volts or Over). The ampacity of single conductors or single conductors bundled together, if one circuit (triplexed or quadraplexed), must comply with the following:

(1) Tables 310.69 and 310.70 cover ampacities for cables 250 kcmil or larger if they are single conductors and in uncovered cable trays. They shouldn't exceed 75 percent of the allowable ampacities shown in these tables. If the cable trays are covered more than 6 feet, the same type of conductor shouldn't exceed 70 percent of the ampacity allowed in the aforementioned tables.

(2) In single-conductor cables, provided the proper spacing is maintained, which is one cable diameter, the ampacities of 250 kcmil and larger cables shall be as shown in Table 310.69 and 310.70, and shouldn't be more than those shown.

(3) Where single conductors of a circuit are in a triangular form, the spacing between the different triangular bundled circuit conductors shouldn't be less than 2.15 times the conductor diameter between the bundled circuit conductors, and the ampacities of 250 kcmil or larger cables shouldn't be allowed to exceed 105 percent of the ampacities covered in Tables 310.71 and 310.72.

Article 394—Concealed Knob-and-Tube Wiring

A number of years ago knob-and-tube wiring was the main method used in wiring houses and commercial buildings. Many are still wired that way, and if proper installation was made, it is a very safe wiring method. The insulators, and the like, are now very hard to obtain, and you will be required to make additions to knob-and-tube wiring by adding NM cable, which needs much less time for installation and incorporates an equipment-grounding conductor. In other words, changes in wiring methods have meant changes in the materials used since this method was popular.

394.1: Definition

Concealed knob-and-tube wiring uses knobs and tubes (usually porcelain) to support and insulate conductors as they pass through hollow wall and ceiling spaces.

394.2: Other Articles

As with open wiring in the preceding Article 394, other parts of the *Code* are applicable, especially Article 300, "Wiring Methods—General."

394.3: Uses Permitted

This type of wiring may be used only in extensions to existing installations, and may be used in other places only with special permission—if the following conditions are met:

See the *NEC* for (1) and (2).

394.4: Uses Not Permitted

(1) In commercial garages.

(2) In theaters and similar locations.

(3) In motion-picture studios.

(4) In hazardous (classified) locations.

394.5: Conductors

(A) **Type.** Knob-and-tube conductors have to be as specified in Article 310.

(B) **Ampacity.** The ampacity of conductors used for knob-and-tube systems must be determined by Section 310.15.

394.6: Conductor Supports

The conductors must be supported by insulated supports and may not contact any other materials. Supports must be placed within 6 inches from a splice or tap, and not exceeding 4½ feet thereafter.

Exception

When providing supports in dry locations is especially difficult, the conductors may be fished in and enclosed in flexible nonmetallic tubing. No splices or taps are allowed in such runs.

394.7: Tie Wires

If solid knobs are used, the conductors shall be mounted on these solid supports by means of tie wires made of the same material as the conductors.

394.8: Conductor Clearance

Conductors shall be separated at least 3 inches and maintained at least 1 inch from the surface wired over. Notice that this is different from Article 394. At distributing centers, meters, outlets, switches, or other places where space is limited and the 3-inch separation can't be maintained, each conductor shall be encased in a continuous length of flexible nonmetallic tubing. Where practical, conductors shall run singly on separate timbers or studding.

Exception

If the space is so limited as not to provide the minimum clearances described above for meters, panelboards, outlets, switch boxes, and so on, the means of protecting conductors shall be a flexible nonmetallic tubing, and this shall extend from the last support continuously to the box and the terminal point.

394.9: Through Walls, Floors, Wood Cross Members, and so on

See Section 394.11 for conductors passing through these points. Also, when conductors pass through cross members in plastered partitions, the conductors must be protected by a nonabsorbent, noncombustible, insulating tube. These tubes shall extend not less than 3 inches beyond the wood member.

394.10: Clearance from Piping, Exposed Conductors, and so on

The same conditions of Section 394.12 are applicable to knob-and-tube wiring.

394.11: Unfinished Attics and Roof Spaces

Conductors in unfinished attics and roof spaces shall comply with the following:

Accessible by Stairway or Permanent Ladder. Conductors in unfinished attics and roof spaces shall be run through or on the sides of joists, studs, and rafters, except in attics and roof spaces that have headroom at all points of not less than 3 feet in buildings completed before the wiring was installed (see Figure 394-1).

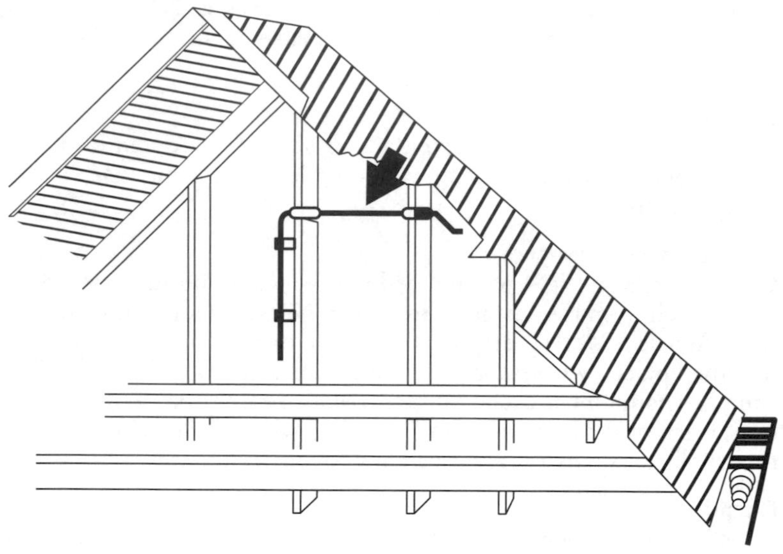

Figure 394-1 Running conductors in attics.

Where conductors in accessible unfinished attics or roof spaces reached by stairway or permanent ladder are run through bored holes in the floor joists or through bored holes in the studs or rafters within 7 feet of the floor or floor joists, such conductors shall be protected by substantial running boards extending 1 inch on each side of the conductors and securely fastened in place (see Figure 394-2).

Where carried along the sides of rafters, studs, or floor joists, neither running boards nor guard strips are required.

Exception

An exception covers buildings being wired after completion that have headroom at all points of less than 3 feet.

394.12: Splices

Splices under strain or in-line are not permitted. Where splices are permitted, they may be soldered (don't use acid core when soldering), or other approved splicing means may be used.

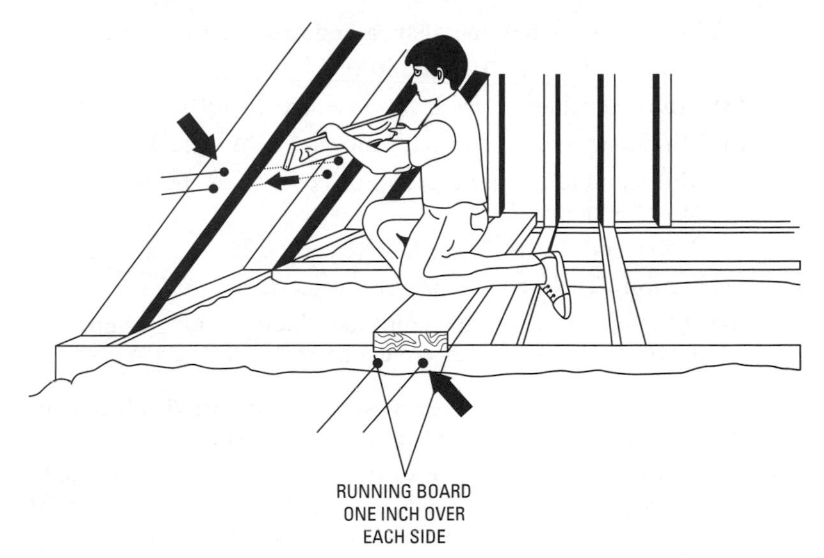

Figure 394-2 Protection by running boards in attics accessible by stairs or permanent ladder.

394.13: Boxes

See Article 314 for outlet boxes.

394.14: Switches

See Sections 404.4 and 404.10(B) for switches that shall comply with this type of wiring.

Article 396—Messenger-Supported Wiring

396.1: Definition

Messenger-supported wiring is a system of exposed wiring that uses a messenger wire to support other conductors. Triplex cables are one type of messenger-supported cables.

396.2: Other Articles

Not only must messenger-supported wiring comply with this article, but it must also comply with any of the provisions that may be applicable in Articles 225 and 300, plus any other applicable articles in the *Code*.

396.3: Uses Permitted

(A) Cable Types. Cable-type wiring may be installed as messenger-supported wiring if it meets all of the conditions of the following articles for each type of cable listed below:

(1) Mineral insulated, metal-sheathed cable (Article 332).

(2) Metal-clad cable (Article 330).

(3) Armored cables in dry locations (Article 320).

(4) Multiconductor service-entrance cable (Article 338).

(5) Multiconductor underground feeder and branch-circuit cable (Article 340).

(6) Power and control tray cable (Article 336).

(7) Section 725.40 covers power-limited tray cable.

(8) Other factory-assembled multiconductor control, signal, or power cable that has been identified for the use.

The use of any of the above for messenger-supported cables shall meet all requirements of the applicable article involved.

(B) In Industrial Establishments. If competent individuals only will be servicing installed messenger-cable wiring in industrial establishments, the following conditions shall be met:

(1) Conductor types in Tables 310.13 or 310.62 may be used.

(2) MV cable may be used.

If the messenger-supported conductors are exposed to a wet location or to sunlight, they shall be listed for wet locations or sunlight or both, if both conditions are prevalent.

(C) Hazardous (Classified) Locations. The only hazardous (classified) locations where messenger-supported wiring shall be used—if they are approved—are those covered in Sections 501.4, 502.4, and 503.3.

The use of messenger-supported cable in hazardous (classified) locations is very limited.

396.4: Uses Not Permitted

In hazardous locations such as hoistways or other places where damage may occur, messenger-supported wiring shouldn't be used.

396.5: Ampacity

Section 310.15 is used to determine the maximum ampacity permissible.

396.6: Messenger Support

Proper and adequate support shall be used for dead-ending the messenger cable. At any location, you must eliminate the conductors being put under tension. Conductors shouldn't come in contact with the messenger cable, points of support, or structure members such as walls or pipes.

396.7: Grounding

Grounding is addressed by Sections 250.32 and 250.33 for enclosures that cover the means of grounding the messenger.

396.8: Conductor Splices and Taps

If approved means are used, splices and taps in conductors, if properly insulated, are permitted for messenger-supportsed wiring.

Article 398—Open Wiring on Insulators

In this article you are referred to the *NEC*. The author's opinion is that this type of wiring is so seldom used that it need not be covered here section-by-section. However, Figures 398-1 through 398-11 show how open wiring is installed.

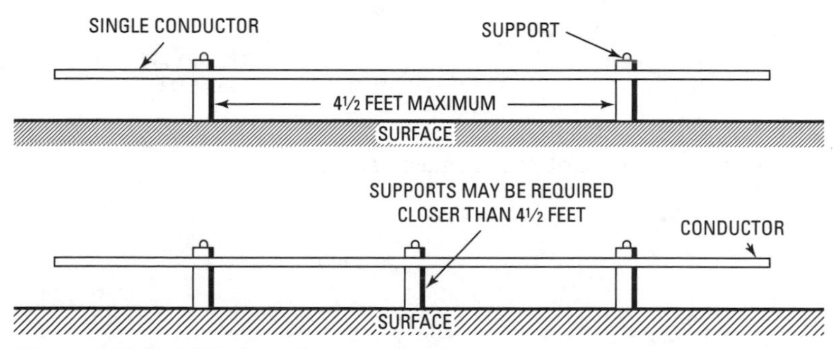

Figure 398-1 Spacing of supporting knobs.

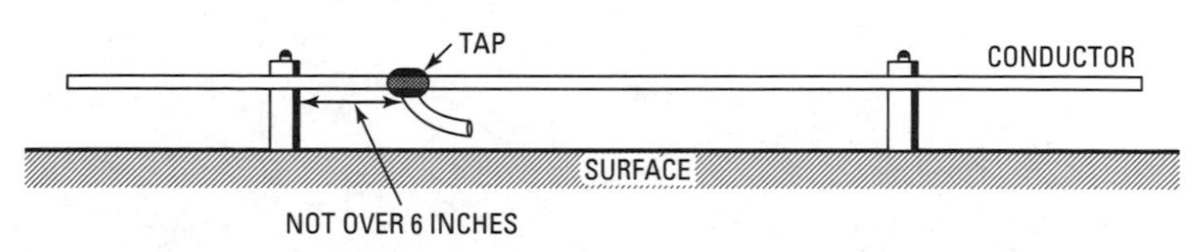

Figure 398-2 Distance between a support and a tap should never exceed 6 inches.

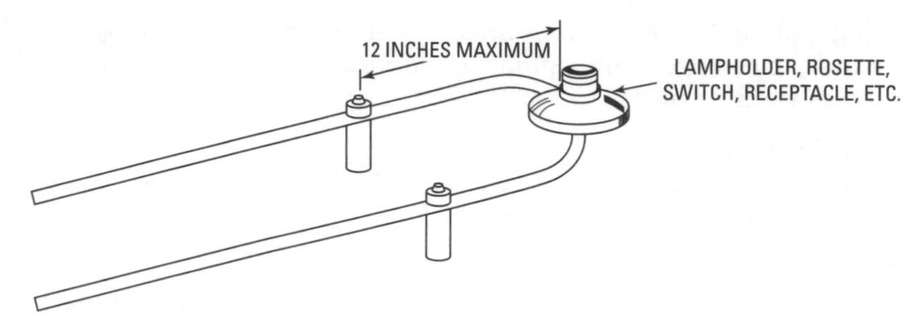

Figure 398-3 Distance of supports when connecting receptacles, switches, rosettes, and other devices.

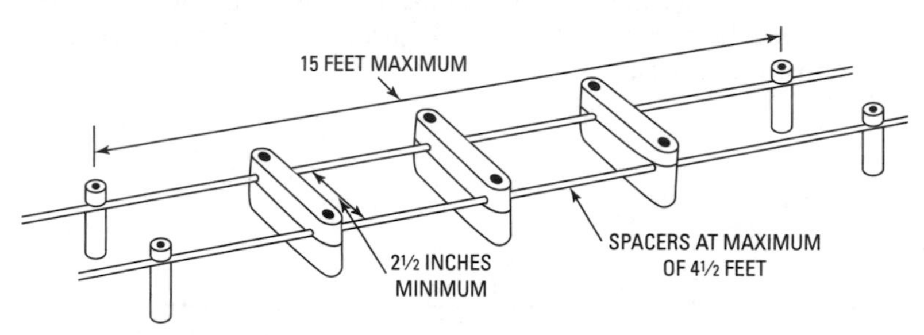

Figure 398-4 A 15-foot span requires spacers.

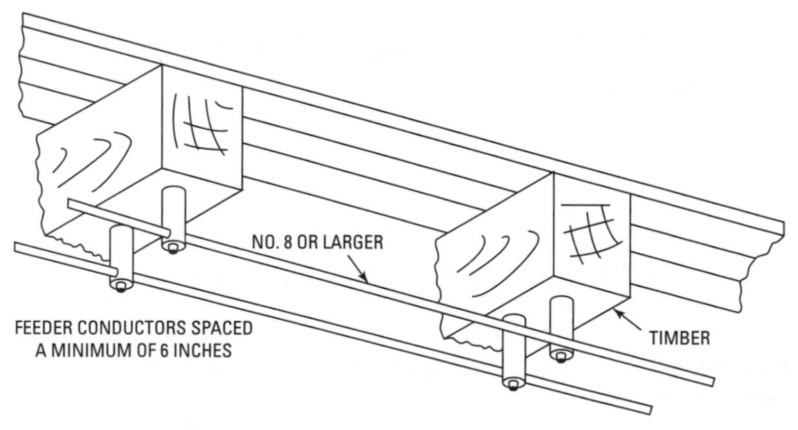

Figure 398-5 Mill construction installation.

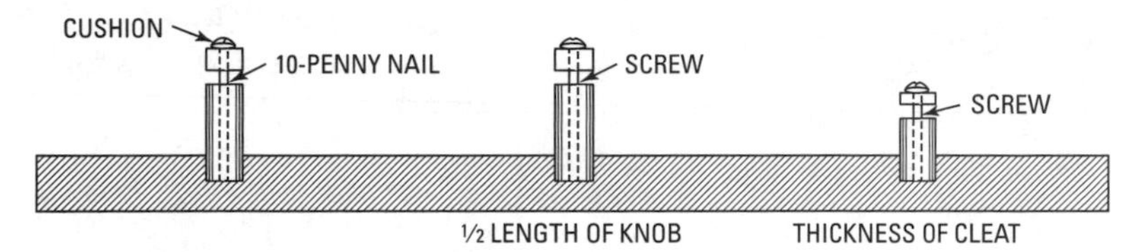

Figure 398-6 Length of nails and screws for knobs and cleats.

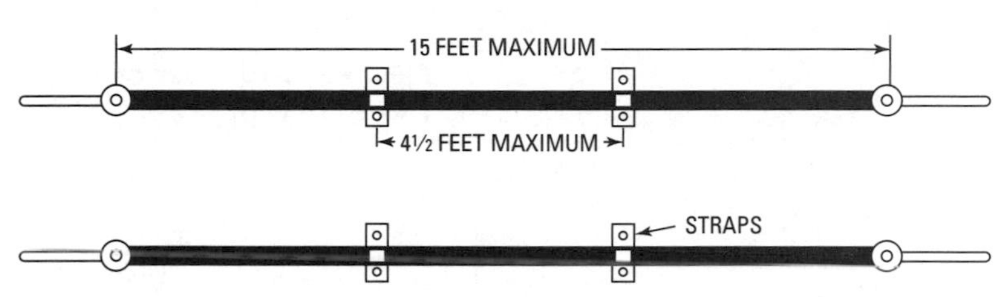

Figure 398-7 Use of approved nonmetallic tubing.

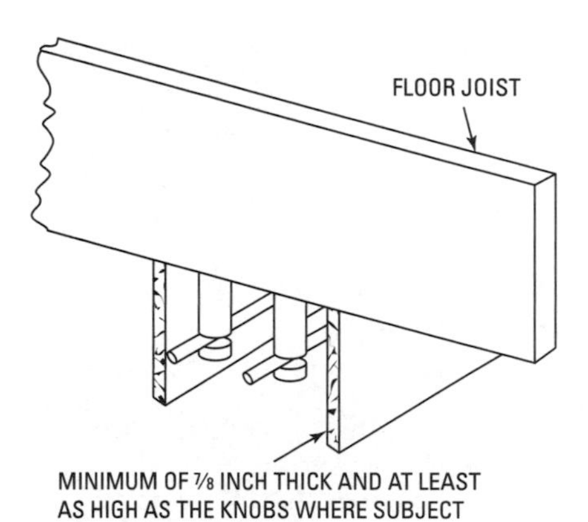

Figure 398-8 Protecting conductors from physical damage.

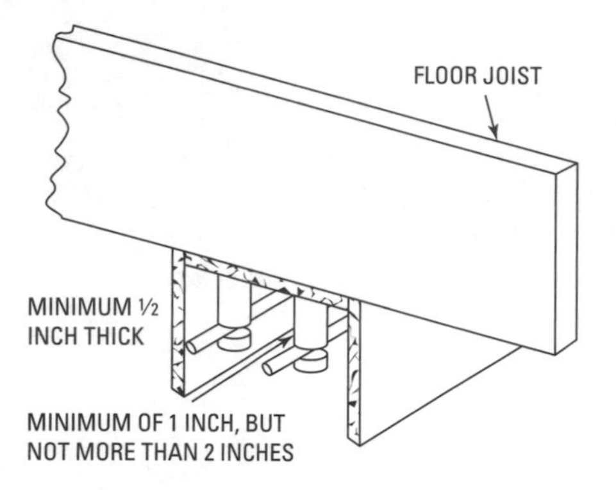

Figure 398-9 Use of running board and sides for protection.

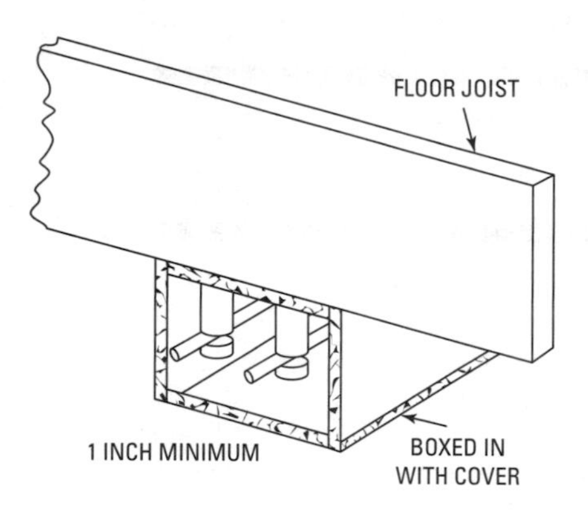

Figure 398-10 Use of enclosure with cover.

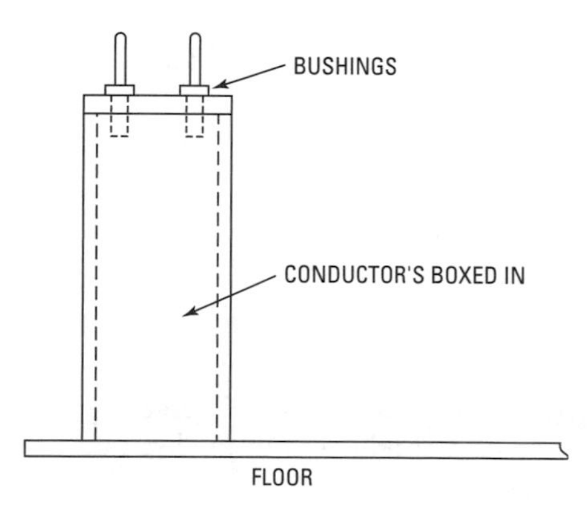

Figure 398-11 Vertical enclosure for conductors.

Chapter 4

Equipment for General Use

Article 400—Flexible Cords and Cables

I. General

400.1: Scope

This article covers the general requirements for flexible cords and flexible cables, uses permitted, and the specifications of construction.

400.2: Other Articles

The requirements of this article are complemented by some of the applicable provisions of other articles of the *Code.*

400.3: Suitability

Cables, cords, and fittings shall be used only in the proper locations and under proper conditions of usage.

Considerable information on flexible cords is included in this chapter of the *NEC,* a greater part being in the form of tables. Flexible cords are essential to the full utilization of electrical energy but are probably the most abused items in use today. A large part of the abuse comes from the use to which they are put by the general public, not by the electrical contractor or wirer.

400.4: Types

Cords of the several types shall conform to the descriptions in *NEC* Table 400.4. Types of flexible cords other than those listed in Table 400.4 and uses for types listed in the table shall be the subject of special investigation and shouldn't be used before being approved. This is rather plain in that it is recognized that flexible cords are subject to many and varied usages, and that the responsibility of usages other than those specified by the *NEC* is given to the inspection authority involved. It would be impossible for a book such as the *NEC* to cover any and all usages to which flexible cords are put.

Also refer to the notes under Table 400.4.

400.5: Ampacity of Flexible Cords and Cables

The allowable ampacity for cords and cables that don't have more than three current-carrying copper conductors is given in Table 400.5

of the *NEC.* If there are more than three current-carrying conductors in a cord or cable, the ampacity shall be reduced from the three current-carrying conductor-cord ratings, and this is shown in the table following this paragraph in the *NEC.*

Be sure to take into consideration the notes that follow Table 400.5.

Refer to subheading C in Table 400.5. Here we find the ampacity for Type W cable, single-conductor, where the individual conductors are installed in raceways, and they shall also not be in physical contact with each other unless they don't exceed 24 inches in length when they pass through an enclosure wall.

In normal circumstances the neutral conductor carries only unbalanced current from the other hot conductors. If properly installed, the currents normally will be balanced in the three or more conductors, and if this applies, the neutral is not considered a current-carrying conductor.

In a four-wire, three-phase wye-connected system, if a three-wire circuit consists of only two-phase wires and a neutral, the common conductor or neutral will carry approximately the same amount of current as the two-phase conductors, and shall therefore be considered a current-carrying conductor.

If in a four-wire, three-phase wye circuit a major part of the load consists of electric discharge lighting or any data processing or similar equipment, there will always be a third harmonic set up, causing currents in the neutral conductor. In these cases the neutral is always to be considered a current-carrying conductor.

Equipment-grounding conductors are not considered current-carrying conductors, as they take care of only grounds that appear from phase conductors.

As was provided in Section 250.60, if a single conductor is used only for equipment grounding and unbalanced current (neutral) from the other conductors, it shouldn't be considered a current-carrying conductor when supplying such appliances as electric clothes dryers and electric ranges.

400.6: Marking

Marking is required for cords and cables on printed tags that are attached to the coil or carton containing the cords or cables. As required by Section 310.11(A), the tag shall designate the type of cord or cable as SJ, SJE, SJO, SJEO, or other types of flexible cord, and G and W flexible cables shall be marked several times on the cable not more than 24 inches apart. The type, size, and number of conductors shall also be designated.

400.7: Uses Permitted

(A) Uses. Flexible cords and cables can be used for the following:

(1) For pendants.

(2) For wiring of fixtures.

(3) For connecting portable appliances or lamps.

(4) As elevator cables.

(5) For wiring cranes and hoists.

(6) For connecting stationary equipment that is frequently changed.

(7) To prevent the transmission of vibration or noise.

(8) To connect appliances that must be removed for service.

(9) As data processing cables.

(10) For connecting moving parts.

(11) For temporary wiring.

(B) **Attachment Plugs.** In the uses permitted in (A)(3), (A)(6), and (A)(8) of this section, the flexible cord must have an attachment plug for energizing from an outlet. There are many types of attachment plugs for use, such as twisting-type plugs, different-voltage–type plugs, and so on. Refer to NEMA (National Electrical Manufacturers Association) standards for configurations of the proper attachment plug to be used, and the proper place.

400.8: Uses Not Permitted

(A) Flexible cords shouldn't be substituted for fixed wiring in a structure.

(B) They shouldn't be run through holes as the wiring between walls, ceilings, or floors.

(C) They shouldn't be run through doorways, windows, or other openings. If electrical connection is required, it shall be done by proper wiring methods.

(D) Only proper wiring methods shall be used for attaching to proper wiring surfaces, not cords.

(E) Cords shouldn't be used instead of proper wiring methods for concealed wire between building walls, ceilings, or floors.

(F) They can't be installed in raceways, unless specifically allowed by the *Code* for certain special installations.

400.9: Splices

This is important: During initial installation, flexible cord shall be installed in a continuous length and shall contain no splices or taps.

Repairs on hard-service cords (flexible), No. 14 or larger, are permitted if the following conditions are met: (1) They are spliced in accordance with Section 110.14(B), and (2) the completed splice retains the outer sheath properties, original flexibility, usage characteristics of the cord being spliced, and the insulation. These above restrictions are plain, and in essence mean that the cord must retain its original characteristics.

400.10: Pull at Joints and Terminals

When flexible cords are used, they shall be connected and installed in such a way that there will be no tension on joints or terminals.

Note

There are many methods of preventing pull on cords. Some are by approved clamps on the attachment plug, using tape or fittings designed for the purpose.

400.11: In Show Windows and Show Cases

Flexible cords used in show windows and show cases shall be Types S, SO, SE, SEO, SOO, SJ, SJE, SJO, SJEO, SJOO, ST, STO, STOO, SJT, SJTO, SJTOO, or AFS, which are heavy-duty cord types. There are two exceptions: where used for the wiring of chain-supported fixtures, and for supplying current to portable lamps and other merchandise for exhibition purposes, the cord may be other than those listed, but care should be exercised to secure safe cord of the proper type.

400.12: Minimum Size

The minimum size of flexible cord conductors is covered in Table 400.4.

400.13: Overcurrent Protection

As described in Section 240.4, you will often find cords smaller than the wires covered in Table 310.16, which may be protected against overload by the installed overcurrent device. Some of these should be flexible cords smaller than No. 18 or tinsel cords. They may also include cords of similar characteristics or smaller size if approved for use with specific appliances.

400.14: Protection from Damage

If outlet boxes or similar enclosures are provided with holes for a cord to pass through the covers, these holes shall be protected with proper bushings or other approved fittings.

II. Construction Specifications

400.20: Labels

Factory testing and labeling shall be done before shipping flexible cords. Also look for the UL listing. The public very often interprets the UL label on a flexible cord as meaning that the appliance they are purchasing has been tested by UL. This is not the case. The appliance will also have a UL label attached whether it has been tested or not.

400.21: Nominal Insulation Thickness

The thickness of insulation covered in Table 400.4 for flexible cords and cables shall meet the specifications thereof.

400.22: Grounded-Conductor Identification

When one conductor of a flexible cord is used as a grounded circuit conductor (neutral), it must be identified as such by one of the following methods:

(1) A colored braiding.

(2) A tracer in the braid.

(3) Colored insulation, or by three white stripes 120 degrees apart.

(4) By a colored separator.

(5) By tinning the conductor.

(6) Surface marking.

400.23: Grounding-Conductor Identification

The aim of this section is to point out the coloring methods used for the grounding conductor in flexible cords. The grounding conductor shall be identified by a continuous green covering, or by a green braid with a yellow stripe, so as to not mistake its purpose as being for grounding only. This conductor is never to be used as a current-carrying conductor.

400.24: Attachment Plugs

Section 250.138(A) and (B) cover attachment plugs that are equipped with a grounding terminal. If the flexible cord has an equipment-grounding conductor, it shall be attached to this type of plug.

III. Portable Cables Over 600 Volts, Nominal

400.30: Scope

This part applies to multiconductor cables for connection to mobile and portable machinery or equipment that operates at over 600 volts, nominal.

400.31: Construction

- **(A)** Conductors. The minimum size of conductor that is permitted is No. 8 AWG copper, and the conductors shall be of the flexible type (stranded).
- **(B)** **Shields.** To confine the voltage stresses to the insulation, cables that operate at over 2000 volts shall be of the shielded type.
- **(C)** **Grounding Conductor**(s). Grounding conductor(s) are to be provided. This grounding conductor shouldn't be less than the size required in Section 250.122.

400.32: Shielding

The shields mentioned in Section 400.31(B) shall be grounded.

400.33: Grounding

Part XI of Article 250, "Grounding Conductor Connections," shall be followed when connecting grounding conductors.

400.34: Minimum Bending Radii

The minimum bending radii, especially on shielded cables, must be taken into account during installation and handling in service in order to prevent damage to both the cable and the shielding.

400.35: Fittings

Cables may have to be connected together; if so, the connectors shall be of the locking type and provisions should be taken to prevent the opening or closing of these connectors while the cables are energized. Care also must be taken to eliminate tension at cable connections and terminations.

400.36: Splices and Terminations

Portable cables are not to be used with splices except where the splices are permanently molded and vulcanized types as covered in Section 110.14(B). Only qualified persons shall have access to terminals of high-voltage portable cables.

Portable high-voltage cables are usually used on sites where they may be easily damaged, and thus every precaution must be taken to prevent shorts, grounds, and accidents.

Article 402—Fixture Wires

402.1: Scope

This article deals with the general requirements and specifications for the construction of fixture wires.

402.2: Other Articles

Fixture wire complies not only with this section, but with sections elsewhere dealing with fixture wires. Article 410 covers the usage of flexible wire in lighting fixtures.

402.3: Types

Table 402.3 in the *NEC* applies to fixture wires and their usage shall comply with the applicable parts of this table. Fixture wires are intended for fixture wiring and the application of the same is covered for lighting fixtures in Article 410. In Table 402.3, fixture wires are usable for 600 volts, nominal, unless not permissible as otherwise provided.

Note

As we have discovered before, thermoplastic insulation at minus 10°C (plus 14°F) requires special care and installation. Thermoplastic insulation is subject to deformation at normal temperatures where subjected to pressure. Thus, extra care must be exercised during installation, especially at points of support.

402.5: Ampacity of Fixture Wires

See Table 402.5 in the *NEC* for the ampacity of fixture wires and the note following.

The conductor insulation temperature rating shouldn't be less than specified in Table 402.3 of the *NEC* covering types of insulation.

Note

Temperature limitations of conductors is covered in Section 310.10.

402.6: Minimum Size

No conductor smaller than 18 AWG shall be used for fixture wires.

402.7: Number of Conductors in Conduit

Table 2 of *NEC* Chapter 9 gives the number of fixture wires that may be pulled into a single run of conduit.

402.8: Grounded-Conductor Identification

The grounded conductor for fixture wires shall comply with the same requirements as the grounded conductor for flexible cords and cables, which are covered in Article 400, Section 400.22(A) through (E).

402.9: Marking

Fixture wires require the same types of marking as was covered in Section 310.11(A).

- **(A)** **Required Information.** The information marked on the conductor must conform to the requirements of Section 310.11(A).
- **(B)** **Method of Marking.** Thermoplastic-insulated fixture wire shall be marked every 24 inches on the surface of the wire, and the marking shall be durable. All other fixture wire may be marked on the reel, coil, or carton.
- **(C)** **Optional Marking.** Fixture wires in Table 402.3 can have surface markings that indicate special characteristics.

402.10: Uses Permitted

Fixture wires may be used:

- **(1)** For installation in light fixtures. They must be protected, and can't be subject to being twisted or bent.
- **(2)** For connecting light fixtures to branch circuit conductors.

402.11: Uses Not Permitted

Branch circuits shouldn't use fixture wiring for the conductors.

Exception

There are some exceptions to the above. Section 725.16 allows it to be used in Class I circuits, and in fire-protective signal circuits it may be used as covered in Section 760.16.

402.12: Overcurrent Protection

Section 240.4 specifies overcurrent protection that may be used for the protection of fixture wires.

Refer to *NEC* Tables 400.5 and 402.5.

Article 404—Switches

I. Installation

404.1: Scope

This article applies to certain breakers used as switches, and to all switches, including switching devices.

404.2: Switch Connections

- **(A)** **Three-Way and Four-Way Switches.** With three-way and four-way switches, the connections shall be made in such a way that the neutral or grounded conductor is not switched. If a

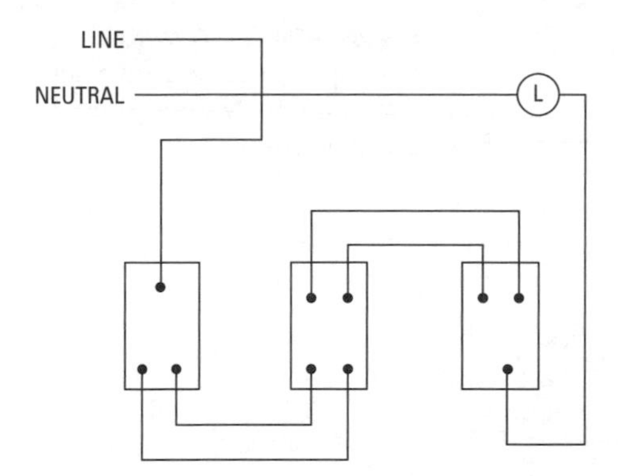

Figure 404-1 The neutral on three- and four-way switching shouldn't be opened.

metal raceway is used, the conductors shall be run so that both polarities are in the raceway in order to counteract inductance (see Figure 404-1).

Exception

A grounding conductor shouldn't be required on switch loops. It is the author's opinion that a grounding conductor could be very appropriate where switches have plates mounted with metal screws, and metal plates shall be used in places that must be grounded. This is not a *Code* requirement.

(B) Grounded Conductors. A grounding conductor must never be opened unless all other conductors associated with it are opened at the same time or before disconnecting the neutral or grounding conductor. In most instances, the grounded conductor is not disconnected, but with circuits going to or passing through gasoline-dispensing islands, it is required that the neutral also be disconnected at the same time as the ungrounded conductor. See Section 514.5 in the *NEC.*

Figure 404-2 shows two circuits; in one the grounded conductor is not disconnected, and in the other the grounded conductor is disconnected simultaneously with the ungrounded conductors. Note that the ground is on the supply side in each case so that the system is still grounded.

404.3: Enclosure

It is required that all switches and circuit breakers be enclosed in metal cabinets with externally operable handles, and the enclosure

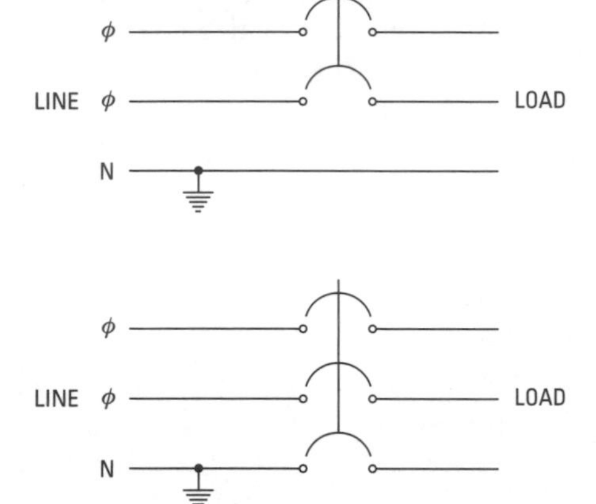

Figure 404-2 A switch that opens the neutral shall simultaneously open the phase wires.

marked to indicate the position of the switch, "off" or "on." Exceptions to this are pendant switches, surface-type snap switches, and open-knife switches mounted on an open-face switchboard or panelboard.

Section 312.6 gives the minimum bending space of terminals, and also includes the minimum gutter space provided in switch enclosures. See this section in the *NEC* for the space required.

404.4: Wet Locations

Section 312.2 covers cabinets and cutout boxes in damp or wet locations, and requires a ¼-inch spacing between the cabinet and the mounting surface. The same applies to switches in wet locations and, if mounted out of doors, they must be in a weatherproof enclosure. Rain-tight is commonly acceptable if prevailing weather conditions are not such that they are endangered by the elements.

No switches may be installed in showers or similar wet locations unless they are part of a listed assembly.

404.5: Time Switches, Flashers, and Similar Devices

These devices must be either enclosed in cabinets and cutout boxes or of the enclosed type. Barriers must be provided so that operators can't be exposed to live parts when working on the equipment.

Exception

If mounted in an enclosure that is only accessible to qualified people, a barrier is needed only if energized parts are within 6 inches of the manual adjustments or switches.

404.6: Position of Knife Switches

Single-throw knife switches shall be mounted in such a position that gravity won't tend to close the switch, but preferably so that they will tend to open. This ruling does not apply to double-throw switches, which may be mounted either vertically or horizontally as required.

(A) **Single-Throw Knife Switches.** Single-throw knife switches shall be mounted in such a position that gravity won't tend to close the switch. If single-throw switches are approved for mounting in an inverted position, they must be provided with a locking device, which ensures that they will stay in the open position.

(B) **Double-Throw Knife Switches.** Double-throw knife switches may be mounted vertically or horizontally, but if mounted vertically, they shall have a locking device to maintain the blades in the open position.

(C) **Connection of Knife Switches.** Knife switches, except those of the double-throw type, shall be connected in such a way that when open, the blades are dead (Figure 404-3).

Author's Note

If this is a fusible switch, the fuses shall also be dead when the switch is open. This is not a problem with modern switches, as they would not be approved unless they contained this feature. However, some older switches are not arranged in this manner.

Exception

There may be times when the load side of the switch is connected to surface or equipment so that there may be feedback from another source of power, thus energizing the open knifeblades or switch. If so, you must install immediately adjacent to the switches a sign reading, "WARNING—LOAD SIDE OF SWITCH MAY BE ENERGIZED BY BACKFEED."

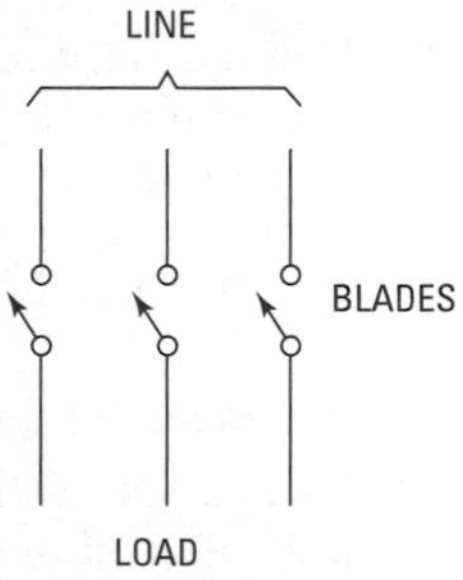

Figure 404-3 Position of knife switches.

404.7: Indicating

Switches and circuit breakers that are mounted enclosures for general use in motor circuits as described in Section 404.3 shall be clearly marked to indicate when they are in the OFF position and when they are in the ON position.

For handles of switches and circuit breakers that are operated vertically and not rotationally or horizontally, the switch shall be in the ON position when the handle is up. This does not apply to double-throw switches.

404.8: Accessibility and Grouping

(A) Location. Switches and circuit breakers that are used as switches shall be mounted at a height that is readily accessible from the floor or working platform, and shall be installed such that the grip's center, when it is at its highest point, doesn't exceed 6 feet 7 inches.

Take note of the above. The handle must be such that it may be reached by hand for opening. There are three important exceptions to the above to take care of impossible cases, where it is impossible to mount the handle as specified above.

Exception No. 1
Actually, this exception is covered early in the *NEC*. The point is that when switches are located on busways that are out of reach from the floor or platform, a suitable device such as a rope, chain, or an adequate stick shall be provided so that they can be accessed from the floor.

Exception No. 2
Motors and appliances and other equipment supplied by switches installed adjacent thereto may be located as specified above, if they are provided with portable means of operation for accessibility.

Exception No. 3
Isolating switches that are higher than 6½ feet may be operated by a hookstick.

(B) Voltage Between Adjacent Switches. Snap switches in outlet boxes may be ganged only when the voltage between live metal parts of adjacent switches does not exceed 300 volts, or if they are installed in boxes equipped with permanently installed barriers between adjacent switches.

404.9: Provisions For General-Use Snap Switches

(A) Faceplates. Whenever flush snap switches are mounted on ungrounded boxes, the plates shall be of a nonconducting material in any location where they may be touched at the same time that contact can be made with grounded surfaces or any other conductive surface. Ferrous faceplates shall be at least 0.030 inch (0.762 mm) thick, while nonferrous faceplates shall be at least 0.040 inch (1.016 mm) thick. Insulated faceplates shall be noncombustible and shall be at least 0.10 inch (2.54 mm) thick. If they are less than 0.10 inch (2.54 mm) thick, they shall be reinforced or formed to provide additional strength. In other words, be certain that approved plates are used. Faceplates shall completely cover the opening in the wall surface.

2005

(B) **Grounding.** Snap switches and dimmers must be grounded. They must also provide a means for grounding metal faceplates, which may be done either with a metal yoke that is in contact with a grounded box, or by using a separate grounding conductor.

If no grounding means is available at the switch (for existing installations only), such a switch may be replaced, but a nonconducting faceplate must be used. Alternatively, a GFCI may be used to provide protection.

404.10: Mounting of Snap Switches

(A) **Surface-Type.** When these switches are used with open wiring on insulators, they must be mounted on insulating material so that the conductors are separated by at least ½ inch from the surface on which they are mounted.

(B) **Box-Mounted.** Boxes for snap switches as permitted in Section 314.10 are to be set back from the wall surface not more than ¼ inch, and shall use the plaster ears so that they sit firmly against the surface of the wall. Boxes that are mounted so that they are flush with the wall or at a slight projection from the wall shall be installed so that the mounting yoke or strap of the switch sits firmly against the box.

In installing boxes, wherever possible there shall be no extension of the box out from the wall surface. This is not a *Code* requirement, just good workmanship.

This is quite plain, but it might be a good idea to return to Section 370.10. In walls of noncombustible materials, the boxes may be set back not to exceed ¼ inch, but on walls that are of combustible material, the boxes are to be set flush.

404.11: Circuit Breakers As Switches

A circuit breaker may be used as a switch provided that it has the same number of poles required by the switch to do the job. Thus, a single-pole breaker with a switched neutral may be used as a switch for simultaneously opening the hot and the neutral conductor feeding or passing through a gasoline-dispensing island. A power-operated circuit breaker capable of being opened by hand in the event of power failure may serve as a switch if it has the required number of poles. (See Section 404.8.) Also, the circuit breakers to be used as switches shall be approved for the purpose.

404.12: Grounding of Enclosures

It is mandatory that enclosures for switches or circuit breakers be grounded, except where accessible to qualified operators only. Article 250 sets forth the manner of grounding enclosures. This does not imply that enclosures for switches or circuit breakers of less than 150 volts-to-ground are not grounded, as in most cases this has been taken care of elsewhere. But it does make it mandatory to ground the enclosures of switches or circuit breakers on circuits of more than 150 volts-to-ground.

404.13: Knife Switches

The following sets forth the requirements of knife switches and the interrupting capacities as to sizes for amperages and voltages. Notice that nothing is stated about interrupting currents larger than the amperage of the switch. Unless marked, a knife switch is not intended to interrupt more than the rating of the switch. For example, a 200-ampere switch will interrupt 200 amperes, but a 200-ampere switch might be purchased that has an interrupting capacity of 8000 amperes. The intent is to show that there is a difference between the current-carrying ratings and the interrupting ratings of switches. Therefore, a switch that will interrupt more than the normal current-carrying rating of the switch will be so marked.

- **(A) Isolation Switches.** Isolation switches are often used in faces of motors that are not in sight of their controller. For safety's sake, they are usually knife switches with no fuses. If they are rated at over 1200 amperes at 250 volts or less, or at 600 amperes at 251 to 600 volts, they are not to be opened while under load. They are for isolating only when working on machinery.
- **(B) To Interrupt Currents.** Only circuit breakers or switches of special design shall be used for current interruption greater than 1200 amperes at 250 volts or less, or at 600 amperes at 251 to 600 volts.

- **(C) General-Use Switches.** Knife switches rated lower than 1200 amperes at 250 volts or less, and 600 amperes at 251 to 600 volts, may be used as general-purpose switches and may be opened under load. Nothing, however, is mentioned about opening under fault currents. See Article 100 for the definition of a general-use switch.
- **(D) Motor-Circuit Switches.** Reference is made to Article 100 for the definition of a motor-circuit switch in which percentages of load and horsepower ratings are mentioned. Knife-type switches may be used for motor switches.

404.14: Rating and Use of Snap Switches

General-use snap switches are switches such as the general-purpose switches in wall boxes or handy boxes. These switches may be used within their rating, and as follows:

- **(A) AC General-Use Snap Switch.** The following covers what form of general-use snap switch is suitable to use only on alternating currents when it controls any of the following:
  - **(1)** If the ampere and voltage rating of the switch involved is not exceeded, it may be used for resistive and inductive loads, including electric-discharge lamps.
  - **(2)** If it is for 120 volts and the ampere rating is not exceeded, it may be used with tungsten-filament lamps.
  - **(3)** If it does not exceed 80 percent of the operating-ampere rating of the switch, it may be used on motor loads.
- **(B) AC-DC General-Use Snap Switch.** A form of general-use snap switch suitable for use on either AC or DC circuits for controlling the following:
  - **(1)** Resistive and inductive loads, if used where the ampere rating of the switch and the voltage supplied is not exceeded by the amperage and voltage of the resistive load. Resistive loads are similar to DC loads and don't have an inductive kickback, which causes the arcing common with inductive loads.
  - **(2)** Inductive-type loads are not to load a switch over 50 percent of the ampere rating of the switch at the applied voltage. Switches when rated in horsepower may be used to control motor loads that are within the horsepower rating of the switch and the voltage applied. Note the percentage and the restrictions because of the inductive loads.

(3) Only "T"-rated switches shall be used on tungsten-filament lamp loads, and they shouldn't exceed the rating of the "T"-rated switch.

For noninductive loads other than tungsten-filament lamps, switches shall have an ampere rating not less than the ampere rating of the load. Noninductive loads are loads that have a 100-percent power factor. Induction is basically a load with a lagging power factor. However, the same effect would result when a capacitive load presented itself, but this would be a rather unusual case. Tungsten-filament lamps draw a heavy current on start.

Tungsten filaments draw heavy initial current for energizing the tungsten filament. Inductive loads include discharge lighting and all other inductive loads. Switches controlling signs and outline lighting should be selected according to the requirements of Section 600.2, and switches controlling motors, according to Sections 430.83, 430.109, and 430.110.

Inductive loads, when opened, create a high-voltage kick caused by the collapsing flux, with a resultant tendency to burn the contacts. High-amperage switches are the answer to this problem.

(C) **CO/ALR Snap Switches.** Snap switches of 20 amperes or less may be directly connected to either copper or aluminum conductors, but they shall be marked CO/ALR.

II. Construction Specifications

404.15: Marking

Switches shall be marked with the current rating, voltage rating, and maximum horsepower rating (if so rated).

404.16: 600-Volt Knife Switches

This is fundamentally a design factor, but it is important. When breaking currents over 200 amperes with 600-volt-rated knife switches, the switchblades shall have auxiliary contacts of a renewable or quick-break type, or they shall be of a type that will serve the equivalent purpose.

404.17: Fused Switches

Switches that are fused are not to have parallel fuses.

Note

The *NEC* refers to Section 240.8.

404.18: Wiring Bending Space

Wiring bending spaces in switches shall meet the requirements of Section 380.3, and shall also meet the requirements of Table 312.6(B), where spacings to the enclosure wall opposite the line and load terminals are given.

Article 408—Switchboards and Panelboards

408.1: Scope

Covered in this article are:

(1) All types of switches that are installed for lighting and power circuits; among these are distribution boards, panelboards, and so on.

(2) Panels supplied from light or power circuits used for charging batteries.

Exception

Switchboards, if used only to control battery-operated signaling circuits.

408.2: Other Articles

As with many articles, the requirements of this article are often supplemented by others in the *Code*, such as Articles 240, 250, 370, 380, and any others that might apply. If switchboards or panelboards are to be located in hazardous (classified) areas, they must meet the requirements of Articles 500 through 517.

408.3: Support and Arrangement of Bus Bars and Conductors

(A) **Conductors and Bus Bars on a Switchboard.** Conductors and bus bars on switchboards, panelboards, or control boards must be mounted so that they are physically protected and secure. Service sections of switchboards must be separated from the other parts of switchboards. Each section of the switchboard should contain only those wires that terminate in that section, except where not practical for control or interconnect wiring.

(B) **Overheating and Inductive Effects.** See the *NEC*.

(C) **Used as Service Equipment.** Each switchboard, switchboard section, or panelboard used as service equipment may require a main bonding jumper sized as covered in Section 250.79(C) and placed within the panelboard or one of the sections of the switchboard for connecting the grounded service conductor on the supply side of the disconnecting means. Bonding is required to all sections of a switchboard to bond them together with an equipment-grounding conductor, and this equipment-grounding conductor shall be sized according to Table 250.95.

Exception
High-impedance grounded systems are accepted, according to Section 250.27.

(D) Load Terminals. Terminals in switchboards and panelboards shall be located such that it will be unnecessary to reach over phase bus bars or conductors in order to make up the connections.

(E) High-Leg Marking. When supplied by a four-wire delta system and a mid-tap of one phase is grounded, the bus bar or conductor having the higher voltage-to-ground should be permanently marked by an outer finish of orange color or by any other effective means.

Refer to Figure 210-3, which shows voltage relations on such a system. It will be noted that phase C has 208 volts to the neutral. This is the conductor with the higher voltage-to-ground being referred to, and care must be exercised to keep from connecting 120-volt equipment to this phase and ground. This phase is commonly termed the *wild leg*.

(F) Phase Arrangement. This is a much-needed *Code* requirement to achieve uniformity in phase lettering.

On three-phase buses, phases shall be indicated as A, B, and C, beginning from front to back or top to bottom or left to right, as if viewed from the front of the switchboard or panelboard. The bus bar's B phase is to be the higher voltage (wild leg) to ground the phase with. Bus bar arrangements other than those above are permitted if they are additions to existing installations, and shall be marked.

Exception
On three-phase, four-wire delta connected systems the same configuration of the phase is permitted as the metering equipment when single or multisection switchboards or panelboards are used for metering.

(G) Minimum Wire-Bending Space. The requirements of Section 373.6 cover the bending space at terminals, and also the minimum gutter space to be provided in switchboards and panel boards.

2005

408.4: Circuit Directory
Every circuit and modification of a circuit must be clearly identified at its panel or switchboard. This identification must provide sufficient information to identify the circuit and its purpose, and to differentiate it from all others.

I. Switchboards

408.5: Location of Switchboards

Switchboards that have any exposed live parts shall be installed in dry locations and accessible only to qualified persons and under the supervision of competent persons. The location of switchboards shall be such that the likelihood of damage from equipment or processes is kept to a minimum.

408.6: Wet Locations

Wherever switchboards are in a wet location, in or outside a building, they must be in a weatherproof enclosure and, if necessary, shall conform to Section 312.2, which requires a ¼-inch spacing from the surface on which they are mounted.

408.7: Location Relative to Easily Ignitable Material

All switch and panelboards must be mounted away from readily ignitable materials. This section can be used in conjunction with Section 240.24(D).

If a combustible floor is used for mounting, protection shall be provided.

408.8: Clearances

- **(A)** **From Ceilings.** Unless there is an adequate fireproof ceiling, a switchboard shouldn't be installed closer than 3 feet from the ceiling.
- **(B)** **Around Switchboards.** See Section 110.16, the provisions of which must be complied with for clearances around switchboards.

408.9: Conductor Insulation

When insulated conductors are used in switchboards, the insulation shall be flame retardant and the voltage rating on the insulation shall be not less than the voltage that is applied to it and to any bus bars or conductors with which it may come in contact.

408.10: Clearance for Conductors Entering Bus Enclosures

For conduits or raceways that enter a switchboard or a panelboard standing on the floor or similar enclosures, if they enter the enclosure at the bottom, be sure to install the conductors in the enclosure so as to have sufficient space for the entering conduit or raceways. Following this section in the *NEC* is a table covering where conduit or raceways leave or enter below the bus bars, supports for bus bars, or any other obstruction. The conduit or raceway, including the fittings used, shouldn't rise in the enclosure more than 3 inches above the bottom of the enclosure.

408.11: Grounding Switchboard Frames

The grounding of all switchboard frames is mandatory except on frames of direct-current, single-polarity switchboards that are effectively insulated.

408.12: Grounding of Instruments, Relays, Meters, and Instrument Transformers on Switchboards

This is covered in Sections 250.170 through 250.178. In addition to other facts concerning grounding, the grounding conductor shall be no smaller than No. 12 copper.

II. Panelboards

408.13: General

Article 220 gives the facts necessary for computing the feeder loads to panelboards. After arriving at the feeder load, the panelboard shall have a rating not less than the minimum feeder size as calculated. All of the facts as to voltages, phase, capacity, and so on, shall be plainly visible after installation. Most inspection authorities will also require the UL label. However, a UL label on an enclosure shouldn't be construed to mean that it also applies to the devices installed therein. Each must have its own UL label.

The door of each panelboard must have mounted on it a directory noting the use of each circuit in the panelboard. This has always been a good trade practice, but this new addition to the *Code* makes it a requirement. See also Section 110.22.

408.14: Lighting and Appliance Branch-Circuit Panelboard

It is necessary that a distinction be made between lighting and appliance panelboards and power panelboards. Therefore, in this section, a lighting and appliance branch-circuit panelboard is defined as a panelboard that has more than 10 percent of its overcurrent devices rated at 30 amperes or less and for which neutral connections are provided.

408.15: Number of Overcurrent Devices on One Panelboard

The *Code* limits the number of overcurrent devices that are permitted in a lighting and appliance branch-circuit panelboard to forty-two. This number does not include the devices used as a main. In counting the devices, a double-pole overcurrent device is counted as two, and a three-pole overcurrent unit is counted as three. Each of these units is often counted as only one, which defeats the intent of this section. Older equipment often was made up with wafer-thin and piggyback breakers that made it possible to increase or even double the number of overcurrent devices in a panelboard. Thus, it might have been possible to install eighty-four breakers in a forty-two-breaker panel. A lighting and appliance branch-circuit panelboard, to be approved, must provide physical means to prevent more overcurrent devices being installed than the panelboard was approved and listed for. The newer panels have ratings such as 12-24, which means that it is designed for twelve full-size breakers or twenty-four piggyback breakers with provisions made for a bus ample to supply these. In this way, the number can be limited to the forty-two requirement.

A case of interest: someone wanted to weld two forty-two-circuit panels together and run buses between them. The decision made, and rightly so, was that the UL label on the panelboard was for the panelboard as originally built, and that welding them together and adding the buses was not approved.

408.16: Overcurrent Protection

(A) Lighting and Appliance Branch-Circuit Panelboard Individually Protected. If the combined rating of two panelboards is not exceeded, each lighting and appliance branch-circuit panelboard may be individually protected by two main breakers or two sets of fuses.

This pertains to a feeder panel. As an example, a 100-ampere panel (one with 100-ampere buses) is fed by 200-ampere conductors and protected by 200-ampere overcurrent devices. In this case, the feeder panel must have a 100-ampere maximum main in it in order to give the proper protection.

A similar case involves a feeder panel with 100-ampere buses. The conductors feeding the panel are of 100-ampere capacity or larger, and are protected by 100-ampere or smaller overcurrent protection at the source. In this case, it is not required to have a main in the feeder panel.

Author's Note

This has always been a little confusing to me, as attempts have sometimes been made to interpret it, for instance, where two circuit breakers are used as the mains in the panel. One could be tripped off and the other still energized, causing injury. It is the responsibility of each inspection authority to see that this sort of thing does not happen.

Exception No. 1

Individual protection of the service-entrance panel for lighting and appliance branch-circuit panelboards is not required if the panelboard feeder at the service-entrance equipment has overcurrent protection that is not higher than the overcurrent protection for which the panelboard is rated.

Exception No. 2

This covers existing installations. Individual protection for lighting and appliance panelboards isn't required when the individual occupancy has service equipment supplying its panelboards. Please note that this exception applies to an individual residential occupancy only. Thus, a split-bus panel is permissible on service equipment in a single-dwelling occupancy as long as it is limited to six operations and the 15- or 20-ampere branch-circuit bus is protected by an overcurrent device that will adequately protect these buses.

There is controversy on split-bus panels as to the size of the service-entrance conductors to be used. In a residence, and this is the only place split-bus panels are permitted, we might have a two-pole, 50-ampere service for the range; a two-pole, 40-ampere service for the dryer; a two-pole, 30-ampere service for the water heater; and a two-pole, 50-ampere service for the 15- and 20-ampere branch circuits. This would give us a total of 170 amperes as mains with the use of 100-ampere service-entrance conductors. Would the *Code* be met where there were a total of 170-ampere mains? It would be our interpretation that we have not met the *Code* requirements. Refer to Section 230.90.

Sometimes a conventional panelboard is installed that has lugs on the buses, and a breaker is fed backwards as a main and the lugs are still intact. Inspection authorities will probably not approve this, as there is a chance that the panelboard may be converted back to one without a main. They will probably accept it if the lugs are sawed off or otherwise fixed so that they may not be used later, but this will void the UL approval.

(B) Power Panelboard Protection. Power panelboard (with more than 10 percent of its branch-circuit overcurrent devices rated 30 amperes or less) must have an overcurrent device on its line side rated not more than the rating of the panelboard. An exception may be made for such a panel used as service equipment with multiple disconnecting means. See Section 230.71.

(C) **Snap Switches Rated 30 Amperes or Less.** Only panelboards protected at 200 amperes or less may be equipped with snap switches that are rated 30 amperes or less.

(D) **Continuous Load.** If the load is continuous for 3 hours or more, the load on any overcurrent device that is located in the panelboard shouldn't exceed 80 percent of the rating of the overcurrent device covering that particular circuit.

Exception

If the overcurrent device is approved for 100-percent rating on continuous loads, the load may be up to 100 percent of the overcurrent device rating.

(E) **Supplied Through a Transformer.** When a panelboard is supplied through a transformer, items (A) and (B) in this list shall be located on the secondary side of the transformer.

Exception

If the load is supplied by a single-phase two-wire transformer with a single voltage, you may consider the secondary to be protected by the overcurrent devices in the primary or supply side of the transformer, provided the primary is protected as covered in Section 450.3(B)(1) and at a value that does not exceed the value determined by multiplying the transformer-voltage ratio of the secondary-to-primary by the panelboard rating.

(F) **Delta Breakers.** A three-phase disconnect shall never be connected to the buses of a panelboard that has fewer than three buses.

(G) **Back-Fed Devices.** Main lug assemblies or overcurrent protection devices that plug in and can be back-fed must have an additional fastener to secure them in place. Only plugging them in is not enough support.

408.17: Panelboards in Damp or Wet Locations

Reference is made to Section 312.2, which requires that panelboards in wet places be weatherproof; in damp locations be arranged to drain; and in both locations be installed with a ¼-inch air space back of them.

408.18: Enclosure

Panelboards shall be the dead-front type, and may be mounted in cabinets or cutout boxes or other enclosures designed for the purpose.

Exception

Panelboards other than dead-front panelboards that are externally operated are permitted where accessible only to qualified persons.

408.19: Relative Arrangement of Switches and Fuses

Panelboards that have switches on the load side of any type of fuses shouldn't be installed, except for use as service equipment as provided in Section 250.146(D). With fuses ahead of the switch, they would have to be replaced while still energized.

408.20: Grounding of Panelboards

Panelboard cabinets shall be grounded in the manner specified in Article 250 or Section 408.3(C). An approved terminal bar for equipment-grounding conductors shall be provided and secured inside of the cabinet for the attachment of all feeder and branch-circuit equipment-grounding conductors when the panelboard is used with nonmetallic raceway or cable wiring, or where separate grounding conductors are provided. The terminal bar for the equipment-grounding conductor must be bonded to the cabinet or panelboard frame (if it is metal), or else connected to the grounding conductor that is run with the conductors that feed the panelboard. Grounded conductors can't be connected to a neutral bar, unless it is identified for such use. In addition, it must be located at the connection between the grounded conductor and the grounding electrode.

Exception

When an isolated grounding conductor is used (see Section 250.74, Exception 4), the *equipment-grounding* conductor that is run with the circuit conductors can pass through the panelboard without connecting to the equipment-grounding terminal bar. This is a necessary exception for isolated computer-grounding systems.

III. Construction Specifications

Most of these sections are not very applicable to installations, but apply more for manufacturing equipment.

408.30: Panels

See the *NEC*.

408.31: Busways

See the *NEC*.

408.32: Protection of Instrument Circuits
See the *NEC*.

408.33: Components Parts
See the *NEC*.

408.34: Knife Switches
See the *NEC*.

408.35: Wire Bending Space
See the *NEC*.

408.36: Minimum Spacing
See the *NEC*.

2005

Article 409—Industrial Control Panels

409.2: Definitions
An *industrial control panel* is an assembly of separate components (such as motor controllers, overload relays, fused disconnects, and circuit breakers), combined with push buttons, selector switches, timers, control relays, terminal blocks, and so on.

409.3: Other Articles
While this article contains requirements for the installation of these panels, each of their components are covered in other articles of the *NEC*. The locations for these other articles are shown in *NEC* Table 409.3.

409.20: Conductors–Minimum Ampacity
Supply conductors to these panels must have an ampacity of at least 125 percent of all resistance heating loads, plus the full-load current of the highest rated motor. The balance of the motors may be added to this amount at only 100 percent of their rated full-load. Allowances are permitted for duty cycles.

409.21: Overcurrent Protection

(A) **General.** Industrial control panel installations must comply with the requirements of Article 240, especially Parts I, II, and IX.

(B) **Location.** Overcurrent protective devices must be provided either ahead of the panel or as a single overcurrent protective device inside the panel. If a single device is located in the panel, the conductors feeding it are considered either feeders or taps. See Section 240.21.

(C) Rating. The setting of the overcurrent protective device must be no more than the setting of the largest branch-circuit short-circuit, and ground-fault protective device provided with the panel, plus the 125 percent of the full-load rating of all resistance heating loads, and the full-load current of all other motors and apparatus that could be in operation simultaneously. An exception is made for special circumstances. See the *NEC* for details.

409.30: Disconnecting Means

Motor disconnects must meet the requirements of Article 430, especially Part IX.

409.60: Grounding

Multisection panels must be bonded together with a conductor or bus, sized according to Table 250.122.

Article 410—Lighting Fixtures, Lampholders, Lamps, Receptacles, and Rosettes

I. General

410.1: Scope

Lighting fixtures (also called *luminaires*), lampholders, pendants, receptacles, rosettes, incandescent filament lamps, arc lamps, electric-discharge lamps, and the wiring and equipment forming a part of such lamps, fixtures, and lighting installations shall conform to the provisions of this article, except as otherwise provided in this *Code*.

410.2: Applications to Other Articles

Articles 500 through 517 cover hazardous (classified) locations, and all equipment with, and pertaining to, the fixtures in such locations shall meet the requirements of these articles.

410.3: Live Parts

There shall be no live parts normally exposed on any fixture, lampholder, lamp, rosette, or receptacle, with the exception of cleat-type lampholders, rosettes, and receptacles that may have exposed live parts if they are at least 8 feet above the floor. Lampholders, receptacles, and switches that have live terminals exposed and accessible shouldn't be installed in metal canopies or in the open bases of portable table or floor lamps.

II. Fixture Locations

410.4: Fixtures in Specific Locations

(A) **Wet and Damp Locations.** The installation of fixtures in all damp or wet locations (definitions follow) shall be done in such a way that water can't enter or accumulate in the lampholders, wiring compartments, or any other electrical parts of the installation. Fixtures that are to be installed out of doors in damp or wet locations shall be marked "Suitable for Wet Locations." This should eliminate any haphazard installations that are exposed to moisture. All fixtures installed in wet or damp locations shall be marked "Suitable for Wet Locations" or "Suitable for Damp Locations."

Wet Locations: installations underground, in concrete slabs or masonry in direct contact with the earth, unprotected and exposed to weather, vehicle washing areas, and similar locations.

Damp Locations: interior locations protected from weather, but with moderate degrees of moisture, basements, some barns, some cold-storage warehouses, under canopies, marquees, roofed open porches, and so on.

The *NEC* refers us to Article 680 for lighting fixtures in swimming pools, fountains, and similar locations.

(B) **Corrosive Locations.** If fixtures are installed in corrosive locations, they must be suitable and approved for such locations.

This will, of course, include any installation in atmospheres with corrosive ducts, gases, or liquids.

Note

Receptacles in fixtures are covered in Section 210.7.

(C) **In Ducts or Hoods.** Lighting fixtures may be installed in nonresidential occupancies, provided that they meet the conditions listed below.

(1) The fixtures used in commercial cooking hoods shall be installed so that the temperature limits of the material used are not exceeded by the temperatures encountered. Also, the fixtures have to be listed for such purposes.

(2) The construction of fixtures shall be such that the lamps and wiring compartments are not exposed to exhaust vapors, cooking vapors, grease, or oil. Thermal shock to diffusers shall be resisted.

(3) Parts of the surface of the fixture shall be smooth so that deposits don't readily collect on them, and they can be easily cleaned. They shall also be resistant to corrosion, or additional protection shall be installed to protect them against corrosion. In order to facilitate cleaning and to prevent collection of grease, they must be of the smooth-surface type.

(4) The wiring methods used to light these fixtures shall be exposed outside of the hood.

Note

The *NEC* refers to Section 110.11.

(D) Bathtub and Shower Areas. In bathroom areas, no parts of pendants, hanging fixtures, or cord-connected fixtures shall be within an area of 3 feet measured horizontally and 8 feet vertically from the rim of the bathtub. This encompasses the entire bath area, including that directly over the tub.

410.5: Fixtures Near Combustible Material

The construction of fixtures shall be such, or they shall be so installed or equipped with shades or guards that any combustible materials in the immediate vicinity of the fixture won't be subject to a temperature more than 90°C (194°F). Combustible materials that are subject to higher temperatures will change in composition to substances that ignite very readily.

410.6: Fixtures Over Combustible Material

Lampholders that are installed over highly combustible materials shouldn't have a switch as a part of the lampholder, but shall be switched elsewhere and, unless an individual switch (located elsewhere) is used for each fixture, the lampholder shall be located at least 8 feet above the floor or otherwise located or guarded so the lamp may not be readily removed or damaged. See Figures 410-1 and 410-2.

410.7: Fixtures in Show Windows

Fixtures that are externally wired (except chain-mounted fixtures) can't be used in show windows.

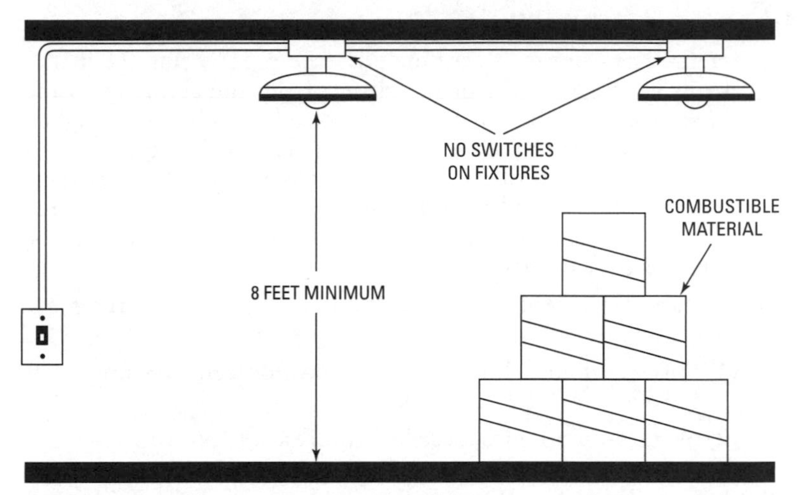

Figure 410-1 Fixtures over combustibles must not have a switch as part of the fixture but must be switched elsewhere.

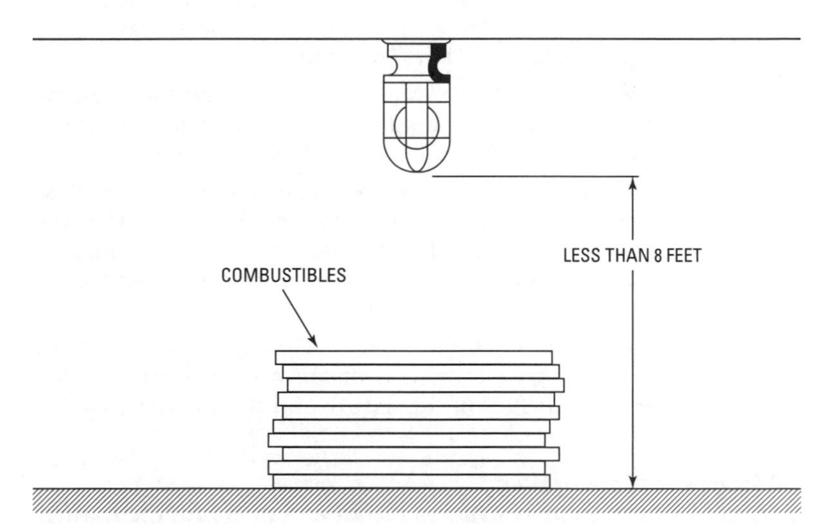

Figure 410-2 Guarding and switching of fixtures over combustibles when fixture is less than 8 feet from the floor.

410.8: Fixtures in Clothes Closets

(A) Definition. *Storage space* includes the area within 24 inches from back and walls of the closet, from the floor up to the highest clothes hanging rod or 6 feet, whichever is higher. Above the highest rod, or 6 feet (whichever is higher), the area within 12 inches of the back or walls is considered storage space. If shelves wider than 12 inches are in this area, the area above the shelves (whatever the width of the shelves) must be considered a storage area.

(B) Fixture Type Permitted. The following types of fixtures are permitted in closets:

(1) Surface-mounted or recessed incandescent fixtures that have a completely enclosed bulb.

(2) Surface-mounted or recessed fluorescent fixtures.

(C) Fixture Types Not Permitted. Incandescent fixtures with open or partially exposed lamps may not be used in closets. Pendant fixtures and lampholders (pull-chain or keyless fixtures) are also prohibited.

(D) Location. Fixtures in clothes closets must be installed as follows:

(1) Surface mounted incandescent fixtures can be mounted on the wall above a door, as long as there is a distance of at least 12 inches between the bulb and the storage area, which was detailed in part (A) of this article.

(2) Surface-mounted fluorescent fixtures can be mounted on the wall over a door, as long as there is a distance of at least 6 inches between the bulb(s) and the storage area.

(3) Recessed incandescent fixtures that completely enclose the lamp (the fresnel-type may be used, but the baffle-type cannot) can be installed in the wall or ceiling, as long as there is a distance of at least 6 inches between the surface of the fixture and the storage area.

(4) Recessed fluorescent fixtures can be mounted in the wall or ceiling, as long as there is a distance of at least 6 inches between the surface of the fixture and the storage area.

410.9: Space for Cove Lighting

Coves must have enough space to install and maintain the fixtures properly.

III. Provisions at Fixture Outlet Boxes, Canopies, and Pans

410.10: Space for Conductors

Basically, this section refers to Article 370, which provides for the proper capacity in outlet boxes, canopies, and pans. The fundamental idea is that an adequate space must be provided so as not to crowd the fixture wires, the branch-circuit conductors, and their connecting devices.

410.11: Temperature Limit of Conductors in Outlet Boxes

A great many fixtures are of the enclosed-type, which trap considerable heat. This section provides that the fixtures be constructed such that the conductors are not subjected to a temperature higher than their rating. Often an outlet box is separated from the fixture proper by a short piece of flexible tubing through which high-temperature wire is run. This is done so that the wire with ordinary temperature insulation may be connected to the fixture conductors and not be overheated.

A common violation of the *Code* is in the use of an outlet box, which is an integral part of an incandescent lighting fixture, for the purpose of passing branch-circuit conductors through this outlet box. The only time that this is permissible is if the fixture and outlet box have been listed for this purpose. Look for the UL label.

410.12: Outlet Boxes to Be Covered

When an installation is complete, all outlet boxes shall be covered unless covered by a fixture canopy, lampholder, receptacle, rosette, or similar device. An exception is made in Section 410.14(B).

410.13: Covering of Combustible Material at Outlet Boxes

Whenever a combustible wall or ceiling is exposed between the outlet box to which a fixture is connected and the canopy or pan of the fixture, the exposed part of the combustible material shall be covered with a noncombustible material (Figure 410-3).

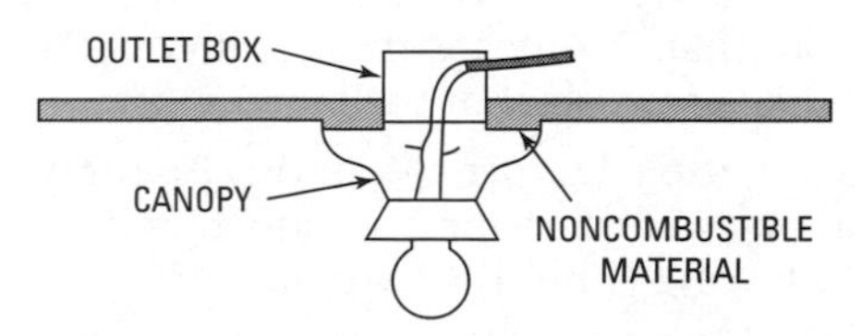

Figure 410-3 Method of installing a canopy larger than the outlet box.

410.14: Connection of Electric-Discharge Lighting Fixtures

As a general rule, fluorescent fixtures that are supported independently of the outlet box are to be connected to the outlet box by metal raceways, metal-clad cable, or nonmetallic-sheathed cable, not by flexible cords. When supported directly below the outlet box, however, this ruling may be waived and flexible cord used if there is no strain on the cord and if it won't be subjected to physical damage. When thus connected, the outer end of the cord shall terminate in an approved grounding-type plug or busway plug. See Section 410.30(C).

(A) Independently of the Outlet Box. Sometimes a fluorescent fixture may be located away from the outlet box to which it is connected. If this occurs, the connection shall be made with metal or nonmetallic raceways, Types AC, MC, MI, or nonmetallic-sheathed cables.

Exception

Sections 410.30(B) and (C) allow some exceptions for connecting these fixtures.

2005

(B) Access to Boxes. If electric-discharge fixtures are mounted on unconcealed boxes, and not designed to be supported completely by the outlet box, there shall be suitable openings in the back of the fixtures, or some other means of access to the box.

IV. Fixture Supports

410.15: Supports

(A) General. Secure supports shall be provided for all fixtures, lampholders or rosettes, and receptacles. Any fixture that weighs over 6 pounds and is more than 16 inches in any dimension must not be supported by the screw shell of the lampholder.

(B) Metal Poles Supporting Lighting Fixtures. Lighting fixtures may be installed in metal poles. Their supply conductors may also be enclosed, if the following conditions are met:

(1) The metal lamp pole shall have a hand-hole that is not less than 2 inches by 4 inches. The cover over the hand-hole shall be rain-tight. This will provide access within the pole or pole base for the supplying of raceway or cables. If the raceway risers or cable are not installed within the pole, a threaded fitting or nipple must be brazed, welded, or tapped to the pole, and this shall be opposite the hand-hole.

Exception

A hand-hole is not required for poles less than 20 feet with hinged bases. Both parts of the pole must be bonded, and there must be a grounding terminal accessible.

(2) A place for grounding the pole shall be provided that is accessible from the hand-hole.

Exception No. 1

The hand-hole and grounding terminal required by this section can be omitted when the wiring that supplies the fixture runs without splice or pulling point to a fixture mounted on a metal pole that is no taller than 8 feet from grade. The interior of the pole and any splices must be accessible by removing the fixture.

Exception No. 2

The hand-hole required above can be omitted for poles 20 feet tall or less, if the connections and the grounding terminal are accessible through a hinged base.

(3) As provided in Section 250.91(B), any metal raceway or any other equipment-grounding conductor must be bonded to the pole by means of equipment-grounding conductors, thereby tying raceways, and so on, of the pole together. They shall be sized according to Section 250.95.

(4) Section 300.19 covers the supporting of conductors that are run vertically. If the conductors are run vertically in a metal pole used as a raceway, they shall be supported as required by the aforementioned section.

Section 410.15(A) might cause some discussion, but the final decision is left up to the inspection authority. The one point that must be satisfied is that the fixture must be secure and not subject to a failure of the supporting device.

410.16: Means of Support

(A) Outlet Boxes. Section 314.23 covers the support and methods of supporting outlet boxes. When a box is properly supported, it may be used to support fixtures weighing up to 50 pounds. Where not supported adequately, or where the fixture exceeds 50 pounds, the fixture shall be supported independently from the box.

(B) **Inspection.** Fixtures must be installed such that any connections between fixture conductors and circuit-supplying conductors may be inspected.

Exception

Any lighting fixtures connected by cord and plug to an outlet.

(C) **Suspended Ceilings.** Lighting fixtures may be supported from the frames of suspended ceilings. The fixture shall be securely fastened to the frame by means of bolts, screws, or nuts. Clips that are made especially for clipping the lighting fixture to the ceiling frame may be used. The ceiling-framing members must be attached to the building structure at appropriate intervals.

(D) **Fixture Studs.** If fixture studs are not a part of the fixture, it is necessary to install hickeys, tripods, and crowfeet, and these are to be made of steel or other material that has been deemed suitable for such usage.

(E) **Insulating Joints.** Where there are insulated joints that are not designed to be mounted with screws or bolts, they shall have a metal casing insulating both screw connections.

(F) **Raceway Fittings.** When raceway fittings are used to support lighting fixtures, they must be capable of supporting the weight of the fixture assembly and the lamps.

(G) **Busways.** Section 368.12 covers fixtures that may be connected to busways.

V. Grounding

410.18: Exposed Fixture Parts

(A) **With Exposed Conductive Parts.** When light fixtures that have exposed conductive parts and equipment that is directly wired to the fixture or attached thereto are supplied by a wiring method that has an equipment-grounding conductor, these exposed parts, and so on, shall be grounded.

(B) **Made of Insulating Material.** When a fixture is directly wired or attached to a wiring method that has no means of equipment grounding, the fixture shall be made of nonconducting material and shall have no exposed conducting parts.

2005

Replacement fixtures are permitted to be connected to an equipment-grounding conductor from a nearby receptacle, or they may be protected with a GFCI.

410.19: Equipment over 150 Volts-to-Ground

(A) Metal Fixtures. Metal fixtures and transformers, and the enclosures used with a fixture on circuits operating at over 150 volts-to-ground, shall be grounded. This grounding shall be by means of an equipment-grounding conductor tying in with the system-grounding conductor.

(B) Other Exposed Metal Parts. Any exposed metal parts must be grounded, or, if insulated from ground or any conductive surface, they shall be accessible only to qualified persons.

Exception

For any tie wires for lamps, mounting screws, clips, or any other decorative metal bands that might be on the glass lamp space, there shall be at least 1½ inches of clearance from these and the lamp terminals. If this is the case, grounding of these parts is not required.

410.20: Equipment-Grounding Conductor Attachment

A place shall be provided for connecting the equipment-grounding conductor to light fixtures that have exposed metal parts.

410.21: Methods of Grounding

As specified in Section 250.118, fixtures are to be considered grounded where they are mechanically connected to an equipment-grounding conductor. The grounding conductor size shall be as covered in Section 250.122.

VI. Wiring of Fixtures

Sections 410.22 through 410.25 are basically used for the manufacture of fixtures. In the unlikely event that you need specific information, you may refer directly to the *Code*.

410.22: Fixture Wiring—General

See the *NEC*.

410.23: Conductor Size

See the *NEC*.

410.24: Conductor Insulation

See the *NEC.*

410.25: Conductor for Certain Conditions

See the *NEC.*

410.27: Pendant Conductors for Incandescent Filament Lamps

(A) **Support.** Pendant lampholders with permanently attached leads (commonly termed pigtail sockets) that are used for other than festoon wiring are to be hung by separate stranded rubber-covered conductors that are separately soldered to the circuit conductors. These rubber-covered conductors are not to support the lampholder; other means of support shall be used. See Figure 410-4.

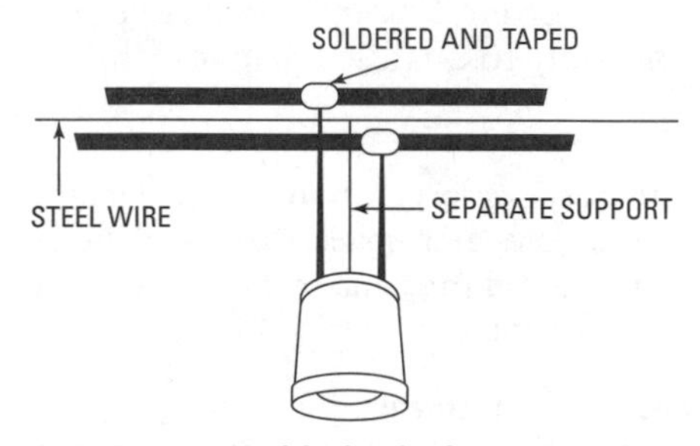

Figure 410-4 Method of attaching pigtail sockets.

(B) **Size.** Conductors shouldn't be smaller than No. 14 for mogul-base or medium-base, screw-shell lampholders when used as pendant conductors. They shouldn't be smaller than No. 18 where used as pendant conductors for intermediate or candelabra-base lampholders.

Exception

Conductors smaller than No. 18 shouldn't be used for Christmas trees and decorative-lighting systems. Conductors feeding these systems can't, of course, be smaller than No. 14.

(C) **Twisted or Cabled.** If pendant conductors are longer than 3 feet, they shall be twisted together where not cabled in a listed assembly.

410.28: Protection of Conductors and Insulation

(A) **Properly Secured.** The conductors shall be secured properly so there will be no cutting or abrading of the insulation of the conductors.

(B) **Protection Through Metal.** Whenever the conductors pass through a metal plate, protection from abrasion is required. This has been discussed above.

(C) **Fixture Stems.** There shall be no splices or taps when fixture stems or arms are used to carry the fixture conductors to the lamp.

(D) **Splices and Taps.** Splices and taps shouldn't be made on a lighting fixture, except where absolutely necessary.

Note

Section 110.14 gives approved methods for making connections.

(E) **Stranding.** Any movable or flexible parts, including fixture chains, shall be wired using stranded conductors.

(F) **Tension.** The weight of the fixture or movable parts shouldn't put tension on the conductors. When installing these conductors, fixtures, or movable parts, therefore, they shall be arranged so as not to cause tension on these conductors.

410.29: Cord-Connected Showcases

Fixed showcases shall be permanently wired. Movable showcases may be wired by flexible cord running to a permanently installed receptacle. A group consisting of not more than six such showcases may be connected to each other with flexible cord that has a lock-type separable connection.

The installation of these showcases shall comply with (A) through (D) of this section:

(A) **Cord Requirements.** The flexible cord used shall be only of the hard-service type, and the conductors in this cord shouldn't be smaller than the branch-circuit conductors serving the receptacle. The flexible-cord conductors shall have an ampacity equal to that of the branch-circuit overcurrent protection and shall be equipped with an equipment-grounding conductor.

Note

Table 250.95 gives the size of the equipment-grounding conductor that shall be used.

(B) Receptacles, Connectors, and Attachment Plugs. All receptacles, connectors, and attachment plugs shall be rated 15 or 20 amperes, shall be of a type used for grounding, and shall be listed.

(C) Support. Flexible cord shall be securely fastened to showcases on the underside so that:

(1) None of the wiring is exposed to mechanical damage.

(2) Where more than one case is involved, they are not more than 2 inches apart and the distance from the first case to the receptacle is not more than 12 inches.

(3) At the end of the group the free end of the cord in showcases has a female connection that does not extend beyond the case.

(D) No Other Equipment. No equipment, other than the showcase-lighting equipment, shall be connected to the showcase wiring.

(E) Secondary Circuit(s). Where showcases are cord-connected, the secondary circuit(s) or discharge lighting shall be limited to one showcase.

410.30: Cord-Connected Lampholders and Fixtures

(A) Lampholders. When a lampholder is attached to a flexible cord, it must be equipped with an insulated bushing. If, by chance, it is equipped with a pipe nipple and threaded, this pipe shouldn't be smaller than ⅜-inch pipe size. It must be large enough for the cord to enter, and it shall be reamed or otherwise have all the burrs on the inside removed so that there will be no chance of damaging the cord insulation.

With plain pendant cord, 9⁄32-inch bushings may be used. Bushing holes 13⁄32 inch in diameter may be used with reinforced cord.

(B) Adjustable Fixtures. Hard-usage cord or extra-hard-usage cord may be used on fixtures requiring adjusting or aiming after installation (such as playing-field lighting). They are not required to have attachment plug- and cord-connector if the exposed cord is of the hard-usage or extra hard-usage type, and the cord is not longer than that which would be required for maximum adjustment and is not subject to strain or physical damage.

(C) Electric-Discharge Fixtures. Permission is granted to use cord-equipped electric-discharge fixtures below the outlet box, provided that the cord is entirely visible for its entire length outside the fixture. The cord shouldn't feed more than one fixture; neither shall it be subject to strain or physical damage. When these conditions are met, such cord-equipped fixtures shall terminate at the outer end of an attachment plug that is of the grounding-type or a busway plug.

If the requirements of Section 240.4 are met, electric-discharge fixtures that use mogul-base screw-shell lampholders may be connected to 50-ampere or less branch circuits or cords complying with the above. If these fixtures are connected by receptacles and attachment plugs, they may have an ampere rating less than that of the branch circuit supplying them, but in no case less than 125 percent of the full-load current of the fixture.

Cord connectors may supply electric-discharge lighting fixtures that have flanged surface inlets. The inlet and connectors may be lower than the rating of the branch circuit, but the pendant shouldn't be lower than 125 percent of the full-load-current rating of the fixture.

410.31: Fixtures As Raceways

Fixtures in general are not intended to be used as raceways for circuit conductors unless they are specifically designed and listed for that purpose. There are three exceptions to this, but it must be remembered that the exceptions cover only the conductors of a single two-wire or multiwire branch circuit supplying these fixtures.

Exception No. 1

This exception is permitted only when the fixture has been listed for use as a raceway.

Exception No. 2

If fixtures are so designed as to be mounted in a continuous row, that is, end to end, or if they are connected by fittings that are recognized wiring methods, the conductors of a two-wire or multiwire branch circuit may carry the conductors straight through.

Exception No. 3

In Exception No. 2 above, one additional branch circuit that separately supplies one or more of the connected fixtures is permitted.

Note

The definition of multiwire circuits is covered in Article 100.

Branch-circuit conductors that come within 3 inches of the ballast of an electric-discharge lighting fixture shall be recognized for use at temperatures not lower than 90°C (194°F). Some of these types are RHH, THW, THHN, THHW, FEP, FEPB, SA, XHHW, and AVA.

VII. Construction of Fixtures

See the *NEC*.

VIII. Installation of Lampholders

410.47: Screw-Shell Type

It has long been a common practice to use screw-shell type lampholders for purposes other than for lamps, such as the female end of an attachment plug, or a pigtail socket as a fuse holder, and so on. These uses are all prohibited by the *NEC*. Screw-shell sockets are to be used as lampholders only.

410.48: Double-Pole Switched Lampholders

Where a lampholder is connected to the hot conductors of a multiwire branch circuit, the switching device of the lampholder shall simultaneously disconnect both conductors of the circuit. This can be readily understood considering that whenever hot wires feed a circuit, if only one is opened, it leaves the other one hot. The *Code* has provided for this dangerous condition by requiring that both hot conductors be opened simultaneously.

410.49: Lampholders in Wet or Damp Locations

Lampholders in these locations must be the weatherproof type.

IX. Construction of Lampholders

See the *NEC*.

X. Lamps and Auxiliary Equipment

410.53: Bases, Incandescent Lamps

Incandescent lamps for general use on lighting branch circuits are limited to not over 300 watts when used in medium-base lampholders. Mogul-base lampholders are limited to 1500-watt incandescent lamps. Special bases or other devices shall be used for incandescent lamps above 1500 watts.

410.54: Enclosures for Electric-Discharge Lamp Auxiliary Equipment

Resistors and regulators used with mercury-vapor lamps are sources of heat and shall be so treated by being installed in noncombustible cases and wired accordingly.

The switching device of auxiliary equipment when supplied by the hot legs of a circuit must disconnect all conductors simultaneously.

410.55: Arc Lamps

See the *NEC*.

XI. Receptacles, Cord Connectors, and Attachment Plugs (Caps)

410.56: Rating and Type

(A) **Receptacles.** Receptacles for attaching portable cords should be rated at not less than 15 amperes, 125 volts, or 15 amperes, 250 volts. These are to be portable only for lampholders. Notice that receptacles shall be of such a type that lamps can't be screwed into them.

Exception

In nonresidential occupancy only, receptacles of 10 ampere/250-volt rating used to supply equipment other than portable hand tools, portable hand lamps, or extension cords are permitted.

Author's Note

I have yet to find where a 10-ampere receptacle can be purchased; this does not mean, however, that they cannot be found.

(B) **CO/ALR Receptacles.** Any receptacles 20 amperes or less to which are connected aluminum conductors must be permanently marked CO/ALR. These, of course, may also be used for copper.

(C) **Isolated Ground Receptacles.** Receptacles intended for this use [permitted by Section 250.146(D)] must be orange, and marked with a triangle.

(D) Faceplates. See the *NEC*.

(E) **Position of Receptacle Faces.** Upon installation, receptacle faces shall be flush, or if mounted on insulating material, the faceplate of the insulating material project a minimum of .015 inch from metal faceplates. Faceplates shall be mounted after installation so that they completely cover the opening and sit against the mounting surface. If the mounting boxes for receptacles are set back of the wall surface, as was covered in Section 314.10, the receptacle shall be installed in such a way that the yoke or strap on the receptacle is be held rigidly at the surface of the wall. When boxes are mounted flush with the wall surface or project somewhat from the wall surface, the receptacle shall be mounted so that the yoke or strap on the receptacle will be well set against the box or raised box cover.

(F) **Receptacle Mounting.** When mounted on boxes that are set back in a wall, the mounting strap of receptacles must be held rigidly against the wall's surface. When mounted on boxes flush with or protruding from a wall, receptacles must be mounted rigidly to the box. Receptacles mounted on covers must be secured with more than one screw or must be part of an assembly that is listed for such use.

(G) **Attachment Plugs.** Fifteen- and 20-ampere attachment plugs and connectors shall be constructed such that no current-carrying parts will be exposed, with the exception of the blades or pins. If by necessity the connections for the wires must be a part of the plug or connector, they must be covered.

A separate insulated disk that is mechanically secured won't be permitted. This applies to the fiber-insulating disk that is slipped on the prongs of the plug.

(H) **Attachment Plug Ejector Mechanisms.** If attachment plugs have ejector mechanisms, they shall in no way affect the proper contact between the blades of the plug and the contacts of the receptacle.

(I) **Noninterchangeability.** This refers to NEMA plug-and-connector configurations, which prohibit interchangeable plugs, connectors, and receptacles, so that only the proper parts will fit for the proper voltage and currents for which they were designed. Thus, if properly installed, the wrong voltage equipment can't be plugged into higher or lower voltages, and so on.

Exception

T-slot receptacles or cord connectors that are rated 20 amperes may accept attachment plugs of the same voltage rating and a 15-ampere rating.

(J) **Receptacles in Raised Covers.** Receptacles that are mounted in raised covers must be supported by more than a single screw.

410.57: Receptacles in Damp or Wet Locations

(A) **Damp Locations.** The enclosure for receptacles installed outdoors and protected from the weather, or installed in any other damp location, must be weatherproof when the receptacle is covered. This means that after the attachment plug is removed, the cover is automatically closed.

Receptacles that can be installed in wet locations are also suitable for damp locations. In damp or wet locations, the only sensible thing to do is to use GFCIs.

Receptacles located under roofs, open porches, canopies, or marquees, and the like, and not subjected to a beating rain or water runoff, including blowing snow, are considered to be in a location protected from weather.

(B) **Wet Locations.** When receptacles are installed outdoors, exposed to the weather, or in any other wet location, they shall be in a weatherproof enclosure that shouldn't be affected when the receptacle is in use (that is, attachment plug-cap inserted).

Exception

An enclosure that is weatherproof only when the self-closing receptacle cover is closed may be used where the receptacle is installed outdoors. This is only when the receptacle is used for portable tools, or other portable equipment, that are only for occasional use and not left connected to the outlet.

(C) **Bathtub and Shower Space.** No receptacles are permitted in such areas.

(D) **Protection for Floor Receptacles.** Floor receptacle standpipes must be high enough that floor-cleaning equipment can be used without damage to receptacles. It should be noted that Section 370.17 requires listed boxes where receptacles are located in the floor.

(E) **Flush Mounting with Faceplate.** An enclosure for outlets, if the box is installed flush with the wall surface, may be made weatherproof by using a weatherproof plate-receptacle assembly that provides a watertight connection between the plate and the wall in which the box is recessed.

(F) **Installation.** If a receptacle outlet is installed outdoors, it shall be located so that accumulation of water is not likely to occur in the outlet cover or plate.

410.58: Grounding-Type Receptacles, Adapters, Cord Connectors, and Attachment Plugs

(A) **Grounding Poles.** In addition to the circuit poles, on grounding-type receptacles, plugs, and connectors there shall also be a pole for grounding purposes.

(B) Grounding-Pole Identification. There shall be a means of connecting the grounding conductor to the grounding pole of grounding-type receptacles, adapters, cord connectors, and attachment plugs.

The grounding-pole terminal shall be designated by one of the following:

(1) The grounding screw or nut shall be hexagonal and painted green, and shall be made so that it can't be easily removed.

(2) The wire connector body shall be pressure-connected in a green color.

(3) Where adapters are used, similar green devices for connection shall be used. The grounding adapter and ground terminal shall be a green-colored rigid ear, lug, or similar device. The grounding connection shall be designed in such a way that it won't come into contact with a current-carrying conductor and either parts of the receptacle adapter or attachment plug. The adapter shall be polarized.

(4) When the terminal for the equipment-grounding conductor is not visible and the wire is inserted into a hole where contact is made, the hole where the wire is to be inserted shall be marked with the word "green," the word "ground," the letter "G," the letters "GR," or the grounding symbol (see Figure 250-12).

(C) Grounding Terminal Use. The grounding terminal or grounding-type device is intended only for grounding purposes and shall never be used for any other purpose.

(D) Grounding-Pole Requirements. The shape and/or length of grounding-type devices shall be designed such that the grounding poles of attachment plugs can't be brought into contact with current-carrying parts of receptacles or cord connectors. The design of grounding-type attachment plugs shall be such that the ground connection is made before the current-carrying connections are made.

(D) Use. Grounding-type plugs shall be used only where an equipment-grounding conductor is provided.

XII. Special Provisions for Flush and Recessed Fixtures

410.64: General

Fixtures installed in cavities of walls or ceilings shall be of a listed type. Sections 410.65 through 410.72 give information on the *Code* requirements.

410.65: Temperature

- **(A) Combustible Material.** It is very important that combustible materials adjacent to fixtures not be subjected to temperatures over 90°C (194°F).
- **(B) Fire-Resistant Construction.** Fixtures that are recessed in fire-resistant material in a fire-resistant building may have allowable temperatures of more than 90°C (194°F) but not more than 150°C (302°F). In order that fixtures thus used are acceptable they shall be listed for this type of service.
- **(C) Recessed Incandescent Fixtures.** Recessed incandescent fixtures must be identified (listed) as having thermal protection. The heat produced by the incandescent fixture may start a fire, so the thermal protection shuts them off before this temperature is reached.

Exception No. 1

If the recessed fixture is identified for installation in poured concrete, it may be used.

Exception No. 2

If recessed fixtures are listed and identified for contact with thermal insulation, they may be so used.

410.66: Clearance and Installation

- **(A) Clearance.** On recessed lighting, the recessed part of the fixture enclosures, except at the point of support, must be spaced at least ½ inch from combustible materials.

Exception

Recessed fixtures can be installed in direct contact with thermal insulation, provided they are listed for this use.

- **(B) Installation.** Thermal insulation is not permitted within 3 inches of the fixture-enclosure wiring compartment, and may not be placed above the ballast. The insulation shall be placed so as not to stop free circulation of air.

Exception

Recessed fixtures can be installed in direct contact with thermal insulation, provided they are listed for this use.

410.67: Wiring

- **(A) General.** Only conductors with insulation suitable for the temperatures that will be encountered shall be used.

(B) Circuit Conductors. Branch-circuit conductors that have an insulation-temperature rating suitable for the temperature they will encounter may be terminated in the fixture.

(C) Tap conductors. Tap connection conductors that have a temperature rating for that which will be encountered shall be run from the fixture terminal connection to an outlet or junction box placed at least 1 foot away from the fixture. Take note of "1 foot away," and don't forget: These tap conductors shall be in a suitable raceway, Type AC cable, or Type MC cable that is at least 4 feet but not more than 6 feet long. This means that the box may be not less than 1 foot away from the fixture, but that a raceway shall be run to it that is not less than 4 feet or more than 6 feet long.

The reason for the 4-foot minimum length is to prevent the heat from traveling through too short a raceway. Type AF wire is not allowed as a branch-circuit conductor, but would be allowed for the fixture tap from a branch circuit into the fixture where the above method of connection is used.

XIII. Construction of Flush and Recessed Fixtures

These sections pertain mostly to the manufacturing of flush and recessed fixtures. However, attention is called to Section 410.70, which requires marking the fixture in ¼-inch letters or larger to indicate the maximum-wattage bulb that may be used. Larger wattage lamps will produce a higher temperature than the fixture is designed for. Also refer to Section 410.71, which prohibits the use of solder. See Part XVI of this article in the *NEC* for complete instructions.

XIV. Special Provisions for Electric-Discharge Lighting Systems of 1000 Volts or Less

410.73: General

(A) Open-Circuit Voltage of 1000 Volts or Less. With electric-discharge lighting that is designed for an open-circuit voltage of 1000 volts or less, the equipment used with such systems shall be of a type suitable for the service.

(B) Considered As Alive. The discharge-lamp terminals are to be considered alive when any lamp terminal is connected to a circuit of over 300 volts.

(C) Transformers of the Oil-Filled Type. Oil-filled–type transformers are not permitted.

(D) **Additional Requirements.** In addition to the requirements we have already covered, lighting fixtures must comply with Part XVII of this article.

(E) **Thermal Protection.** Fluorescent fixtures that are installed indoors must have thermal protection (Class P) built into the ballasts. If replacement of the ballasts is required, they shall be replaced only by ballasts that have thermal (Class P) protection within them.

Exception

Fluorescent fixtures with reactance ballasts don't require thermal protection.

(F) **Recessed High-Intensity Discharge Fixtures.** These fixtures must be thermally protected, and must be identified as thermally protected. When operated by remote ballasts, the ballasts must also be thermally protected.

Exception

Recessed HID fixtures can be installed in poured concrete when identified for such use.

410.74: Direct-Current Equipment

All discharge-lighting equipment in this class that is designed for use on direct current shall be so marked and the resistors and other equipment used shall be designed for direct-current operation. A fluorescent fixture designed for alternating current won't work on direct current.

410.75: Voltages—Dwelling Occupancies

(A) **Open-Circuit Voltage Exceeding 1000 Volts.** Lighting equipment for dwelling occupancies is limited to an open-circuit voltage of not more than 1000 volts. Sometimes a request is made to use neon lighting in a residence. According to the *Code*, this type of lighting would be prohibited because the voltage will exceed 1000 volts in practically all cases.

(B) **Open-Circuit Voltage Exceeding 300 Volts.** Equipment that has an open-circuit voltage over 300 volts should not be installed in dwelling occupancies unless it is designed in such a way that there will be no exposed parts when lamps are being replaced or removed.

410.76: Fixture Mounting

(A) Exposed Ballasts. There are times when exposed ballasts or transformers are used. In such installations, they shall be mounted such that they don't come in contact with any combustible material.

(B) Combustible Low-Density Cellulose Fiberboard. Where a fixture containing a ballast is to be installed on combustible low-density cellulose fiberboard it shall, where surface-mounted, be listed for direct mounting on this type of combustible material, or it shall be 1½ inches from the surface of the fiberboard. Where such a fixture is partially or wholly recessed, the provisions of this article (Sections 410.64 through 410.72) apply.

Note

This covers combustible low-density cellulose fiberboards. Included are sheets, panels, and tiles, the density of which is 20 pounds-per-cubic-foot or less. These fiberboards are formed of bonded plant fibers. Solid or laminated wood is not included, nor is fiberboard that has a density of over 20 pounds-per-cubic-foot. It can be a material that has been properly treated with fire-retarding chemicals, and these fire-retarding chemicals shall be treated so that the flame spread in any place on this material shouldn't exceed 25. This will be determined by tests for surface burning characteristics of building materials, which appear in ANSI/ASTM E84-2000, Surface Burning Characteristics of Building Materials.

410.77: Equipment Not Integral with Fixture

(A) Metal Cabinets. Auxiliary equipment (remote reactors, capacitors, resistors, and so on) must be enclosed in permanent metal cabinets.

(B) Separate Mounting. Separately mounted ballasts that are designed for direct connection to the wiring system do not have to be enclosed.

(C) Wired Fixture Sections. Wired fixture sections are assemblies of two fixtures that share a common ballast. These paired units can be connected with ⅜-inch flexible metal conduit in lengths of 25 feet or less. Fixture wire operating at line voltage (not at the ballast output voltage) supplying power to only the ballast of one of the paired fixtures can be installed in the same raceway as the conductors supplying the lamps.

Only one of the paired fixtures can be used when using other specified wiring methods.

410.78: Autotransformers

This section is very similar to the requirements for autotransformers supplying branch circuits as covered in Section 210.9, except that this section covers lighting. Autotransformers that are used as a part of a ballast for supplying lighting units and that raise the voltage to more than 300 volts shall be supplied only from a grounded system.

410.79: Switches

Section 404.14 covers snap switches. They are to meet the requirements of this section.

XVI. Special Provisions for Electric-Discharge Lighting Systems of More Than 1000 Volts

410.80: General

(A) **Open-Circuit Voltage Exceeding 1000 Volts.** Only equipment that is of a type intended for such service shall be used on electric-discharge lighting systems with an open-circuit voltage in excess of 1000 volts.

(B) **Considered As Alive.** All terminals are to be considered alive when connected to over 1000 volts and shouldn't be installed in or on dwellings.

(C) **Additional Requirements.** Not only do the general requirements for lighting fixtures apply here, but you must also comply with Part XVII of this article.

Signs and outline lighting are covered in Article 600.

410.81: Control

(A) **Disconnection.** Fixtures and lamp installations may be controlled individually or in groups by switches or circuit breakers, but either shall open all of the ungrounded conductors of the primary circuit (the branch circuit supplying the fixtures or lamps).

(B) **Within Sight or Locked Type.** For protection, the switches or circuit breakers shall be located within sight of the fixtures or lamps, or shall be capable of being locked in an open position.

410.82: Lamp Terminals and Lampholders

Any fixture parts that must be removed for lamp replacement shall be hinged or fastened in an approved manner, and the lamps and lampholders, or both, shall be so designed that there will be no exposed live parts when lamps are being removed or replaced.

410.83: Transformer Ratings

The open-circuit rating of ballasts and transformers shouldn't exceed 15,000 volts, with an allowable 1000 volts on test in addition. The secondary current rating shouldn't exceed 120 milliamperes (0.120 ampere) when the open-circuit voltage to the ballast or transformer exceeds 7500 volts, and the secondary current shouldn't exceed 240 milliamperes (0.240 ampere) when the open-circuit voltage is 7500 watts or less.

410.84: Transformer Type

Transformers must be listed and enclosed.

410.85: Transformer Secondary Connections

- **(A) High-Voltage Windings.** The high-voltage windings of transformers shouldn't be connected in parallel or series, with the exception that two transformers, each of which has one end of the high-voltage winding grounded and attached to the enclosure, may have the high-voltage windings connected in series to form the equivalent of a midpoint grounded transformer. See Figures 410-5 and 410-6.
- **(B) Grounded Ends of Paralleled Transformers.** When transformers are paralleled, the grounded ends shall be connected by an insulating conductor that is no smaller than No. 14 AWG.

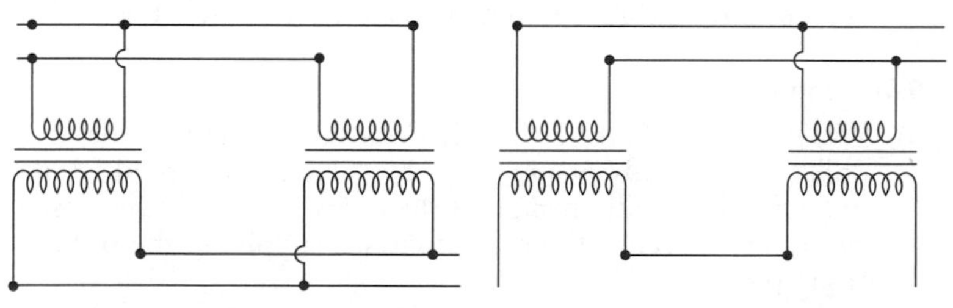

(A) High-voltage secondaries shall not be paralleled.

(B) High-voltage secondaries shall not be connected in series.

Figure 410-5 These connections are prohibited on high-voltage lighting transformers.

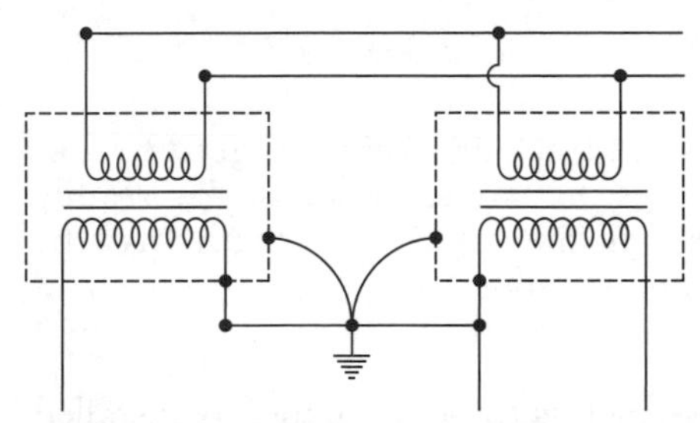

Figure 410-6 A connection that is permissible on high-voltage lighting transformers.

410.86: Transformer Locations

(A) **Accessible.** Transformers must be accessible after their installation.

(B) **Secondary Conductors.** Transformers must be installed close to the lamps so that the wires may be kept as short as possible.

(C) **Adjacent to Combustible Materials.** Transformers can't be installed close enough to combustible materials to subject them to temperatures higher than 90°C (194°F).

410.87: Transformer Loading

The loading of the transformer should be such that there won't be a continuous over-voltage condition. At first glance this sounds odd, but these transformers are of the high-leakage reactance type and as the load increases the voltage increases to maintain the proper current. The lamp load should be properly balanced for the transformer being used.

410.88: Wiring Method—Secondary Conductors

See the *NEC*.

410.89: Lamp Supports

See Section 600.33.

410.90: Exposure to Damage

See the *NEC*.

410.91: Marking

See the *NEC*.

410.92: Switches

See the *NEC*.

XVII. Lighting Track

410.100: Definition

A lighting track is a manufactured track assembly designed to support and energize lighting fixtures that are capable of being readily moved to different positions on the track. Its length may be changed by adding or subtracting sections.

410.101: Installation

(A) Lighting Track. Lighting track is to be permanently installed and permanently connected to a branch circuit. Only fittings made for lighting track shall be installed thereon. There shall be no general-purpose receptacles installed on lighting track.

(B) Connected Load. As with any conductor the load on the lighting track shouldn't exceed the rating (ampacity) of the track. The rating of the branch-circuit supplying the lighting track shouldn't be greater than the rating of the lighting track.

(C) Locations Not Permitted. Lighting track may not be installed in the following locations:

- **(1)** Where it may be subjected to physical damage.
- **(2)** In wet or damp locations.
- **(3)** Where corrosive vapors exist.
- **(4)** In battery-storage rooms (the vapors there are corrosive).
- **(5)** In hazardous (classified) locations—sparks may occur when fixtures are moved on the track.
- **(6)** In concealed places.
- **(7)** Where it penetrates walls or partitions.
- **(8)** Unless protected from physical damage, less than 5 feet above finished floors, except if the track operates at no more than 30 volts.
- **(9)** Within 3 feet horizontally or 8 feet vertically of a bathtub rim.

(D) Support. Only identified fittings shall be used on lighting track and they shall be designed for the specific lighting track on which they will be used. The fittings shall be securely fastened to the track. Polarization and grounding shall be maintained. They shall be designed to be suspended directly from the track.

410.102: Track Load

When calculating the branch circuits for the lighting track, each 2-foot or less length of lighting track shall be considered as 150 VA.

If multiwire circuit track is installed and each section of the track is supplied by one portion of the multiwired circuit, the load should be considered evenly divided.

Exception

When installed in dwelling units or hotel rooms, lighting track can be considered to be loaded differently.

410.103: Heavy-Duty Track

Heavy-duty lighting track is considered and identified as exceeding 20 amperes. Each fitting used for attachment of heavy-duty lighting track must have individual overcurrent protection.

The above indicates to the author that when circuits are over 20 amperes to the lighting track, each individual lighting track over 20 amperes shall be supplied by an individual branch circuit.

410.104: Fastening

The lighting fixture track must be securely mounted so that it will handle the maximum load put on it after installation. Unless it is listed for supports at greater intervals of 4 feet or shorter length, it must have two supports. When installed in a single length of track, the individual supports shall be not more than 4 feet apart, and there shall be one additional support.

410.105: Construction Requirements

See the *NEC*.

Article 411—Lighting Systems Operating at 30 Volts or Less

411.2: Definition

This article concerns lighting systems operating with isolated power supplies producing 30 volts (42.4 volts peak) under any load condition. More than one circuit, each limited to 25 amperes, may operate the lighting.

411.3: Listing

All equipment must be listed for the purpose.

2005

411.4: Locations Not Permitted

These sytems may not be concealed or extended through a wall, except if using one of the wiring methods listed in Chapter 3, or if supplied by a Class 2 source per Section 725.52. They are also prohibited within 10 feet (3.0 m) of pools, spas, and similar locations.

411.5: Secondary Circuits

(A) **Grounding.** Secondary circuits may not be grounded.

(B) **Isolation.** The secondary circuits must be isolated from the branch circuits by an isolating transformer.

2005

(C) Bare Conductors. Bare conductors and parts are permitted for these systems. However, bare conductors may not be installed less than 7 feet (2.1 m) above a finished floor, unless specifically listed for the use. Bare conductors may be used indoors only.

411.6: Branch Circuit

Twenty-ampere branch circuits are the maximum permitted.

411.7: Hazardous Locations

If installed in hazardous locations, these systems must also comply with Articles 500 through 517.

Article 422—Appliances

I. General

422.1: Scope

This article covers the installation of electric appliances in any occupancy. Equipment shall be of a type approved for the purpose and location.

422.2: Live Parts

The live parts of appliances shall be normally enclosed to avoid exposure to contact. This, however, can't always be done, as in the case of toasters, grills, and the like.

422.3: Other Articles

Article 430 covers motors. The requirements of this article amend and supersede portions of Article 430. Some other articles in the *Code* may also be applicable to this article. Hazardous (classified) locations must comply with Articles 500 through 517.

Article 440 applies to appliances that contain hermetically sealed refrigeration motor-compressor(s), unless amended within this article.

II. Branch-Circuit Requirements

422.4: Branch-Circuit Sizing

This section is not intended to apply to the conductors that are an integral part of an appliance. This section applies to the sizes of branch-circuit conductors that may be used without heating under the conditions specified in Article 210.

(A) **Individual Circuits.** The rating marked on the appliance shall be used in determining the individual branch-circuit rating to which it is connected. The branch-circuit rating shouldn't be lower than the rating on the appliance or an appliance that has a combination of loads, such as those you will find in Section 422.32.

Exception No. 1

Motor-operated appliances that don't have the ampacity of the branch circuit marked on the appliance shall be treated according to Part II of Article 430, which covers this type of operation.

Exception No. 2

Appliances that are not motor-operated shall have an overcurrent-device rating of less than 125 percent of the marked rating on the appliance, if used for continuous loading. If the branch-circuit devices are rated for use at 100 percent of the rating, then the branch-circuit rating shall be not less than 100 percent of the marked rating.

Exception No. 3

Branch circuits to household cooking appliances have a special concession on rating. This is covered in accordance with Table 220.19 and the notes therewith.

(B) **Circuits Supplying Two or More Loads.** Section 210.23 gives percentages to use on branch-circuit loading, so branch circuits supplying appliances and other loads shall conform to Section 210.23.

422.5: Branch-Circuit Overcurrent Protection

Section 240.3 gives us the necessary information for protecting branch circuits. When there is a protective-device rating marked on an appliance, the branch-circuit overcurrent-device rating shouldn't exceed the protective-device rating marked on the appliance.

III. Installation of Appliances

422.6: General

Care must be taken to install all appliances in an approved manner.

422.7: Central Heating Equipment

Central heating or fixed space-heating equipment must be supplied with an individual branch circuit. Nothing else can be connected to that circuit.

Exception

Pumps, valves, humidifiers, and the like may be connected to the circuit.

422.8: Flexible Cords

Flexible cords used to connect heating appliances shall comply with the following:

- **(A)** **Heater Cords.** This part defines the types of cords required for all smoothing irons and portable electrically heated appliances rated at more than 50 watts that produce temperatures in excess of 121°C (250°F) on surfaces that the cord is likely to come in contact with. Table 400.4 lists cords for heating appliances, along with their temperature ranges and usages.
- **(B)** **Other Heating Appliances.** All other portable heating appliances shall have cords that are listed in Table 400.4 and meet the requirements for the use to which they are put.
- **(C)** **Other Appliances.** Flexible cord may be used for the following purposes: (1) if appliances are frequently interchanged or transmit noise or vibration; (2) if appliances are fastened in place, but have to be moved frequently for maintenance and repair. A good example of this is a garbage disposal. With a flexible cord connection, it merely has to be unplugged and the danger of a wrong connection upon reinstallation is largely avoided.
- **(D)** **Specific Appliances.**
  - **(1)** Kitchen-waste disposals household use that are supplied by types of cords identified for that purpose may be connected with an attachment plug, if all of the following conditions are met:
    - **(a)** The cord attached to the appliance is not less than 18 inches in length, nor more than 36 inches in length.
    - **(b)** Receptacles are placed such that they won't cause any damage to the flexible cord.
    - **(c)** The receptacles are placed such that they are accessible.
  - **(2)** For dwelling purposes, built-in dishwashers and trash compactors supplied by types of cords that are identified for

that purpose may be connected with an attachment plug. The following conditions must be met:

(a) The length of the cord allowed shall be 3 to 4 feet.

(b) See (1)(b).

(c) The receptacle shall be mounted in the space where the appliance is located or immediately adjacent thereto.

(d) See (1)(c).

Exception

Listed kitchen waste disposals, dishwashers, or trash compactors, when protected by double insulation and so marked, are not required to have an equipment-grounding conductor.

(3) Both the external cord and the internal wiring supplying high-pressure spray washing machines must have GFCI for protection for persons. The GFCI is to be listed for use with portable equipment, and may also be part of the attachment plug.

Exception

High-pressure spray washers operating at over 125 volts or double insulated don't have to be provided with a GFCI, but must be marked with a note stating that it must be connected to an outlet that is GFCI protected.

422.9: Cord- and Plug-Connected Immersion Heaters

Portable immersion heaters are to be constructed and installed so that the current-carrying parts don't come into contact with the substance into which they are immersed. This prohibits the type of immersion heaters that have open coils in contact with the liquid.

422.10: Protection of Combustible Material

All electrically heated appliances that, because of size, weight, or service, are intended to be used in fixed position, shall be provided with ample protection from combustibles.

422.11: Stands for Cord- and Plug-Connected Appliances

All portable heating appliances, such as irons, that are intended to be used by applying them to combustible substances shall be provided with stands for support when not actually being used. These stands may be part of the appliance or a separate item.

422.12: Signals for Heated Appliances

This does not apply to appliances intended for residential use, but other electrically heated appliances that are applied to combustible articles shall have a signaling device unless the appliance has a temperature-limiting device.

422.13: Flatirons

All electrically heated smoothing irons shall be equipped with an identified temperature-limiting means, such as a thermostat.

422.14: Water Heaters

(A) **Storage- and Instantaneous-Type Water Heaters.** In addition to control thermostat, temperature-limiting means to disconnect all ungrounded conductors are to be used on storage- or instant-type water heaters. They shall be installed in such a way that they sense the maximum water temperature. The water heater must be either trip-free or of a type having replacement elements. This type must have a temperature- and pressure-relief valve, and shall be marked for such.

Exception

If the water heater supplies water at 180°F (82°C) or higher, or has a capacity of 60 kW or more, it shall be listed for this use; also, water heaters of a capacity of one gallon or less should be listed as being satisfactory for that use.

It is almost universally accepted that there should be a pressure- and temperature-relief valve installed on water heaters. The requirements above are in addition to this requirement. Some heaters in the past had provisions for opening only one ungrounded conductor by means of a thermostat. It is now required that all ungrounded conductors be opened by the thermostat. Often, if an element is shorted to ground, one side will open but the other side of the circuit will stay energized. Thus, current will go to ground, still heating the water. By opening all ungrounded conductors, this problem will be eliminated.

(B) **Storage-Type Water Heaters.** With fixed storage-type water heaters that have a capacity of 120 gallons or less, the branch circuit must be rated at not less than 125 percent of the water heater's nameplate rating.

Note

Sizing of the branch circuit is covered in Section 422.5(A).

422.15: Infrared Lamp Industrial Heating Appliances

See the *NEC.*

422.16: Grounding

Metal frames of portable, stationary, and fixed electrically heated appliances, operating on circuits above 150 volts-to-ground, shall be grounded in the manner specified in Article 250, provided, however, that where this is impractical, grounding may be omitted by special permission, in which case the frames shall be permanently and effectively insulated from the ground.

In rural areas especially, grounding the water-heater frame to the same ground as the service entrance will cause many burnouts due to lightning. This provision of grounding that requires the cold-water pipe to be bonded to a made electrode (when a made electrode is required) will automatically do much to prevent the troubles that used to develop. Even when (by special permission) the equipment is not required to be grounded, the *Code* recommends that the frames be grounded in all cases. It will be recalled that the frames of ranges and dryers may be grounded to the neutral if the neutral is sized at No. 10 or larger. These are two cases in which this type of grounding is permitted. It does not apply to mobile homes or travel trailers.

Refrigerators and freezers shall comply with the requirements of Sections 250.42, 250.43, and 250.45. Electric ranges, wall-mounted ovens, counter-mounted cooking units, and clothes dryers shall comply with the requirements of Sections 250.57 and 250.60. This, of course, won't apply to mobile homes or travel trailers.

422.17: Wall-Mounted Ovens and Counter-Mounted Cooking Units

(A) **Permitted to Be Cord- and Plug-Connected or Permanently Connected.** Wall-mounted ovens and counter-mounted cooking units are considered fixed appliances, and this includes the provisions for mounting and connection of the wiring. They may be cord- and plug-connected, or permanently connected.

(B) **Separable Connector or Plug-and-Receptacle Connector.** A separable connector or plug-and-receptacle combination in the supply-line to a wall-mounted oven or counter-mounted cooking top shouldn't be considered a disconnecting means, but should only be used as a means for ease of servicing. This means that it is not to be used as a disconnecting means as covered in Section 422.20. Also, it shall be approved for the temperature to which it might be subjected, especially since the space is limited in most cases.

422.18: Support of Ceiling Fans

Ceiling fans that are listed and don't weigh over 35 pounds—with or without added accessories—may be supported by outlet boxes that have been identified and listed for such use. These boxes shall be supported as outlined in Sections 314.13 and 314.17.

422.19: Other Installation Methods

If appliances use methods of installation other than those covered in this article, they will require special permission. You will recall that "special permission" is permission granted in writing by the authority that has jurisdiction.

IV. Control and Protection of Appliances

422.20: Disconnecting Means

Each appliance shall have some means for disconnecting all ungrounded conductors. These means for fixed, portable, and stationary appliances will be covered in the next few sections.

422.21: Disconnection of Permanently Connected Appliances

(A) **Rated at Not Over 300 Volt Amperes or 1/8 Horsepower.** The branch-circuit overcurrent device may be used as a disconnecting means for appliances not over 300 volt amperes or 1/8 horsepower.

(B) **Permanently Connected Appliances of Greater Rating.** If the overcurrent-protection device is readily accessible to the user, it may be used as the disconnecting means for permanently connected appliances that have greater ratings.

Note

Section 422.26 provides for motor-driven appliances using motors over 1/8 horsepower.

Exception

Section 422.24 covers disconnecting means when unit switches are used for appliances.

The above shouldn't prevent the addition of a separable connection or a plug and receptacle approved for being installed as a means of serving the fixed appliance. They are not intended merely to be the disconnecting means.

422.22: Disconnection of Cord- and Plug-Connected Appliances

(A) **Separable Connector or an Attachment Plug and Receptacle.** Separable connectors or plugs and receptacles can be used as the required disconnecting method for these appliances, but only where the cord-and-plug connection is accessible.

(B) **Connection at the Rear Back of a Range.** Cord-and-plug connections are permitted for residential ranges.

(C) **Rating.** Receptacles or separable connectors must be rated no less than the appliances they feed.

(D) **Requirements for Attachment Plugs and Connectors.** These devices must conform to the following:

(1) They must be built to avoid the user contacting live parts.

(2) They must be able to interrupt their rated current without hazard to the operator.

(3) They must not be able to be interchanged with devices of a lesser rating.

422.23: Polarity in Cord- and Plug-Connected Appliances

The attachment plug must be polarized if the appliance has an Edison-base lampholder or a single-pole snap switch.

422.25: Unit Switches As Disconnecting Means

See the *NEC*.

422.26: Switch and Circuit Breaker to Be Indicating

Switches or circuit breakers used as disconnecting means for appliances shall indicate whether they are in the open or closed position.

422.27: Disconnecting Means for Motor-Driven Appliances

When a switch or circuit breaker serves as the disconnecting means for a stationary or fixed motor-driven appliance of more than ⅛ horsepower, it shall be located within sight of the motor controller or shall be capable of being locked in the open position. It will be found that most inspection authorities will consider any distance over 50 feet as being out of sight. The intent of this part is to make motor-driven appliances of over ⅛ horsepower safe during service work.

Exception

As required in Section 422.24(A), (B), (C), or (D), a switch or circuit breaker serving as the other disconnecting means may be out of sight from the appliance motor controller provided with unit switch(es). There shall be a marked OFF position on all such means of disconnecting the ungrounded conductor.

422.28: Overcurrent Protection

(A) Appliances. See Sections 422.5 and 422.6. Appliances shall be protected against overcurrent as provided in (B) through (F) below and these two sections.

Exception

Part III of Article 430 requires that motor-operated appliances shall have overload protection provided. Part VI of Article 440 covers the motors of hermetic refrigerant motor-compressors in air-conditioning or refrigerating equipment that shall be provided with overcurrent protection. When separate overcurrent devices are provided for an appliance, the data covering these overcurrent devices must be marked on the appliance. Sections 430.7 and 440.4 cover the minimum marking on such appliances.

(B) Household-Type Appliance and Surface-Heating Elements. Table 220.19 lists the demand factors for residential ranges. Any residential range that has a demand factor of over 60 amperes (as calculated from this table) must have its power supply divided into two or more circuits, each of which is provided with overcurrent protection of not more than 50 amperes.

(C) Infrared Lamp Commercial and Industrial Heating Appliances. Infrared heating appliances shall be protected by overcurrent protection of not more than 50 amperes. This covers commercial and industrial heating appliances.

(D) Open-Coil or Exposed Sheathed-Coil Types of Surface-Heating Element in Commercial-Type Heating Appliances. Exposed sheathed-coil or open-coil heating elements of surface-heating–type appliances shall be protected by overcurrent protection, and this heating device shall not be over 50 amperes.

(E) Single Nonmotor-Operated Appliance. When a branch circuit supplies a single, nonmotor-operated appliance, the overcurrent protection can't exceed the rating marked on the appliance. If the rating is not marked on the appliance, and the appliance uses more than 13.3 amperes, the overcurrent protection can't be more than 150 percent of the appliance's current. If the rated current is not marked, and the appliance uses less than 13.3 amperes, the overcurrent protection can't exceed 20 amperes.

Exception

When 150 percent of the appliance rating does not equal a standard rating, the appliance may be protected at the next higher standard rating.

(F) **Electric Heating Appliances Employing Resistance-Type Heating Elements Rated More Than 48 Amperes.** When electric heating appliances use resistive-type heating elements and they are rated more than 48 amperes, the heating elements must be subdivided. When subdivided, each subdivision shall be rated not more than 48 amperes and shall be protected by not over 60 amperes.

The supplementary overcurrent devices must be: (1) supplied by the factory and in or on the heater enclosure, or they may be a separate assembly furnished by the heater manufacturer; (2) accessible, but need not be readily accessible; and (3) suitable for branch-circuit protection.

The conductor that supplies the overcurrent protection for these heaters is a branch circuit.

Exception No. 1

In household equipment that is supplied by surface-heating elements as covered in Section 422.27(B), and also in commercial-type heating appliances that are covered in Section 422.27(D).

Exception No. 2

Any commercial kitchen or cooking appliances using sheath-type heating elements that are not covered in Section 422.27(D) may be subdivided into circuits that shouldn't exceed 120 amperes, and the protection for such shouldn't be over 150 amperes if one of the following conditions is met: (a) The heating elements is a part of the equipment and is enclosed in the cooking element; (b) the enclosure is listed, suitable, and the elements are completely contained in the enclosure; or (c) the vessel is ASME-rated and stamped.

Exception No. 3

If resistive-type elements are mounted in water heaters or steam boilers and use resistive-type commercial heating elements, and the vessel that contains them is marked with an ASME rating and is stamped, a surface may be divided into a circuit not exceeding 120 amperes and protected at not more than 150 amperes.

V. Marking of Appliances

See the *NEC*. This part concerns primarily manufacturer's requirements. The markings must be on the appliances, so look for the UL label.

Article 424—Fixed Electrical Space-Heating Equipment

I. General

424.1: Scope

This article covers fixed equipment for space heating. The equipment may be heating cables, unit heaters, boilers, central systems, or other approved fixed heating equipment. Not covered in this article are process heating and room air conditioning.

424.2: Other Articles

This is basically the same requirement that appears in other articles of the *Code* to indicate primarily that heating equipment intended for use in hazardous (classified) locations must also comply with Articles 500 through 517. It also indicates that heating equipment employing a hermetic refrigerant motor-compressor must comply with Article 440.

Fixed, electric space-heating equipment that incorporates a sealed (hermetic-type) motor-compressor shall comply with Article 440.

All other applicable portions of the *Code* apply.

424.3: Branch Circuits

(A) Branch-Circuit Requirements. Branch circuits supplying two or more outlets for fixed heating equipment are limited to 15, 20, or 30 amperes. An individual branch circuit may supply any load.

There is an exception that says that branch circuits for fixed infrared heating equipment shouldn't be rated at more than 50 amperes. This covers other than residential occupancies.

(B) Branch-Circuit Sizing. The sizing of overcurrent devices and branch-circuit conductors for electrical space-heating equipment (fixed) shall be calculated on the basis of 125 percent of the total load of the heaters and motors, if equipped with motors. Contactors, relays, thermostats, and so on that are rated at continuous load ratings of 100 percent may be used to supply the full rated load. See Section 250.19, Exception.

Sections 440.34 and 440.35 shall be used in computing the size of the branch-circuit conductors if the electric space-heating equipment consists of a heat pump that is mechanical refrigeration with or without resistive elements.

The provisions of the above paragraph don't apply to conductors that supply part of an approved, fixed, electric space-heating equipment.

II. Installation

424.9: General

Only approved methods of installation shall be used for all electrical space-heating equipment.

424.10: Special Permission

Special permission shall be secured from the inspection authority to install fixed space-heating systems that require methods of installation other than those covered in this article.

424.11: Supply Conductors

Some heating equipment will require high-temperature wiring for the connections. Any equipment that is not suitable for connection to conductors with 60°C insulation shall be clearly and permanently marked, and this marking shall be plainly visible after installation.

424.12: Location

(A) **Exposed to Severe Physical Damage.** Fixed electrical space-heating equipment shall never be installed where it will be subject to physical damage. However, if subject to physical damage, it has to be adequately protected.

(B) **Damp and Wet Locations.** Only approved heaters and related equipment shall be installed in damp or wet locations. The construction and installation of these shall be such that water can't enter or be allowed to accumulate in or on the electrical parts or duct work.

A fine-print note calls attention to Section 110.11, which explains where equipment shall be covered during installation for protection.

424.13: Spacing from Combustible Materials

All fixed electrically heated appliances (heating equipment) that, because of size, weight, or service, are intended to be used in a fixed position shall be amply protected from combustibles, that is, unless they have been approved for installation in direct contact with combustible materials.

424.14: Grounding

Metal parts that are exposed and are a part of fixed electrical heating equipment shall be grounded as required in Article 250.

III. Control and Protection of Fixed Electric Space-Heating Equipment

424.19: Disconnecting Means

All fixed electric space-heating equipment shall have means provided to disconnect all of the ungrounded conductors to the heater, controller(s), and overcurrent device(s). Where heating equipment is supplied from more than one source, the disconnecting means shall be grouped and identified.

- **(A) Heating Equipment with Supplementary Overcurrent Protection.** If supplementary overcurrent protection is in the enclosure of the disconnecting means for fixed electric space-heating equipment, this disconnection equipment must be in sight on the supply side of a supplementary overcurrent device(s). It must also comply with either (1) or (2):
  - **(1) Heater Containing No Motor Rated Over 1/8 Horsepower.** The disconnecting means mentioned above or unit switches complementing Section 424.19(C) may be used as the required disconnecting means to both the heater and the motor controller(s) as permitted either under (a) or (b) below: (a) The disconnecting means must be in sight of the motor controller(s) and the heater itself; or (b) The disconnection means is the type that can be locked in the open position.
  - **(2) Heater Containing a Motor(s) Rated Over 1/8 Horsepower.** This covers motors used with heaters that are rated over 1/8 horsepower.
    - **(a)** If the disconnecting means is in sight of both the motor controller(s) and the heater, it may serve as a disconnecting means for both.
    - **(b)** Switches for disconnecting means shall comply with Section 424.19(C). If they are not visible from the heater, a separate disconnecting means shall be installed, or the disconnecting means shall be capable of being locked in the open position.
    - **(c)** When the disconnecting means is not in sight of the motor controller location, the disconnecting means used shall comply with Section 430.102.

(d) Section 430.102(B) shall apply when the motor disconnect is not visible from the location of the motor controller.

(B) Heating Equipment Without Supplementary Overcurrent Protection. This covers heating equipment that does not have overcurrent protection.

(1) Without Motor or with Motor Not Over 1⁄8 Horsepower. When fixed heating equipment does not have a motor over 1⁄8 horsepower, it may use the branch-circuit switch or circuit breaker for the disconnecting means, if it is readily accessible for servicing.

(2) Over 1⁄8 Horsepower. If a motor over 1⁄8 horsepower is used in motor-driven electric space-heating equipment, the disconnecting means shall be visible from the motor control.

Exception

Section 424.19(A)(2) provides an exception for the above.

(C) Unit Switch(es) As Disconnecting Means. If a disconnecting switch(es) for a unit is plainly marked with the opposition that will disconnect all ungrounded conductors to the fixed heater, it may be the disconnecting means that this article requires, or other disconnecting means may be provided in the following occupancies:

(1) Multifamily Dwellings. In multifamily dwellings, the disconnecting means must be in the dwelling unit involved or on the same floor as the dwelling unit where the fixed heater is installed. The disconnecting means may also have branch-circuit breakers supplying lamps and appliances.

(2) Two-Family Dwellings. In a two-family dwelling the disconnecting means may be installed on the inside or outside of the dwelling unit, supplying the dwelling in which the fixed heating is installed. In this case the disconnecting means may have overload protection, not only for the fixed heating protection, but also as protection for lamps and appliances.

(3) One-Family Dwellings. In a one-family dwelling, the disconnecting means for the dwelling may be the other disconnecting means. This disconnecting means, it is assumed, disconnects all the service to the one-family dwelling.

(4) Other Occupancies. In other occupancies, the fixed heater may have the branch-circuit switch or breaker. If it is readily accessible for providing service to the fixed heating, it may be used as the other disconnecting means.

424.20: Thermostatically Controlled Switching Devices

(A) **Serving As Both Controllers and Disconnecting Means.** A thermostatically controlled switch for controlling the temperature may be a combination thermostat, capable of being manually controlled, with a switch incorporated. This switch may be used if the following conditions are met:

(1) The thermostat is marked with an OFF position and is able to open all current-carrying conductors.

(2) When placed in the OFF position, it manually opens all ungrounded conductors.

(3) When the switch on the thermostat is set to the OFF position, the design is such that the circuit can't be energized automatically.

(4) It is located as covered in Section 424.19.

(B) **Thermostats That Don't Directly Interrupt All Underground Conductors.** If the thermostat does not directly open all the underground conductors, and if the thermostat is operable by remote control, it does not have to meet the requirements of (A). Such devices shouldn't be used as the disconnecting means of the fixed heating.

424.21: Switch and Circuit Breaker to Be Indicating

Switches or circuit breakers used for disconnecting means for heating equipment shall indicate whether they are in the open or closed position.

424.22: Overcurrent Protection

(A) **Branch-Circuit Devices.** Article 210 covering branch circuits goes into considerable detail on overcurrent protection. Heating equipment is generally considered to have overcurrent protection when the conditions of Article 210 are met, with the exception of motor-driven heating equipment, which comes under Articles 430 and 440.

(B) Resistance Elements. Any electric space-heating equipment that draws a total of more than 48 amperes shall have the heating units subdivided so that each subdivided section doesn't draw more than 48 amperes; if it draws 48 amperes, it shall be protected by not more than 60 amperes overcurrent protection.

Exception

Section 424.72(A) covers the exception involved here.

(C) Overcurrent Protective Devices. Supplementary overcurrent-protective devices that are used on subdivided loads specified in (B) of this list must comply with the following: (1) They shall be either factory-installed within, on the heater enclosure itself, or supplied as a separate assembly for use with that heater by the manufacturer; (2) Although required to be accessible, they are not required to be readily accessible; (3) These devices for overcurrent protection shall be suitable for use on branch circuits.

Note

This refers to Section 240.10.

If cartridge fuses are used to provide the overcurrent protection, several subdivided loads may be served by the single disconnecting means.

Note

Refer to Section 240.40.

(D) Branch-Circuit Conductors. Any conductors supplying the supplementary overcurrent protection devices are branch-circuit conductors.

Exception

If the conditions following this paragraph are met for heaters rated at 50 kW or more, conductors used to supply the supplementary overcurrent devices, as covered in (C) of this list, may not be sized at less than 100 percent of the nameplate rating that is supplied on the heater. For this purpose, the following conditions must be met: (a) The heater shall have a marking giving the minimum size of conductors that may be used; (b) these conductors may not be smaller than the size of the conductors as marked on the nameplate of the heater; and (c) there must be a device that is actuated by temperature to control the cycles of the operation of heating equipment.

(E) Conductors for Subdivided Loads. With wiring that is installed in the field, the conductors between the supplementary overcurrent devices and the heater are to be sized at not less than 125 percent of the load that is being served. Section 240.3 covers supplementary overcurrent protective devices that are specified in (C) of this list, and these conductors shall be protected in accordance with the aforementioned section.

Exception

With heaters that are rated at 50 kW or more, the ampacity of field-wired conductors allowed between the feeder and the supplementary overcurrent devices in no case shall be permitted to be less than 100 percent of each subdivided circuit. This is permitted if all of the following conditions are met:

- **(a)** See (a) of the exception to (D).
- **(b)** See (b) of the exception to (D).
- **(c)** See the *NEC*.

IV. Marking of Heating Equipment

424.28: Nameplate

Every electric space-heating unit must be marked, showing the manufacturer's name, operating voltage, and current (or voltage and wattage), and must be marked if it is to be used with AC only, or DC only. Units with motors of more than ⅛ horsepower must be marked, showing the frequency, volts and amps of motor load, and the volts and watts, or volts and amps, of the heating load. All markings must be visible after installation.

424.29: Marking of Heating Elements

Heating elements that are replaceable in the field shall have the following markings on each heating element: volts and amperes, or volts and watts.

V. Electric Space-Heating Cables

424.34: Heating Cable Construction

Heating cables are supplied by the manufacturer in a complete unit with 7-foot (2.13 m) nonheating leads. The nonheating leads shouldn't be cut off or shortened in any manner. See the next section. Also see Figure 424-1.

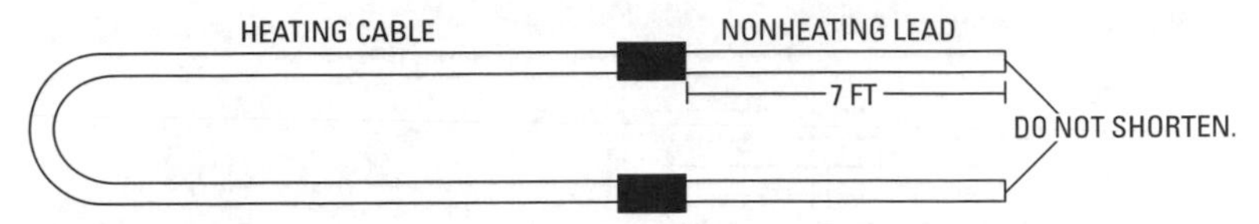

Figure 424-1 Nonheating leads shall be a minimum of 7 feet.

424.35: Marking of Heating Cables

Identifying name or identifying symbol, catalog rating number, the rating in volts and watts or in volts and amperes shall be marked on each heating cable.

Every heating cable length shall have a permanent marking on the nonheating lead, and this shall be located within 3 inches of the terminal end. The lead wires shall have distinctive colors to indicate the voltage at which they are to be used:

- 120 volts, nominal, yellow.
- 208 volts, nominal, blue.
- 240 volts, nominal, red.
- 277 volts, nominal, brown.

424.36: Clearances of Wiring in Ceilings

Wiring located above heated ceilings and in thermal insulation must be sized according to the temperature that it might be exposed to. Therefore, wiring above heated ceilings shall be located not less than 2 inches above the heated ceiling and, if it is in thermal insulation at this height, it shall be considered as being operated at an ambient temperature of 50°C. It will be necessary to refer to the correction factors that accompany Tables 310.16 through 310.31 to find the correction factors for the ampacity of conductors at 50°C. Wiring located above heated ceilings or above thermal insulation that has a minimum thickness of 2 inches requires no correction factor.

424.37: Location of Branch-Circuit and Feeder Wiring in Exterior Walls

The methods of wiring covered in Article 300 and Section 310.10 must be used for branch-circuit and feeder wiring installed in exterior walls.

424.38: Area Restriction

(A) Shouldn't Extend Beyond the Room or Area. Heating cables shouldn't, in any circumstance, extend beyond the room or area that is being heated. See Figure 424-2.

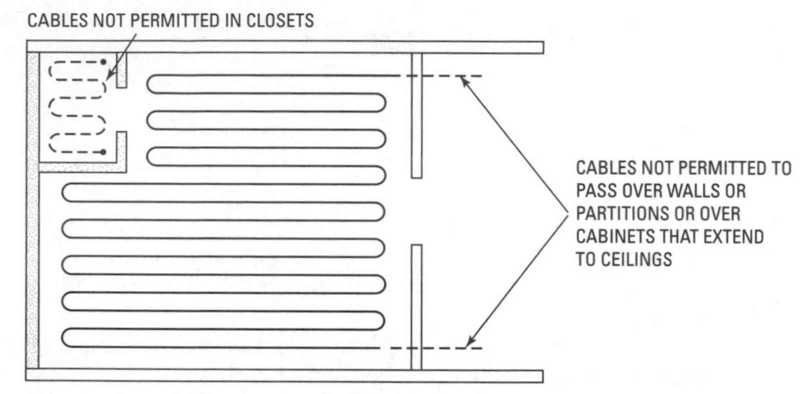

Figure 424-2 Illustration showing where heating cables may and may not be installed in ceilings.

(B) **Uses Prohibited.** Heating cables shouldn't be installed in closets, over walls or partitions that extend to the ceiling, or over cabinets. They may be installed in front of cabinets unless the ceiling above the cabinets is equal to their minimum horizontal dimensions. This takes into account the nearest cabinet edge to the room or area.

Exception

If heating cables are embedded, a single run may pass over partitions.

(C) **In Closet Ceilings As Low-Temperature Heat Source to Control Relative Humidity.** There are climates where humidity control is required in closets, but the prohibition of cables in closets is not intended to prohibit the use of low-temperature humidity controls in closets. See the *NEC*.

424.39: Clearance from Other Objects and Openings

There shall be a clearance of at least 8 inches from the edge of outlet and junction boxes that are to be used for mounting lighting fixtures.

Two inches (50.8 mm) shall be provided from recessed lighting fixtures and their trims, ventilating openings, and other such openings in room surfaces. Sufficient area shall be provided to ensure that no heating cable or panel will be covered by other surface-mounted lighting units. The temperature limits may be exceeded and the cable may overheat in this instance. Therefore, the requirements of the section should be very carefully followed and, if in doubt, a little extra clearance should be given.

424.40: Splices

Splicing of cables is prohibited except where necessary due to breaks. Even then the length of the cable shouldn't be altered, as this will change the characteristics of the cable and the heat. It will be necessary occasionally to splice a break, but only approved methods shall be used.

424.41: Installation of Heating Cables on Dry Board, in Plaster, and on Concrete Ceilings

(A) Shouldn't Be Installed in Walls. Heating cable shouldn't be installed in walls. It is not designed for this purpose and is strictly forbidden, with the exception that isolated runs of cable may run down a vertical surface to reach a drop ceiling.

(B) Adjacent Runs. Adjacent runs of heating cable shall be spaced not closer than 1½ inches on centers and have a wattage not to exceed 2¾ watts-per-square-foot (Figure 424-3).

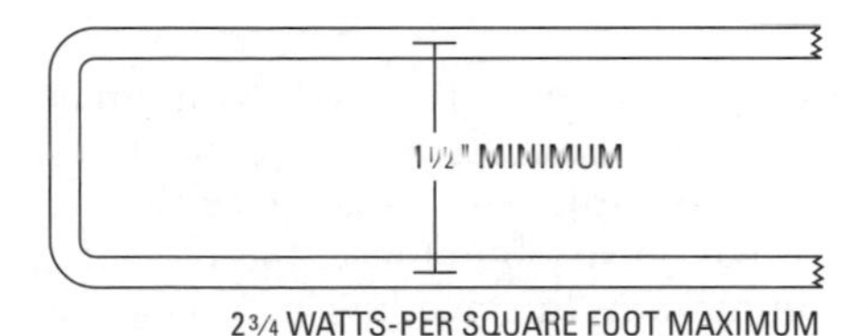

Figure 424-3 Heating cables installed in ceilings shall be placed at least 1½ inches apart.

(C) Surfaces to Be Applied. Heating cables shall be applied only to gypsum board, plaster lath, or similar fire-resistant materials. If applied to the surface of metal lath or any other conducting material, there shall be a coat of plaster commonly known as a brown- or scratch-coat applied before the cable is installed. This coating of plaster shall entirely cover the metal lath or conducting surface. Also see Section 424.41(F).

(D) Splices. Splices to the nonheating leads shall have a 3-inch minimum after the splice is embedded in plaster or in dry board. The embedding shall be done in the same manner as with the feeding cable.

(E) Ceiling Surfaces. On plastered ceilings, the entire surface shall have a finish coat of thermally noninsulating sand plaster or other approved coating that shall have a nominal thickness of ½ inch. Insulation (thermal) plaster shouldn't be used.

(F) Secured. The cable shall be fastened at intervals not to exceed 16 inches by means of taping, stapling, or plastering. Staples or metal fasteners that straddle the cable shouldn't be used with metal lath or other conducting surfaces. The fastening devices shall be of an approved type.

Exception

Heating cable shall be mounted so that there is no more than 6 feet between mountings.

(G) Dry Board Installations. When dry board ceilings are used, the cable shall be installed and the entire ceiling below the cable shall be covered with gypsum board not exceeding ½ inch (12.7 mm) in thickness, but the voids between the two layers and around the cable shall be filled with a conducting plaster or other approved thermal conducting material so that the heat will readily transfer.

(H) Free from Contact with Conductive Surfaces. Heating cables shouldn't come in contact with metal or other conducting materials.

(I) Joists. When dry board applications are used, you must install the cable parallel to joists and leave a clearance of 2½ inches between centers of adjacent runs of cable. At times it is necessary to cross joists, but the cable crossing joists shall be kept to a minimum. Surface layers of heating cable and gypsum board shall be mounted in such a way that nails or other types of fasteners don't penetrate into the heating cable.

Note

When practical, heating cable shall cross joists only at the ends of the joist. This will reduce the chances of nails being driven into the heating cables.

424.42: Finished Ceilings

The question often arises as to whether wallpaper or paint may be used over a ceiling that has heating cable. These materials have been used as finishes over heating cable since cables first were used in the late 1940s. This section gives formal recognition to painting or papering ceilings.

The above does not permit ceilings to be covered with decorative panels, or beams constructed of materials that have thermal insulating properties, such as wood, fiber, or plastic.

424.43: Installation of Nonheating Leads of Cables

(A) **Free Nonheating Leads.** Only approved wiring methods shall be used for connecting the nonheating leads from a junction box at the heating leads to a location in the ceiling.

In these installations, single leads in raceways (conductors) or single- or multiconductor Type UF, Type NMC, Type MI, or other approved conductors may be used. Please note the absence of Type NM.

(B) **Leads in Junction Box.** Where the nonheating leads terminate in a junction box, there shall be not less than 6 inches of the nonheating leads free within the junction box. Also, the markings of the leads shall be visible in the junction box. This is highly important so that the heating cable can be identified.

(C) **Excess Leads.** Nonheating leads shouldn't be shortened. They shall either be fastened to the underside of the ceiling or embedded in plaster or other approved material, leaving a length sufficient to reach the junction box with not less than 6 inches of free lead remaining in the box.

424.44: Installation of Cables in Concrete or Poured Masonry Floors

This section is for fixed indoor space heating and is not to be confused with Part IX or ice and snow melting, which is covered in Article 426.

(A) **Watts per Linear Foot.** The wattage per foot of heating cable shouldn't exceed 16½ watts.

(B) **Spacing Between Adjacent Runs.** Runs of heating cable shall be not less than 1 inch on centers to adjacent heating cables.

(C) **Secured in Place.** Cables have to be secured in place while concrete or other finish material is being applied. Approved means, such as nonmetallic spreaders or frames, shall be used. Concrete floors often have expansion joints in them. Cables shall be so installed that they don't bridge an expansion joint unless they are protected so as to prevent damage to the cables due to expansion or contraction of the floor.

(D) **Spacings Between Heating Cable and Metal Embedded in the Floor.** There shall be some distance between heating cable and any metal embedded in the floor.

Exception

If the cable is metal-covered, it may come into contact with metal embedded in the floor. This includes MI heating cable, which is used extensively.

(E) **Leads Protected.** Sleeving of the leads by means of rigid metal conduit, rigid nonmetallic conduit, intermediate metal conduit, or EMT shall be used for protection where the leads leave the floor.

(F) **Bushings or Approved Fittings.** The sleeves mentioned in (C) shall have bushings or other approved means used where the leads emerge within the floor slab to prevent damage to the cable.

(G) **GFCI Protection for Conductive Heated Floors of Bathrooms, Hydromassage Bathtubs, Spas, and Hot Tub Locations.** GFCI protection must be provided for all such electric heating systems.

424.45: Inspection and Test

Great care shall be exercised in the installation of embedded heating cables to prevent any damage thereto. In addition, the installation must be inspected and approved before cables are covered or concealed. If MI cable is run, the insulation resistance to the metal covering should be Megger tested.

After plastering ceilings or pouring floors, the cables shall be tested to see that the insulation resistance is within safe limits. A Megger test should be made at about 500 volts.

VI. Duct Heaters

424.57: General

Part VI covers heaters mounted in air ducts where they furnish a stream of air through the duct and where the air-moving unit is not a part of the equipment design.

424.58: Identification

If heaters are installed in air ducts, they must be suitable for the installation. Look for the UL label, and check the installation instructions closely.

424.59: Air Flow

Air flow over the face of the heater must be adequate and uniform in accordance with the manufacturer's instructions. If the heaters are mounted within 4 feet of an air-moving device, heat pump, air conditioner, elbows, baffle plates, or other obstructions that might be present in the ductwork and which might interfere with adequate and uniform air flow, it may be required to use turning vanes, pressure plates, or some other device on the inlet side of the duct heater to ensure that the air flow will be uniform over the face of the heater.

424.60: Elevated Inlet Temperature

Many duct heaters are used with heat pumps or other sources of air that are above room temperature. The operation of the heater may not be the same with an elevated air temperature at the inlet as it would be with the inlet temperature at room temperature. Therefore, the heater shall be identified as suitable for use at the elevated temperatures.

424.61: Installation of Duct Heaters with Heat Pumps and Air Conditioners

A duct heater installed immediately adjacent to an air conditioner or heat pump under abnormal operating conditions could adversely affect electrical equipment in the heat pump or air conditioner. It has been found that the duct heater shall be mounted not less than 4 feet from such equipment, unless the duct heater has been identified as suitable for closer mounting and is so marked.

424.62: Condensation

Duct heaters that are also used with air conditioners or with other air-cooling equipment (such equipment will cause condensation of moisture) shall be identified as suitable for use with air conditioners (for example, a furnace that had air conditioning installed within it).

424.63: Fan Circuit Interlock

A duct heater that operates in such a manner that the fan circuit is not energized with the first heating element might cycle on the high limit. This can cause an undue number of operations with the high limit to occur. Conceivably, in the course of years, the high limit might operate enough times to exceed the 100,000-cycle requirement and become erratic in its operation. By providing means to energize the fan circuit with the first heater circuit, we eliminate this potential problem.

424.64: Limit Controls

In order to prevent overheating, each duct heater must be provided with an integral, approved automatic-reset temperature-limiting control or controllers to deenergize the circuits if the temperature rises too high.

Also, an integral independent supplementary control must be provided with each duct heater that will disconnect a sufficient number of duct heaters and conductors to stop the flow of current. Such a device must be manually resettable or replaceable.

424.65: Location of Disconnecting Means

All control equipment for duct heaters shall be accessible and a disconnecting means provided that shall be installed at or within sight of the controllers for the heating units.

Exception
Section 424.19(A) gives an exception that is permitted.

424.66: Installation
See NFPA pamphlets *90A* and *90B*. Also see the *NEC*.

VII. Resistance-Type Boilers

424.70: Scope
Part VII covers resistance-type heating elements used in boilers. Electrode-type heating is not included here; that topic will be found under Part VIII.

424.71: Identification
Resistance-type boilers must be identified for this type of use.

424.72: Overcurrent Protection

(A) **Boiler Employing Resistance-Type Immersion Heating Elements in an ASME-Rated and Stamped Vessel.** Resistance-type immersion heating elements, covered by ASME, are to be stamped vessels. The heating elements shall be protected at not more than 150 amperes. If boilers are rated at more than 120 amperes, the heating units shall be subdivided so that each heating element load does not exceed 120 amperes.

Section 424.3(B) covers subdivided loads that are less than 120 amperes. The ratings of the overcurrent-protection devices are covered in that section.

(B) **Boiler Employing Resistance-Type Heating Elements Rated More Than 40 Amperes and Not Contained in an ASME-Rated and Stamped Vessel.** If the boiler is not an ASME-rated and stamped vessel employing resistance-type heating elements, it shall have elements subdivided into loads of not more than 48 amperes and they shouldn't be protected by more than 60 amperes.

If such a boiler is rated at more than 48 amperes, the heating elements shall be subdivided into loads not to exceed 48 amperes.

If the load, when subdivided, is less than 48 amperes, the overcurrent-protection device shall comply with Section 424.3(B).

(C) **Supplementary Overcurrent Protective Devices.** The supplementary overcurrent-protective devices as required by (A) and (B) of this list shall be:

(1) Factory-installed within or on the boiler enclosure or provided as a separate assembly by the boiler manufacturer.

(2) Accessible, but need not be readily accessible.

(3) Suitable for branch-circuit protection.

See Section 240.40. If fuses are used to provide the overcurrent protection, a single disconnecting means may be used for the several subdivided circuits.

(D) Conductors Supplying Supplementary Overcurrent Protective Devices. The conductors supplying the supplementary overcurrent protection are considered branch-circuit conductors.

Exception

If all of the following conditions are met, where feeders are rated at 50 kW or more, the conductors that supply the protective overcurrent device, as specified in (C) of this list, may be sized at not less than 100 percent of the nameplate rating of the heater: (a) The minimum conductor size is marked on the heater; (b) The conductors used are not smaller than those covered on the marking on the heater for minimum size; and (c) There is a temperature- and pressure-operative device to control the operating cycles of the equipment.

(E) Conductors for Subdivided Loads. When field wiring of the conductors is done between the heater and the supplementary overcurrent device, the conductors shall be sized at not less than 125 percent of the load that they are serving. Section 240.3 covers the protection of these conductors with the supplementary overcurrent-protective devices specified in (C) of this list.

Exception

When heaters are 50 kW or more, the field-wired conductor ampacity from the feeder to the supplementary overcurrent device shall be not less than 100 percent of their individual subdivided circuits if all of the following conditions are met: (a) The heater is marked with a minimum conductor size; (b) The conductor is not smaller than the minimum marked size on the feeder; and (c) There is a temperature device that controls the cycles of the equipment operation.

424.73: Over-Temperature Limit Control

In addition to a temperature-regulating system and other devices protecting the tank against excessive pressure, each boiler design shall be equipped with a temperature-sensitive limiting means so there is no change in state of the heat transfer medium. This shall be installed in such a way that it limits the maximum liquid temperature and shall directly or indirectly disconnect all ungrounded conductors to the heating elements; thus, a temperature sensitive relay device may open a magnetically controlled disconnecting means.

424.74: Over-Pressure Limit Control

Means shall be provided for a pressure-regulating system and other devices protecting the tank against excessive pressure on each boiler so that in normal operation there is a change in state of the heat transfer medium from a liquid to a vapor. In addition to this, the boiler shall be equipped with a pressure-sensing and pressure-limiting device that is installed to limit pressure and directly or indirectly disconnect all ungrounded conductors to the heating elements.

424.75: Grounding

All metal parts that don't carry current shall be grounded as required by Article 250. A connection means shall be provided for the equipment grounding, and it must be sized in accordance with Table 250.95.

VIII. Electrode-Type Boilers

424.80: Scope

Part VIII covers boilers heated by means of electrodes in the water. The voltage of the current passing between these electrodes shouldn't exceed 600 volts, nominal.

424.81: Identification

Electrode-type boilers must be identified where used as indicated, and shall be suitable for the installation involved.

424.82: Branch-Circuit Requirements

The overcurrent protection for conductors must be calculated as having 125 percent of the total load, with the exception of motors. A contactor or relay, if it has been approved for continuous service at 100 percent of its rating, can supply its full-current-rated load. Refer to Section 210.22(C). This section shouldn't apply to conductors when they are an integral part of an approved boiler.

Exception

Conductors can be sized at 100 percent of the nameplate rating for electrode boilers rated 50 kW or more, as long as they are marked showing a minimum conductor size, the conductors are no smaller than the marked size, and the operation of the equipment is controlled by either a temperature- or pressure-activated device.

424.83: Over-Temperature Limit Control

In addition to a temperature regulating system and other devices protecting the tank against excessive pressure, each boiler is designed so that there is no change in state of the heat transfer medium, and shall be equipped with a temperature-sensitive limiting means. This shall be installed in such a way that it limits the maximum temperature and shall directly or indirectly interrupt all current flow through the electrodes.

424.84: Over-Pressure Limit Control

Boilers used for making steam shall be equipped with pressure-sensitive limiting means, in order to limit the maximum pressure created in the boiler by the steam, and also directly or indirectly to interrupt the flow of electricity to the electrodes. This shall be in addition to a pressure-regulating system and other means of protecting the boiler from excessive pressure.

424.85: Grounding

On those boilers that are designed so that fault currents won't pass through the pressure vessel or boiler, and if the pressure vessel is isolated from the electrodes, all exposed metal parts, including the pressure vessel and supply and return piping, shall be grounded as required in Article 250.

For all other designs, the pressure vessel or boiler that contains the electrodes must be isolated and insulated from ground.

424.86: Markings

Marking will be required on all electrode-type boilers. The marking shall consist of the following:

- **(A)** The name of the manufacturer.
- **(B)** The normal rating in volts, amperes, and kW.
- **(C)** The type of electrical supply of frequency, the number of phases required, and the number of wires required.
- **(D)** The designation "Electrode-Type Boiler."
- **(E)** The warning: "ALL POWER SUPPLIES SHALL BE DISCONNECTED BEFORE SERVICING, INCLUDING SERVICING THE PRESSURE VESSEL."

The location of the nameplate shall be such that it is visible after installation.

IX. Electric Radiant Heating Panels and Heating Panel Sets

Cable and cable sets are not generally the same as heating panels and heating panel sets. The installations and wiring methods can't be covered by the same *Code* requirements; the *Code* addresses this problem in this part. Previously, the inclusion of both radiant-cable heating and radiant-panel heating in Part V caused considerable confusion. The solution—adding Part IX—helps to clarify the requirements for each type of equipment.

424.90: Scope

This part applies to radiant heating panels and sets of heating panels.

424.91: Definitions

(A) **Heating Panel.** Heating panels consist of a complete assembly. The assembly shall be provided with a junction box or a length of flexible conduit so that it may be connected to the branch circuit.

(B) **Heating Panel Set.** A heating panel set may be of a rigid or nonrigid assembly and shall be provided either with nonheating leads or a terminal junction assembly that has been identified as being suitable for the wiring system to which it is connected.

424.92: Markings

Such units must be clearly marked, showing their operating characteristics, and the like.

424.93: Installation

(A) General.

- **(1)** Heating panels and heating panel sets shall be installed in accordance with the manufacturer's instructions.
- **(2)** The heating portions (a) shouldn't be installed in or behind locations where they would be subject to physical damage; (b) shouldn't be run above walls, partitions, cupboards, or any other similar portion of a structure that extends to the ceiling; and (c) shouldn't be run in or through thermal insulation. However, they may be in contact with the surface of insulation. If run in or through insulation, the heating qualities of the panels will be reduced and the panels will overheat.
- **(3)** They shouldn't be installed less than 8 inches from any outlet box or junction box that will be used for the mounting of surface lighting fixtures. If recessed fixtures and trims

are used, heating panels shall have a clearance of 2 inches from these or from ventilating openings and other such openings in the rooms. There shall be sufficient area allowed so that no heating panel or heating panel set will be covered by any surface-mounted unit causing overheating of the panel or panel sets.

If a heating panel is covered by a surface-mounted heating unit, the situation may be dangerous because of overheating, resulting in fire.

Exception

If heating panels are listed to be mounted closer to the items described above, they may be installed at the distance marked in the listing.

(4) After installation and inspection of the heating panel or sets of heating panels, you may install a surface that has been identified and listed in the manufacturer's instructions as suitable for the installation. The surface shall be secured so that no nails or other fastenings can damage the heating panel or heating panel sets.

(5) Surfaces that are permitted in Section 424.93(A)(4) may be covered with paint, wallpaper, or other surface coverings that have been listed and identified in the manufacturer's instructions as suitable for use.

It should be mentioned that there is an uncontrollable future danger of tenants or homeowners driving nails into walls with heat panels and contacting or breaking the heat cables in the heating panel.

(B) Heating Panel Sets.

(1) In installing heating panel sets, you may secure them to the lower face of the joists, or they may be mounted between joists, headings, or nailing strips.

(2) Heating panel sets must be mounted parallel to joists or nailing strips.

(3) Only the unheated portion of heating panel sets may be nailed or stapled in place. You shouldn't cut heating panel sets or nail them at points closer than ¼ inch to the elements. Nails, staples, and so on must never be used where they penetrate current-carrying parts.

(4) Unless identified for cutting in an approved manner, heating panel sets must be installed as complete units.

424.94: Clearances of Wiring in Ceilings

Any wiring above ceilings shall be spaced not less than 2 inches above the heated ceiling, and the ambient temperature at which they are to be operated shall be 50°C or 122°F. The ampacity of the ambient temperature shall be computed with the correction tables for ambient temperatures, as in Tables 310.16 through 310.31.

Exception

When wiring above the ceilings is located above thermal insulation, and the thermal insulation is 2 inches or more in thickness, the wiring need not be corrected for temperature.

Author's Note

The material in Section 424.94 is, for all practical purposes, the same as covered previously for heat cables in ceilings.

424.95: Location of Branch-Circuit and Feeder Wiring in Walls

(A) **Exterior Walls.** Wiring methods installed in exterior walls must comply with Article 300 and Section 310.10.

(B) **Interior Walls.** Wiring installed in interior walls behind heating panels and heating panel sets shall be considered to be operating at an ambient temperature of 40°C or 104°F. The ampacity shall be computed as covered in Tables 310.16 through 310.31. Besides interior walls, this section also covers interior partitions.

424.96: Connection to Branch-Circuit Conductors

(A) **General.** The manufacturer's instructions must be complied with when heating panels or heating panel sets are assembled in the field for installation in one room or area.

(B) **Heating Panels.** Only wiring methods that are approved shall be used when connecting heating panels to branch-circuit wiring.

(C) **Heating Panel Sets.**

(1) Only identified wiring methods that are considered suitable for the purpose shall be used for connecting heating panel sets to branch-circuit wiring.

(2) The manufacturer's instructions shall be used on heating panel sets that are provided with terminal junction assemblies that have had the nonheating leads attached at the time of installation.

424.97: Nonheating Leads

It is permissible to cut nonheating leads of heating panels or heating panel sets to the required length. They shall meet the requirements for the installation of the wiring used, as covered in Section 424.96. Nonheating leads are part of the heating panel or heating panel set and are not subject to the ampacity requirements covered in Section 424.3(B) for branch circuits. Remember too that there is a requirement in the *Code* for a minimum of 6 inches of conductors, including these leads, in junction boxes.

424.98: Installation in Concrete or Poured Masonry

- **(A)** **Maximum Heated Area.** The maximum wattage for heating panels or heating panel sets installed in concrete or poured masonry is 33 watts-per-square-foot.
- **(B)** **Secured in Place and Identified As Suitable.** The manufacturer's instructions shall specify the means of securing heating panels or heating panel sets in concrete or poured masonry. The means of securing shall also be identified for that use.
- **(C)** **Expansion Joints.** Unless protected from expansion and contraction, heating panels and heating panel sets shouldn't bridge expansion joints.
- **(D)** **Spacings.** Heating panels and panel sets shouldn't come in contact with metal embedded in the floor.

Exception

If the heating panels have a metal-clad covering that is grounded, they may be in contact with metal embedded in the floor.

- **(E)** **Protection of Leads.** Rigid metal conduit, intermediate metal conduit, rigid nonmetallic conduit, EMT, or other approved means shall be used for leads to protect them where they leave the floor.
- **(F)** **Brushings or Fittings Required.** Leads emerging within the floor shall have bushings or fittings that are approved.

424.99: Installation Under Floor Covering

- **(A)** **Identification.** Any heating panels or heating panel sets to be installed under floor covering shall be identified and listed as suitable for that use.

(B) **Maximum Heated Area.** The maximum heated area in wattage for heating panels or heating panel sets installed under floor covering is 15 watts-per-square-foot.

(C) **Installation.** Panels or panel sets that are listed for installation in the floor covering are to be installed only on a smooth and flat floor area as called for by the manufacturer's instructions, which shall also comply with the following:

See the *NEC* for the five conditions that must be met for the above.

Article 426—Fixed Outdoor Electric De-icing and Snow-Melting Equipment

I. General

426.1: Scope

This article covers electrically energized heating systems and their installation.

(A) **Embedded.** Driveways, steps, and various other areas may have heating cable embedded.

(B) **Exposed.** Heating cable may be used for drainage systems, bridges, structures, and roofs, and they may be installed on other structures.

It is the author's opinion that where they are mounted in downspouts, in roofs, and so on, they should be MI-heating cables, either copper or stainless steel, and the copper continuity of the equipment-grounding system should be maintained. The author would also go so far as to require that GFCIs supply these branch circuits.

426.2: Definitions

Four definitions are given in this section: heating systems, resistance heating element, impedance heating system, and skin-effect heating system. These definitions are sometimes helpful in understanding this article.

426.3: Application of Other Articles

As is true throughout the *Code*, this article must be in accordance with other parts of the *Code* unless specifically amended by this article. In particular, applications of the coverage of this article for use in hazardous (classified) locations must comply with Articles 500 through 517.

When the outdoor heating, de-icing, and snow-melting equipment is used and is cord- and plug-connected, it shall be identified and installed as described in Article 422.

It is the author's opinion that heating cable used for roofs, downspouts, steps, and so on should be controlled by a thermostat mounted where it is not exposed to direct heat from the sun, so that you need not depend upon people to turn it on and off when it is needed and not needed.

426.4: Branch-Circuit Sizing

The ampacity of branch-circuit conductors used for these purposes and the rating of the overcurrent-protection devices supplying this fixed outdoor de-icing and snow-melting equipment is to be not less than 125 percent of the maximum load to the heaters to which they are connected. See Section 240.3.

II. Installation

426.10: General

Equipment and materials used for installing fixed outdoor electric de-icing and snow-melting systems may be subject to many troubles and possible hazards, so extreme care should be taken in the installation and inspection to ensure that the equipment and other installation materials are identified as being suitable for the use and installed according to the *Code* for the following conditions:

- **(A)** Chemical, thermal, and physical environments must all be considered; and
- **(B)** The manufacturer's drawings and instructions must be followed in making the installation.

An actual incident that the author was involved in was the use of stainless steel MI heating cable. It was laid on the earth surface of steps, and then precast concrete step tops were added. During the time that the cable was exposed, it drew moisture, which was found by using a Biddle Megger. Upon further examination of the MI cable, it was found that there were places in the stainless steel covering that were not completely closed, so the installation was not approved until the MI cable was replaced with cable that didn't have leaks in the outer covering.

426.11: Use

The installation of de-icing and snow-melting heating equipment must be done in such a manner as to provide protection from physical damage.

426.12: Thermal Protection

De-icing and snow-melting electrical equipment that operate at temperatures exceeding 60°C (140°F) shall be guarded, thermally insulated, or otherwise isolated to prevent persons coming into contact with it.

426.13: Identification

De-icing and snow-melting equipment used outdoors shall be identified for the location by posting signs or markings that are appropriate and clearly visible.

426.14: Special Permission

Special types of de-icing or snow-melting equipment that employ methods of construction or installation other than those covered by this article shouldn't be used or installed unless special permission is granted in writing.

III. Resistance Heating Elements

426.20: Embedded De-icing and Snow-Melting Equipment

(A) **Watt Density.** The heater area shall have not more than 120 watts per square-foot.

(B) **Spacing.** The spacing of the cable is dependent upon the rating of the de-icing cable, but it shouldn't be less than 1 inch on centers.

(C) **Cover.** Heating panel or cables shall be installed as follows:

- **(1)** If installed in asphalt or concrete or other masonry, you must have at least 2 inches of masonry or asphalt, and the cable shall be at least 1½ inches below the surface of the masonry or asphalt or the installation of panels or other units.
- **(2)** They may be installed over other approved bases of masonry or asphalt and embedded with 3½ inches of the material, but they shouldn't be installed less than 1½ inches from the top surface.
- **(3)** If other forms of heating equipment have been specifically listed for other forms of installation, then the installation shall be done only in the manner for which it has been investigated and listed.

(D) **Secured.** Frames or spacers shall be installed between cables for support while the masonry or asphalt is being applied over the cables.

(E) Expansion and Contraction. Added protection for expansion and contraction is required for cables, units, and panels that are installed on expansion joints on bridges.

426.21: Exposed De-icing and Snow-Melting Equipment

(A) Secured. Approved means must be used when installing heating unit assemblies so that they will be securely fastened to the surface.

(B) Over-Temperature. In any place where heating elements are not in direct contact with the surface that they are heating, the design of the heating assembly shall be such that its temperature limitations are not exceeded.

(C) Expansion and Contraction. Expansion joints shouldn't be crossed with heating elements and assemblies unless provision is made for expansion and contraction.

(D) Flexible Capability. When heating is installed where structures are flexible, the heating assemblies must have a flexibility that is compatible with the structure.

426.22: Installation of Nonheating Leads for Embedded Equipment

(A) Grounding sheath or Braid. If the nonheating leads have a grounding sheath or braid, they may be embedded in masonry or asphalt in the same manner as heating cable without the need for additional protection.

(B) Raceways. All but 1 to 6 inches of nonheating leads of Type TW, or cables with other approved types of insulation that don't have a grounding sheath or braid, shall be installed in conduit, intermediate metal conduit, EMT, or other raceway, and this conduit or other protection shall extend a minimum of 1 inch from the factory splice to the heating part of the cable to not more than 6 inches from the factory splice.

(C) Bushings. Insulated bushings shall be installed on the conduit, EMT, and so on in the masonry or asphalt where the cable emerges from the conduit.

(D) Expansion and Contraction. All leads shall be protected where buried in expansion joints; where they emerge from masonry or asphalt, they shall be installed in conduit, intermediate metal conduit, EMT, or other raceway.

(E) Leads in Junction Boxes. There shall be a minimum of 6 inches of nonheating leads left in junction boxes.

426.23: Installation of Nonheating Leads for Exposed Equipment

(A) Nonheating leads. Power supplies to nonheating leads that are cold leads, supplying resistive elements, must be suitable for the temperature that they will encounter. Nonheating leads attached to approved heaters may be shortened, provided the markings that were covered in Section 426.25 are kept on the nonheating leads. There must be no less than 6 inches of nonheating leads in the junction box.

(B) Protection. Rigid metal conduit, intermediate metal conduit, EMT, or any other approved means shall be supplied over the nonheating leads from the power supply.

426.24: Electrical Connection

(A) Heating Element Connection. Only insulated connectors that are listed for the purpose shall be used to make connections to heating elements where factory connections are not supplied when heating elements are embedded in masonry or asphalt or are on exposed surfaces.

(B) Circuit Connections. The nonheating leads may be spliced or terminations made according to Sections 110.14 and 300.15. This does not apply to connection to the heating element itself.

426.25: Marking

Factory-assembled heating units must be marked within 3 inches from the heating element on the nonheating lead at both ends. This permanent marking shall have identification symbol, catalog number, and rating in volts, watts, or in volts and amperes.

426.26: Corrosion Protection

Raceways made of ferrous or nonferrous cable, cable armor, cable sheaths, boxes, fittings, supports, and support hardware may be installed in concrete or in direct contact with the earth. If the areas are subject to severe corrosive conditions, the material must be suitable for the conditions. Keep in mind that conduits and the like are not corrosion-resistant in many types of earth.

It is up to the installer to prove to the authority that has jurisdiction that the materials are adequately protected from corrosion.

426.27: Grounding

(A) Metal Parts. All exposed metal parts of fixed outdoor electric de-icing and snow-melting equipment, raceways, boxes, and so on likely to become energized shall be grounded as required in Article 250.

(B) Grounding Braid or Sheath. The heating section of the cable, panel, or unit shall have a means for grounding, such as copper braid, lead, copper sheath, or other approved means.

(C) Bonding and Grounding. Any metal parts involved in cable or cable installations that are likely to become energized shall be grounded. Table 250.95 shall be used to find the equipment-grounding conductor size.

Author's Note

A very dangerous situation can exist when heat cables are installed in drainage gutters or downspouts, such as on residences. This is done for the most part by do-it-yourselfers. Attention should be called to the fact that unless downspouts are properly grounded, as per Article 250, a dangerous situation may occur.

426.28: Equipment Protection

Equipment on branch circuits that supply fixed outdoor electric de icing and snow-melting equipment must have ground-fault protection.

IV. Impedance Heating

426.30: Personnel Protection

All exposed impedance heating elements must be guarded or insulated to prevent personnel from contacting the elements.

426.31: Isolation Transformer

The heating system must be isolated from the distribution system by using shielded isolation transformer.

426.32: Voltage Limitation

These elements can't be operated at more than 30 volts, except when GFCIs are used; in such instances, they can be used with up to 80 volts.

426.33: Induced Current

Section 300.20 covers current-carrying components.

426.34: Grounding

If impedance-operated heating systems are operated at voltages over 30 and under 80 volts, they shall be grounded at specified points.

V. Skin-Effect Heating

426.40: Conductor Ampacity

An insulated conductor within the ferromagnetic envelope can carry more current than its rating per Article 310, as long as it is identified for such use.

426.41: Pull Boxes

All pull boxes must be accessible.

426.42: Single Conductor in Enclosure

Single conductors in the ferromagnetic enclosure are exempt from the requirements of Section 300.20.

426.43: Corrosion Protection

Components used for skin-effect heating systems (ferromagnetic envelopes, metal raceways, boxes, supports, and so on) are allowed to be installed in concrete, as long as they are made of a suitable material, or coated with a corrosion-resistant substance.

426.44: Grounding

The ferromagnetic envelope of a skin-effect heating system must be grounded at each end, and may also be grounded at additional points in between.

VI. Control and Protection

426.50: Disconnecting Means

(A) **Disconnection.** Disconnecting means that open all ungrounded conductors shall be used for de-icing and snow-melting equipment. Branch-circuit switches or circuit breakers may be used for this purpose if readily accessible and marked so as to indicate what position they are in. Never trust any circuit as being deenergized unless proper tests have been performed to assure you that they are open.

(B) **Cord- and Plug-Connected Equipment.** If attachment plugs or cord- and plug-connected equipment are factory-installed and rated at 20 amperes or less and 150 volts-to-ground or less, the attachment plug may be used as a disconnecting means.

426.51: Controllers

(A) **Temperature Controllers with OFF Position.** Temperature controllers with switching devices that indicate an OFF position and open all ungrounded conductors are not usable as a disconnecting means unless they have a positive lockout in the OFF position.

(B) **Temperature Controllers Without OFF Position.** Temperature-control switching devices that don't open all ungrounded conductors are not required to serve as the disconnecting means.

(C) **Remote Temperature Controllers.** Remote-control temperature-actuated devices do not have to meet the requirements of Section 426.51(A) in order to meet all of the specifications in this section. Also, they may not be used as a disconnecting means.

(D) **Combined Switching Devices.** The following conditions apply when combined temperature-actuated devices and manually controlled switches serve both as the controller and the disconnecting means:

(1) When placed in the OFF position, the manually operated device will open all ungrounded conductors.

(2) The design is such that the heating elements can't be energized when the manually operated device is in the OFF position.

(3) There is to be a positive lockout in the OFF position.

426.52: Overcurrent Protection

Section 426.4 of this article covers branch-circuit requirements. Fixed outdoor electric de-icing and snow-melting equipment shall be considered protected against overcurrent when the preceding section is adhered to, remembering that this is considered a continuous load and is subject to the 80-percent factor.

Article 427—Fixed Electric Heating Equipment for Pipelines and Vessels

I. General

427.1: Scope

This article covers the application of electrically energized systems of various types and how to install these systems on pipelines and/or vessels.

427.2: Definitions

For the purpose of this article, the following definitions are used:

Pipeline. The length of pipe, including the valves, pumping, strainers, control devices, and other equipment that carries fluids. The purpose is to maintain the fluid condition of fluids that may tend to congeal at lower temperatures.

Vessels. The portion of the system that holds the volume of the fluid that is to be carried by pipeline. They may be barrels, drums, or tanks. They may contain not only fluids, but also other materials.

Integrated Heating Equipment. This comprises the complete system and all components and vessels used by the heating elements, medium of transferring the heat, thermal insulation, barriers to keep out moisture, the nonheating leads for connection to the source of supply, and any temperature controllers, warning signs, and all the electrical raceway fittings, and so on, to connect the system to the source.

Resistance Heating Element. A separate element that generates the heat that is applied to the pipeline or vessel, either externally or internally.

Note

Resistive-type heaters may come in many forms: tubular heaters, strip heaters, heating cable, heating blankets, and immersion heating.

From the above definition it may readily be seen that there is a certain tie-in between Articles 426 and 427.

Impedance Heating System. A system in which the pipeline and vessels are fed from a dual-winding transformer to the pipeline of vessel wall, causing current to flow therein by direct connection to an AC source of voltage.

See Part IV in Article 426.

Induction Heating System. Heat is generated in the pipeline of vessel walls by inductive current and the hysteresis effect of the alternating current in the pipeline of vessel wall. The AC current must be supplied from an external and isolated AC field.

Skin-Effect Heating System. The heating effect of a ferromagnetic envelope attached to the pipeline and/or vessel. The heat is generated on the inner surface of the ferromagnetic envelope.

Note

Electrically insulated conductors are routed through and connected to the envelope at the other end. The envelope and the insulated conductor must be connected to an AC voltage source that originates from a dual-winding transformer.

See Part V of Article 426.

427.3: Application of Other Articles

Article 422 covers cord-connected pipe-heating assemblies intended for specific uses and approved for the purpose. Hazardous (classified) locations shall comply with Articles 500 through 516 for fixed pipeline- and vessel-heating equipment in such hazardous locations.

427.4: Branch-Circuit Sizing

As with other branch-circuit requirements for continuous loads, conductors and overcurrent devices that supply fixed electric pipeline- and vessel-heating equipment shall be calculated on the basis of 125 percent of the total load of the heaters. Section 240.3 covers rating or setting of overcurrent devices.

II. Installation

427.10: General

The chemical, thermal, and physical environment by necessity requires that the equipment for pipelines and vessels shall be compatible with the problems listed in Part I of this article.

The manufacturer's drawings and instructions shall be followed in the installation.

427.11: Use

Care must be taken in the installation of feeding equipment to protect it from physical damage.

427.12: Thermal Protection

Pipelines or vessels with external heating equipment that operate with a surface temperature of 60°C or 140°F shall protect personnel from contacting it by thermal insulation.

427.13: Identification

The presence of electrically heated pipelines or vessels shall be made known by posting caution signs or by markings placed at intervals along the pipeline or vessel.

III. Resistance Heating Elements

427.14: Secured

Thermal insulation shouldn't be used as the method of securing the heating element assemblies to the surface that is being heated. Other means must be used.

427.15: Not in Direct Contact

Unless the heater assemblies are such that the temperature limits can't be exceeded when the heating element is not in direct contact with the pipeline or vessel that is being heated, other means must be provided to prevent the temperature of the heating element from rising higher than that at which it should be operated.

427.16: Expansion and Contraction

Unless provision is made for expansion and contraction of the heating source, heating elements shouldn't be allowed to bridge expansion joints.

427.17: Flexural Capability

When installed on flexible pipelines, the flexibility of the heating element shall be compatible with the flexibility of the pipeline.

427.18: Power Supply Loads

(A) **Nonheating Leads.** The nonheating leads from the power supply, which are also known as the cold leads, when used with resistance-type elements shall have insulation to withstand the temperatures with which they may come in contact. Nonheating leads that are factory assembled may be shortened, as was covered in Section 427.20, but there shall be no less than 6 inches of lead extending into the junction box.

(B) **Power Supply Lead Protection.** Rigid metal conduit, intermediate metal conduit, EMT, or other raceways that are listed as being suitable for this application shall be supplied for the protection of nonheating supply leads where they emerge from the electrically heated pipeline or vessel.

(C) **Interconnecting Leads.** The nonheating leads interconnecting with the heating leads may be covered by the thermal insulation in the same way as the heaters.

427.19: Electrical Connections

(A) **Nonheating Interconnections.** Insulated and identified connectors, when properly installed, shall be used to make nonheating interconnections under thermal insulation.

(B) Circuit Connections. Splices and terminations outside thermal insulation shall be in a box or fitting in accordance with Sections 110.14 and 300.15.

427.20: Marking

Each factory-assembled heating unit shall be legibly marked within 3 inches of the end of each power-supply nonheating lead with the permanent identification or symbol, catalog number, rating in volts and watts, or rating in volts and amperes.

427.21: Grounding

All exposed noncurrent-carrying metal parts of the electrical heating equipment shall be grounded, as required by Article 250, if there is any possible chance of such parts becoming energized.

427.22: Equipment Protection

Heating equipment that is not metal covered must have ground-fault protection for the branch circuit supplying it.

427.23: Metal Covering

Heating wires and cables must have grounded metal coverings. Heating panels must have a grounded metal cover on the side of the panel opposite the heated surface.

IV. Impedance Heating

427.25: Personnel Protection

Personnel protection, either by guarding or by insulation, shall be provided for all external portions of the pipeline and/or vessel that is being heated. If the insulation is exposed to the outside, it must be weatherproof.

427.26: Isolation Transformer

Dual-winding transformers that have a shield between the primary and secondary windings shall provide isolation between the heating system and the distribution system.

2005

427.27: Voltage Limitations

The voltage use for the heating shouldn't be greater than 30 volts AC. Voltages up to 80 volts may be used when ground-fault protection is provided. Voltages up to 132 volts to ground are permitted if the following conditions are met: (a) The system will be supervised only by qualified persons, (b) ground-fault protection is provided, (c) the heated vessel is contained entirely in a grounded metal enclosure, and (d) the transformer secondary connections are enclosed completely in a grounded metal enclosure.

427.28: Induced Currents

Refer to Section 300.20, which covers the installation of all current-carrying parts.

427.29: Grounding

When a heating system for a pipeline or vessel operates at not less than 30 and not more than 80 volts, it shall be grounded at designated points, which should be described in the installation instructions supplied by the manufacturer.

427.30: Secondary Conductor Sizing

The ampacity of the current-carrying conductors from the transformer shall be rated at least 30 percent of the load the heater will be using.

V. Induction Heating

427.35: Scope

Covered in this section is the installation of line-frequency induction-heating equipment that is used to heat pipelines and vessels and all their accessories.

Note

You are referred by the *Code* to Article 665 for other applications of inductive heating.

427.36: Personnel Protection

The induction coils used for heating pipelines and vessels that operate at voltages greater than 30 volts AC must be enclosed in nonmetallic raceways or in split metallic enclosures. These shall be either isolated or made inaccessible to protect people in the area. The purpose of this split metallic enclosure is to keep heat being induced in the metallic enclosure, thereby allowing it to go to the pipeline or vessels.

427.37: Induced Current

To prevent induction into the metal that surrounds the area of mechanical equipment supports or structures, there shall be shielding, isolation, or insulation of the path that the induced currents might take other than the pipelines or vessels. The stray current paths shall be bonded to prevent arcing between them.

VI. Skin-Effect Heating

427.45: Conductor Ampacity

The ampacity of conductors that are electrically insulated inside ferromagnetic envelopes may be permitted to exceed values that appear in Article 310, provided that they are identified or listed for the purpose for which they are used.

427.46: Pull Boxes

Pull boxes used for pulling the electrically insulated conductors into ferromagnetic envelopes may be buried under the thermal insulation. This may be done only if these locations are permanently marked on the insulation jacket and on drawings. Pull boxes thus mounted outdoors shall have watertight construction.

427.47: Single Conductor in Enclosure

Section 300.20 doesn't apply where a single conductor is enclosed in ferromagnetic envelopes or other metal enclosures.

427.48: Grounding

Not only shall the ferromagnetic envelope be grounded at both ends, but it may be grounded at points in between the two ends, as might be required in the design. All joints of ferromagnetic envelopes or enclosures shall be bonded so that the electrical continuity is ensured.

Skin-effect heating systems are not required to comply with Section 250.26.

Note

Refer to Section 250.26(D).

VII. Control and Protection

427.55: Disconnecting Means

(A) **Switch or Circuit Breaker.** You must have a disconnecting means that opens all conductors to fixed electrical pipeline- or vessel-heating equipment. For this purpose, the branch-circuit switch or circuit breakers may be used as the disconnecting means if they are readily accessible. However, on either means the position must be identified, and there shall be a positive means of locking it in the OFF position.

(B) **Cord- and Plug-Connected Equipment.** If the circuit does not exceed 20 amperes or 150 volts-to-ground or less, the factory-installed attachment cord-and plug-connection may be used as the disconnecting means.

427.56: Controls

(A) Temperature Control with OFF Position. If the temperature-controlling means has an OFF position that opens all ungrounded conductors and the OFF position is plainly indicated, these may not be used as the disconnecting means unless they are provided with a positive lockout on the OFF position of the control.

(B) Temperature Control Without OFF Position. Temperature-controlled devices that have an OFF position need not open all ungrounded conductors; neither shall they serve as disconnecting means.

(C) Remote Temperature Controller. Temperature control devices located in a remote position don't have to meet the requirements of Sections 427.56(A) and (B), and shouldn't be used as a disconnection means.

(D) Combined Switching Devices. Switching devices that are a combination of temperature-actuated devices and manually controlled switches that serve both controllers and the disconnecting means must comply with the following conditions:

(1) When manually placed in the OFF position, they must manually open all ungrounded conductors.

(2) The design shall be such that when thrown in the manually OFF position they won't automatically energize the system.

(3) A positive means must be provided to lock it in an OFF position.

427.57: Overcurrent Protection

If Section 427.4 is complied with, heating equipment supplied by a branch circuit shall be considered adequate protection by the overcurrent-protection device.

Article 430—Motors, Motor Circuits, and Controllers

I. General

430.1: Scope

Article 430 covers motors and all associated branch circuits, overcurrent protection, overload protection, and so on. The installation of motor-control centers is covered in Article 408, and air-conditioning equipment is covered in Article 440.

430.2: Adjustable-Speed Drive Systems

The power-conversion equipment, if fed from a branch circuit or feeder and including part of an adjustable-speed drive system, is considered a part of the adjustable-speed drive system, and the rating is to be based upon the power required by the conversion equipment. When the power-conversion equipment supplies overcurrent protection for the motor, no additional overload protection is required.

Adjustable-speed drive systems may have the disconnecting means in the line coming into the conversion equipment, and the rating of the line shall be not less than 115 percent of the input-current rating of the conversion unit.

Conversion units may be solid-state units to change the cycles or chop part of the wave forms to vary the speed of squirrel cage motors as necessary for the application.

430.3: Part-Winding Motors

Induction or synchronous motors that have a part-winding start are arranged so that at starting they energize this part at the primary (armature) winding. After starting, the remainder of the winding is energized in one or more steps. The purpose of this is to reduce the initial insurge of current until the motor accumulates some speed, developing a counter electric motor force. The inrush current at start is locked-rotor current, and is quite high. A standard part-winding-start induction motor is arranged so that only one-half of its winding is energized at start; then, as it comes up to speed, the other half is energized, so that both halves are energized and carry equal current. Hermetically sealed refrigerating motors are not considered standard part-winding-start motors.

Separate overload devices are used on the standard part-winding-start induction motor. This requires that each half of the motor winding be individually protected. This is covered in Sections 430.32 and 430.37. Each half of the winding has a trip current that is one-half of the specified running current.

As specified in Section 430.52, each of the two motor windings shall have branch-circuit, short-circuit, and ground-fault protection that is to be not more than one-half the currents mentioned in Section 430.52.

Exception

A single device with this one-half rating is permitted for both windings, provided that it will allow the motor to start. If a time-delay (dual-element) fuse is used as a single device for both windings, its rating is permitted if it does not exceed 150 percent of the motor full-load current.

430.5: Other Articles

The equipment covered in this article must also comply with other sections of the *Code*. A thorough list of such other sections is given in this section of the *NEC*.

430.6: Ampacity and Motor-Rating Determination

Ampacities shall be determined as follows:

(A) General Motor Applications. This part is very important and should be carefully read. Whenever the current rating of a motor is used for figuring the ampacity of conductors, switches, branch-circuit overcurrent devices, and so on, the actual motor nameplate current rating shouldn't be used. Instead, Tables 430.147 through 430.150, which give full-load currents for direct-current motors, single-phase AC motors, two-phase AC motors, and three-phase AC motors, are used.

Separate motor overload protection is to be figured from the actual nameplate current rating. When a motor is marked in amperes instead of horsepower, the horsepower rating shall be assumed to correspond to the values given in Tables 430.147 through 430.150. At times, it may be necessary to interpolate to arrive at an intermediate horsepower rating. Listed motor-operated appliances (such as garage-door openers) that are marked with both horsepower and full-load current may have their circuits based on the full-load current.

Exception No. 1

Sections 430.22(A) and 430.52 cover multispeed motors.

Exception No. 2

When shaded-pole or permanent-split-capacitor-type fan or blower motors are applied on equipment, if the motor type is marked, the full-load current for the motor marked on the nameplate of the equipment where the fan or blower is being used shall be used instead of the horsepower rating of the motor where the amperage is being used to determine the rating of the disconnecting means and branch-circuit conductors, as well as the controller, the short-circuit branch-circuit and ground-fault protection and the separate overload protection. The amperage marked on the nameplate shouldn't be less than that marked on the fan or blower nameplate.

(B) Torque Motors. Torque motors are commonly designed for applications that require prolonged stalled torques or special running-torque characteristics such as being turned against the direction of rotation. Direct-current, single-phase or polyphase induction wound-rotor, polyphase induction, repulsion, repulsion-induction, universal, and other motors can be designed as torque motors. A torque motor develops its maximum torque at locked rotor or stalled conditions. This information will make the following clearer to you.

The locked-rotor current is the rated current of a torque motor. Thus, the nameplate current shall be used to determine the ampacity of the branch circuit conductors (see Sections 430.22 and 430.24), the ampere rating of the motor overcurrent protection, and ground-fault protection.

Note

See Section 430.83 and Section 430.110.

(C) **AC Adjustable Voltage Motors.** For motors used on adjustable alternating current, adjustable voltage, variable torque drive systems, and so on, the maximum current marked on the motor and the controller nameplate shall be used for the amperage of the conductors or the amperage rating of the switches, branch-circuit short circuits, and any ground-fault protection used. Should the amperage not appear on the nameplate, the ampacity should be determined on the basis of 150 percent of the values found in Tables 430.149 and 430.150.

430.7: Marking on Motors and Multimotor Equipment

(A) **Usual Motor Applications.** The following information shall be marked on a motor:

(1) Manufacturer's name.

(2) The maximum motor amperes on shaded-pole and permanent-split-capacity motors are to be used when the motors run only at maximum speed. On multispeed motors the full-load current for each speed shall be marked.

(3) If it is an alternating-current motor, the frequency and number phases shall be marked.

(4) The rated full-load speed is to be marked.

(5) The rated ambient temperature and the rated temperature rise of the insulation system class shall be marked.

(6) The time-rating of use shall be marked: for example, 5, 15, 30, or 60 minutes, or continuous if it is for continuous use.

(7) If a ⅛-horsepower motor or more, then the horsepower rating shall be given. On multispeed motors that are ⅛ horsepower or more, the horsepower rating for each speed shall be given, except for shaded-pole or permanent-split-capacitor motors at ⅛ horsepower or more. Arc-welding motors are not required to be marked in horsepower.

(8) On alternating-current motors that are rated ½ horsepower or more, the code letter shall be given if they are alternating current motors, so that the motor can be replaced with a similar motor. Code letters shall be omitted on polyphase wound-rotor motors. See (B) of this list.

(9) B, C, D, or E motors must be marked with the appropriate design-letter designation.

(10) If it is a wound-rotor induction motor, the full-load current and secondary voltage shall be given.

(11) On synchronous motors that are DC excited, the field current and voltage shall be given.

(12) The type of winding, whether straight-shunt, stabilized-shunt, compound, or series-wound, if they are direct-current motors, shall be marked. Direct-current motors that are 7 inches or less in diameter and are fractional horsepower motors are not required to be marked.

(13) When a motor is provided with thermal protection that meets the requirements of Section 430.32(C)(2), it shall be marked "Thermally Protected." The abbreviated marking "T.P." may be used on motors that are thermally protected with a rating of 100 watts or less, provided they meet the requirements of Section 430.32(C)(2).

(14) When a motor complies with Section 430.32(C)(4), it shall be marked "Impedance Protected." If impedance-protected motors are 100 watts or less and meet the requirements of Section 430.32(C)(4), the marking "Z.P." may be used.

(B) Locked-Rotor Indicating Code Letters. The code letters marked on the motor nameplate show the kilovolt-ampere input per horsepower when the rotor is locked. These code letters are given in Table 430.7(B) and also used with Table 430.152 in determining branch-circuit overcurrent protection as provided in Section 430.52.

(1) Code letters shall be marked on multispeed motors and shall designate the locked-rotor kVA per horsepower. This will be for the locked-rotor current at the highest speed at which the motor may be started.

Exception

The locked-rotor current for kVA and horsepower shall be marked on multispeed motors.

(2) Single-speed motors that start on Y connection but run on delta connection must be marked with the locked-rotor kVA per horsepower for the Y connection. Y-started motors are often used for large motors to cut down on the locked-rotor current to start them; as the current drops, they are automatically cut over to a delta connection, and this is where they run when in use.

(3) Dual-speed motors will have two locked-rotor current ratings and kVA per horsepower on the two different voltages. They should be marked with the code letter for the voltage that gives the highest locked-rotor rating and kVA per horsepower.

(4) Motors that have both a 60- and 50-Hertz rating shall be marked with the code letter designating the locked-rotor current, and the kVA per horsepower will be for the 60 Hertz.

(5) Motors that start on part winding shall have the code letter that designates the locked-rotor kVA per horsepower that is based on the locked-rotor current for the full winding of the motor.

Refer to Table 430.7(B), "Locked-Rotor Indicating Code Letter."

(C) Torque Motors. Torque motors have a standstill-rated operation, so all of the markings as covered in (A) of this section apply except that locked-rotor torque shall replace horsepower.

(D) Multimotor and Combination-Load Equipment. See the *NEC*. This marking is needed for multimotor and combination-load equipment because often the individual nameplates are visible after mounting. It may be very difficult for both the inspector and the installing contractor to determine which loads may or may not be in operation at the same time, and this is necessary to determine the minimum circuit ampacity and maximum rating of the circuit-protective device required. Such information can best be furnished by the equipment builder who knows the conditions of operation, thereby minimizing the chance of errors in field calculations where the specific conditions may not be known.

A considerable portion of the above applies to the manufacturer, but it is also very essential that the trade and inspectors know what and where to look. See Section 430.25 for the computation of the ampacity of the conductors.

430.8: Marking on Controllers

Controllers shall be marked as follows: (1) manufacturer's name or identification; (2) voltage; (3) current or horsepower rating; (4) any other data that might be necessary to indicate the proper application of the motor.

If a controller includes motor overload protection that satisfies the requirements for group application, the controller shall be marked with the following information: motor overload protection, ground-fault protection, and the maximum branch-circuit short circuit for use with such applications.

When instantaneous circuit breakers are used in combined controllers that have instantaneous trip, they are to be clearly marked on the circuit breaker to show the ampere settings on the trip-adjustable elements.

Where a controller is built in as an integral part of a motor or of a motor-generator set, the controller need not be individually marked when the necessary data is on the motor nameplate.

430.9: Marking at Terminals

(A) Markings. Terminals of both motors and controllers must be marked, so that correct terminations can be made.

(B) Conductors. Motors and controller terminals must be connected to copper wiring, unless the terminals are clearly marked to accept different conductors.

(C) Torque Requirements. Screw terminals that receive wires No. 14 or smaller must be torqued to at least 7 pound-inches (0.79 N-m), except if they are clearly identified as having a different torque rating.

430.10: Wiring Space in Enclosures

(A) General. In Article 312, it is stated that switches are not intended to be used as junction boxes or wireways. This ruling also includes enclosures for controllers and disconnecting means for motors. Neither shall they be used as junction boxes or raceways for conductors feeding through or tapping off to other apparatus, unless they are specifically designed for this purpose. Other means, such as auxiliary gutters, junction boxes, and so on, shall be used instead. See Article 312 for switch and overcurrent-device enclosures.

(B) Wire-Bending Space in Enclosures. Table 430.10(B) gives the minimum bending space allowable in motor controllers. The measured distance, in the direction in which the wire leaves the terminal, is in a straight line from the end of a connector or lug to the wall or barrier. If alternate wire terminating means is substituted for other than that supplied by the manufacturer of the controller, it must be identified *by the manufacturer* for use in that controller, and the minimum wire-bending space shouldn't be reduced. [See the *NEC* for Table 430.10(B).]

When a motor control center is used as the enclosure, Article 312 applies to the minimum wire-bending space.

430.11: Protection Against Liquids

Motors that are mounted directly under or in locations where dripping or spraying oil, water, or other injurious liquids may occur shall be either suitably protected or designed for the existing conditions.

430.12: Motor Terminal Housings

(A) Material. Motor terminal housings shall be made of substantial constructed metal.

Exception

Nonmetallic, nonburning housings are permitted in other than hazardous (classified) locations. They shall be of substantial construction, and a grounding means on the motor inside the terminal housing shall be provided so that the equipment-grounding conductor will properly ground the motor.

(B) Dimensions and Space—Wire-to-Wire Connections. Table 430.12(B) shall be used in terminal housing where wire-to-wire connections will be made. The minimum dimensions and usable volumes shall be in accordance with Table 430.12(B).

See the *NEC* for this table and the note thereto.

(C) Dimensions and Space—Fixed Terminal Connections. See the *NEC* for Tables 430.12(C)(1) and (C)(2), which describes terminal housings that enclose rigidly mounted motor terminals. These tables give the usable volumes so that the terminal housing is of sufficient size to provide minimum terminal spacings.

Incorporated in this section is Table 430.12(B), which gives the horsepower, minimum cover-opening dimension, and minimum usable volume. Please note, however, that nothing is stated concerning the voltages of the motors. Part (C) has two tables. Table 430.12(C)(1) gives the information necessary for terminal housings, enclosing rigidly mounted motor terminals, while Table 430.12(C)(2) lists the usable volumes. All of this is information pertinent to the manufacturer and to the trade, as well.

(D) Large Wire or Factory Connections. The foregoing provision covering the volumes of terminal housings is not to be considered applicable when the following is encountered: During the installation of motors at larger ratings and greater number of leads, if they have large wire sizes or if the motors are installed as part of factory-wired equipment so that no additional connections have to be made at the motor terminal housing, it is required that the motor terminal housing be of adequate size for making the connections.

(E) Equipment-Grounding Connections. With motor terminal housings or wire-to-wire connections, or fixed terminal connections, a means shall be provided as covered in Section 250.8 for attaching the equipment-grounding conductor so as to ground the motor properly.

Exception

When the motor is part of factory-wired equipment that must be grounded, and is without additional terminal connection provided in the motor terminal housing during the installation of such motor equipment, a separate means for motor grounding at the motor terminal housing won't be required.

430.13: Bushings

All conductors to the motor, including leads, shall be properly bushed to prevent abrasion and deterioration by oils, greases, or other contaminants. If conductors are exposed to deteriorating agents, you are referred to Section 310.8

430.14: Location of Motors

(A) Ventilation and Maintenance. Motors shall be located so as to give adequate ventilation and provide room for normal maintenance, such as greasing bearings and changing brushes.

(B) Open Motors. Open motors that have commutators or slip rings shall be sò located that sparks from the brushes won't ignite combustibles. This does not prohibit the mounting of motors on wooden floors or supports.

430.16: Exposure to Dust Accumulations

Motors in locations where dust or flying materials will accumulate within the motor in such quantities as to interfere with the cooling and thereby cause dangerous temperatures shall be replaced with types that are suitable for the conditions. There are many types of motors designed to eliminate the accumulation of dust, dirt, and the like. If extremely severe conditions exist, pipe-ventilated motors should be used or the motors located in reasonably dust-free rooms.

430.17: Highest-Rated or Smallest-Rated Motor

Refer to the following sections to determine how to comply with them: 430.24, 430.53(B), and 430.53(C). The highest full-load current rating shall be considered the highest-rated motor. To determine the full-load current used to determine the highest-rated motor, the equivalent value shall correspond to the motor horsepower rating, and shall be selected from Tables 430.147 through 430.150.

You are advised that it might be possible in calculations to become fooled by the horsepower rating and find that you should, in fact, have taken the full-load currents instead.

430.18: Nominal Voltage of Rectifier Systems

The value of the rectified voltage derived from a system will be determined from the nominal voltage (AC) rectified.

Exception

If the nominal DC voltage from the rectifier exceeds the peak value of the AC voltage being rectified, it shall be used.

II. Motor Circuit Conductors

430.21: General

Part II gives us the sizes of the conductor specified and capable of carrying the current encountered under specified conditions.

Exception

For volts over 600, nominal, you are referred to Section 430.124.

The provisions covering grounding in Article 250 and conductors in general in Article 310 are not intended to apply to conductors that form an integral part of equipment such as motors, motor controllers, and the like. See Sections 300.1(B) and 310.1.

430.22: Single Motor

(A) **General.** Branch-circuit conductors that supply a single motor shall have an ampacity of not less than 125 percent of the full-load current of the motor. Remember that the ampacity in Tables 430.147 through 430.150 will govern the branch-circuit conductor calculations and not the nameplate amperes.

When multispeed motors are used, the ampacity of branch-circuit conductors from the controller to the motor is to be based on the highest full-load current ratings. This will be shown on the motor nameplate. In selecting the ampacity of the branch-circuit conductors running between the controller and the motor, when they are to be energized for some particular speed of that multispeed motor, the current rating speed shall be used to determine selection of branch circuits between the controller and the motor. Wye-start/delta-run motors must have their branch-circuit conductor between their controller, and the motor can be rated at 58 percent of the motor's full-load current.

Note

See Diagram 430.1.

Exception No. 1

Table 430.22(A), Exception, shows the motor nameplate current rating. This shall be used for conductors to motors that are run for short periods of time, or for intermittent, periodic, or varying duty. This is to be followed except when the authority that has jurisdiction gives permission for use with smaller conductors. Notice that there is a variation of from 85 percent to as high as 200 percent in Table 430.22(A). This is why it is necessary to use this table instead of the usual 125 percent.

Exception No. 2

When direct current supplies motors on a rectified source derived from a single-phase power supply, the ampacity of the conductors between the controller and the motor shall be not less than the full-load current rating. The percentage is covered below: (a) 190-percent ampacity is derived from a rectifier bridge of a half-wave, single-phase AC (b) when a full-wye rectifier is used from a single-phase, the percentage of ampacity is 150 percent.

Refer Table 430.22(A) in the ***NEC***.

(B) Separate Terminal Enclosure. The conductors between a stationary motor rated 1 horsepower or less and the separate terminal enclosures permitted in Section 430.145(B) may be smaller than No. 14 but not smaller than No. 18, provided they have an ampacity as specified in Table 430.22(A).

430.23: Wound-Rotor Secondary

(A) Continuous Duty. When a wound-rotor motor is used for continuous duty, the conductors that connect the wound-rotor to its controller shall have a rating of 125-percent ampacity for the full-load current of the secondary of the motor.

(B) Other Than Continuous Duty. When wound-rotor motors are not used for continuous duty, the ampacity of the conductors from the secondary shall be not less than those specified in Table 430.22(A), Exception.

(C) Resistor Separate from Controller. Table 430.23(C) gives the ampacity of the conductors where the controller is mounted separately from the secondary resistor. See Table 430.23(C), which gives the percentage of full-load secondary current from which the ampacity of the conductor that may be used can be figured. Always remember that an oversized conductor is preferable to a conductor that is too small.

430.24: Conductors Supplying Several Motors

Conductors that supply several motors sometimes cause confusion. The basic rule is that 125 percent of the ampacity of the largest motor is taken, plus the ampacity (100 percent) of the other motors connected. This applies when there are two or more motors. It will be recalled that Tables 430.147 through 430.150 give the ampacity used for figuring conductor sizes.

Example

For four 3-phase, 230-volt motors, the ratings in Table 430-1 would be used to calculate the service-entrance conductors:

The preceding example won't apply when one or more motors are used on short-time, intermittent, periodic, or varying duty. In this case, 125 percent of the nameplate full-load current of the largest continuous-duty motor, or the highest current obtained by multiplying the applicable percentage of Table 430.22(A), Exception, by the nameplate full-load current of the noncontinuous-duty motor (whichever is larger), plus the nameplate full-load currents of the other motors, each multiplied by 100 percent or the applicable percentage of Table 430.150, whichever is smaller.

Table 430-1 Ampacity for Service-Entrance Conductors of Four 3-Phase, 230-Volt Motors

Motor Size	*Full-Load Current (from NEC Table 430.150)*
25 hp	64
25 hp	64
5 hp	15
10 hp	28
Total amperes	171
Plus 25% of 64	16
Ampacity to use in figuring the conductors	187

Exception

If there is interlocking in the circuitry that will prevent starting a second motor or a group of motors, the size of the conductor shall be determined from the largest motor or group of motors that is to be operated at any one time. See Chapter 9, Example B.

Cases of interlocking, where only part of the motors may be operated at a given time, are often encountered. In such a case, this is taken into consideration, as the conductors won't be called upon to serve all motors at the same time.

430.25: Conductors Supplying Motors and Other Loads

(A) **Combination Load.** When one or more motors is served from a circuit that also has a lighting and appliance load, the lighting and appliance load is computed from Article 220 or other applicable sections. The conductors shall have sufficient ampacity to carry this appliance and lighting load as calculated, plus the motor-load ampacity as calculated for a single motor in Section 430.22 or for two or more motors in accordance with Section 430.24.

Exception

Section 424.3(B) governs the ampacity of conductors that supply a motor that is operated with electrical space-heating equipment.

(B) Multimotor and Combination-Load Equipment. Section 430.7(D) covers the ampacity of conductors that supply multimotors and combination-load equipment. The ampacity shouldn't be less than the minimum ampacity that is marked on the equipment. Also, you must conform to the section mentioned.

430.26: Feeder Demand Factor

The authority enforcing the *Code* may grant special permission for feeder conductors to be of less capacity than specified in Sections 430.24 and 430.25, which covers motors that operate on duty-cycle, intermittently, or where all motors don't operate at the same time.

The conductors shall have sufficient ampacity for the maximum load determined in accordance with the sizes and numbers of motors supplied and the characteristics of their loads.

430.27: Capacitors with Motors

Capacitors are used for power-factor correction and will change the current that the conductor is required to carry. (Refer to Section 240.21). In addition, they shall also be required to meet the requirements of Sections 460.8 and 460.9.

430.28: Feeder Taps

The ampacity of feeder taps shall be not less than that required in Part II, and they shall terminate in a protective device for the branch circuit. In addition, they shall also be required to meet one of the following:

(A) They are to be enclosed either in a raceway not exceeding 10 feet or in a controller.

(B) The ampacity of these conductors shall be at least one-third the ampacity of the feeder conductors. They shall be protected from physical damage and are not allowed to be over 25 feet in length.

(C) They may have the same ampacity as the feeder conductors.

Exception

Feeder Taps Over 25 Feet Long. If the feeder taps are supplying, for example, a manufacturing area that is of a high-bay type (over 35 feet), conductors from the tap to a feeder may be over 25 feet in length, horizontally. The overall length to the feeder shouldn't be over 100 feet if the following conditions are met: (a) The feeder taps are at least one-third the size of the feeder conductors. (b) A single circuit breaker or switch with fuses is used for termination of the tap conductors conforming with (1) Part IV, if the tap is a branch circuit or (2) Part V, if the tap is a feeder. (c) The conductors are installed in raceways and must not be subjected to physical damage. (d) The taps are continuous from end to end without splices. (e) The tap conductors are not smaller than No. 6 AWG copper, or if aluminum, not smaller than No. 4 AWG. Of course, they may be larger. (f) At no point do the conductors penetrate walls, floors, or ceilings. (g) The tap is at least 30 feet above the floor.

430.29: Constant-Voltage DC Motors—Power Resistors

When the conductors are connected to the motor controller, or to breaking resistors, and they are in the armature circuit, the ampacity shall not be less than that shown in Table 430.29, and the motor full-load current shall be used. When an armature-shunting resistor is used and the power resistor is used for acceleration conductors, you must calculate the ampacity by using the total of both the full-load motor current and the amperage of the armature-shunting resistor.

The ampacity of armature-shunt resistor conductors can be calculated from Table 430.29. Use the shunt-resistor current as the full-load current.

See Table 430.9, "Conductor Rating Factors for Power Resistors."

III. Motor and Branch-Circuit Overload Protection

430.31: General

This part gives the overload protection that is intended to protect motors, motor controllers, and branch-circuit conductors from overheating due to motor overloads. When a motor controller continually trips, don't merely increase the size of the overload protection—something is causing the trouble, such as low voltage, high voltage, high ambient temperatures, unbalanced voltages, or an overload on the motor. Take time to evaluate the cause and remedy the trouble.

Overload is defined as any current operating for a length of time such that it would cause damage or overheating of the motor involved. This does not include short circuits or ground faults. The *Code* sets up the maximum overcurrent protection that is permitted for various motors. This is a proven value and should be followed.

NFPA Standard for Centrifugal Pumps (No. 20) sets up provisions for overcurrent protection for equipment such as fire pumps, where the overcurrent would be the lesser of the hazards. The reference to *NFPA No. 20* makes it a part of the *NEC*.

The provisions of motor circuits over 600 volts are covered in Part IX. Part III does not apply to them.

430.32: Continuous-Duty Motors

(A) **More than 1 Horsepower.** One of the following means shall be used to protect motors rated more than 1 horsepower and used for continuous duty:

(1) Separate overload devices for the motor shall be selected to trip by being rated at no more than the following percent of the current that appears on the motor nameplate for full-load current:

See the *NEC* for the percentage ratings.

Some modification of these values is permitted by Section 430.34.

Each section of a multispeed motor shall be considered separately.

If a separate overcurrent device is located so as not to carry the full current as stated on the motor nameplate (such as for wye-delta starting), the proper percentage of current appropriate for either the selection or the setting of the overcurrent device must be clearly marked on the equipment involved, or come from a table from the manufacturer selecting the overcurrent value taken into consideration.

(2) A thermal integral protector supplied with the motor and approved for use with the motor it protects will be acceptable if it prevents damage or overheating of the motor in case of overload and failure-to-start. In the case where there is a separate current-interrupting device apart from the motor, and this device is actuated by the integral thermal device, it shall be arranged so that the interruption of the control circuit will interrupt the current to the windings. An example of this is a thermal protector built into the winding of a motor that interrupts the control current to a magnetic coil.

The trip current on thermally protected motors shouldn't be greater than the percentage of those that are given for full-load current in Tables 430.180, 430.149, and 430.150. See the *NEC* following this part for the percentages.

If the device opening the current is separate from the motor and its control circuit is operated by a protective device integral with the motor, it shall be such that the opening of the control circuit will interrupt the current to the motor.

(3) An overcurrent device, as part of the motor, that protects the motor from damage due to failure-to-start is permitted if the motor is part of an approved assembly that normally is not subject to overloads.

(4) With motors larger than 1500 horsepower, a protective device shall be included that has embedded detectors that will cause the current to the motor to be interrupted if the temperature rises higher than that marked on the nameplate when the motor is in an ambient temperature of 40°C.

(B) One Horsepower or Less, Nonautomatically Started.

(1) A motor of 1 horsepower or less used for continuous duty that is not permanently installed and that is nonautomatically started, if it is in sight from the controller location, may be protected by a device that has overcurrent protection on the branch circuit that will protect it from overload, short-circuit, and ground-fault. The branch-circuit protective device is not to be larger than specified in this Article 430.

Exception

With any motor just described above operating at 120 volts, overload protection that is not greater than 20 amperes is permitted.

(2) This portion tells us that any motor located out of sight of the controller, and any motor rated at 1 horsepower or less that is permanently installed, has to be protected as in the next part (C).

(C) One Horsepower or Less, Automatically Started. Where a motor of 1 horsepower or less is started automatically, it is to be protected by one of the following means:

(1) It may have a separate overload device that responds to the motor current. Such a device shall have a trip rating that is no more than the following percentages applied to the motor nameplate full-load current rating: see the *NEC*.

On multispeed motors, each winding is considered separately, and modification of this rating is permitted as covered in Section 430.34.

(2) An integral thermal protector approved for use with the motor is considered acceptable. This type of protector prevents damage or overheating of the motor in case of overload or failure-to-start. In the case where a separate current-interrupting device, apart from the motor, is actuated by the integral thermal device, it shall be arranged so that the interruption of the control circuit will interrupt the current to the motor windings.

(3) Overcurrent devices, when integral with the motor and protecting it against damage due to failure-to-start, are permitted (a) if the motor is part of an approved assembly that won't permit the motor to be subject to overloads; or (b) when the assembly is provided with other safety controls (such as the controls for safety from combustion on an oil burner) that protect the motor against damage if it fails to start. If the motor assembly has safety controls for this protection of the motor, it shall be marked so on the nameplate of the assembly, where it will be visible after the assembly is installed.

(4) Motors that have a winding impedance high enough to prevent overheating due to failure-to-start will be considered protected when complying with (B) of this section if they are manually started.

Note

A number of AC motors are less than 1⁄20 horsepower, including clock motors, series motors, and the like, and also some larger motors such as torque motors all come within this classification. Split-phase motors have an automatic switch to disconnect the starting windings, and are not included.

(D) Wound-Rotor Secondaries. The secondary of wound-rotor motors for alternating current that includes conductors, controllers, resistors, and so forth may be protected by means of the motor overload protection.

430.33: Intermittent and Similar Duty

If a motor is used in a situation where it can't be run continuously [see Table 430.22(B)], it is not required to have a motor started. Instead, this motor's overload protection can simply be supplied by the branch circuit or ground-fault protective device.

Note that it must be absolutely certain that the motor will only see intermittent duty for it to come under this section.

430.34: Selection of Overload Relay

This section allows for overload protection where the proper size devices are not available. Where the values specified for overload protection for a motor are not built to the standard sizes or ratings of fuses, nonadjustable circuit breakers, thermal cutouts, thermal relays, heating elements of thermal-trip motor switches, or possible settings of adjustable circuit breakers adequate to carry the load, the next higher size, rating, or setting may be used. Other types of motors shouldn't exceed 130 percent of their full-load current rating. In cases where the overcurrent protection is not shunted during starting time, it shall have sufficient delay to allow starting and acceleration of the load. See the *NEC*, which gives the percentages allowable for overcurrent protection.

430.35: Shunting During Starting Period

(A) Nonautomatically Started. For a nonautomatically started motor the overload protection may be shunted or it is allowable to shunt or cut out the overload protection during start, providing that the shunting device can't be left in use after starting and that the fuses or time-delay circuit breakers are rated or set so as not to exceed 400 percent of the full-load motor current and that these fuses or circuit breakers are not so located as to be in the circuit during starting.

(B) Automatically Started. It is not permissible to shunt or cut out overload protection on automatically started motors.

Exception

The overload protection may be shunted or cut out when starting automatically during the starting period where (1) the motor starting period is greater than the time delay of the available motor overload protective device, and (2) where a listed means is provided that (a) senses the motor rotation and will automatically prevent the shunting or cutout if the motor fails to start, (b) limits the time of shunting of the overload protection or cutout to a point that is less than the locked-rotor rating of the motor that you are protecting, and (c) causes shutdown. You will have to restart the motor manually if running position has not been reached.

430.36: Fuses—In Which Conductor

If fuses are used for motor overload protection, a fuse shall be inserted in the ungrounded conductor.

Fuses shall also be inserted in the grounded conductor if the supply system is a three-wire, three-phase AC system with one phase grounded.

In the past there has been much discussion over what is now covered in the Exception. This will clarify the matter. For instance, if the fuse were not in the grounded phase, serious results could occur when one phase opened on a three-phase irrigation pump motor on a grounded-phase delta system.

Not only does this apply to the above discussion, but also to a delta or wye where one phase conductor is grounded.

430.37: Devices Other Than Fuses—In Which Conductor

Whenever devices other than fuses are installed in series with a motor conductor, the installation must be in accordance with Table 430.37. Usually, this applies to motor-starter overload units. In these cases, the requirements are taken care of by the manufacturer of the starter, who builds starters that require the number of overload units specified in Table 430.37. The buyer needs only to choose the rating of the overload units.

430.38: Number of Conductors Opened by Overload Device

Motor-running overload devices such as are located in the controller, other than fuses, thermal cutouts, or thermal protectors, must open simultaneously a sufficient number of ungrounded conductors so as to stop the flow of current into the motor.

430.39: Motor Controller As Overload Protection

Motor controllers may serve as overload devices for motors under the following conditions: (1) where the number of overload devices comply with Table 430.37; (2) where these units are operable both during starting and running when used with direct-current motors; and (3) when in the running position and used with alternating-current motors. When nonautomatic motor controllers serve as the overload protection for motors, it is recommended that all ungrounded conductors be opened. Always follow *Code* recommendations.

It is clear that practically all three-phase motor installations shall have three overload units or other approved means for overload protection. Many inspection authorities will accept dual-element motor fuses as the other approved means if they are sized properly and the fuse box marked with a warning as to what the replacement is to be. In fact, these dual-element fuses are a very economical form of motor insurance in addition to the three overload units.

430.40: Overload Relay

If motor-running overload protection consists of overload relays and other devices for motor-running overload protection are not capable of opening short circuits, then protection shall be required by means of fuses or circuit breakers, and the ratings or settings shall comply with Section 430.52 or where the same section covers motor short-circuit protection.

Exception No. 1
This allows for approved group installation and marking for the maximum size of fuse or inverse time circuit breaker by which they may be protected.

Note
See Section 430.52, which covers instantaneous-trip circuit breakers for protection against motor short circuits.

430.42: Motors on General-Purpose Branch Circuits
Article 210 covers the connection of certain motors on general-purpose branch circuits but the overload protection of motors on general-purpose branch circuits shall conform to the following:

(A) **Not Over 1 Horsepower.** One or more motors may be connected to general-purpose branch circuit without individual overload protection of the motor or motors provided the requirements of Sections 430.32(B) and (C) and Sections 430.53(A)(1) and (A)(2) are complied with.

(B) **Over 1 Horsepower.** Motors rated at over 1 horsepower, as covered in Section 430.53(A), may be connected to general-purpose branch circuits only where each motor has its own overload protection that meets the overload ratings specified in Section 430.32. Both the controller and the motor overload device shall be approved for group installation with the protective device of the branch circuit to which they are connected. See Section 430.53.

(C) **Cord- and Plug-Connected.** When a motor is connected to a branch circuit by means of a plug and receptacle and it has no individual overload protection as allowed in Section 430.42(A), the rating of the plug and receptacle shouldn't exceed 15 amperes at 125 volts or 250 volts.

Where a motor is connected to a branch circuit by means of a plug and receptacle and is used as a motor or motor-operated appliance and also meets the requirements of Section 430.42(B), the overload device shall be an integral part of the motor or appliance. The rating of the plug and receptacle shall be assumed to determine the rating of the circuit to which the motor may be connected. See Section 210.

(D) **Time Delay.** Motors take considerable current on start. Therefore, it is required that branch-circuit overcurrent devices for motors have sufficient time delay to allow them to accelerate their load.

430.43: Automatic Restarting

Care should be taken in installing motors that automatically restart after the overload protection tips. Such a protection device shouldn't be installed unless it is approved for use with the motor that it protects. When a motor can restart automatically after tripping, it shall be installed in such a way that an injury to persons can't result from its automatic starting.

430.44: Orderly Shutdown

Where a number of motors are used in conjunction with one another, if immediate shutdown occurs, motor overload protective devices could—and often do—introduce additional or increased hazards to a person(s), thus necessitating continued motor operation for safe shutdown of the equipment or process. Motor overcurrent-sensing devices may be installed if they conform to provisions that were covered in Part III of this article, and may be installed through a supervised alarm so that immediate interruption to the motor circuits won't result and corrective action may be taken for an orderly shutdown.

IV. Motor Branch-Circuit, Short-Circuit, and Ground-Fault Protection

430.51: General

This part is intended to specify protection for branch-circuit conductors that supply motors, motor control apparatus, and the motors against overcurrent due to short circuits and ground faults. These provisions are in addition to or amendatory to the provisions of Article 240.

Devices provided for by Sections 210.8, 230.95, and 305.6 are not included in Part IV.

See Part IX, as the provisions of Part IV don't apply to motor circuits over 600 volts, nominal.

430.52: Rating or Setting for Individual Motor Circuit

The motor branch-circuit overcurrent device shall be capable of carrying the starting current of the motor. Short-circuit and ground-fault current will be considered taken care of when the overcurrent protection does not exceed values in Table 430.152. An instantaneous-trip circuit breaker (without time delay) shall be used only if it is adjustable, if it is a part of a combination controller that has overcurrent protection in each conductor, or if the combination has been approved. An instantaneous-trip circuit breaker may have a damping device to limit the inrush current when the motor is started.

In case the values for branch-circuit protective devices determined by Table 430.152 don't correspond to the standard sizes or ratings of fuses, nonadjustable circuit breakers, thermal devices, or possible settings of adjustable circuit breakers adequate to carry the load, the next higher size rating or setting may be used. See Section 240.6 for standard ratings.

There are exceptions where the overcurrent protection as specified in Table 430.152 won't take care of the starting current of the motor.

Exception No. 1

If the values of the branch-circuit, short-circuit, and ground-fault protection devices determined from Table 430.152 don't conform to standard sizes or ratings of fuses, nonadjustable circuit breakers, or possible settings on adjustable circuit breakers—and they are capable of adequately carrying the load involved—then the next higher setting or rating will be permitted.

Exception No. 2

If the ratings shown in Table 430.152 are not sufficient for the starting current of the motor, then (a) when no time delay fuses are used and don't exceed 600 amperes in rating, it is permitted to increase the fuse size up to 400 percent of the full-load current, but not over 400 percent; (b) time-element fuses (dual-element) are not to exceed 225 percent of full-load current, but they may be increased to this percentage; (c) inverse time-element breakers may be increased in rating, but they (1) shouldn't exceed 400 percent of full-load current or 100 amperes or less, or (2) may be increased to 300 percent to a full-load current greater than 100 amperes; and (d) fuses of ratings from 601 to 6000 amperes may be increased, but in no case shall they exceed 300 percent of the full-load rating of the motor involved.

Exception No. 3

In accordance with Section 240.3, Exception No. 1, torque-motor branch circuits are to be protected at the motor nameplate current rating.

Note

Standard rating of fuses and circuit breakers is given in Section 240.6 of the *NEC*.

If instantaneous-trip breakers are not adjustable and are a part of the motor controller-covering overload and also short-circuit and ground-fault protection for each of the conductors, a motor short-circuit protector is permitted in lieu of the listed devices in Table 430.152. This is where short-circuit protectors are a part of a combination controller that has both motor-overload protection and short-circuit and ground-fault protection. These must be present in each conductor. There is a provision that they shall operate at not more than 1300 percent of the full-load motor current. Circuit breakers with instantaneous-trip or motor short-circuit protectors shall be used only as part of the combination motor controller that provides protection that is coordinated to the motor branch-circuit overload short-circuit and ground-fault protection.

Exception

If the specified setting in Table 430.152 is found to be insufficient for the starting current of the motor, the setting on an instantaneous-trip circuit current may be increased, provided that in no instance will it exceed 1300 percent of the motor full-load current rating for types B, C, and D motors, and 1600 percent for type E motors. Engineering evaluation is required for B, C, and D motors over 800 percent, and type E motors above 1100 percent.

For multispeed motors, a single overcurrent-protective device may be used. The size of the protective device must be based on the motor that has the highest current rating. In addition to this protection, each winding of the motor must be provided with running overload protection. Also, the conductors to each winding must be sized based on the winding with the highest current rating.

When the short-circuit and ground-fault protective device ratings are shown in the manufacturer's overload relay table for use with the motor controller involved or as otherwise marked on the equipment, they shouldn't be allowed to exceed the highest values shown above.

Note

Refer to Diagram 430.1.

In lieu of devices listed in Table 430.152, suitable fuses may be used instead for adjustable-speed drive systems. This is only if markings for replacement are provided adjacent to the fuses.

Where maximum protective device ratings are shown in the manufacturer's heater table for use with a marked controller or are otherwise marked with the equipment, they shouldn't be exceeded even if higher values are allowed as shown above.

Torque Motors

Branch circuits that supply torque motors must be protected at the motor nameplate current rating. (See Section 240.3). Note also that the full-load current of a torque motor is the same as the locked-rotor current, as torque motors are designed to operate at speeds far below synchronous speed.

430.53: Several Motors or Loads on One Branch Circuit

Under conditions specified in (A), (B), or (C) of the following list, two or more motors or one or more motors may be run from the same branch circuit.

(A) Not Over 1 Horsepower. Two or more motors of 1 horsepower or less that don't have a full-load rating of more than 6 amperes may be used on branch circuits that are protected at not more than 20 amperes at 120 volts or less, or on branch circuits rated at not more than 15 amperes at 600 volts or less. Individual running-overcurrent protection for these motors is not required unless specifically called for in Section 430.32.

Short-circuit and ground-fault protective devices for branch circuits shall have the ratings marked on any of the controllers, and this rating is not to be exceeded.

(B) If Smallest Motor Protected. If it is determined that the branch-circuit, short-circuit, and ground-fault protective device won't open under the most severe normal operating conditions that might be encountered, and the branch-circuit protective device is not larger than allowed in Section 430.52, then two or more motors, each having individual running-overcurrent protection may be connected to one branch circuit.

(C) Other Group Installation. On two or more motors of any rating, or one or more motors plus other load(s), each motor is to have individual overcurrent protection, but they may be connected to one branch circuit where (1) the motor controller(s) and the overcurrent device(s) are a factory assembly and the short-circuit ground-fault protective devices may be a part of the assembly or may be specified by marking the assembly and installing separately; or (2) the branch-circuit overload protection, short-circuit and ground-fault protection devices, and the motor controller(s) and motor overload device(s) are field-installed by use of a separate assembly that has been listed for this purpose, and are to be provided with the instructions from the manufacturer where the use with each other is provided; or (3) all the following conditions are met:

(1) Each motor overload device (listed for group installation) is tested with a specific maximum rating of the inverse circuit breaker or fuse.

(2) The motor controller used is listed for group installation, and there is a specific maximum rating for the fuse or circuit breaker.

(3) Each circuit breaker is listed for group installation and is of the inverse-time-type breaker.

(4) The branch circuit using fuses or inverse-time circuit breakers has a rating not exceeding that covered in Section

430.52. This rating is for the largest motor that is connected to the branch circuit; to this must be added the sum of the full-load current ratings of the other motors involved, and is for the rating of other loads that are to be connected to the branch circuit. It is permitted to increase the maximum rating of fuses or circuit breakers to a value that is not over that permitted by Section 240.3, Exception No. 1, if the results of the calculation come to less than that of the supply conductors.

(5) These branch-circuit fuses or inverse-time circuit breakers are not larger than Section 430.40 allows for thermal cutouts or overload relay that is protecting the smallest motor of the group.

The *NEC* refers to Section 110.10.

(D) Single Motor Taps. With group installations of motors as described above, the conductors of any tap supplying a single motor shall comply with the following: (1) All of the conductors to the motor shall have an ampacity of at least that of the branch-circuit conductors; and (2) conductors to a motor shouldn't have an ampacity of less than one-third that of the branch-circuit conductors, and should meet the requirements of Section 430.22, which requires an ampacity of at least 125 percent that of the full-load current of the motor. The conductors to the motor-running protection shouldn't exceed 25 feet in length and shall be protected from physical damage.

Any tap supplying a single motor shouldn't be required to have individual branch-circuit, short-circuit, and ground-fault protective devices, provided they comply with the conditions outlined above in this section.

430.54: Multimotor and Combination-Load Equipment

Multimotor and combination-load equipment is covered in Section 430.7(D). The rating of the branch-circuit, short-circuit, and ground-fault protective devices for this purpose shouldn't exceed the rating on the equipment provided for this purpose.

430.55: Combined Overcurrent Protection

The overcurrent protection may be combined in one piece of equipment with the branch-circuit overcurrent protection, ground-fault protection, and motor-running overcurrent protection when the setting or rating of the devices provides the 115-percent and

125-percent overcurrent-running protection as specified in Section 430.32.

430.56: Branch-Circuit Protective Devices—In Which Conductor

Overcurrent devices shall open all ungrounded conductors of the circuit as required in Section 240.20. They may also open the grounded conductor if it is opened simultaneously with the ungrounded conductors.

430.57: Size of Fuseholder

Where fuses are used in branch circuits for motor protection, the fuseholders shouldn't be smaller than required to accommodate fuses as specified by Section 430.152. An exception to this would be when time-delay fuses are used that have appropriate characteristics for motors, in which case the fuseholders may be smaller, but check to make certain that the switch and fuseholders are rated in a horsepower rating to accommodate the motor. Fuseholder adapters may be used to reduce the size of the fuseholder to adapt to the time-delay motor fuses, but this is not always the best policy, as they may be easily removed and a larger size inserted. Check the horsepower rating of fusible switches. See the *NEC* for the Exception.

430.58: Rating of Circuit Breaker

Sections 430.52 and 430.110 cover the current rating of a circuit breaker that is used for motor branch-circuit, short-circuit, and ground-fault protection.

V. Motor Feeder Short-Circuit and Ground-Fault Protection

430.61: General

The provisions of Part V specify overcurrent devices intended to protect feeder conductors supplying motors against overcurrents due to short circuits or grounds.

Note

Refer to Chapter 9, Example No. 8.

430.62: Rating or Setting Motor Load

(A) Specific Load. A feeder that supplies a fixed motor load must be calculated as far as the conductor ampacity from Section 430.24, which allows a carrying capacity of 125 percent of the rating of the full-load current of the largest motor plus 100 percent of the full-load current rating of the other motors. This conductor shall further be protected by overcurrent protection that shouldn't be greater than allowed for the protection of the largest motor [as figured from Table 430.152, or Section 440.22(A) for hermetic refrigerant motor-compressors] plus the sums of the full-load current of the other motors.

If two motors of the same rating are used, only one is considered the larger. This motor is then used to calculate the excess rating and the other motor the 100 percent rating.

Where two or more motors are started simultaneously, the feeder sizes and overcurrent protection must be figured accordingly and will require higher ratings. See Example No. 8 in Chapter 9.

If instantaneous-trip circuit breakers are used for motor short-circuit or ground-fault protection, the rating of that device can be the same as the rating of a dual element fuse. (Such fuses interrupt short circuits at many times their stated rating. For example, a 100-amp dual-element fuse may not interrupt a very brief short circuit unless it reaches a current level of 1200 amps). Since a 100-amp instantaneous-trip circuit breaker (which would trip at exactly 100 amps in the same situation) would not allow its motor to start (it would trip immediately, because of the starting surge), a 1200-amp instantaneous-trip circuit breaker can be used.

(B) Other Installations. On large-capacity installations (for example, large industrial plants), heavy-capacity feeders are usually or should be installed to provide for future additions of load or changes that might be made. The rating or setting of the feeder-protective devices may be based on the rated ampacity of the feeder conductors.

430.63: Rating or Setting—Power and Light Loads

Where a feeder carries a motor load in addition to lighting and/or appliance loads, the ampacity of the lighting and/or the appliance loads must be figured as in Articles 210 and 220. To this is added the capacity for the motor as figured in Section 430.52 or Section 430.62. These totals are combined to determine the ampacity of the feeder conductors and the overcurrent protection for the feeders.

VI. Motor Control Circuits

430.71: General

Part VI applies to special conditions of motor control circuits, and thus applies to the general requirements.

Note
The *NEC* refers to Section 430.9(B).

Definition of Motor Control Circuit: The control circuit of a motor apparatus controlling the operation of the motor or system carries only the electrical signals that direct the controller to stop, go, jog, and so on, and does not contain the main power of the circuit.

430.72: Overcurrent Protection

(A) General. The control circuit that originates at the motor-branch circuit by being tapped there from the load side and short-circuit and ground-fault protective device(s) functioning to control the motor(s) connected to that branch circuit shall be protected against overcurrent, complying with Section 430.72. The motor control circuit is not considered a branch circuit, and supplementary overcurrent protection separate from the branch-circuit overcurrent protection is used. A motor control circuit other than tapped control circuit is to be protected against the overcurrents covered in Section 725.12 or Section 725.35, whichever is applicable.

(B) Conductor Protection. The values specified in Column A of Table 430.72(B) shouldn't be exceeded for the overcurrent protection of the conductors.

Exception No. 1
When the control circuit conductors don't extend beyond the motor-control-equipment enclosure, they must have only short-circuit and ground-fault protection, and this can be covered by the motor branch-circuit, short-circuit, and ground-fault protective device(s), if the rating of the protective devices does not exceed the value as determined from Column B of Table 430.72(B).

Exception No. 2
When the control circuit conductors extend beyond the controller, they will be required to have only short-circuit and ground-fault protection. They may also be protected by the motor branch-circuit, short-circuit, and ground-fault protective devices when the ratings of the protective devices don't exceed those of Column C of Table 430.72(B).

Exception No. 3
When control conductors are supplied by a single-phase transformer on the secondary side and have only a two-wire, single-voltage secondary, they may be protected by the overcurrent protection on the primary side of the transformer. This is if they are provided by

overcurrent protection not exceeding the value obtained by multiplying the appropriate minimum rating of the overcurrent device for the secondary conductor as provided in Table 430.72(B) by the secondary-to-primary voltage ratio. If the secondary conductors are other than two-wire, they are not permitted to be protected by the primary overcurrent protection.

Exception No. 4
Control conductor circuits are required to have only short-circuit and ground-fault protection, and may be protected by the motor short-circuit and ground-fault protection device. Also, they may be protected by the branch-circuit protection to the motor that is short circuit and ground fault protected if the opening of the control circuit would cause a hazardous condition (as in a control circuit to a fire pump). See Table 430.72(B) and the notes following.

(C) Control Circuit Transformer. When a control circuit transformer is used, this transformer protection is covered in Article 450.

Exception No. 1
If the control circuit transformer is rated at less than 50 VA and is a part of the motor control and located in the enclosure for the motor controller, it can be protected by primary overcurrent devices, by impedance-limiting devices, or by other inherently protective means.

Exception No. 2
If the control-circuit transformer current is less than 2 amperes, the overcurrent device on the primary side is not to exceed 500 percent of the primary current rating and may be used.

Exception No. 3
Refer to Section 725.11(A). This covers transformers that supply Class 1 power-limited circuits and transformers that supply Class 2 and Class 3 remote-control circuits. See Article 725, Part III, in your *NEC*.

Exception No. 4
If other means of protection are provided.

Exception No. 5
Where opening of the control circuit would create a hazard, such as for a fire pump, the overcurrent protection may be omitted.

430.73: Mechanical Protection of Conductor
Where damage to a control circuit would constitute a hazard, all conductors of such remote-control circuit shall be installed in a

raceway or be otherwise suitably protected from physical damage outside the control device itself. Damage to control wiring might cause an important operation to stop, or might cause a machine to unintentionally start, causing a hazard.

Where one side of a motor control circuit is grounded, the control circuit shall be arranged in such a way that an accidental ground other than the intended ground elsewhere in a circuit won't start the motor. Most circuits have a grounded conductor along with the ungrounded conductor. Extreme care should be exercised to connect the stop and start buttons of a magnetic starter so that it won't energize the magnetic coil and start the motor should one of the conductors become grounded. The same could very easily happen on an ungrounded 480-volt, three-phase circuit that uses a 115-volt grounded system to energize the controls. Grounding one side could cause the motor to start. This item is very important, but so often little attention is paid to the matter.

430.74: Disconnection

(A) **General.** Motor control circuits are to be arranged in such a way that if the disconnecting means is in open position, the control circuit will also be open. Two or more separate disconnecting means will be used. One will open the power to the motor controller itself, and the other disconnecting means will open the motor controller circuit(s) permitting power supply. Where more than one disconnecting means is used, they shall be mounted adjacent to one another.

Author's Note

It is the author's recommendation that the two disconnecting means be marked indicating that both supply sources are disconnected when either of the disconnecting devices is opened.

Exception No. 1

If all the following conditions are met and if there are more than 12 motor-control circuit conductors that must be disconnected, this exception permits the disconnecting means to be located elsewhere than immediately adjacent to each other: (a) Part X of this article permits the live parts to be accessible only to qualified persons. (b) There shall be a warning sign posted on the door or enclosure cover on the outside of each disconnecting means that permits access to live parts of the motor control circuit(s). This warning shall indicate that the disconnecting means of the control are remote from the controller of the motor, giving the location and identification of each disconnect. If, as permitted in Sections 430.132 and 430.133, live parts are not in equipment enclosures, an additional warning sign(s) is to be located so as to be visible to persons working in the area of the live parts.

Exception No. 2

Conditions (a) and (b) of Exception No. 1 above shall be adhered to where the opening of one or more of the motor-control circuit disconnects might result in an unsafe condition for personnel or property.

(B) Control Transformer in Controller. A transformer or other device is often used to obtain a reduced voltage for the control circuit and is located in the controller. Such a transformer or other device shall be connected to the load side of the disconnecting means for the control circuit, so that when the motor disconnecting means is opened, the control circuit is also deenergized.

VII. Motor Controllers

430.81: General

The intent of Part VII is to cover suitable controllers for all motors.

(A) Definition. For the definition of *controller*, the *NEC* refers to Article 100. In this portion, the purpose of a controller is any switch or other device that is used to start or stop a motor.

(B) Stationary Motor of ⅛ Horsepower or Less. On motors less than ⅛ horsepower, where mounted stationary or continuously, the construction must be such that if they fail to start, they don't burn out and the branch circuit to which they are connected will be the overload protection for them.

(C) Portable Motor of ⅛ Horsepower or Less. Motors of ⅛ horsepower or less may use a plug -and-cord connection for a disconnecting means.

430.82: Controller Design

(A) Starting and Stopping. Every motor controller shall be capable of both starting and stopping the motor that it controls. Controllers for alternating-current motors shall be capable of interrupting the stalled-rotor current. Stalled-rotor current was covered in the first part of this article. Recall that this current is much higher than the running current and, unless the controller is capable of interrupting this larger current, damage may result to the controller. Controller listings in catalogs show the horsepower rating and the voltage.

(B) Autotransformer. Autotransformer controllers are alternating-current devices that incorporate an autotransformer to reduce the voltage and increase the current at start. This will reduce the current drawn from the branch circuit at start.

Autotransformer controllers shall have an OFF, RUN, and at least one START position. More than one start position may be incorporated to accelerate the motor in steps. The design shall be such that the controller can't remain in the start position, which would render the overcurrent protection inoperative.

(C) Rheostats. Rheostats shall conform to the following:

(1) Motor-starting rheostats shall be designed in such a way that they won't remain in any of the starting positions or segments when the starting handle is released, but will return to the off position. In addition, the design should be such that the first contact won't engage any part of the rheostat.

(2) Rheostats used as motor-starts on direct current that operate from a constant voltage supply are to be equipped with a device that, should the voltage drop, will release the starter before the motor speed has dropped to less than one-third of its normal speed.

430.83: Rating

Controllers shall be marked in horsepower and shouldn't be rated at less horsepower than the motor being served.

Exception No. 1

The controllers for type E motors rated more than 2 horsepower must be marked with a locked-rotor rating no less than that shown in Table 430.51(B). In addition, it must be marked as suitable for controlling type E motors and must have a rating at least 1.3 times that of the motor it controls.

Exception No. 2

A general-use switch may be used as a controller for motors rated at 2 horsepower or less and 300 volts or less, provided that it has a current rating of at least twice the full-load current rating of the motor with which it is to be used.

On AC circuits it is permissible to use an AC general-use snap switch, but not a general-use AC/DC snap switch, as the controller for motors of 2 horsepower or less and 300 volts or less, providing that the full-load current rating of the motor does not exceed 80 percent of the current rating of the snap switch.

Exception No. 3
Only inverse-time circuit breakers rated in amperes may be used in branch circuits supplying a controller. If this circuit breaker is also used for overload protection, it must conform to the parts of this article that govern overload protection.

Exception No. 4
Torque motors shall have a continuous-duty and also a full-load current rating that is not less than the full-load rating marked on the motor. If the motor controller is rated in horsepower and not marked as covered above, you shall determine the ampere or the horsepower rating by using Tables 430.147, 430.148, 430.159, and 430.150.

Exception No. 5
Horsepower ratings are not required when devices allowable under 430.81(B) and (C) are used as controllers.

430.84: Need Not Open All Conductors
Section 430.111 covers controllers that serve as both the controller and the disconnecting means, and these shall open all ungrounded conductors. Where the controller does not serve as both the controller and the disconnect, it need not open all of the conductors, but only those necessary to start and stop the motor.

430.85: In-Grounded Conductors
There is nothing that prohibits the opening of the grounded conductor if one is used to supply a motor. However, if the grounded conductor is opened, all ungrounded conductors shall be simultaneously opened.

430.87: Number of Motors Served by Each Controller
Each motor must have a separate controller.

Exception
Under one of the following conditions, a single controller of 600 volts or less shouldn't be rated less than the sum of the horsepower ratings of all the motors that arc in a group, and they may be served by the group of motors as mentioned in the beginning of this paragraph: (a) A single machine or a piece of apparatus, such as metal or woodworking equipment, crane, hoist, and so on, might require several motors in its

operation, in which case a single disconnecting means may serve all the motors. Other conditions of this article, such as overcurrent protection, must also be met. (b) In Section 430.53(A), more than one motor was permitted to be protected by one overcurrent device if certain requirements as to current and voltage were met. (c) If a group of motors is located in a single room, and all are within sight of the disconnecting means, then a single disconnecting means may be used.

430.88: Adjustable-Speed Motors

Some motors, especially shunt and compound-wound DC motors, have speed adjustment controlled by the regulation of the field current. Where such is the case, the controller shall be designed such that the motor can't be started with a weakened field unless the motor is specifically designed for such starting. Starting a motor with a weakened field is dangerous and may cause a current high enough to burn out the armature.

430.89: Speed Limitation

There are three types of motors that must be provided with speed limitation.

(A) Separately excited DC motors.

(B) Series motors.

(C) Motor generators and converters.

Exception No. 1

When the motor is attached to a load or device that will effectively limit the speed.

Exception No. 2

If the machine will always be under manual control by a qualified operator.

430.90: Combination Fuseholder and Switch As Controller

When a combination fuseholder and switch is used as a motor controller, it shall be such that the fuseholder will permit only the size of fuse specified in Part III of this article covering overload protection. A fuseholder of larger capacity is permissible if fused down to proper value.

Exception

Where time-delay fuses that have the appropriate motor-starting characteristics are used, fuseholders that are smaller than those covered in Part III of this article will be permitted.

430.91: Motor Controller Enclosure Types

Table 430.91 gives the basis for the selection of enclosures in nonhazardous locations for specific purposes. These enclosures are not intended to be used for protection for the conditions such as condensation, icing, corrosion, or contamination that might be likely to occur in the enclosures or to enter through openings that are not sealed, or through the conduit. When any of these conditions exists, the installer or user must give it special consideration.

See the *NEC* for Table 430.91, "Motor Controller Enclosure Selection Table."

VIII. Motor Control Centers

430.92: General

A motor control center is an assembly of motor controllers having a common power bus.

430.94: Overcurrent Protection

Overcurrent protection shall be provided for these units based upon the current rating of the power bus and the requirements of Article 240. This must be provided either by an overcurrent device located ahead of the motor control center or by a main overcurrent protection device that is within sight of the center.

430.95: Service-Entrance Equipment

When motor control centers are used as service-entrance equipment, they must have a main disconnect that disconnects all ungrounded conductors. If necessary, a second service disconnect may be used to feed additional equipment. If a grounded conductor is used, a main bonding jumper must be installed.

430.96: Grounding

All sections of motor control centers must be bonded together with a conductor or bus sized per Section 250.122. All equipment-ground conductors must be connected to this conductor or bus. (A bus is almost always used).

430.97: Bus Bars and Conductors

- **(A)** **Support and Arrangement.** Bus bars must be protected and held rigidly in place. Conductors shouldn't be installed in vertical sections of the motor control center unless necessary or when protected from the bus bars by a barrier.
- **(B)** **Phase Arrangement.** The phase arrangement for three-phase systems must be A, B, C from front to back, top to bottom, or

left to right. An exception is made for back-to-back units with vertical buswork.

(C) **Wire Bending Space.** As per Article 312.

(D) **Spacings.** No less than shown in Section 430.97.

(E) **Barriers.** These must be used to isolate service bus bars in all service-entrance centers.

430.98: Marking

See the *NEC*.

IX. Disconnecting Means

430.101: General

The intent of Part IX of this article is the requirement of the disconnecting means for the disconnection of motors and controllers from the circuit.

Note

The *Code* refers us to Diagram 430.1.

Note

Identification of the disconnecting means can be found in Section 110.22.

430.102: Location

(A) **Controller.** The disconnecting means shall be located in sight of the controller.

Exception No. 1

In circuits using over 600 volts for a motor, the disconnecting means may be out of sight of the controller, if the controller is plainly marked with a warning label that gives the location and identifies the disconnecting means that can be locked in an open position.

Exception No. 2

Where a multimotor continuous process machine is used, a single disconnecting means may be located by the group of coordinated controls that are mounted and adjacent to each other.

(B) **Motor.** The disconnecting means must be mounted within sight of the motor and driven machinery.

Exception

As provided in Section 430.102(A), if the disconnecting means is capable of being locked in an OFF position, it may be out of sight of the motor.

A great many inspectors will probably require, and rightly so, that the disconnecting means be marked as to what motor-driven machinery it disconnects. They might also require a sign at the motor or machinery stating that the disconnecting means is to be locked in an open position. This is only common sense, in that the main purpose of this ruling is safety to anyone who might perform mechanical or electrical work on the motor or the driven machinery.

(C) Phase Converter. Any phase converter must have a disconnecting means within sight of its location.

430.103: To Disconnect Both Motor and Controller

All ungrounded conductors to the motor and controller supply conductors shall be energized when the disconnecting means is in the OFF position. The disconnecting means shall be designed so that no pole can be operated independently. The disconnecting means may also be located in the control enclosure. See Section 430.113. This section covers those installations where energy is received from more than one source.

430.104: To Be Indicating

Any means of disconnection shall readily indicate whether it is OFF or ON.

430.105: Grounded Conductors

One pole of the disconnecting means may open the grounded conductors if one is used in the circuit. However, if it opens the grounded conductor, it shall also simultaneously open all conductors of the circuit.

430.106: Service Switch As Disconnecting Means

If the installation consists of only one motor, the service switch may be used as a disconnecting means if it meets all of the requirements of this article.

430.107: Readily Accessible

There shall be no obstructions in the way of ladders needed in the operation of the disconnecting means.

430.108: Every Switch

This section requires that every disconnect switch in the motor branch circuit (between the feeder and the motor) comply with the requirements stated in Sections 430.109 and 430.110.

2005

430.109: Type

The disconnecting means may be a listed motor switch rated in horsepower, a circuit breaker, or a molded-case switch.

(A) General.

(1) Motor Circuit Switch. These must be listed and rated in horsepower. For Design E motors, these switches must be marked as listed for use with Design E motors, or must be rated 1.4 times the motor rating up to 100 horsepower, and 1.3 times the motor rating above 100 horsepower.

(2) Molded Case Circuit Breakers. Listed breakers may be used.

(3) Molded Case Switch. Listed switches may be used.

(4) Instantaneous-Trip Circuit Breakers. Listed devices may be used.

(5) Self-Protected Combination Controller. Listed devices may be used.

(6) Manual Motor Controllers. Devices listed and marked "Suitable as motor disconnect" may be used between the final branch-circuit short-circuit ground-fault protective device and the motor.

(7) System Isolation Equipment. This special equipment may serve as a disconnecting means if it is listed and located on the load side of the disconnecting means and overcurrent protection.

(B) Stationary Motors ⅛ Horsepower or Less. The branch-circuit protective device is permitted to be the disconnecting means for these motors.

(C) Stationary Motors of 2 Horsepower or Less. For such motors operating at 300 volts or less, the disconnecting means can be one of the following:

(1) A general-use snapswitch with twice the full-load rating of the motor.

(2) An AC general-use snapswitch (on AC circuits only) that is rated at least 125 percent of the motor full-load current.

(3) Or, a listed manual motor controller that is marked "Suitable as motor disconnect" and rated to at least the rating of the motor.

(D) Autotransformer-Type Controlled Motors. General-use snapswitches may be used for these motors rated between 2 and 100 horsepower, provided that the motor drives a generator with overload protection, the controller is able to interrupt the locked-rotor current of the motor, the overload protection does not exceed 125 percent of the motor's full-load current, and separate fuses or an inverse-time breaker are provided, and are rated or set at no more than 150 percent of the motor's full-load current.

(E) Isolating Switches. These devices may serve as the disconnecting means for motors rated over 40 horsepower DC or 100 horsepower AC. Such switches must be clearly marked "Do Not Operate Under Load."

(F) Cord- and Plug-Connected Motors. Cord-and-plug connections may serve as a disconnecting means if the plug and receptacle are horsepower rated and are rated to at least the motor's horsepower rating. For Design E motors over 2 horsepower, the plug and receptacle must have a rating 1.4 times the motor's rating. Horsepower-rated plugs and receptacles are not required for room air conditioners (Section 440.63), appliances (Section 422.33), or for portable motors rated ⅓ horsepower or less.

(G) Torque Motors. A general-use switch may serve as the disconnecting means for these motors.

430.110: Ampere Rating and Interrupting Capacity

(A) General. The disconnecting means must have an interrupting capacity of at least 115 percent of the full-load rating of the motor, for motors, and circuits rated 600 volts, nominal, or less.

(B) For Torque Motors. The disconnecting means for a torque motor shall be at least 115 percent of the nameplate rating on the motor.

(C) For Combination Loads. The following will give a method of determining the disconnecting means size where two or more motors are used together or where one or more motors are in combination with other types of load such as resistance heaters. If the combined loads can come on at the same time, the signal disconnecting means shall be calculated by the ampere and the horsepower ratings of the combined loads.

(1) In determining the size of the disconnecting means, the sum of all currents, including the resistance loads at the full-load condition, and also at the locked-rotor condition and the combined full-load current condition and combined locked-rotor current so obtained shall be considered as follows:

From Table 430.148, 430.149, or 430.150, select the full-load current of each motor. The combination of these full-load currents plus the other loads added to them in amperes will determine the full-load current of the combined motor.

From Tables 430.151(A) or (B), select the locked-rotor current equivalent for the horsepower rating of each motor. Locked-rotor current ratings are to be added to the other combined loads. If two or more motors and/or other loads can't be started simultaneously, the appropriate combinations will determine the combination locked-rotor rating and may be used in making a decision on the locked-rotor current for the loads that come on simultaneously.

Exception

Where the total concurrent load or part of the load is a resistance load, and if the disconnecting means is a switch that is rated in horsepower and amperes, the switch used may have a horsepower rating of not less than the total combined load of the ampere rating of the switch that is not less than the sum of the locked current and the motor(s) plus the resistance load.

(2) The ampere rating of the disconnecting switch is not allowed to be less than 115 percent of the total combination of all currents at the full-load condition as was determined above in (C)(1).

(3) For small motors that are not covered by Tables 430.147, 430.148, 430.149, or 430.150, the locked-rotor current may be assumed to be six times the full-load currents of the small motors.

(D) Phase Converters. To properly size the single-phase supply disconnecting means for phase converters, the following steps must be taken:

(1) Add the locked-rotor currents (not full-load currents) of each motor that will be connected to the converter.

(2) To this amount, add the ampere rating of all other loads.

(3) Multiply this amount by 1.73, which will give you the required rating.

430.111: Switch or Circuit Breaker As Both Controller and Disconnecting Means

A switch or circuit breaker can be used as both the controller and the disconnecting means, as long as it has overcurrent protection, opens all ungrounded conductors, and is of one of the following types:

(A) An air-brake switch, operated by a lever or handle.

(B) An inverse-time circuit breaker, operated by a lever or handle.

(C) An oil switch, rated at less than 600 volts or 100 amps, except when special permission is given.

430.112: Motors Served by Single Disconnecting Means

Every motor must be provided with its own disconnecting means.

Exception

A single disconnect may supply more than one motor when it is sized according to Section 430.110(C) and falls under one of these conditions: (a) where several motors drive separate parts of a machine, such as metal or wood-working machines; (b) as allowed under Section 430.53(A); or (c) where the motors are all within one room, and within sight of the disconnecting means.

430.113: Energy from More Than One Source

If equipment receives its energy from more than one source, such as the control circuit from an isolation transformer of a different voltage from that of the motor or from a DC source and the motor from an AC source, the disconnecting means for each source (and there must be a disconnecting means for each) shall be located adjacent to each other. Each source may have a separate disconnecting means.

Exception No. 1

If a motor receives power from more than one source, the disconnecting means that is used for the main power supply to the motor is not required to be located immediately next to that motor, but it must be capable of being locked in the open position.

Exception No. 2

If the control circuit is derived from a Class 2 circuit that is remote and conforms with Article 725, and this Class 2 circuit has not more than 30 volts and is isolated and ungrounded, a separate disconnecting means is required.

X. Over 600 Volts, Nominal

430.121: General

This part covers motors and controllers operating at more than 600 volts. There are special hazards that are encountered at these voltages and must be taken into consideration. Other requirements for circuits and equipment that operate at over 600 volts are covered in Article 490.

430.122: Marking on Controllers

The control voltage shall be marked on the controller in addition to the marking that is required in Section 430.8.

430.123: Conductor Enclosure Adjacent to Motors

Flexible metal conduit including liquid-tight flexible metal conduit with listed fittings not exceeding 6 ft in length may be used for raceway connection to the motor terminal housing. Be certain that if required, the proper size equipment-grounding conductor is also in the wiring methods mentioned above and size it according to Table 250.95. In many places in industry, the equipment-grounding conductor may be required regardless of the *Code* requirement.

430.124: Size of Conductors

The setting at which the motor overload-protective device(s) is set will determine the ampacity of the conductors supplying the motor.

430.125: Motor Circuit Overcurrent Protection

(A) General. High-voltage circuits for each motor must include protection that is coordinated to interrupt automatically any overloads or fault currents in the motor, and also the control apparatus for the motor and the conductors to the motor.

Exception

When one particular motor is vital to the operation of the entire plant, and if the motor's failing to operate means greater hazards to persons, then there will be a sensing device(s) that will be permitted to operate a supervised annunciator alarm. These are permitted instead of interrupting the motor circuit.

(B) Overload Protection.

(1) A thermal protector within the motor, or external devices sensing the current, shall be used in each motor to prevent damage by overheating caused by motor overloads or failure-to-start.

(2) The overcurrent device protecting the motor is considered to protect the secondary circuits of wound-rotor, alternating-current motors, including the controllers and resistors that are used in the secondary circuit and rated for that application.

(3) All ungrounded conductors shall be opened at the same time by the overload-interrupting device.

(4) Devices to sense overload shouldn't automatically reset after tripping unless the resetting of the overload-sensing device won't cause automatic restarting of the motor, or unless there can be no hazards to persons from the restarting of the motor or by the machinery it runs.

(C) Fault-Current Protection.

(1) The following means shall be used to provide fault-current protection in each motor circuit: (a) The circuit breaker may be used if it is of suitable type and proper rating, and is so arranged that it can be serviced without causing a hazard. All ungrounded conductors must be opened at the same time by the circuit breaker. Internal or external sensing devices shall be permitted for the circuit breaker to sense fault currents. (b) Fuses arranged so that they can't be serviced while energized may be used if of suitable type and rating and if they are placed in each ungrounded conductor. There shall be a suitable disconnecting means for the fuses, although they can also serve as the disconnecting means if listed for that use.

(2) Circuits shouldn't be automatically closed by fault-current interrupters.

Exception

This permits automatically closing if the opening was caused by a transient fault current, or if, in closing the circuit, it does not create a hazard to people.

(3) The same device may be used for overload and fault-current protection.

430.126: Rating of Motor Control Apparatus

The ultimate trip current of overcurrent relays or other protective devices can't be more than 115 percent of the controller's continuous-current rating. If the disconnecting means is separate from the controller, its rating must be at least as high as the ultimate trip current of the overcurrent relays.

430.127: Disconnecting Means

The disconnecting means for a controller must be capable of being locked in the open position.

XI. Protection of Live Parts—All Voltages

430.131: General

Part XI requires that live parts be adequately protected from possible hazards.

430.132: Where Required

Any live parts exposed on motors or controllers that operate at 50 volts or more between terminals must be guarded from accidental contact by an enclosure, or by location as follows:

- **(A)** They may be enclosed in a room or enclosure that is accessible only to qualified persons. A "*qualified person*", as defined in Article 100, is someone who is familiar with the operation and construction of the equipment and with any hazards that might be involved.
- **(B)** They may be installed on a suitable balcony, gallery, or platform that is so elevated and arranged as to keep all unqualified persons away. An open stairway that is not barred or that does not have a gate would not comply.
- **(C)** They shall be elevated 8 feet or more above the floor. This 8-foot requirement appears often in the *Code*.

Exception

When commutators, collectors, and brush rigging on stationary motors are inside of the motor end brackets and are not conductively connected to a voltage of 150 volts to ground from the supply source. An example might be a wound-rotor motor.

430.133: Guards for Attendants

In certain instances (when live parts of more than 150 volts to ground are capable of being touched by a guard or attendant), insulating mats or platforms must be installed in such a way that the live parts can't be touched, except from the mat or platform.

XII. Grounding All Voltages

430.141: General

Part XII covers the grounding of motors and controller frames to prevent voltages above ground in the event of accidental contact of the live parts and the frames. There are permissible conditions where insulation, isolation, or guarding will suffice instead of grounding the frames.

430.142: Stationary Motors

Under any of the following conditions, frames of stationary motors shall be grounded: (1) if connected to metal enclosed wiring; (2) if they are in wet locations and are not isolated or guarded; (3) if, as covered in Articles 500 through 517, they are in a hazardous (classified) location; and (4) if any motor operates with terminals over 150 volts to ground.

If the frame of the motor is not grounded, it shall be effectively and permanently isolated from ground.

Other suitable precautions should be taken to keep persons from coming in contact with the ground, should a fault occur.

430.143: Portable Motors

Frames of motors (portable) that operate at more than 150 volts to ground must be either guarded or grounded. In Section 250.45(D), portable motors in other than residential occupancies were to be grounded where the location was of a damp or wet nature, or where persons might come into contact with grounded objects. An exception to this was that motors that operate at not more than 50 volts or are supplied by an isolation transformer did not need to be grounded. There was also a recommendation that motors that operate at more than 50 volts be grounded.

Note

Portable appliances and other than residential occupancies will be found in Section 250.45(D).

Note

The grounding conductor color is covered in Section 250.59(B).

430.144: Controllers

Controller cases shall be grounded, regardless of voltage, except for controllers attached to ungrounded portable equipment and lined covers of snap switches, that is, where the cover is properly insulated.

430.145: Method of Grounding

Grounding of motors and controllers shall conform to Article 250.

(A) Grounding Through Terminal Housings. In Article 250, Type AC metal-clad cable or metal raceways was, in most cases (except in some hazardous installations), acceptable as a ground if properly installed. Where these types of wiring are used, a junction box shall be provided for the terminals of the motor, and the cable or raceway attached to this junction box as specified in Article 250.

Note

See *NEC,* Section 430.12(E) for grounding and terminal housings.

(B) Separation of Junction Box from Motor. The junction box, as provided for in (A) of this list, may be separated from the motor by a distance not to exceed 6 feet (1.83 m), provided that the leads to the motor are one of the following: (1) Type AC metal-clad cable; (2) armored cable; (3) rigid nonmetallic conduit; or (4) stranded leads enclosed in flexible metal conduit, liquid-tight flexible nonmetallic conduit, rigid or intermediate metal conduit, or in electrical metallic tubing no smaller than 3⁄8-inch trade-size. Where stranded leads that are protected by one of the means above are used, they shall be no larger than No. 10, and all other requirements for conductors, as provided by the *Code*, shall be met.

(C) Grounding of Controller-Mounted Devices. Instrument transformer secondaries and exposed noncurrent-carrying metal or other conductive parts or cases of instrument transformers, meters, instruments, and relays shall be grounded as specified in Sections 250.121 through 250.125.

See *NEC* Part XIII for the following tables:

Table 430.147, "Full-Load Current in Amperes, Direct-Current Motors" (the values of current at full load are the average direct-current quantities for motors running at base speed).

Table 430.148, "Full-Load Current in Amperes for Single-Phase Alternating-Current Motors".

Table 430.149, Full-Load Current for Two-Phase Alternating-Current Motors, Four-Wire".

Table 430.150, "Full-Load Current, Three-Phase Alternating-Current Motors" (induction-type, squirrel cage, wound-rotor type, and synchronous types at unity power factor).

Table 430.151, "Locked-Rotor Current Conversion" (determined from horsepower and voltage rating for use only with Sections 430.110, 440.12, and 440.41).

Table 430.152, "Maximum Rating or Setting of Motor Branch-Circuit Protective Devices" (for all types of motors and also the code letter on motors; gives the percent of full-load current for non–time-delay fuses, dual-element (time-delay) fuses, instantaneous-trip breakers, and inverse-time breakers; be sure to observe the notes at the bottom of this table).

Article 440—Air-Conditioning and Refrigerating Equipment

I. General

440.1: Scope

This article covers electric motor-driven air-conditioning and refrigeration equipment, and the controllers and branch circuits that supply this equipment. Included will be special considerations necessary to supply hermetic refrigerant motor-compressors. It takes into consideration an individual branch circuit that supplies hermetic refrigerant motor-compressors, and any air-conditioning and/or refrigeration equipment supplied from an individual branch circuit.

440.2: Definitions

Hermetic Refrigerant Motor-Compressor. Here the compressor and the motor are both enclosed in the same housing; therefore, they don't have a shaft extending out from a compressor to an external motor.

440.3: Other Articles

(A) Article 430. Provisions in this article are in addition to or amendatory to Article 430 and other articles in the *NEC* that apply unless modified by this article.

(B) Article 422, 424, or 430. Air-conditioning and refrigeration equipment that employ conventional motors, instead of hermetic-type motors, come under Article 422, 424, or 430. This includes the following units when driven by conventional motors: refrigeration compressors, furnaces with air-conditioning equipment, fan-coil units, remote forced-air-cooled condensers, and remote commercial refrigerators.

(C) Article 422. The following items are considered appliances and come under Article 422: room air conditioners, household refrigerators, household freezers, drinking-water coolers, and beverage dispensers.

(D) Other Applicable Articles. The applicable conditions follow for the circuits, controllers, and equipment as well as the hermetic refrigerant motor-compressor. Refer to the *NEC*, which lists places that require special installations.

440.4: Marking on Hermetic Refrigerant Motor-Compressor and Equipment

(A) Hermetic Refrigerant Motor-Compressor Nameplate. On the hermetic refrigerant motor-compressor, a marking must be provided on the nameplate giving the manufacturer's name, trademark, or symbol and designating the identification, number of phase, voltage, and frequency. The rated-load current is to be marked in amperes for the motor-compressor. It can be marked on the manufacturer's nameplate and the nameplate of the equipment on which the motor-compressor is to be used. Also marked on the motor compressor nameplate shall be the locked-rotor current for a single-phase motor-compressor that has a rated-load current of more than 9 amperes at 115 volts, or for 230-volt operation if more than 4.5 amperes. Polyphase motor-compressors shall be marked as such on the motor-compressor nameplate. Where there is a thermal protector that complies with Section 440.52(A)(2), the motor-compressor nameplate or the equipment nameplate shall be permanently marked "Thermally Protected." If a protective system complying with Sections 440.52(A)(4) and (B)(4) is used and is also furnished as part of the equipment, the equipment nameplate shall be marked "Thermally Protected System." Where the protective system complies with Section 440.52(A)(4) and (B)(4), the equipment nameplate shall be marked to indicate that it is not supplied with the equipment, but must be used.

Note

The rated-load current of the refrigerant motor-compressor that is marked thereon is the result of full-load current when the motor compressor is operated at its rated load and the rated voltage and frequency that is marked for the equipment it serves.

(B) Multimotor and Combination-Load Equipment. Multimotor and combination-load equipment shall be provided with a visible nameplate marked with marker's name, rating in volts, frequency, number of phases, minimum circuit ampacity, and maximum rating of the branch-circuit short-circuit and ground-fault protective-device rating. The ampacity calculations will follow in Part IV. Part III covers ground-fault protection calculation. See the *NEC* for further details.

Exception No. 1
When suitable under the provisions of this article, multimotor and combination-load equipment that is equipped for a single 15- or 20-ampere connection at 120 volts, or connected to a 15-ampere source or a 208- or 240-volt single-phase branch circuit, shall be marked and is permitted.

Exception No. 2
Part VII of Article 440 covers room air-conditioners.

(C) Branch-Circuit Selection Current. Sealed (hermetic-type) motor-compressors or equipment containing such compressor(s) in which the protective system, approved for use with the motor-compressor that it protects, permits continuous current in excess of the specified percentage of nameplate-rated-load current given in Section 440.52(B)(2) or (B)(4) shall also be marked with a branch-circuit selection current that complies with Section 440.52(B)(2) or (B)(4). This marking shall be on the nameplate(s) where the rated-load current(s) appears.

Note
Definition: Value of amperes to be used in the selection of branch-circuit current instead of a rated full-load current shall be determined by the rating of the motor branch-circuit conductors, the controllers, the disconnecting means, and also the branch-circuit, short-circuit, and ground-fault protective devices, wherever the running overload protective device will permit sustained current that is more than is specified as the percentage of load current. The value of the branch-circuit selected current will always be equal to or greater than the marked rated full-load current.

440.5: Marking on Controllers
Controllers shall be marked with the following: maker's name, trademark, or symbol; identifying designation; voltage; phase; full-load current; locked-rotor current (or horsepower) rating; and any other pertinent information.

440.6: Ampacity and Rating

The rating of the equipment and the ampacity of the conductors selected from Tables 310.16 through 310.19, or according to Section 310.15, must be arrived at as follows:

(A) Hermetic Refrigerant Motor-Compressor. For a sealed (hermetic- type) motor-compressor, the rated-load current marked on the nameplate of the equipment in which the motor-compressor is employed shall be used in determining the rating or ampacity of the disconnecting means, branch-circuit conductors, controller, ground-fault protection, and separate motor overload protection. If there is no rated-load current on the equipment nameplate, you shall use the rated-load on the compressor nameplate. For disconnecting means and controllers, see Sections 440.12 and 440.41.

Exception No. 1

Permits us to use the branch-circuit selection current when shown, instead of the rated-load current for determining the rating or ampacity of disconnecting means, branch-circuit conductors, controller size, branch-circuit protection, short-circuit protection, and ground-fault protection.

Exception No. 2

As covered in Section 440.22(B), it is permitted for the branch-circuit, short-circuit, and ground-fault cord- and plug-connected equipment to be used for the protection.

(B) Multimotor Equipment. If multimotor equipment employs shaded-pole or permanent split-capacitor-type fans that run blowers, the current for such motors that is marked on the nameplate of the equipment for which the fan or blower is used shall be the current used instead of the horsepower rating in determining the ampacity that is required for disconnecting means, branch-circuit conductors, controllers, the branch-circuit, short-circuit, and ground-fault protection, and also the separate overcurrent protection that is needed for this equipment. The marking that is on the nameplate or equipment is to be not less than the current marked on the fan- or blower-motor nameplate.

440.7: Highest-Rated (Largest) Motor

In compliance with this article and with Sections 430.24, 430.53(B) and (C), and 430.62(A), the motor with the highest load-rated current is the largest motor. If there should be two or more motors of the same load rating, only one is used. Any motors other than (hermetic-type) compressor-motors, as were covered in Section 440.6(B), shall have their full-load current determined by using Tables 430.148, 430.149, or 430.150, using horsepower to determine full-load current, not the nameplate current-rating.

Exception

If it is so marked, then the branch-circuit selection current is to be used instead of the rated-load current in determining the largest motor compressor.

440.8: Single Machine

Section 430.87 considers air-conditioning systems a single machine. Section 430.87 and Section 430.112 stipulate that the motors need not be mounted adjacent to each other, but may be mounted remote to each other.

II. Disconnecting Means

440.11: General

The *NEC* refers to Diagram 430.1. The intent of Part II is to require disconnecting means for air-conditioning and refrigeration equipment, including the motor and the compressor and controllers to be supplied from a circuit feeder.

440.12: Rating and Interrupting Capacity

(A) Hermetic Refrigerant Motor-Compressor. In considering the rating of the disconnecting means that serve a hermetic refrigerant motor-compressor, it is to be considered as selected from the nameplate rated-load current or the branch-circuit current, whichever is greater, and locked-rotor current, respectively, of the motor-compressor as follows:

(1) The rating of the ampere shall be at least 115 percent of the rated-load current on the nameplate or the branch-circuit current, whichever is larger. This gives us a minimum, but on large units the 115 percent won't be of sufficient ampacity for opening under load.

(2) To determine the equivalent horsepower in complying with the requirements of Section 430.109, select the horsepower rating from Table 430.148, 430.149, or 430.150 corresponding to the rated-load current or branch-circuit selection current, whichever is greater, and also the horsepower rating from Table 430.151 corresponding to the locked-rotor current. See the *NEC* for the balance of this part.

The first three tables mentioned, Tables 430.148, 430.149, and 430.150, cover horsepower to full-load current. These may also convert full-load current to horsepower. Table 430.151 gives us the locked-rotor conversion table.

Hermetic-type motor-compressors usually don't have a horsepower rating on the nameplate, so we use the ampere rating and if amperes and horsepower rating don't correspond when we convert to horsepower, we use the next larger horsepower in the tables.

(B) Combination Loads. If one or more hermetic refrigerant motor-compressors are operated together, or other combinations of motors or loads such as resistive heaters are used with them, if the combination load can come on at the same time on a single disconnecting means, then the rating for the combined load shall be as follows:

(1) The disconnecting means shall be determined by the sum of the currents, such as current of resistive load at rated load-current plus locked-rotor currents of the motor(s). The combined rated-load current and the combined locked-rotor current are obtained by the summation of (a) and (b) to convert the ampacity as needed as if the combination were just one motor:

(a) Other than hermetic refrigerant motor-compressors and fans for blower motors, the full-load current to a horsepower rating to each motor, as covered in Section 440.6(B), is to be selected from Tables 430.148, 430.149, or 430.150. Of the sum of these currents and the motor-compressor rated-load current(s), or branch-circuit current(s), you are to use the greater. Also, amperes of other loads, to obtain equivalent, will give the full-load current of the combined loads.

(b) Other than the hermetic refrigerant motor-compressor, the locked-rotor current equivalent to the horsepower rating of each motor shall be selected from Table 430.151, and for the fan and blower motors of the shaded-pole or permanent split-capacitor type that are marked with the locked-rotor current, the marked values shall be used. The locked-rotor current values of the motor-compressor locked-rotor current(s) and the rating in amperes of other loads shall be added to obtain the locked-rotor current for the combined load. Where two or more motors and/or other loads can't be started at the same time, the appropriate locked-rotor and rated-load currents and branch-circuit selection of current, whichever is greater, will be acceptable as the means for determining the equivalent locked-rotor current combinations that will operate at the same time.

Exception

If part of the load is resistive and if the disconnecting means is rated in horsepower and amperage, the switch may have a horsepower rating that is not less than the combined load of the motor-compressor(s) and the other motors at locked-rotor current. If the amperage is equal to or greater than the locked-rotor plus the resistance load, it may be used.

(2) As referred to in Section 440.12(B)(1), the disconnecting means shall have an ampere rating that is at least 115 percent of the combination of all currents at the rated-load conditions.

(C) **Small Motor-Compressors.** For small motors that don't have locked-rotor current rating marked on the nameplate, or for small motors covered in Table 430.147, 430.148, 430.149, or 430.150, the locked-rotor current shall be assumed to be six times the full-load current rating.

(D) **Every Switch.** The requirements of Section 440.12 shall be met wherever the disconnecting means in the refrigerant motor-compressor circuit is between the point of attachment to the heater and a point of connection to the refrigerant motor-compressor.

(E) **Disconnecting Means Rated in Excess of 100 Horsepower.** The provisions of Section 430.109 apply when, as determined above, the rated-load or locked-rotor current indicates that the rating is in excess of 100 horsepower.

440.13: Cord-Connected Equipment

When room air conditioners, home refrigerators and freezers, drinking-water coolers, and beverage dispensers are cord-connected, a cord, plug, or receptacle, which may be separable, is permitted as the disconnecting means. For additional information on this, see Section 440.63.

440.14: Location

This is slightly different from some of Article 430, but refers to Parts VII and VIII of Article 430 for additional requirements. The disconnecting means shall be located within sight of the air-conditioning or refrigeration equipment and shall also be readily accessible. The disconnecting means may be installed within or on the refrigerating or air-conditioning equipment.

Note

Additional requirements may be found in Parts VII and IX of Article 430.

III. Branch-Circuit, Short-Circuit, and Ground-Fault Protection

440.21: General

Part III covers devices that are intended to protect the branch-circuit conductors, control apparatus, and motors that serve hermetic refrigerant motor-compressors against overcurrent caused by short circuits and grounds. They are also supplementary or amendatory to Article 240.

440.22: Application and Selection

(A) **Rating or Setting for Individual Motor-Compressor.** The branch-circuit, short-circuit, and ground-fault protective device of the motor-compressor shall be capable of handling the motor starting current, but shouldn't be over 175 percent of the motor's rated-load current or branch-circuit selection current, whichever is the greater; 15 amperes is the minimum size.

Should the above 175 percent be too small to take care of the starting current, this 175 percent may be increased to a maximum of 225 percent with the same stipulations.

(B) **Rating or Setting for Equipment.** The equipment branch-circuit, short-circuit, and ground-fault protective device shall be capable of carrying the starting current of the equipment. Where the hermetic refrigerant motor-compressor is the only load on the circuit, the protection shall conform with Section 440.22(A). Where the equipment incorporates more than one hermetic-refrigerant motor-compressor or a hermetic refrigerant motor-compressor and other motors or other loads, the equipment protection shall conform with Section 430.53 and the following: (Article 430 and Article 440 are closely tied together. Refer to

Section 430.53, which provides the information required.) Section 430.53 and the following cover short-circuit and ground-fault protection:

(1) If the hermetic refrigerant motor-compressor is the larger load, the rating or setting of the protective device shall conform to Section 440.22(A) plus the rated-load current or branch-circuit selection current, whichever is the greater of the other smaller motor-compressor(s) and/or the rating of any other loads covered by this protective device. The values specified in Section 440.22(A) are not to exceed the rating and setting of the branch-circuit, short-circuit, and ground-fault protective device.

(2) Where a hermetic refrigerant motor-compressor is not the largest load connected to the circuit, the rating or setting of the protective device shouldn't exceed a value equal to the sum of the rated-load current or branch-circuit selection current, whichever is greater, rating(s) for the motor-compressor(s) plus the value specified in Section 430.53(C)(4), where other motor loads are supplied in addition to the motor-compressor(s), or the value specified in Section 240.3, which describes only nonmotor loads that are supplied in addition to the motor-compressor(s).

Exception No. 1

In many cases where the equipment will start and operate on a 15- or 20-ampere, 120-volt or a 15-ampere, 208- or 240-volt single-phase branch circuit, a 15- or 20-ampere overcurrent device may be used to protect the branch circuit. Should the maximum branch-circuit, short-circuit, and ground-fault protective device rating marked on the equipment be less than these values, then the values of the overcurrent device in the branch circuit shouldn't exceed the values marked on the nameplate of the equipment.

Usually such equipment will have protective devices built into the unit itself.

Exception No. 2

On equipment not rated over 250 volts, such as household refrigerators, drinking-water coolers, and beverage coolers, that has a nameplate marking for cord-and-plug connection, this nameplate rating is to be used for the branch-circuit device rating and, unless the nameplate is marked otherwise, this equipment is to be considered single-motor equipment.

(C) Protective-Device Rating Not to Exceed the Manufacturer's Values. Where the maximum protective-device ratings that will be shown on the manufacturer's heater table for use with a motor controller are less than the rating or setting as determined by Section 440.22(A) and (B), the manufacturer's values marked on the equipment shouldn't be exceeded by the protective-device rating.

IV. Branch-Circuit Conductors

440.31: General

Conductor sizes are specified in Part IV and in Articles 300 and 310. Conductors shall be of large enough ampacity to carry the motor current without heating. Exception No. 1 of Section 440.5(A) modifies this specification.

The provision of Articles 300 and 310 don't cover the following: integral conductors of motor, motor controllers, and the like; and conductors that form an integral part of approved equipment. See Sections 300.1(B) and 310.1.

440.32: Single Motor-Compressors

This is the same for a single motor as covered in Article 430, that is, not less than 125 percent of either the motor-compressor rated-load current or the branch-circuit selection current, whichever is greater.

440.33: Motor-Compressor(s) with or Without Additional Motor Loads

The ampacity of conductors supplying one or more motor-compressors, with or without additional motor loads shall be figured from the rated-load or branch-circuit selection current ratings, whichever is the larger. The rated load shall be 125 percent of the full-load current of the highest rating of the largest motor or motor-compressor rating of the group, plus the full-load ratings of all of the other motors or motor-compressors in the group.

Exception No. 1

If the circuitry is interlocked to prevent starting and running of a second motor-compressor or group of motor-compressors, you will use the current value of the largest motor-compressor or, if a group of compressors can run together at a given time, the conductor size will be for this largest group of motor-compressors.

Exception No. 2

Part III of Article 440 covers room air conditioners.

440.34: Combination Load

For conductors that supply a motor-compressor load and also a lighting or appliance load, the computation was given in Article 220, and there are other articles that might apply. The conductors shall have sufficient ampacity to handle the lighting or appliance load in addition to the required ampacity for the motor-compressor load. This shall meet the requirements of Section 440.33, or, if for a single motor-compressor load, the requirements of Section 440.32.

Exception

Should the circuitry be so designed that the motor-compressor and other loads on the circuit can't be operated at the same time, the conductor sizes will be determined by the motor-compressor(s) and any other loads that may be operated at the same time.

440.35: Multimotor and Combination-Load Equipment

For such equipment we use the minimum circuit ampacity marked on the equipment in accordance with Section 440.3(B), which need not be repeated here as it has just been covered.

V. Controllers for Motor-Compressors

440.41: Rating

(A) **Motor-Compressor Controller.** The motor-compressor controller must have not only the continuous-duty full-load current, but also must have the locked-rotor current rating. This shouldn't be less than the load indicated on the nameplate-rated load current or the selected current for a branch-circuit, whichever is greater, and the locked-rotor current (you are referred to Sections 440.6 and 440.7) tied in with the locked-rotor current of the compressor. In case the controller for the motor is rated in horsepower but is without one or both of the foregoing current ratings required, the equivalent currents of the horsepower may be determined from ratings by using Table 430.148, 430.149, or 430.150 to determine the equivalent full-load current. Table 430.151 will be used to determine the equivalent locked-rotor current.

This is basically no different from that for the conventional type of motor, and the ampacity of the controller is determined in the same manner and with the use of the same tables.

(B) **Controller Serving More Than One Load.** Where there is more than one motor-compressor or a motor-compressor and other loads, the controller rating shall be determined by the continuous-duty full-load current rating, and a locked-rotor rating of not less than the combined loads, as determined in accordance with Section 440.12(B).

VI. Motor-Compressor and Branch-Circuit Overload Protection

440.51: General

See Section 240.3. Part VI is intended to cover devices that give protection to the motor-compressor, the apparatus to control the motor, and the branch-circuit conductors. In this part, the main idea is to protect the motor from overload and excessive heating and/or failure-to-start.

Note

Overloading on electrically driven equipment can cause damage or overheating of the electrical system. Not included here are short circuits or ground faults.

440.52: Application and Selection

(A) Protection of Motor-Compressor. Overloads or failure-to-start may be prevented by one of the following means:

(1) This tells us the separate overload relay shall be responsive to not over 140 percent of the motor-compressor rated-load current.

(2) A thermal protector built into the compressor-motor is acceptable, if it prevents damage from overheating due to overload or failure to start.

The current-interrupting device may be external and operated by a thermal device in the motor, provided that it is arranged in such a way that the control circuit from the thermal device will cause interruption of current to the motor-compressor.

(3) Fuses or inverse-time circuit breakers that respond to the motor current are permitted if they are not rated at more than 125 percent of the motor-compressor rated-load current. Time-delay shall be sufficient to take care of the starting current.

Either the motor-compressor or the equipment shall be marked with the maximum size of branch-circuit fuse or inverse-time circuit-breaker rating.

(4) The protective system we are covering, whether furnished or supplied with the equipment, is intended to prevent overheating of the motor-compressor due to overload and failure-to-start. When the current interrupting device is not a part of the motor compressor and the control circuit is operated by a device that is not a part of the current-interrupting equipment, it shall be arranged so that the control circuit results in the interruption of the current to the motor-compressor.

(B) Protection of Motor-Compressor Control Apparatus and Branch-Circuit Conductors. This portion explains that the motor-compressor controller(s), the disconnecting means, and branch-circuit conductors are to be protected from overcurrent due to motor overload or failure-to-start. It also gives us the means that we may use to accomplish this and tells us that these means may be the same device or system that protects the motor-compressor as called for in Section 440.52(A).

Exception

As provided in Sections 440.54 and 440.55, motor-compressors and equipment that operate from a 15- or 20-ampere single-phase branch-circuit will be permitted if: (1) An overload relay that meets the requirements of (A)(1) of this section. (2) As provided in Section 440.52(A)(2), the thermal protector applied in accordance with this section that prevents continuous current in excess of 156 percent of the marked rated-load current or branch-circuit selection current. (3) A fuse or inverse-time circuit breaker that meets the requirements of (A)(3) of this section. (4) A protective system that meets the requirements of Section 440.52(A)(4) and that won't permit the current to be over 156 percent of the marked rating of load current or the selection of branch-circuit current.

440.53: Overload Relays

Relays or other devices for motor protection that are not able to open short circuits must be protected by fuses or inverse-time circuit breakers, and the ratings of these shall be in accordance with Part III—that is, unless approved for group installation of part-winding motors. They shall be marked to indicate the maximum size of the fuse or of the inverse-time circuit breaker that is used to protect them.

Exception

Fuses or inverse-time circuit breakers shall permit the size marking on the nameplate of the approved equipment in which the overload relay or other overload device is used.

440.54: Motor-Compressors and Equipment on 15- or 20-Ampere Branch-Circuit—Not Cord- and Attachment Plug-Connected

As permitted by Article 210, (A) and (B) of this section cover overload protection for motor-compressors and equipment used on 15- or 20-ampere, 120-volt or 15-ampere, 208- or 240-volt, single-phase branch circuits.

(A) **Overload Protection.** Motor-compressor overload protection shall be selected as required by Section 440.52(A). The controller and motor overload protective device must be approved for the installation of short-circuit and fault-current protective devices. This is for the branch circuit that supplies the equipment.

(B) **Time Delay.** The branch-circuit, short-circuit, or ground-fault protective device (usually fuse or circuit breaker) shall have a sufficient time delay to take care of the starting current.

440.55: Cord- and Attachment Plug-Connected Motor-Compressors and Equipment on 15- or 20-Ampere Branch Circuits

As permitted by Article 210, (A), (B), and (C) of this section cover overload protection for motor-compressors and equipment that are cord- and attachment plug-connected and used on 15- or 20-ampere, 120-volt or 15-ampere, 208- or 240-volt, single-phase branch circuits.

(A) **Overload Protection.** See (A) in Section 440.54.

(B) **Attachment Plug and Receptacle Rating.** Plug-and-cord connection to the receptacle shouldn't exceed 20 amperes at 125 volts or 15 amperes at 250 volts.

(C) **Time Delay.** See (B) in Section 440.54.

VII. Provisions for Room Air Conditioners

440.60: General

Electrically energized air conditioners that control both temperature and humidity are covered in Part VII. Room air conditioners (which may or may not have provisions for heating) shall be considered to be of the alternating-current type of appliance that may be a window-mounted, console, or in-the-wall type of installation located in the room that is to be conditioned and that uses a hermetic refrigerant motor-compressor(s). Part VII covers equipment not over 250 volts and single-phase. This type of equipment may be cord- and attachment plug-connected.

If the room air conditioner is three-phase or operated at over 250 volts, it must be directly connected (not cord- and plug-connected) by methods covered in Chapter 3 and to follow in this Part VII.

440.61: Grounding

Sections 250.110, 250.112, and 250.114 cover the grounding of this type of equipment, and room air-conditioners shall be grounded in accordance with these sections.

440.62: Branch-Circuit Requirements

(A) Room Air Conditioner. If the following conditions are met, a room air conditioner shall be considered a single-motor unit when determining its branch-circuit requirements provided all the following conditions are met:

(1) It is cord- and attachment plug-connected.

(2) Its rating is less than 40 amperes, single-phase and 250 volts.

(3) A nameplate rating of the air-conditioner is shown rather than individual currents.

(4) The rating of the branch-circuit, short-circuit, and ground-fault protective equipment doesn't exceed the ampacity of the branch-circuit conductors or of the receptacles from which it is supplied.

As we are aware, a room air conditioner usually has the hermetic-type motor-compressor, as well as another fan motor or two.

(B) Where No Other Loads Are Supplied. If no other loads are supplied, the marked rating of cord-and-attachment-plug connection may not exceed 80 percent of the branch-circuit.

(C) Where Lighting Units or Other Appliances Are Also Supplied. The marked rating of the air conditioner is not to exceed 50 percent of the rating of the branch-circuit conductors, where lighting or other appliances are supplied from the same branch circuit.

440.63: Disconnecting Means

The room air conditioner operating at 250 volts or less may be cord- and plug- and receptacle-connected for single-phase air conditioners if (1) the manual controls for the air conditioner are readily accessible and within 6 feet of the floor, or (2) a separate manually operated switch to control the air conditioner is mounted in sight of it.

440.64: Supply Cords

Flexible cords supplying room air conditioners shall be no longer than (1) ten feet for a nominal 120-volt rating, or (2) six feet for a nominal 208- or 240-volt rating.

Article 445—Generators

445.1: General

In addition to the requirements of this article, generators must comply with the requirements of other articles of the *Code*, most notably Articles 230, 250, 695, 700, 701, 702, and 705.

445.2: Location

The first requirement is that the generator be suitable for the location where it is installed. The requirements of Section 430.14 must also be complied with. If generators are to be installed in hazardous locations, the requirements of Articles 500 through 503, 510 through 517, 520, 530, and 665 must also be complied with.

445.3: Marking

Every generator must have a nameplate that contains the following information:

- The manufacturer's name
- The operating frequency
- Number of phases
- Power factor
- Rating in kVA or kW, with the corresponding volts and amperes
- Rated RPMs
- Insulation type, ambient temperature, and time rating

445.4: Overcurrent Protection

(A) **Constant-Voltage Generators.** Generators that produce a constant voltage (almost all the generators we commonly use are of this type) have to be protected from overloads. This may be done by overcurrent protective devices (fuses, circuit breakers, and the like), or by inherent design.

(B) **Two-Wire Generators.** Two-wire DC generators are required to have overcurrent protection in only one wire, if the overcurrent device is set off by the entire current (not the current in the shunt coil). The overcurrent device may not, under any circumstances, open the shunt coil.

(C) **65 Volts or Less.** A generator that operates at 65 volts or less and is driven by only one motor may be protected at 150 percent of its full rated current.

(D) **Balancer Sets.** Two-wire DC generators are often used with balancer sets, which convert their output to a three-wire system with a neutral. When this is done, the system must be provided with overcurrent devices that disconnect the three wires in the event of an excessive imbalance.

(E) **Three-Wire, Direct-Current Generators.** These generators (either shunt-wound or compound-wound) must be provided with overcurrent protection in each armature lead, which must sense the entire armature current. These devices (which must be multipole devices) must open all the poles in the event of an overcurrent condition.

Exceptions to (A) Through (E)

There are some instances when a generator failure would be less of a hazard than disconnecting it when an overcurrent occurs (the proverbial lesser of two evils). In such circumstances, the authority that has jurisdiction (usually an electrical inspector) may allow the generator to be connected to a supervision panel and annunciator or alarm, instead of overcurrent protection.

445.5: Ampacity of Conductors

Phase conductors from a generator must be rated at no less than 115 percent of the nameplate current rating. Neutrals can be sized according to Section 220.22. Conductors carrying ground-fault currents must be sized in accordance with Section 250.24(B).

Exception No. 1

The conductors can be protected at 100 percent of the rated current if the generator's design will prevent overloading.

Exception No. 2

If there are internal overcurrent devices in the generator, the conductors can be rated at the nameplate current.

445.6: Protection of Live Parts

No live parts of generators that are connected to voltages of more than 50 volts to ground can be exposed, except where only qualified people are ever allowed in the area.

445.7: Guards for Attendants

The requirements of Section 430.133 apply also to generators.

445.8: Bushings

Bushings must be used to protect wires that pass through openings in enclosures, boxes, and the like. The bushings must be properly designed for the environment in which they are used.

445.9: Generator Terminal Housings

These shall comply with Section 430.12.

445.10: Disconnecting Means Required for Generators

A means must be supplied to disconnect completely the generator and all protective or controlling devices from the circuits supplied by the generator. Exceptions are made when the driving means (prime mover) of the generator can't be easily shut down and when the generator is arranged to operate in parallel with another generator or power source.

Article 450—Transformers and Transformer Vaults (Including Secondary Ties)

450.1: Scope

This article applies to the installation of all transformers with the following exceptions:

- **(A)** Current transformers.
- **(B)** Dry-type transformers that are a component part of apparatus and that conform to the requirements of this apparatus.
- **(C)** Transformers for use with X-ray and high-frequency or electrostatic-coating apparatus.
- **(D)** Transformers used with Class 2 and Class 3 circuits that comply with Article 725.
- **(E)** Transformers for sign and outline lighting conforming to Article 600.
- **(F)** Transformers for discharge lighting as covered in Article 410.
- **(G)** Transformers used for power-limited fire-protective signaling circuits that comply with Part III of Article 760.

(H) With transformers used for research, development, or testing, whether liquid-filled or dry type, proper safeguards must be provided for unqualified persons coming in contact with high-voltage terminals or conductors that are energized.

As covered in Articles 501 to 503, this article also covers transformers in hazardous (classified) locations.

I. General Provisions

450.2: Definitions

This article covers the following:

Transformer. *Transformer* means a single-transformer or polyphase transformer identified by a single nameplate, unless otherwise covered elsewhere in this article. See Section 450.6, for example.

450.3: Overcurrent Protection

In this section, a *transformer* is either a single transformer or a polyphase bank of two or three single-phase transformers operating together.

A secondary overcurrent device can be made of up to six circuit breakers or fuses in one location. In such cases, the total of the overcurrent device ratings can't be higher than what the rating would be for one overcurrent device.

(A) **Transformers Over 600 Volts, Nominal.**

(1) **Primary and Secondary.** Every transformer of over 600 volts, nominal, shall have primary and secondary overcurrent protective devices rated or set at not over the value shown in Table 450.3(A)(1). See the *NEC.*

Electronically actuated fuses that can be set to variable current levels must be set according to the setting required for circuit breakers.

Exception No. 1

If the overcurrent fuse of a circuit breaker does not meet the standard rating or settings, the next higher rating or setting is to be used.

Exception No. 2

See (A)(2) in this list.

Exception No. 3

Transformers that are dedicated to fire pumps don't require secondary overcurrent protection. The primary overcurrent protection must be sufficient to carry the total of all locked-rotor and associated currents fed by it for an indefinite period of time. This is necessary since fire pumps operate only under extreme emergency conditions, and in such cases, the circuit tripping an overcurrent device could lead to an unrestrained fire. Since the risks due to a fire pump not functioning are greater than the risks from locked-rotor currents, it is better to allow for the locked-rotor currents, and guarantee that the fire pumps will function without interruption.

(2) Supervised Installations. If maintenance and supervision is done only by qualified persons who will be monitoring and servicing the transformers and installation, overcurrent protection as provided in (A)(2)(a) is permitted.

(a) Primary. *Primary* is often inferred in the field as being the high side and the term *secondary* as being the low side of the transformer. This is not the case, however. The primary is always the input side of a transformer and the secondary is always the output side. Thus, voltage has nothing to do with the terms. Each transformer is to be protected by an overcurrent device in the primary side. If the overcurrent protection is fuses, they shall be rated at not more than 250 percent of the rated primary current of the transformer. When circuit breakers are used, they shall be set at not more than 300 percent of the rated primary current. This overcurrent protection may be mounted in the vault or at the transformer if identified for the purpose. It may also be mounted to protect the circuit supplying the transformer. Thus, in the case of a vault, it could be mounted out of doors on a pole, but a disconnecting means will have to be provided in the vault (Figure 450-1).

(b) Primary and Secondary. If a transformer with more than 600 volts, nominal, has an overcurrent device on the secondary, meets the rating, or is set to open at a value noted in Table 450.3(A)(2)(b), or if the transformer is equipped with a thermal overload protection device installed by the manufacturer, overcurrent devices won't be required in the primary connection. The primary feeder overcurrent device shall be rated or set to open at no more than required in Table 450.3(A)(2)(b). See the *NEC*.

Author's Note

The inrush current in charging a transformer primary winding wire exceeds the maximum current the transformer is designed to normally carry.

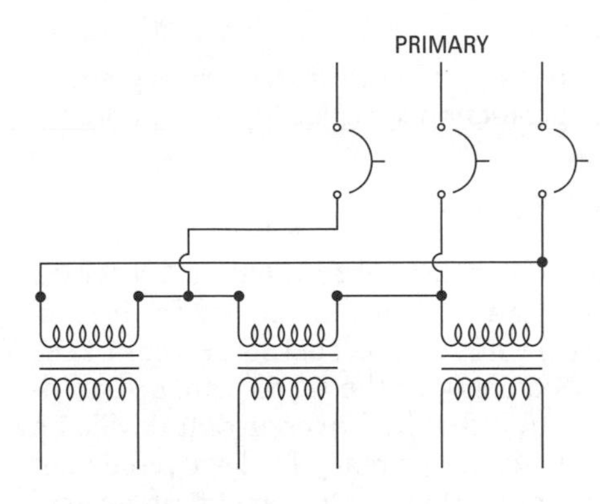

Figure 450-1 Overcurrent protection may be at the transformer or in the circuit supplying the transformer.

Exception No. 1
If the primary current of a transformer is 9 amperes or more, and 125 percent of this current does not show up in standard rating of fuse or nonadjustable circuit breakers, according to Section 240.6, the next higher standard rating may be used. If the primary current is less than 9 amperes, the overcurrent device, if set at not more than 167 percent of the primary current, may use the standard rating.

The primary overload protection device may have a setting at not more than 300 percent. This is permitted only where the primary current is less than 2 amperes.

Exception No. 2
If the primary circuit overcurrent protection provided in this section is used, an individual overcurrent device won't be required. This applies to where the feeder or branch circuit originates.

Exception No. 3
See (B)(2) in this list.

(B) Transformers 600 Volts, Nominal, or Less.

(1) Primary. An individual overcurrent device on the primary side of each transformer of 600 volts, nominal, or less shall be installed to protect the transformer. The rating or setting of this overcurrent device shall be rated or set at not more than 125 percent of the rated primary current of the transformer.

Exception No. 1
See Exception No. 1 of 450.3(A)(1).

Exception No. 2
Individual overcurrent devices are not required if the primary circuit overcurrent device provides the protection specified in this section.

Exception No. 3
See (A)(2)(b) in this list.

(2) Primary and Secondary. A transformer rated at 600 volts, nominal, or less, having an overcurrent device in the secondary that is rated or set not to exceed 125 percent of the current rating of the secondary of the transformer, shouldn't be required to have an individual overcurrent device on the primary side, provided the primary feeder overcurrent is rated or set at not more than 250 percent of normal operating current of the primary of the transformer.

Many of these installations will be separately derived systems, which is still not as clear as the author feels it should be. First, we have a 10-foot-tap rule in Section 240.21 covering transformer secondaries and applicable rules governing the same. Then, in the same section, we have an Exception, which governs the 15-foot rule. See (E) of Section 240.21. In Exception No. 2, covering the 10-foot-tap rule, a note states "lighting and branch-circuit panelboards are covered in Section 408.16(A)." This requires secondary overcurrent devices. Consider all of these statements in making your decision on overcurrent secondary devices.

If a transformer of 600 volts, nominal, or less, has been equipped with thermal overload protection by the manufacturer and is arranged to disconnect the primary current, the transformer primary won't be required to have an individual overcurrent device, provided that the feeder overcurrent device is rated or set at a current value of not over six times the rated current for the transformer with an impedance of not over 6 percent. If the transformer has an impedance rating of more than 6 percent but not over 10 percent, the rating of the overcurrent shouldn't be over four times the load of the rated current of the transformer.

There is more inrush current upon energizing a transformer, with lower impedance transformers.

Exception
If the secondary current rating of a transformer is 9 amperes or more and if 125 percent of the full-load current does not correspond to the

rating of a fuse or a nonadjustable circuit breaker, you may use the higher standard rating described in Section 240.6.

When the secondary current is less than 9 amperes, and an overcurrent device of the circuit is not rated over 167 percent of this full-load current rating, the overcurrent device may be used.

(C) **Potential (Voltage) Transformers.** When installed indoors or enclosed, potential transformers are to be protected by primary fuses. To the author, it is just good practice to install protection for all potential transformers.

Note

The *NEC* refers to Section 408.22.

450.4: Autotransformers 600 Volts, Nominal, or Less

(A) **Overcurrent Protection.** Potential transformers of 600 volts or less must have individual overcurrent devices installed, and so must each ungrounded conductor on the input side of the autotransformer. These overcurrent devices shall be set at not more than 125 percent of the full-load current of the autotransformer. An overcurrent device is not to be installed in the shunt winding of an autotransformer (the winding that is common to both the primary and secondary circuits).

See Diagram 450.54 in the *NEC*.

Exception

If the input rating of the current to the autotransformer is 9 amperes or more, and if 125 percent of this current does not meet the standard ratings for fuses and nonadjustable circuit breakers as covered in Section 240.6, the next higher standard rating will be permitted. If the rating of the input current to an autotransformer is less than 9 amperes, the overcurrent device rating or setting shall be not more than 167 percent of the input current rating for a full-load current.

(B) **Transformer Field-Connected as an Autotransformer.** An autotransformer shall be identified as such and for the use at higher voltage if the autotransformer is field-connected.

Note

The use of autotransformers is covered in Section 210.9.

2005

450.5: Grounding Autotransformers

Many electrical systems are not grounded, especially older systems of 600 volts or less, and many 2400-, 4800-, and 6900-volt systems. Also when it is desired to ground existing ungrounded delta systems, grounding autotransformers may be used to obtain a neutral. Grounding autotransformers such as the zigzag- type may be used. The type most generally used is the 3ϕ zigzag-type transformer with no secondary winding. See Figure 450-2 for the schematic connections of such an autotransformer. Zigzag transformers may not be connected on the load side of any system grounding connection, including connections made per Sections 250.24(B), 250.30(A)(1), or 250.32(B)(2).

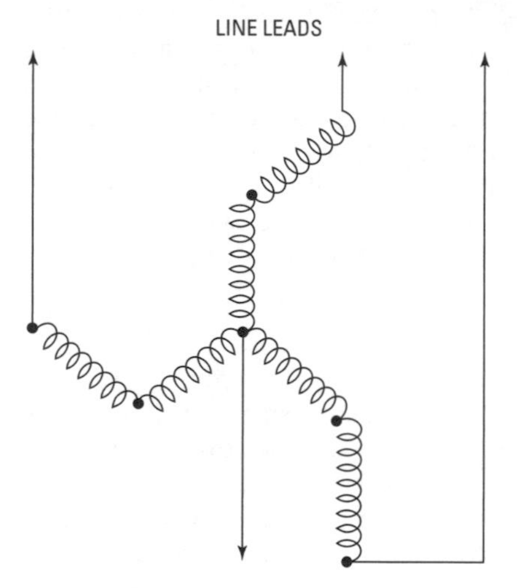

Figure 450-2 Schematic diagram of a 3ϕ zigzag grounding autotransformer.

The impedance of the autotransformer to 3ϕ currents is high so that there is no fault on the systems; only a small magnetizing current flows in the transformer winding. The transformer impedance to ground current, however, is low so that it allows high ground current to flow. The transformer divides the ground current into three equal components; these currents are in phase with each other and flow in the three windings of the grounding autotransformer.

Due to the direction of the actual windings, this tends to have equal division of the three lines and accounts for the low impedance of the transformer to ground currents. The preceding will give an idea of the operation and results of what Section 450.4 covers.

Grounding autotransformers are to be zigzag- or T-connected transformers connected to three-phase, three-wire systems that are ungrounded, the purpose of which is creating a three-phase, four-wire distribution system that will provide a neutral reference for grounding purposes. This kind of system shall have continuous phase currents and neutral current ratings.

Note

The current and the phase of an autotransformer shall be one-third of the neutral current.

(A) Three-Phase, Four-Wire System. When a three-phase, three-wire autotransformer system is converted to a three-phase, four-wire system, the following conditions shall be met:

(1) Connections. The ungrounded phase conductors shall be connected directly and shouldn't be switched or provided with overcurrent protection that is independent of the main switch with a common-trip overcurrent protection for the three-phase, four-wire system.

(2) Overcurrent Protection. A device to provide sensing of overcurrent and a phase or phases is permitted if it creates main-switch or common-trip overcurrent protection, as covered in (A)(1) of this list, to open should the load on the transformer reach or exceed 125 percent of a full-load continuous current per phase or neutral rating. The sensing device may be used to delay tripping for temporary overloads that are being sensed. The purpose of this is to allow proper operation of branch or feeder protection in the four-wire system.

(3) Transformer Fault Sensing. A sensing device for faults that will cause the opening of a main switch or common tripping of the overcurrent device on the three-phase, four-wire side of the system is to be provided to protect against single phasing or internal faults.

Note

You can accomplish this by using two subtractive-connected donut-type current transformers installed such that they sense and signal any unbalanced currents in the line current of 50 percent or more of the rated current of the autotransformer.

(4) **Rating.** The neutral current rating on an autotransformer system or four-wire shall be sufficient to handle the unbalanced load.

(B) Ground Reference for Fault Protection Devices. The following requirements apply where a grounding autotransformer is used to make available the intensity of ground-fault current for a ground-fault protective device on an ungrounded three-phase, three-wire system.

(1) **Rating.** The rating of the ground-fault autotransformer is to have continuous neutral current rating to cover specified ground-fault current.

(2) **Overcurrent Protection.** An overcurrent protection device of sufficient short-circuit rating to open all ungrounded conductors instantaneously when it operates shall be applied to the grounding transformer branch circuit and rated at a current not exceeding 125 percent of the load rating of autotransformers or 42 percent of the continuous current rating of any operated in the series-connected device and the neutral of the autotransformer connection. Delayed tripping is permitted for temporary overcurrents where the proper operation of the ground-tripping device that is responsible for tripping devices on the main system is permitted, but the values shouldn't exceed the short-time rating of either the grounding autotransformer or any devices connected in series with a neutral connection.

(C) Ground Reference for Damping Transitory Overvoltages. Grounding is always advantageous in damping and limiting transitory overvoltages. Autotransformers used for this purpose shall be rated and connected in accordance with (A)(1) of this list.

450.6: Secondary Ties

A secondary tie is a circuit operating at 600 volts, nominal, or less, connected between two phases and connecting two power sources or supplies: for example, the secondary for two transformers. The connections will be permitted to use one or more conductors per phase.

Note
Transformer, as used in this section, means either a single transformer or transformers operated as a bank.

This shouldn't be confused with the definition of transformer given in Section 450.2.

(A) Tie Circuits. Article 240 covers overcurrent protection in general, and the circuits shall be provided with overcurrent protection at each end as provided in Article 240.

Exceptions are under the conditions described in *NEC* Sections 450.6(A)(1) and (A)(2), and in these cases the overcurrent protection may be in accordance with *NEC* Section 450.6(A)(3).

(1) Loads at Transformer Supply Points Only. Where transformers are tied together (parallel) and connected by tie conductors that don't have overcurrent protection as per Article 240, the ampacity of the ties (connecting conductors) shouldn't be less than 67 percent of the rated secondary current of the largest transformer in the tie circuit. This applies where the loads are at the transformer supply points.

The paralleling of transformers is rather common, but great care should be taken to ensure that the transformers are similar in all characteristics. If they are not, one transformer will attempt to take all of the load. If the transformers are of equal capacity and similar characteristics, they would theoretically each take 50 percent of the load. The 67 percent allows for any difference in transformer sizes (See Figure 450-3).

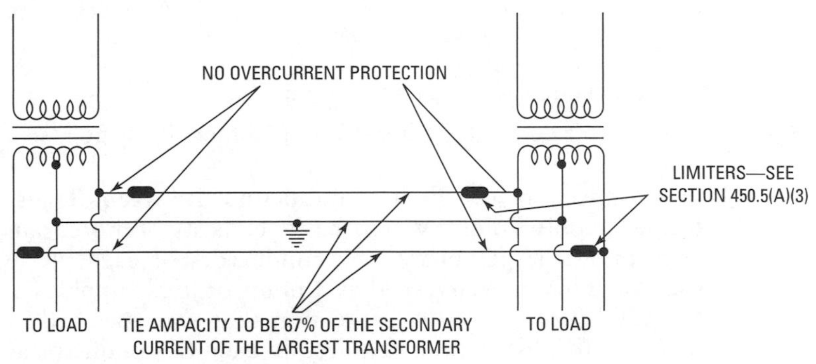

Figure 450-3 Secondary ties.

(2) Loads Connected Between Transformer Supply Points. If Article 240 is not used where load is connected to the tie points between the supply points to the transformer, and overcurrent protection has not been provided as in the article stated above, the ampacity rating of the ties is to be not less than 100-percent current rating of the largest transformer connected in the secondary tie system. See (A)(4) in this list.

(3) Tie Circuit Protection. In *NEC* Sections 450.6(A)(1) and (A)(2), both ends of each tie connection shall be equipped with a protective device that will open at a predetermined temperature of the tie conductor. This is to prevent damage to the tie conductor and its insulation, and may consist of (a) a limiter, which is a fusible link cable connector (not a common fuse) designed for the insulation, conductor material, and so on, on the tie conductors (ordinarily a copper link enclosed in a protective covering); or (b) a circuit breaker actuated by devices having comparable characteristics to the above (See Figures 450-3 and 450-4).

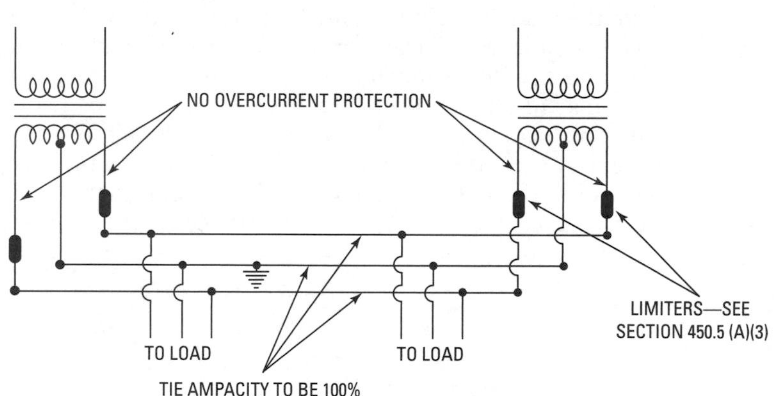

Figure 450-4 Loads connected between transformer supply points.

(4) Interconnection of Phase Conductors Between Transformer Supply Points. When the tie consists of more than one conductor per phase, the conductors of each phase shall be interconnected so that a point of load supply will be established. Also, the protection that was specified in (A)(3) in this list shall be required in each tie conductor at this point.

Exception
Loads may be connected to each conductor of a paralleled conductor tie, but it is not required to interconnect the conductors from each phase, or, without protection specified in (A)(3) in this list, at points for the load connection, provided the tie conductors of each phase have a combined ampacity of not less than 133 percent of the rating of the secondary current using the load of the largest transformer for this purpose that is connected to the tie system. The total load of these taps is not to exceed the rated current of the largest transformer; the loads must be equally divided on each phase and on the individual conductors of each phase as far as practical.

The use of multiple conductors on each phase and the requirement that loads don't have to tap both of the multiple conductors of the same phase could possibly set up unbalanced currents in the multiple conductors on the same phase. This is taken care of in the requirements that the combined capacity of the multiple conductors on the same phase be rated at 133 percent of the secondary current of the largest transformer. Limiters are necessary at the tap or connections to the transformers that are tied together.

(5) Tie Circuit Control. If the operating voltage of secondary ties exceeds 150 volts to ground, there shall be a switch ahead of the limiters and tie conductors that will deenergize the tie conductors and the limiters. This switch shall meet the following conditions: (a) The current rating of the switch shouldn't be less than the current rating of the conductors connected to the switch; (b) the switch shall be capable of opening its rated current; and (c) the switch shouldn't open under the magnetic forces resulting from short-circuit current.

(B) Overcurrent Protection for Secondary Connections. When secondary ties from transformers are used, there shall be an overcurrent device in the secondary of each transformer that is rated or set at not more than 250 percent of the rated secondary current of the transformer. In addition, there shall be a circuit breaker actuated by a reverse-current relay, the breaker to be set at not more than the rated secondary current of the transformer. The overcurrent protection takes care of overloads and shorts, while the reverse-current relay and circuit breaker takes care of any reversal of current flow into the transformer.

450.7: Parallel Operation

Section 450.6 covers ties for transformer connections, which, in essence, is a form of paralleling. However, this section covers the actual paralleling of transformers. Paralleled transformers may be switched as a unit provided that the overcurrent protection meets the requirements of Section 450.3(A)(1) or 450.3(B)(2).

Anyone working with paralleled transformers or transformer tie circuits should be extremely cautious that there are no feedbacks or other conditions that would affect safety. In order to secure a balance of current between paralleled transformers, all transformers should have characteristics that are very much alike, such as voltage, impedance, and so on. These pertinent facts may be obtained from the manufacturer of the transformers.

450.8: Guarding

Transformers shall be guarded as follows:

- **(A) Mechanical Protection.** When exposed to physical damage, transformers shall be protected to minimize damage from external causes.
- **(B) Case or Enclosure.** Dry-type transformers are to be mounted in noncombustible and moisture-resistant enclosures. Also, provision must be made to prevent the accidental insertion of foreign material from the outside into the enclosure.
- **(C) Exposed Energized Parts.** Switches or other control items that operate at low voltages and supply only items within a transformer enclosure may be installed in a transformer enclosure, as long as only qualified people will have access. Any energized parts must be protected according to the requirements of Sections 110.17 and 110.34.
- **(D) Voltage Warning.** Signs indicating the voltage of live exposed parts of transformers, or other suitable markings, shall be used in areas where transformers are located.

450.9: Ventilation

The ventilation shall be adequate to prevent a transformer temperature in excess of the values prescribed in *ANSI/IEEE C57.12.00-2000*. Care must be taken that ventilating openings in transformers are not blocked. The proper clearances must be marked on the transformer.

450.10: Grounding

Any exposed metal parts of transformer installations (fencing, guards, and so on) must be grounded in accordance with Article 250.

450.11: Marking

Nameplates with the following information shall be provided for transformers: name of manufacturer, rated kilovolt-amperes, frequency, primary voltage, secondary voltage, amount and kind of insulating liquid (where used), and impedance of transformers 25 kVA and larger.

The nameplate of each dry-type transformer shall include the temperature class for the insulation system.

450.12: Terminal Wiring Space

The bending space of fixed terminals of transformers of 600 volts or less, including terminals for both load and line connections, shall meet the specification as required in Section 312.6. The wiring space is covered in Table 314.6(B).

450.13: Location

Maintenance and inspection of transformers and transformer vaults shall be readily accessible to qualified personnel.

Exception No. 1

Dry transformers of 600 volts, nominal, or less, are not required to be readily accessible if located in the open, on walls, columns, or structures.

This requirement has never been enforced throughout the country simply because transformers where readily accessible have in many cases been damaged. It would be far better to elevate dry-type transformers when they are installed. This would not cause any further problems where vaults are considered or required.

Exception No. 2

See the *NEC*. This clarifies the installation of dry-type transformers above drop ceilings, where the hollow space is fire-resistant. The main requirement is that the ventilation be sufficient to maintain temperatures not in excess of those prescribed in *NFPA 251-1999*. A limitation of 50 kVA and 600 volts is put on transformers thus installed.

It is impossible to cover everything pertaining to transformer locations in this section, so oil-insulated transformers are found in Sections 450.26, 450.27, and 450.71. Dry-type transformers are covered in Section 450.21, and askarel-insulated transformers in Section 450.25.

Note

The *NEC* refers you to *ANSI/ASTM E119.83*, and *Fire Test of Building Construction and Materials, NFPA 251.1999.*

Note

The location of different type transformers is covered in Part II of this article. Transformer vaults are covered in Section 450.41.

II. Specific Provision Applicable to Different Types of Transformers

450.21: Dry-Type Transformers Installed Indoors

(A) **Not Over 112½ kVA.** Dry-type transformers of not over 112½ kVA that are mounted indoors shall be mounted at least 12 inches from combustible material unless there is a fire-resistant, heat-insulating barrier between the transformer and the wall, or unless the transformer has a rating of not over 600 volts, and, except for ventilation openings, is completely enclosed.

(B) **Over 112½ kVA.** Dry-type transformers of over 112½ kVA are to be installed in a fire-resistant room.

Exception No. 1

When the rating of a transformer is 80°C (176°F) or higher, and there is a fire-resistant barrier separating the transformer from combustible materials and the separating barrier is made of a fire-resistant, heat-insulating material, or is installed not less than 6 feet horizontally and 12 feet vertically from combustible material.

Exception No. 2

When the rise of a transformer is 80°C (176°F) or higher rated, and the transformer is completely enclosed, except for ventilating openings.

(C) **Over 35,000 Volts.** Transformers operating at 35,000 volts or higher must comply with Part III of this article, which covers installation in volts.

450.22: Dry-Type Transformers Installed Outdoors

Only weatherproof enclosures shall be used on transformers installed out-of-doors.

If the transformer exceeds 112½ kVA, it shouldn't be located within 12 inches of combustible materials in buildings, unless it has Class 155 insulation and is completely enclosed except for ventilation openings.

450.23: Less-Flammable Liquid-Insulated Transformers

When liquid that is less flammable is used for insulation in transformers and listed as such, they may be installed without being in a vault in Type I and II buildings and in areas that contain no combustible materials, provided arrangements are made for a liquid-tight area so that in the case of leakage the liquid is confined to this area. Liquid shall have a fire point that is 300°C (572°F)or more, and the installation shall comply with all restrictions provided for in the listing of the liquid. Such indoor transformers not meeting the liquid listing, or installed in other than Type I and II buildings, or in areas where combustible materials are stored, (1) must have a fire-extinguishing system and a liquid confinement area, or (2) are to be installed in a vault complying with Part III of this article.

When transformers rated over 35,000 volts are to be installed indoors, a vault for this must be provided.

Section 450.27 covers safeguards for transformers installed out of doors, and they shall comply with the requirements thereof.

Note

In *NFPA 220-1999 (ANSI)*, the word noncombustible refers to Types I and II building construction, and the materials defined in the building are found in this NFPA code.

Note

Listing is defined in Article 100 of the *NEC*.

450.24: Nonflammable Fluid-Insulated Transformers

When transformers are filled for insulating with a dielectric fluid and this fluid is identified or listed as nonflammable, they may be installed indoors or out-of-doors. If over 35,000 volts when installed indoors, they must be installed in a vault. The dielectric fluid that is nonflammable for the purpose of this section is one that does not have a flash point or fire point and is not flammable in air.

450.25: Askarel-Insulated Transformers Installed Indoors

If askarel-insulated transformers are installed indoors and have a rating of over 25 kVA, they must be manufactured with a pressure-relief vent, and if the area is poorly ventilated there must be other means provided for absorbing gases that would be generated by arcing inside the transformer case. The pressure-relief valve may also be connected to a chimney or other flue to carry the gases formed out of the building. Askarel-insulated transformers over 35,000 volts shall be installed in a vault.

Both askarel and the fumes from the same are extremely toxic, so great care must be taken where askarel is used.

450.26: Oil-Insulated Transformers Installed Indoors

All oil-insulated transformers installed indoors are to be in vaults, with the following exceptions:

(A) Not Over 112½ kVA Total Capacity. The specifications for vaults in Part III of this article may be modified so that the vault may be constructed of reinforced concrete not less than 4 inches thick.

(B) Not Over 600 Volts. A vault is not necessary when the transformer rating is 600 volts or less, if the following conditions are met: Proper means are provided to prevent a transformer-oil fire from igniting other materials, and if the total transformer capacity in this location does not exceed 10 kVA if the section of the building in which they are installed is classified as combustible, or if the total transformer capacity is from 10 to 75 kVA and the surrounding structure is classified as fire-resistant.

(C) Furnace Transformers. Electric-furnace transformers that don't exceed 75 kVA need not be installed in a vault if the building or room is of fire-resistant construction and if a suitable drain or other means is provided for preventing an oil fire from spreading to combustible materials.

(D) Detached Buildings. A building detached from other buildings—that is, an entirely separate building or one that may be defined as such by proper walls and the like—may be used for transformers if neither the building nor its contents constitute a fire hazard to any other building or property, and the building where the transformer is located is used only to supply electric service and has an interior that is accessible only to qualified persons.

(E) Oil Transformers Without a Vault. Oil-insulated transformers without a vault are permitted for portable or mobile surface mining equipment (such as electric excavators) provided each of the following conditions is met:

- **(1)** Provision must be made for draining leaking fluid to the ground.
- **(2)** Safeguards must be used for the protection of persons.
- **(3)** A steel barrier of ¼ inch minimum is provided to protect personnel.

450.27: Oil-Insulated Transformers Installed Outdoors

Oil-insulated transformers that are installed outdoors shall be installed so that no hazard to combustible materials, combustible buildings and parts of buildings, fire escapes, and door and window openings will exist in case of fire. These transformers may also be attached to or adjacent to buildings or combustible materials if precautions are taken to prevent fire hazards to the buildings or combustible materials. The following steps are recognized.

Some safeguards for hazards from transformer installations are as follows: space separations, automatic water spray systems, fire-resistant barriers, and enclosures that will confine oil from a rupture in the transformer tank. One or more of these safeguards shall be used, depending upon the degree of hazard involved.

The enclosure for oil from a rupture may consist of fire-resistant dikes, curbed areas or basins, or trenches filled with porous crushed stone. A basin made of concrete and filled with coarse, crushed stone will keep oil from seeping into the ground into underground water. Where the quantity of oil involved is such that removal of the oil is important, oil enclosures shall be provided with trap drains.

Note

For transformers mounted on poles, see the *National Electrical Safety Code, ANSI C2-1997.*

450.28: Modification of Transformers

If transformers are modified in existing installations in a way that changes the type of the transformer in any respect to Part II of this article, said transformer shall be marked to show the type of insulating liquid installed and what modifications were made to the transformer. All modifications shall comply with the requirements of that type of transformer.

III. Transformer Vaults

450.41: Location

Where practical, transformer vaults shall be located so that they will be ventilated to outside air without the use of flues or ducts.

450.42: Walls, Roof, and Floor

In the fine-print note to this section, the *NEC* refers to standards covering building materials, construction, and so on.

Transformer vaults shall have roofs and walls constructed of material adequate for structural conditions and with a minimum fire resistance of 3 hours. Floors of concrete in contact with the earth shall be less than 4 inches thick, but if the floor is mounted above other space below, the structural strength of the floor shall be adequate to handle the load placed on it, and it shall have a minimum of 3 hours' fire resistance. Stud and wallboard construction is not acceptable as a firewall under this section.

Note

NFPA 251 and *ANSI/ASTM E119* contain added information for fire-test of building materials.

Note

A typical 3-hour construction consists of 6 inches of reinforced concrete.

Exception

One-hour fire rating construction may be used, if the transformers are protected by automatic sprinklers, water spray, carbon dioxide, or halon.

450.43: Doorways

The following covers the protection of doorways and vault doorways:

(A) **Type of Door.** Doors for vaults are covered in the *NFPA Standard for the Installation of Fire Doors and Windows, No. 80*, and this is referenced in the fine-print note. They shall be rated at 3 hours and be tight fitting, and the inspection authority may require a door on each side of the wall.
See Exception to Section 450.42.

(B) **Sills.** The door shouldn't extend completely to the floor, but a sill shall be provided that is high enough to take care of the oil from the largest transformer in the vault. In no case shall this doorsill be less than 4 inches high.

(C) **Locks.** All doors shall be provided with locks that shall be kept locked so that access will be to qualified persons only. Latches and locks shall be such that they are readily opened from the inside of the vault. Many authorities require crash bars. All doors to transformer vaults must swing out.

450.45: Ventilation Openings

When required by Section 450.8, openings for ventilation shall be provided as follows:

(A) **Location.** Openings used for ventilation of vaults shall be kept as far away as possible from doors, windows, and combustible material.

(B) **Arrangement.** Where vaults are ventilated by a natural circulation of air, the total area of the openings for ventilation may be divided, with half of the area at the floor level and the remainder in one or more openings in the roof or near the ceiling. All of the area required for ventilation may be supplied by one or more openings near or in the roof.

(C) Size. When vaults are ventilated directly to the outdoor area, and ducts or flues are not used, the net area of the ventilation openings (after deducting the area that the screen, gratings, or louvers occupy) shouldn't be less than 3 square-inches per kVA of transformer capacity in the vault. For any vault with a transformer capacity of less than 50 kVA, the area of the ventilation opening shouldn't be less than 1 square foot.

(D) Covering. The covering for ventilation openings in vaults is left up to the inspection authority for the final decision, but they shall be covered with durable gratings, screens, or louvers. The final decision will be made to avoid unsafe conditions.

(E) Dampers. Dampers that possess a standard fire rating of not less than 1½ hours shall be provided with automatic closing dampers that will respond to a vault fire.

Note

The *NEC* refers to *ANSI/UL555-1995* for the standard for fire dampers.

(F) Ducts. Ventilating ducts are to be made of fire-resistant material.

When foreign systems can't be avoided, access to them shall be other than the entry into the vault. Any leaks or other malfunction of the foreign systems shouldn't cause damage to the transformers and their equipment.

450.46: Drainage

If vaults have more than 100 kVA transformer capacity, they shall be provided with a drain or other means that will carry off any accumulation of water or oil that might accumulate in the vault, unless this is impractical. The floor shall be sloped enough to aid in the drainage.

450.47: Water Pipes and Accessories

Systems foreign to the electrical system, such as pipes and ducts, shouldn't enter or pass through a transformer vault. Piping or other facilities for fire protection or cooling of transformers are not to be considered as being foreign to the electric system.

450.48: Storage in Vaults

Transformer vaults are just what the term implies, and are not to be used for material storage. This means that transformer vaults are only transformer vaults and not warehouses or storage areas. The vault is to be kept clear at all times—there is high voltage involved and safety is a very important factor.

Article 460—Capacitors

460.1: Scope

The installation of capacitors on electric circuits is covered in this article.

Surge capacitors are excluded from these requirements when they are a part of another apparatus and conform to the application for use on this apparatus.

Capacitors in hazardous (classified) locations are included in this article for installation therein, with some modification covered by Articles 501 through 503.

460.2: Enclosing and Guarding

(A) **Containing More Than 3 Gallons of Flammable Liquid.** Capacitors shall be enclosed in vaults or outdoor fenced enclosures complying with Article 710 if they contain more than 3 gallons of flammable liquid.

(B) **Accidental Contact.** Capacitors are to be enclosed or properly guarded to keep persons from coming into contact with them or from bringing conductive materials in contact with them. This covers all energized parts, terminals, or buses associated with them.

Where capacitors are installed where they are accessible only to authorized and qualified personnel, the above precautions may be omitted.

I. 600 Volts, Nominal, and Under

460.6: Drainage of Stored Charge

Capacitors store up a charge of electricity and the larger sizes may be lethal. Therefore, it is necessary to provide some means of draining off this charge.

(A) **Time of Discharge.** The plates of the capacitor retain a charge of voltage. This is called residual charge, and it shall be reduced to 50 volts, nominal, or less. This shall occur in

1 minute after the capacitor has been disconnected from the source of supply.

(B) Means of Discharge. The means of discharging a capacitor (such as a resistor) may be permanently connected to the terminals of the capacitor, or the disconnecting means to the capacitor may be so arranged that, upon opening, the capacitor is automatically connected to some discharging means. A manual discharging means is prohibited. The windings of transformers, motors, or other equipment that are directly connected to capacitors without a switch will act as the discharging means for the capacitor.

460.8: Conductors

(A) Ampacity. The two statements in this part are not to be confused. The ampacity of conductors to capacitors shouldn't be less than 135 percent of the rated current of the capacitor; power factor enters into this rating, as the capacitors have a leading power factor. The ampacity of the conductors connecting the capacitor to the terminals of a motor or to motor circuit conductors shouldn't be less than one-third the ampacity of the motor circuit conductors, and not less than 135 percent of the rated current capacity of the capacitor.

(B) Overcurrent Protection.

(1) Each ungrounded conductor to each capacitor shall be provided with an overcurrent device.

Exception

When a capacitor is connected to the load side of a motor overcurrent protection device, a separate overcurrent device is not required for the capacitor.

(2) The overcurrent device protecting the capacitor shall be of as low an amperage as possible.

(C) Disconnecting Means.

(1) Each ungrounded conductor to each capacitor shall have a means of disconnecting.

Exception

A separate overcurrent device shouldn't be required for a capacitor connected on the load side of a motor overload protective device.

(2) All ungrounded conductors to capacitors shall be opened at the same time.

(3) The capacitors may be disconnected by their disconnecting means as a normal operating procedure.

(4) The disconnecting means shall have a rating not less than 135 percent of the capacitor rated current.

460.9: Rating or Setting of the Motor, Overload Device

Due to the fact that a capacitor connected to a motor through the motor overload devices (running protection devices), and also due to the fact that a capacitor in a motor circuit corrects the lagging power factor of the motor, the current to the motor will be less than that of the same motor if it did not have capacitors for power-factor correction.

Use Section 430.32 for calculating the rating or setting of the motor overload devices, but the actual value of current drawn by the motor must be used instead of the nameplate current, which will be less in value.

Section 430.22 applies to the rating of the conductors. This is not derated because of the capacitor but is used as if the capacitor were not in the circuit.

460.10: Grounding

All capacitors must be grounded in accordance with Article 250, except capacitors that are mounted on structures that operate at potentials other than ground potential.

460.12: Marking

Capacitors must be provided with nameplates, which state the manufacturer's name, frequency, voltage, kilovar or amps, number of phases, and the amount of fill liquid (if combustible). Also, it must be noted whether or not the capacitor has an internal discharge device.

II. Over 600 Volts, Nominal

460.24: Switching

(A) Load Current. The switches for capacitors shall be group-operated, and shall be capable of:

(1) Continuously carrying the current of the capacitor at 135 percent of the current rating of the capacitor installation.

(2) Interrupting the maximum continuous load current of each capacitor, capacitor bank, or capacitor installation when switched as a unit.

(3) Withstanding the inrush current, not only of one capacitor, but including currents contributed from adjacent capacitor installations.

(4) Carrying fault currents that occur on the capacitor side of the switch.

(B) Isolation.

(1) All sources of voltage to each capacitor, capacitor bank, or capacitor installation shall have a means installed to isolate sources of voltage from any of these that will be removed from service as a unit.

(2) The isolating means must have a gap that is visible in the electric circuit and is also adequate for the operating voltage.

(3) Isolating or disconnecting means for capacitors (that don't have the load interrupting rating on the nameplate) must be interlocked with the load interrupting device, or they shall be permanently marked with signs that are prominently displayed as required by Section 710.22. This is to prevent switching of load current.

(C) Additional Requirements for Series Capacitors. The switching sequence shall be done in a proper manner by one of the following: (1) isolating and bypassing switches that are mechanically sequenced; (2) interlocks; or (3) the operating sequence shall be visibly displayed giving the procedures, and this shall be installed at the location of the switching.

460.25: Overcurrent Protection

(A) Provided to Detect and Interrupt Fault Current. There shall be provided a means of detecting and interrupting fault currents, which may cause dangerous pressure within an individual capacitor.

(B) Single-Phase or Multiphase Devices. You may use either single-phase or multiphase devices for this purpose.

(C) Protected Individually or in Groups. Capacitors may be protected either singly or in groups.

(D) Protective Devices Rated or Adjusted. Capacitors or capacitor equipment shall have the protective devices either rated or adjusted so that they will operate within the safety-zone limits for individual capacitors.

Exception

If the rating or adjustment of the protective devices is within the limits of Zones 1 or 2, it is required that the capacitors be isolated or enclosed.

Under no condition shall the maximum limits of Zone 2 be exceeded by the protective devices.

Note

The definition of Zones 1 and 2 may be found in the pamphlet standard for *Shunt Power Capacitors, ANSI/IEEE 18-2002.*

460.26: Identification

There shall be a permanent nameplate on the capacitor giving the maker's name, rated voltage, frequency, kilovar or amperes, number of phases, number of gallons of liquid contained therein, and, if the liquid is flammable, whether listed as flammable.

460.27: Grounding

The grounding of capacitor neutrals and cases, if grounded, shall meet the grounding requirements of Article 250, except where the capacitors are supported on a structure that is not intended to be grounded.

460.28: Means for Discharge

(A) Means to Reduce the Residual Voltage. Capacitors hold a dangerous amount of electrical energy so a means shall be provided to reduce the stored energy of the capacitor to 50 volts or less within 5 minutes after the capacitor is disconnected from the power source.

(B) Connection to Terminals. The discharging as required in (A) may be accomplished by means of a permanently connected discharge circuit to the terminals of the capacitor, or an automatic means of connection to the terminals of the capacitor bank may be used that immediately connects to the terminals after the disconnection of the capacitor bank from the source of supply.

The requirements of (A) above must be met if the windings of motors or other equipment are connected to capacitors without a switching device.

Article 470—Resistors and Reactors

For rheostats, see Section 430.62

These types of installations are very seldom performed anymore. For specific information, refer to the *Code*.

Article 480—Storage Batteries

480.1: Scope

This article applies to all stationary storage battery installations.

480.2: Definitions

Definitions are given for storage batteries, sealed cells, and nominal battery voltage.

480.3: Wiring and Equipment Supplied from Batteries

Wiring and equipment that is supplied by storage batteries must conform to requirements given in other sections of the *Code* that relate to wiring and equipment operating at that specific voltage.

480.4: Grounding

All storage battery installations must be grounded in accordance with Article 250.

480.5: Insulation of Batteries Not Over 250 Volts

- **(A)** **Vented Lead-Acid Batteries.** No additional insulation is required for vented lead-acid batteries that are contained in sealed, nonconductive, heat-resistant material.
- **(B)** **Vented Alkaline-Type Batteries.** No additional insulation support is required for cells with covers sealed to jars made of nonconductive, heat-resistant material. No more than 20 cells in jars of conductive material (24 volts each) can be installed in series in one tray.
- **(C)** **Rubber Jars.** No extra insulating support is required when the voltage of all cells in series is not more than 150 volts. When this voltage does exceed 150 volts, the cells must be placed in groups of 150 volts or less, and the cells installed on trays or racks.
- **(D)** **Sealed Cells or Batteries.** No extra insulating support is required for these batteries when they are made of heat-resistant, nonconductive material. Insulating support is required for this type of battery when it has a conductive container, and if a voltage is present between the container and ground.

480.6: Insulation of Batteries Over 250 Volts

In addition to the requirements of Section 480.5, cells installed in groups of over 250 volts must have insulation between the groups (air may be considered insulation). Battery parts of opposite polarity (not over 600 volts) must be separated by at least 2 inches.

480.7: Racks and Trays

(A) Racks. Racks for storage batteries may be made of:

(1) Metal that is treated so that the electrolyte won't corrode it and has insulating members supporting the batteries.

(2) Fiberglass, or similar suitable materials.

(B) Trays. Trays or frames must be made of materials that won't be corroded by battery electrolytes.

480.8: Battery Location

The installation locations for storage batteries must conform with the following:

(A) Ventilation. Enough ventilation must be provided to dissipate battery gasses, which can be explosive.

(B) Live Parts. Live parts must be guarded. See Section 110.27.

(C) Working Space. Working space surrounding battery installations shall comply with Section 110.26. Measurements are made from the edge of the battery rack.

480.9: Vents

(A) Vented Cells. All ventilated cells must be provided with flame arrestors, which prevent the cell from being destroyed if a spark or flame ignites the battery gasses.

(B) Sealed Cells. Sealed cells or batteries must have a pressure-release vent, preventing excessive gas pressure, which can lead to explosion.

Article 490—Equipment Over 600 Volts, Nominal

I. General

490.1: Scope

This article is intended to give the general requirements to be used on all circuits and equipment operated at more than 600 volts, nominal. (See Figures 490-1 through 490-4).

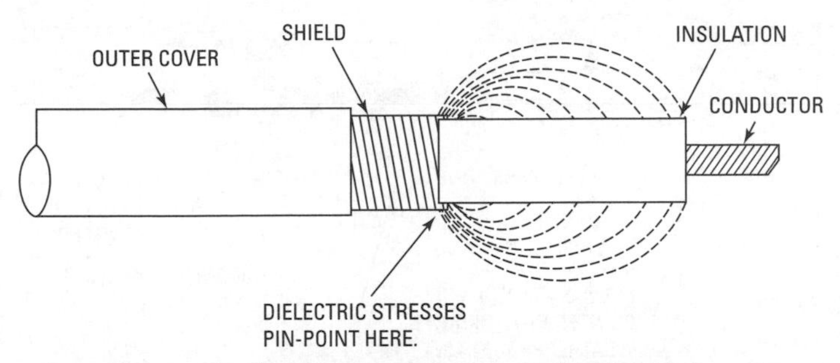

Figure 490-1 Dielectric stresses in a shielded cable.

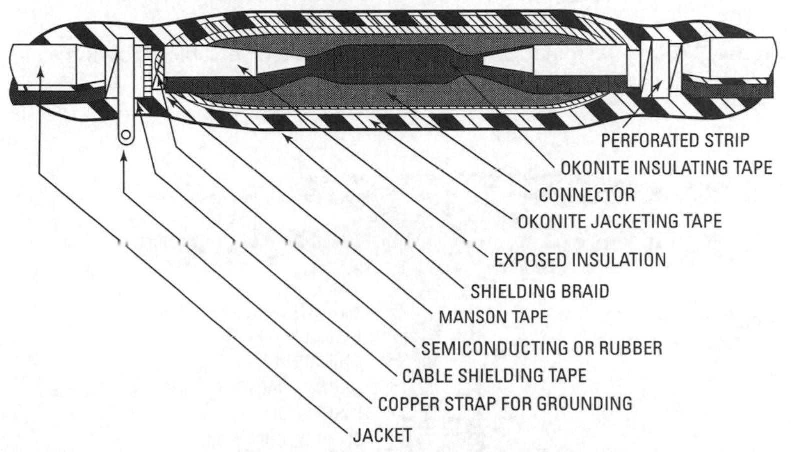

Figure 490-2 A straight-through splice of high-voltage shielded cable.

490.2: Definition

For this article (only), the term *high voltage* refers to voltages over 600 volts.

These stresses pinpoint back to the termination of the shielding and cause undue strain at that point, with accompanying flashovers or breakdowns. The purpose of stress cones, potheads, terminators, and so on, is to spread these stresses out and not let them concentrate at one spot on the cable.

490.3: Oil-Filled Equipment

Except for transformers covered by Article 450, all other devices containing more than 10 gallons of oil must comply with Parts II and III of Article 450.

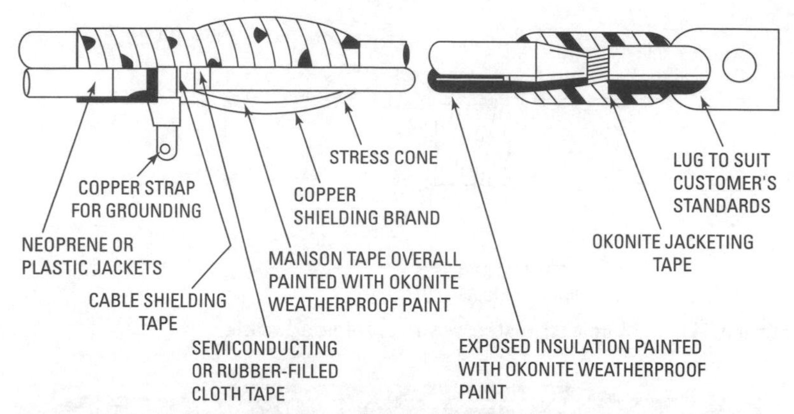

Figure 490-3 Termination of shielded high-voltage cable, showing proper shielding, and so on.

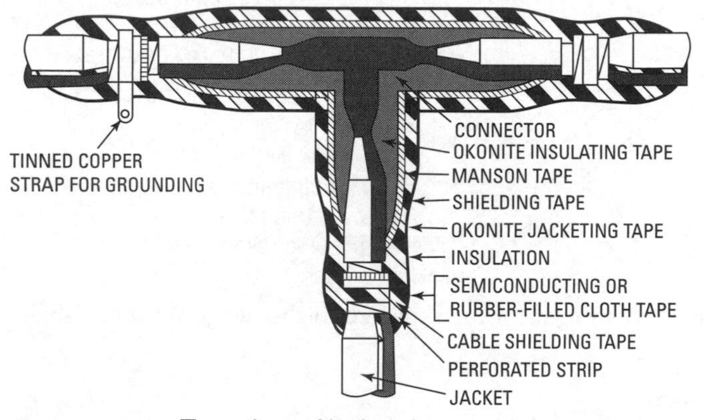

Figure 490-4 Tee-splice of high-voltage cable, showing how properly made and properly shielded.

II. Equipment—Specific Provisions

490.21: Circuit Interrupting Devices

(A) Circuit Breakers.

(1) Indoor installations shall consist of metal-enclosed units or fire-resistant cell-mounted units except that open mounting of circuit breakers is permissible in locations accessible to qualified persons only.

Exception
In locations accessible to qualified persons only, open circuit breakers may be mounted.

(2) When oil-filled transformers are used, circuit breakers shall either be located outside the transformer vault or be capable of operation from outside the vault.

(3) Any adjacent readily combustible structures or materials shall be safe-guarded in an approved manner when oil circuit breakers are used.

(4) The following equipment or operating characteristics shall be used with circuit breakers. Refer to the *NEC* for these, which include (a) through (f).

(5) The continuous current rating of a circuit breaker shall be not less than the continuous circuit that may pass through the circuit breaker.

(6) The maximum fault current that a circuit breaker may be required to interrupt shall be covered by a circuit breaker with a rating of this amount of fault current, and it shall include any contribution to the circuit from other sources of energy.

(7) The circuit breaker that is used for closing a circuit shall have not less than the maximum asymmetrical fault current into which the circuit breaker can be closed.

(8) The momentary rating of the circuit breaker shouldn't be less than the maximum amount of asymmetrical fault current available at the point of installation.

(9) The maximum voltage rating of the circuit determines the maximum voltage of the circuit breaker.

(B) Power Fuses and Fuseholders.

(1) Use. When fuses are used to protect conductors and equipment, a fuse shall be placed in each ungrounded conductor. Two fuses in parallel will be allowed to protect the same load provided that both fuses have the same rating and are installed in an approved common mounting, and the connections between the paralleled fuses shall be such as to divide the current equally. Unless approved for such usage, ventilated power fuses are not to be used indoors, underground, or in metal enclosures.

(2) Interrupting Rating. The interrupting rating of the power fuses shouldn't be less than the maximum fault current that the fuses are intended to interrupt, including all sources of other energy.

(3) Voltage Rating. The maximum voltage of the circuit determines the maximum voltage rating of power fuses. Fuses having a minimum operating voltage are not to be used for supplying circuits below the minimum rating of the fuse.

(4) Identification of Fuse Mounting and Fuse Units. The mountings for the fuses and the fuse units shall be identified by a permanent legible nameplate that shows not only the manufacturer's type or designation, but also the continuous current rating, the maximum voltage rating, and the current interruption rating.

(5) Fuses. If fuses expel flame when the circuit is opened, their design and arrangement shall be such that they will operate properly without causing any harm to persons or property.

(6) Fuseholders. All fuseholders must be so designed that they are in the open position and will be deenergized while replacing a blown fuse.

Exception

If they are designed so that only qualified persons replace them by the use of approved equipment designed for the purpose, they may be replaced without deenergizing the fuseholder.

(7) High-Voltage Fuses. When metal switchgear and substations use high-voltage fuses, they must be designed to be operated by a gang disconnecting switch. Fuses are to be isolated from the circuits either by using a switch between the fuses or by means of a roll-out switch and fuse construction. A switch must be of the load-interrupting type, unless either mechanical or electrical interlocks are provided with load-interrupting devices so as to reduce the interrupting possibility of the switch.

Exception

For the disconnecting means, more than one switch will be permitted for one set of fuses if the switches are provided so that more than one set of supply conductors are connected. An interlocking device shall be provided so that access to the fuses can only be gained when all switches are open. A conspicuous sign shall be placed at the fuses reading, "WARNING—FUSES MAY BE ENERGIZED FROM MORE THAN ONE SOURCE."

(C) Distribution Cutouts and Fuse Links—Expulsion Type. Refer to the *NEC* for (1) through (7).

(D) Oil-Filled Cutouts.

(1) Continuous Current Rating. The oil-filled cutouts shall have a continuous current rating that isn't less than the maximum continuous current that will pass through the cutout.

(2) Interrupting Rating. The interrupting rating of oil-filled cutouts shall meet the maximum fault current, or greater, that the cutout will be required to interrupt. This rating will include contributions from any other sources of energy.

(3) Voltage Rating. The maximum voltage rating of an oil-filled cutout shall be at least the maximum voltage of the circuit it serves.

(4) Fault-Closing Rating. The fault-closing rating of the oil-filled switch rating shall be at least the maximum asymmetrical fault current that can occur at the cutout location unless there are interlocks or suitable procedures that can prevent the possibility of closing into a fault.

(5) Identification. There shall be a permanent and legible nameplate on all oil-filled cutouts that shows the rating of the continuous current, the maximum voltage, and the amount of current it will interrupt.

(6) Fuse Links. There shall be a legible identification that shows the rated continuous current on fuse links. This shall be a permanent rating.

(7) Location. The location of cutouts shall be readily accessible and the top of a cutout shall be not over 5 feet above the floor or platform.

(8) Enclosure. To prevent contact with nonshielded cables or any energized parts of oil-filled cutouts, suitable barriers or enclosures shall be installed.

(E) Load Interrupters. Switches that will interrupt load may be installed, provided suitable fuses or circuit breakers are used along with these devices so that they may interrupt fault currents. Where these devices are used in combination, the electrical coordination between them shall be such that they will safely withstand the effects of closing, carrying, or interrupting all currents. This includes the assigned maximum short-circuit rating.

Where more than one switch is installed and has interconnected load terminals that will provide alternate connections to different supply conductors, it is required that each switch be provided with a conspicuous sign reading, "WARNING: SWITCH MAY BE ENERGIZED BY BACKFEED."

(1) Continuous Current Rating. Interrupting switches shall have continuous current rating that is equal to or greater than the maximum current that can occur at the point of installation.

(2) Voltage Rating. The maximum voltage rating of cutouts shall be equal to or greater than the maximum voltage of the circuit involved.

(3) Identification. Cutouts used for distribution must have on their body or on the door or fuse tube a legible nameplate or other identification showing manufacturer's type or designation, maximum voltage rating, continuous current rating, and also their current interrupting rating.

(4) Switching of Conductors. The mechanism for switching is to be arranged to be operated from a location that won't expose the operator to energized parts, and it must open all ungrounded conductors on the same circuit at the same instant by means of only one operation. Arrangement must be provided to lock switches in the open position. Metal-enclosed switches must be operated from outside the enclosure.

(5) Stored Energy for Opening. The energizing storing apparatus may be left uncharged after the switch has been enclosed, provided that by a single movement of the operating handle it charges the operator and opens the switch.

(6) Supply Terminals. All supply terminals that supply fuse-interrupted switches must be at the top of the enclosure for the switch.

Exception

If there is a barrier provided in the enclosure, so installed as to prevent persons from accidentally contacting the energized parts or dropping tools or fuses into the energized parts, the supply terminals shouldn't be required to be at the top of the switch enclosure.

490.22: Isolating Means

Means shall be provided so that any item of the equipment may be completely isolated. It is not necessary to use isolation switches if there are other means for deenergizing the equipment for inspection and repairs. These could be drawout-type, metal-enclosed switchgear, or removable truck panels.

A sign warning against opening under load will be required on isolating switches not on interlock with circuit-interrupting devices that have been approved.

A fuse and fuseholder will be permitted as an isolating switch if designed for the purpose.

490.23: Voltage Regulators

Proper switching sequence for regulators must be ensured by the use of one of the following: (1) regular bypass switches that are mechanically sequenced; (2) interlocks that are mechanically operated; or (3) instructions for switching procedure that are prominently displayed at the switching location.

490.24: Minimum Space Separation

This section deals with interior wiring design and construction. It does not apply to the space separation provided in electrical apparatus and wiring devices.

Table 490.24 in the *NEC* lists the minimum indoor air separation between bare live conductors and between such conductors and adjacent surfaces. These are the minimum values—the spacing may be greater but not less than the spacing shown in the table. (See Table 490.24, "Minimum Clearance of Live Parts.")

III. Equipment—Metal-Enclosed Power Switchgear and Industrial Control Assemblies

This Part of Article 490 deals with the manufacture of metal-enclosed power switchgear. See the *NEC* for details.

IV. Mobile and Portable Equipment

Much of Part IV is covered in previous sections of this article, so instead of repeating it, we refer you to Part IV in the *National Electrical Code*.

High-voltage equipment is, of course, subject to hazards and therefore only qualified persons should work on it; they will know the proper respect to show it. Grounding is extremely important. The author's suggestion is that all grounds be thoroughly checked for ground resistance and also periodic inspections be made to be certain the grounding is intact.

Chapter 5

Special Occupancies

Articles 500 through 517 refer to hazardous areas and should be well understood by anyone concerned with the wiring in such locations. Such areas are dangerous from many standpoints and each has its own problems that require special methods for taking care of the electrical systems installed in these places.

Article 500—(Classified) Locations

500.1: Scope—Articles 500 Through 505

It is the responsibility of the *Code*-enforcing authority to judge whether or not areas come under Articles 500 through 505, as indicated by the classifications indicated in these articles. These articles cover locations where fire or explosive hazards may exist due to the following: flammable gases, vapors, liquids, combustible dusts, ignitable fibers, or flyings.

500.2: Location and General Requirements

Locations are classified as Class I, Class II, or Class III. The location is classified according to the properties and materials therein, such as flammable vapors, liquids, or gases, or combustible dust or fibers that may be present and could cause flammable concentrations, or the quantity of such. Where pyrophoric materials are the only materials used or handled, they are not to be classified.

Note

A pyrophoric material is one that is capable of igniting spontaneously when exposed to air.

The intent is that each room, section, or area (including motor and generator rooms and rooms for enclosure of control equipment) be individually considered in determining the classification suitable for the conditions.

In judging these areas and what equipment is allowable therein, the inspection authority refers to the *Underwriter's Laboratories* listings for the equipment that is installed in each area. All other wiring methods of the *Code* are applicable except as modified by these articles covering hazardous areas.

The following terms will appear often: approved; dustproof; dust-ignition-proof; dust-tight; and explosion-proof apparatus.

Refer to Article 100 in the *NEC* for the definitions of these terms, except dust-ignition-proof, which is defined in Section 502.1.

Equipment and associated wiring approved as intrinsically safe may be installed in any hazardous location for which it is approved, and the provisions of Article 500 through 517 need not apply to such installation. Intrinsically safe equipment and wiring are incapable of releasing sufficient electrical energy under normal and abnormal conditions to cause ignition of a specific hazardous atmospheric mixture. Abnormal conditions include accidental damage to any part of the equipment or wiring, insulation- or other failure of electrical components, application of overvoltage, adjustment and maintenance operations, and other similar conditions.

In designing the wiring and selecting the equipment to use in a hazardous location, less expensive materials and equipment can often be used by relocation or by adequate ventilation, forced or otherwise. Anyone doing installation in hazardous areas, but more especially inspectors, should become familiar with other codes published by the National Fire Protection Association, such as *Flammable and Combustible Liquids Code, NFPA 30-2000*; *Dry Cleaning Plants, NFPA 32-2000*; *Manufacture of Organic Coatings, NFPA 35-1999 (ANSI)*; *Solvent Extraction Plants, NFPA 36-2001 (ANSI)*; *Storage and Handling of Liquefied Petroleum Gases, NFPA 58-2001*; *Storage and Handling of Liquid Petroleum Gases at Utility Gas Plants, NFPA 59-2001*; and *Classification of I Hazardous Locations for Electrical Installation in Chemical Plants, NFPA 497-1997 (ANSI)*.

All references in the fine-print notes intend to direct users of the *Code* to other NFPA codes and standards. See *NEC* Section 110.1.

Designers and inspectors will find many items pertaining to the subjects covered in this chapter in many of the NFPA Codes. These codes often assist in many decisions that must be made, especially by the inspector.

For electrical installations in hazardous areas where it is necessary to use threaded rigid metal conduit, the *Code* requires that the joints be made up wrench-tight —this applies to threaded joints and connections. Please note that the *Code* states wrench-tight and not plier-tight. In addition, standard conduit dies that provide a taper of ¾ inch per foot shall be used. All of this amounts to the fact that, even if the proper equipment is used and the wiring methods are as required, a loose threaded connection, when subject to fault current, may spark and cause an explosion. Where it is impossible to make a proper threaded connection, the joint shall be bonded with a bonding jumper.

Note
Static electrical hazards are potentially very dangerous in some hazardous locations. See *Recommended Practice on Static Electricity, NFPA 77-2000 (ANSI)* for coverage.

Note
The standard electrical classification for laboratories using chemicals may be found in *NFPA 45-2000*.

500.3: Special Precaution
The intent of Articles 500 through 505 is that more than ordinary precaution must be exercised in the construction of equipment and the installation and maintenance of the entire wiring system for safe operation. This becomes the responsibility of all concerned in the matter.

The atmospheric mixtures of various gases, vapors, and dusts, and the hazard involved, depends upon the concentrations, temperatures, and many other things. It is impossible to cover all these variables here, but they may be found in the various NFPA codes.

Note 1
Inspection authorities and users must use more than ordinary care in the installation or maintenance of wiring in hazardous conditions.

Note 2
Explosive conditions of gas vapors or dust vary greatly with specific materials that are involved in the determination of maximum pressure of explosion, the maximum distances for safe clearances between clamp joints in an enclosure, and the minimum temperature of explosive conditions in the atmospheric mixture with the gases that refers to Class I areas. Referring to Class II locations, including Groups E, F, and G, the classification involves the tightness of the joints of assembly and shaft openings to prevent entrance of dust in the dust-ignition-proof enclosure, the blanketing effect of layers of dust on the equipment that may cause overheating, electrical conductivity of the dust, and the ignition temperature of the dust. It is necessary, therefore, that equipment be approved not only for the class, but also for the specific group of gas, vapor, or dust that will be present.

Note 3
Low ambient temperatures require special consideration. It may not be necessary to use explosion-proof or dust-tight equipment when temperatures are lower than −25°C (−13°F), but equipment must be suitable

for low-temperature service. However, care must be taken, because if the ambient temperature rises, the temperature of flammable concentrations of vapors may not exist in the location that is classified Class I, Division I when the ambient temperature is normal.

Note 4

For testing purposes, approval, and the classification of areas, various air mixtures (not including oxygen-enriched) are grouped according to their characteristics, and facilities are available for the testing of equipment and approval for use in the following atmospheric groups. See the *NEC* for Notes 5 through 18.

(A) Approval for Class and Properties. Equipment in hazardous locations must be approved not only for the class of location in which it is to be used, but also for the explosive, combustible, and ignitable properties of the particular items involved, such as gas, vapor, dust, fiber, or flyings. In addition, Class I equipment may not have any surfaces that are exposed to the igniting temperature of the specific gas or vapor. Later you will find that lighting and painting booths require special installation. It is required that Class II equipment not have a higher temperature than that specified in Section 500.3(D). For Class III equipment, the maximum surface temperature specified by Section 503.1 shouldn't be exceeded.

When equipment is approved for Division 1 locations, permission is granted to use it in Division 2 locations when the class and group are the same.

Articles 501 through 503 cover normal operating conditions, where sources of ignition are not encountered. If specifically permitted, general-purpose equipment or equipment in general-purpose enclosures may be installed in Division 2 locations.

Motors are to be considered as operating steadily under full-load conditions unless otherwise specified.

If either flammable gases or combustible dusts might be present, the presence of both has to be considered when determining what the safe operating temperature of the electrical equipment is.

Note

The various atmospheric mixtures of gases, vapors, and dusts depend upon the characteristics of the materials involved.

(B) **Marking.** Approved equipment shall always be marked showing class, group, and operating temperature. This is referenced to as a 40°C (104°F) ambient temperature.

For the balance of (B) and Table 500.3(B), refer to the *NEC.*

Equipment that is approved for Class I and Class II should be marked with the maximum safe operating temperature. This operating temperature is determined by exposure to the combination of Class I and Class II conditions that occur simultaneously.

Exception No. 1

Operating temperature is not required to be marked on equipment such as junction boxes, conduits, and fittings when the temperature does not exceed 100°C or 212°F.

Exception No. 2

Marking indicating the group is not required for fixed lighting fixtures that are marked either Class I, Division 2, or Class II, Division 2.

Exception No. 3

It is not required that fixed general-purpose equipment be marked with a class, group, division, or operating temperature. In Class I locations this does not include fixed lighting fixtures that are acceptable for Class I, Division 2 locations.

Exception No. 4

Fixed dust-tight equipment, not including fixed lighting fixtures, that is acceptable for use in Class II, Division 2, and Class III locations is not required to be marked with the class, group, division, or operating temperature.

See Table 500.3(B), "Identification Numbers."

Note

For testing and approval, various atmospheric mixtures that are not oxygen-enriched have been grouped according to their characteristics, and facilities have been available for testing and approving equipment for use in the atmospheric groups listed in *Classification of Gases, Vapors, and Dusts if Listed for Use in Hazardous (Classified) Locations, NFPA 497-1997.* Since there is no consistent relationship between explosive properties and ignition temperatures, the two are independent requirements.

(C) **Class I Temperature.** The temperature marking specified in (B) above is not to exceed the ignition point of the specific vapor or gas encountered in that area. See the *NEC* for the balance of (C).

(D) **Class II Temperature.** The surface temperature specified in (B) and so marked is to be less than the ignition temperature of the type of dust, and it shouldn't be permitted in any case to be greater than the temperature given below for Groups E, F, and G. See *Classification of Gases, Vapors, and Dusts for Electrical Equipment in Hazardous (Classified) Locations, NFPA 497-1997*, for the specific minimum temperature of ignition for specific dust.

Exception

When equipment is subject to overloads, it shouldn't exceed 150°C (302°F) in normal operation, and the ignition temperature of the dust shouldn't be in excess of 200°C (392°F), whichever is lower, when the installation is in a location that is classified due to the presence of carbonaceous dusts.

See the table in the *NEC* at this point for maximum surface temperatures that were approved prior to this requirement.

500.4: Specific Occupancies

See Articles 510 to 517, inclusive, for rules applying to garages, aircraft hangars, gasoline-dispensing and service stations, bulk storage plants, spray application, dipping and coating processes, and health care facilities.

500.5: Class I Locations

Locations in which flammable gases and vapors are or may be present in the air in sufficient quantity to produce an explosion or a mixture that is ignitable are Class I locations. Included in Class I locations are those that appear in (A) and (B) below:

(A) **Class I, Division 1.** This is a location (1) in which under normal operating conditions there will be ignitable concentrations of flammable gases or vapors; or (2) in which concentrations of ignitable gases or vapors frequently exist during repair or maintenance or because of leakage; or (3) in which breakdown of operations that are faulty due to faulty equipment or processes could possibly release ignitable concentrations of flammable gases or vapors, and also at the same time cause the electrical equipment to fail.

Class I, Division 1 areas also include (1) locations where volatile flammable liquids or liquefied flammable gases are transferred from one container to another; (2) interiors of spray booths; (3) areas in the vicinity of spraying and painting operations where volatile flammable solvents are used; (4) areas with open tanks or vats of volatile flammable liquids; (5) drying rooms or compartments for the evaporation of flammable solvents; (6) areas containing fat- and oil-extraction apparatus using volatile flammable solvents; (7) portions of cleaning and dyeing plants where hazardous liquids are used; (8) gas generator rooms and other portions of gas manufacturing plants where flammable gas may escape; (9) inadequately ventilated pump rooms for flammable gas or for volatile flammable liquids; (10) interiors of refrigerators or freezers in which volatile, flammable materials are stored in open, lightly stoppered, or easily ruptured containers; (11) any other locations where flammable vapors or gases are likely to occur in the course of normal operations.

(B) Class I, Division 2. A Class I, Division 2 location is a location (1) in which volatile liquids or flammable gases may be handled, processed, or used, but where the liquids, vapors, or gases will be in closed containers or closed systems from which they normally would not escape except in the case of an accidental rupture or breakdown of equipment; (2) in which positive mechanical ventilation would normally prevent concentrations of ignitable gases or vapors, or that may become a hazardous location due to the failure of ventilation equipment; or (3) adjacent to Class I, Division 1 locations where ignitable concentrations of gases or vapors might occasionally pass into that area unless the adjacent area has a positive pressure of clean air and safety guards for ventilation failure are provided.

The locations under this classification usually include those where volatile flammable liquids or flammable gases or vapors are used, but in the judgment of the *Code*-enforcing authority would become hazardous only in the case of accident or unusual operations. In deciding whether or not this would be a Class I, Division 2 location, the *Code*-enforcing authority must take into account the quantity of hazardous materials that might escape in the case of an accident, the records of similar locations with respect to fires and explosions, and any other conditions that might affect the amount of hazard present or the possible amount of hazard present.

Piping that is without valves, checks, meters, and similar devices would not ordinarily be considered capable of introducing a hazard; neither would storage areas for hazardous liquids or liquefied or compressed gases in sealed containers that would normally not be hazardous.

There are special instances where storage areas for material such as anesthetics are considered Class I, Division 1 locations.

If conduits and their fittings have a single seal or barrier separating process fluids, they will be classified as Division 2 if the outside conduit and enclosures are in a nonhazardous area.

500.6: Class II Locations

The presence of combustible dust creates Class II locations and includes the following:

(A) **Class II, Division I.** Class II, Division 1 locations are as follows: (1) where combustible or explosive dusts are in the air when operating normally in quantities sufficient to cause an explosive or ignitable mixture; (2) where mechanical problems or any other abnormal operation of the machinery or equipment could possibly cause an explosive or ignitable mixture to be formed, and could also be the source of ignition through failure of electrical equipment or operational protection devices, or from some other cause; or (3) where mixtures of dusts in the air of an electrical conducting nature may be present in hazardous quantities.

Also included in this classification are (1) grain handling and storage plants; (2) rooms containing grinders or pulverizers, cleaners, graders, scalpers, open conveyors or spouts, open bins or hoppers, mixers or blenders, automatic or hopper scales, packing machinery, elevator heads and boots, stock distributors, dust and stock collectors (except all-metal collectors vented to the outside), and all similar dust-producing machinery and equipment in grain-processing plants, starch plants, sugar-pulverizing plants, melting plants, hay-grinding plants, and any other occupancy of a similar nature; (3) coal-pulverizing plants, unless the pulverizing equipment is essentially tight; (4) all working areas where metal dusts and powders are produced, processed, handled, packed, or stored (except in tight containers); and (5) all other similar locations where combustible dust may, under normal operating conditions, be present in the air in sufficient quantities to produce explosive or ignitable mixtures.

(B) **Class II, Division 2.** This classification covers locations in which, during the normal operation of apparatus and equipment, combustible dust is not likely to be in suspension in the air in quantities sufficient to produce explosive or ignitable mixtures, but where deposits or accumulations of such dust may be of sufficient quantity to interfere with the safe dissipation of heat from electrical equipment and apparatus; or where such deposits or accumulations of dust on or in electrical equipment or apparatus may be ignited by sparks, arcs, or burning material from same.

The decision as to whether an area is a Class II, Division 1 or a Class II, Division 2 location is determined by the inspection authority. Such locations might include (1) rooms containing closed spouts and conveyors; (2) closed bins or hoppers; (3) rooms with machines or equipment from which appreciable quantities of dust might escape under abnormal operating condition; (4) rooms adjacent to Class II locations; (5) rooms in which controls over suspension of dust in explosive or ignitable quantities is exercised; (6) warehouses and shipping rooms where dust-producing materials are stored; and (7) any similar location.

Note

The quantities of dust present and the adequacy of dust removal systems are both factors that should be considered in arriving at the classification of the location; it may result in an unclassified area.

Note

In areas such as where seed is handled, the quantity of dust that is deposited may be so low as not to require classification of that location.

500.7: Class III Locations

Class III locations include those where ignitable fibers or flyings are not likely to be in suspension in high enough quantities to produce explosive or ignitable conditions, but in which they are present. The following are included in Class III locations.

(A) **Class III, Division 1.** Locations in which easily ignitable fibers or materials producing combustible flyings are handled, manufactured, or used. Such locations may include (1) some parts of rayon and cotton or other textile mills; (2) combustible fiber manufacturing and processing plants; (3) cotton gins and cotton-seed mills; (4) flax processing plants; (5) clothing manufacturing plants; and (6) other establishments and industries involving similar hazards.

Included in easily ignitable fibers and flyings are rayon, cotton, cotton linters and cotton waste, sisal or henequen, istle, hemp and jute, tow, cocoa fiber, oakum, baled waste kapok, Spanish moss, excelsior, other similar materials, and sawmills and other woodworking locations.

(B) **Class III, Division 2.** Locations in which easily ignitable fibers are stored or handled (except in process of manufacture).

Article 501—Class I Locations

501.1: General

The general requirements of the *Code* covering wiring and the installation thereof and the provisions of Article 500 as classified under Section 500.5 apply for Class I locations with the modifications covered in this article.

501.2: Transformers and Capacitors

It was stated in Articles 450 and 460 that transformers and capacitors in hazardous locations would also be affected by the articles covering hazardous locations. The installation of transformers and capacitors shall conform to the following:

(A) **Class I, Division 1.** In Class I, Division 1 locations, transformers and capacitors shall conform to the following:

(1) **Containing a Liquid That Will Burn.** Transformers and capacitors containing a flammable liquid shall be installed only in vaults conforming to the provision in Part III of Article 450, Sections 450.41 to 450.48, inclusive, but the following also applies: (a) There shall be no door or other communicating opening between the vault and the hazardous area; and (b) sufficient ventilation shall be provided to remove hazardous vapors or gases; all vent ducts and openings shall lead to a safe location outside of the building; vent ducts and openings shall be of sufficient size to relieve any explosive pressures that might occur in the vault; and all portions of vent ducts within the building shall be of reinforced concrete.

(2) Not Containing a Liquid That Will Burn. Transformers and capacitors that don't contain a flammable liquid shall (a) be installed in vaults conforming to the requirements in (A)(1) above, or (b) be approved for explosion-proof Class I locations.

(B) Class I, Division 2. Transformers and capacitors in Class I, Division 2 must comply with Sections 450.21 through 450.27.

501.3: Meters, Instruments, and Relays

The installation of meters, instruments, and relays shall conform to the following:

(A) Class I, Division 1. Enclosures that are approved for Class I locations shall be used in Class I, Division 1 locations for meters, instruments, and relays, including kilowatt-hour meters, instrument transformers and resistors, rectifiers, and thermionic tubes. It should be determined that the Group Letter is applicable for the location.

Included in enclosures approved for Class I, Division 1 locations are (1) enclosures that are explosion-proof and (2) enclosures that have been purged and pressurized.

Note

The *NEC* refers you to *NFPA 496-2003 (ANSI), Purged and Pressurized Enclosures for Electrical Equipment in Hazardous Locations.*

(B) Class I, Division 2. The installation of meters, instruments, and relays in Class I, Division 2 locations shall conform to the following:

(1) Contacts. Switches and circuit breakers, and make-and-break contacts of push buttons, relays, and alarm bells or horns, shall have enclosures approved for Class I locations, unless general-purpose enclosures are provided, and current interrupting contacts are (a) immersed in oil; (b) enclosed in a chamber hermetically sealed against the entrance of gases or vapors; or (c) in circuits that under normal conditions don't release enough energy to ignite the specific hazardous atmospheric mixture.

(2) Resistors and Similar Equipment. Resistors, resistant devices, thermionic tubes, and rectifiers that are used in or in connection with meters, instruments, and relays shall conform to (A) above, except that enclosures may be of general-purpose type when such equipment is without make-and-break or sliding contacts (other than as provided in (B)(1) above) and when the maximum operating temperature of any exposed surface won't exceed 80 percent of the ignition temperature in degrees Celsius of the gas or vapor involved.

(3) Without Make-or-Break Contacts. When the following don't have sliding or make-and-break contacts, the enclosure may be of a general-purpose type: transformer windings, impedance coils, solenoids, or other windings.

(4) General-Purpose Assemblies. If an assembly is made up of parts for which general-purpose enclosures are acceptable as covered in (B)(1), (B)(2), and (B)(3) above, a single general-purpose enclosure is acceptable for such an assembly. If such an assembly includes the equipment described in (B)(2) above, the maximum temperature at the surface of any component of this assembly must be clearly and permanently indicated or marked on the outside of the enclosure. As an alternative, equipment that is approved may indicate the temperature range for which it is suitable, and shall be so marked. In doing this, use the identification numbers of Table 500.3(B).

(5) Fuses. Where general-purpose enclosures are permitted under (B)(1), (B)(2), (B)(3), and (B)(4) above, fuses for overcurrent protection of the instrument circuits may be mounted in general-purpose enclosures, provided, each such fuse is preceded by a switch conforming to (B)(1) above.

(6) Connections. To make it easier to replace process control instruments, they may be connected by means of a flexible cord, attachment plug, and receptacle, provided that: (1) A switch that complies with (B)(1) above is provided so that the attachment plug won't depend upon the plug for interrupting current; (2) the current doesn't exceed 3 amperes at 120 volts, nominal; (3) the flexible cord isn't over 3 feet and is of a type approved for extra-hard usage, or hard usage if the location provides some protection; also, the attachment plug and receptacle must be of the locking type and grounding type; (4) no more than the necessary receptacles are provided; and (5) the receptacle has a label on it that warns against opening by unplugging while under load.

501.4: Wiring Methods

Wiring methods shall conform to the following:

(A) Class I, Division 1:

(1) Threaded rigid metal conduit.

(2) Threaded steel intermediate metal conduit.

(3) Type MI cable with termination fittings.

(4) Boxes, fittings, and joints shall have a threaded connection to conduit or approved cable terminations.

2005

(5) Threaded joints shall be made up of at least five full threads. This is to cool any escaping gases and to prevent loose joints that may cause sparking or arcs in case of fault. An exception is made for factory threaded NPT entries, which need only be engaged four ½ threads.

(6) MI cable shall be installed and supported in a manner to prevent tensile stress at the termination fittings.

(7) The fittings in (3) shall be explosion-proof.

(8) Flexible fittings shall be explosion-proof and approved for Class I locations. Liquid-tight flexible metal conduit is not permitted for this purpose.

(B) Class I, Division 2:

(1) Threaded rigid metal conduit.

(2) Threaded steel intermediate metal conduit.

(3) Enclosed, gasketed busways and wireways.

(4) Type PLTC cable in accordance with the provisions of Article 725.

(5) Type MI, MC, MV, TC, or SNM cable may be installed in cable-tray systems and shall be installed so as to prevent tensile stress at termination fittings.

(6) Liquid-tight flexible nonmetallic conduit, with approved fittings.

(7) In industrial locations, and where not subject to damage, 1/0 through 4/0 single conductor cables that are listed for use in cable trays may be used.

(8) Boxes, fittings, and joints shouldn't be required to be explosion-proof, except as required in Sections 501.3(B)(1), 501.6(B)(1), and 501.14(B)(1).

Where provisions must be made for a flexible connection, as at motors, the following may be used, but additional grounding must be provided around these flexible connections: flexible metal fittings, flexible metal conduit with approved fittings, extra-hard usage flexible cord with approved bushed fittings and an extra grounding conductor, and liquid-tight flexible metal conduit with approved fittings.

Exception

If there is wiring that under normal conditions can't release sufficient energy to ignite a specific ignitable atmosphere mixture and if the wiring is opening, shorting, or grounding, it may use any of the wiring methods that would be used and allowed by the *Code* in ordinary locations.

For voltages over 600 volts and where adequately protected from physical damage, metallically shielded high voltage cable is acceptable in cable trays when installed in accordance with Article 392.

501.5: Sealing and Drainage

Seals are to be provided in conduit and cable systems to minimize the passage of gases or vapors from one portion of the system to another portion. Type MI cable is inherently sealed, but sealed fittings must be used at terminations to keep moisture and other liquids from entering the insulation of the MI cable. Seals in conduit and cable systems shall conform to (A) through (F) below.

Note

Seals must be provided in conduit and cable that enter from hazardous locations; this is to minimize the passing of gases or vapors and to prevent flames passing from one part of the electrical installation to another through the conduit. Where MI cable is used, such passage of gases or vapors is automatically prevented by the construction of the Type MI cable.

Caution to inspection authorities: Often in the field the wirer installs a seal and then forgets to go back and load it with a sealing compound; therefore, each seal should be thoroughly checked. Used seals shouldn't be reused, but discarded. Seals used to prevent the passage of liquids, gases, or vapors at a continuous differential pressure across the seal must be designed, tested, and listed especially for that purpose. Even with a very small pressure differential across the seal, the equivalent of only a few inches of water pressure, it is possible to have a slow passage of gas or vapor through a seal and through the conductors passing through the seal. Section 501.5(E)(2) is referred to in the *NEC*. Highly corrosive liquids and vapors and temperature extremes can affect the ability of the seals to perform their intended function. Refer to Section 501.5(E)(2).

(A) Conduit Seals, Class I, Division 1. The location of seals for Class I, Division 1 is given in the following:

(1) In every run of conduit that enters enclosures, switches, circuit breakers, fuses, relays, resistors, and any apparatus that may produce sparks, arcs, or high temperature. These seals are to be located as close as practical to the enclosure but in no case shall they be more than 18 inches away from the enclosure. The purpose of the seals is to keep from transmitting an explosion or to keep ignition from traveling between sections of the system. Seals are made for vertical and horizontal installation and are to be used only for the purpose for which they are designed. Thus, a vertical seal is not to be placed horizontally (Figure 501-1). Explosion-proof unions, couplings, reducers, elbows, and condulets can be installed between the enclosure and the sealing fitting.

Exception

When conduit of 1½ inches or smaller enters an explosion-proof enclosure that is used for switches, circuit breakers, fuses, relays, or any

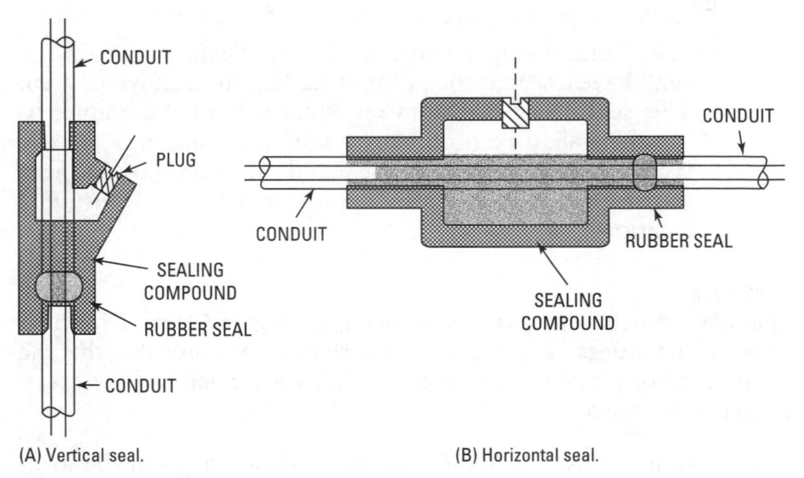

Figure 501-1 Horizontal and vertical seals.

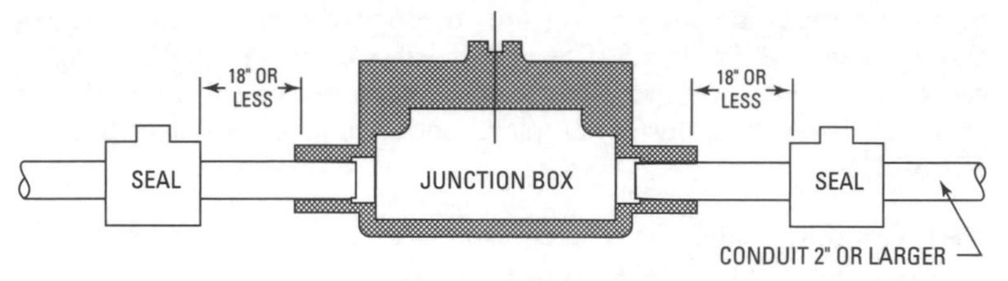

Figure 501-2 Junction boxes with splices or taps shall have seals.

other equipment that may produce arcs or sparks, it won't be required to have seals if the parts that interrupt current are as follows: (a) if the parts that cause the arcs or sparks are in a hermetically sealed chamber to stop the entrance of gas or vapors; or (b) as permitted in Section 501.6(B)(1)(b), and are immersed in oil.

(2) and (3) When a conduit-run of 2-inch conduit or larger enters an enclosure or fitting housing terminals, splices, or taps, there shall be a seal within 18 inches of such enclosure or fitting (see Figure 501-2). When two or more enclosures for which seals are required, as in Section 501.5(A)(1)(b), are connected by a nipple or conduit-run that is no longer than 36 inches in length, only one seal is required (see Figure 501-3). This will fulfill the requirement of not more than 18 inches from the enclosure. The *Code* suggests that you see the notes under Group B in the sixth fine-print note that appears in Section 500.3.

(4) Each conduit-run leaving a Class I, Division 1 location shall have a seal at the point of leaving the hazardous area. This seal may be located on either side of the boundary, but there shall be no fitting, union, coupling, box, or any type of fitting between the seal and the nonhazardous area (Figure 501-4). This also applies to Class I, Division 2 locations.

Exception

Unbroken conduit that passes completely through a Class I, Division 1 area with no fittings 12 inches beyond each boundary, providing that the termination points of the unbroken conduit are in nonhazardous areas, need not be sealed.

(B) Conduit Seals, Class I, Division 2. Below are given the locations for installing seals in Class I, Division 2 locations:

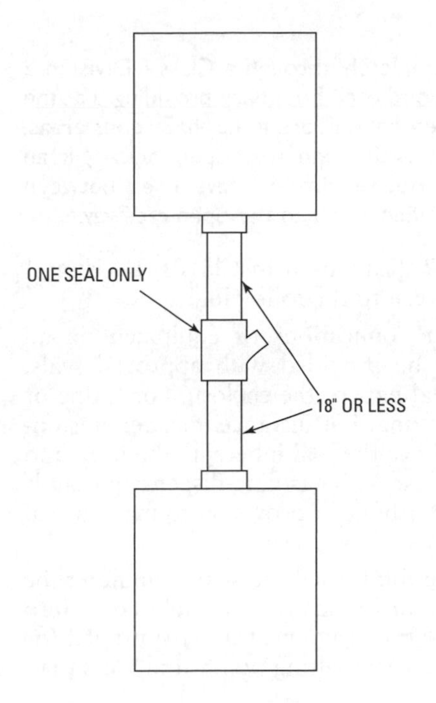

Figure 501-3 One seal only on runs of 36 inches or less.

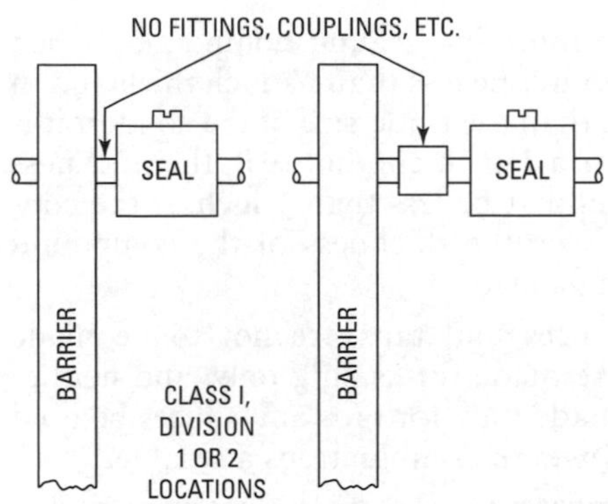

Figure 501-4 Location of seals when leaving Class I locations.

(1) The wiring methods as outlined in Section 501.4(A) apply. Also, seals shall be provided as in (A)(1), (A)(2), and (A)(3) above.

(2) The provisions that were covered in Section 501.5(A)(3) apply to Class I, Division 2 areas where conduit leaves that area and passes into a nonhazardous area.

Exception

Unbroken conduit that passes completely through a Class I, Division 2 area with no fittings 12 inches beyond each boundary, providing that the termination points of the unbroken conduit are in nonhazardous areas, need not be sealed. Conduit systems that end in an open raceway in an unclassified outdoor location are not required to have a seal between where the conduit leaves the classified area and the open raceway.

(C) **Class I, Divisions 1 and 2.** Seals used in Class I, Division 1 and 2 locations shall conform to the following:

- **(1) Fittings.** Enclosures for connections or equipment in all Class I locations shall be provided with approved seals. These may be an integral part of the enclosure or fitting or they may be a sealing fitting. For instance, most explosion-proof lighting fixtures have the seal inherently built in, and curb junction boxes for use with gasoline-dispensing islands may be purchased with a built-in provision to make a seal off from it.
- **(2) Compound.** The compound used in seals shouldn't be affected by temperature or liquids that it might come into contact with. It shall be a compound that is approved for this purpose and shall have a melting point of not less than 93°C (200°F).
- **(3) Thickness of Compound.** The sealing compound, when put into the seal, shouldn't be less than ⅝-inch thick and in no case shall it be less than the trade-size of the conduit it is designed for. Thus, for a 1-inch conduit seal, the thickness of the compound shouldn't be less than 1 inch. If the conduit is ½-inch trade-size, the thickness of the compound shouldn't be less than ⅝ inch.
- **(4) Splices and Taps.** Splices and taps are not to be made within fittings that are made for sealing only, and neither shall boxes that are made only for taps and splices be used as seals. There are, however, combinations available.
- **(5) Assemblies.** When a separate compartment has an assembly of equipment that could produce arcs, sparks, or high temperatures and this compartment is separate from another compartment that contains splices or taps, if a seal is provided for the conductors that pass from one compartment to the other, the entire assembly and the two compartments shall be approved for Class I locations. When conduit enters the compartment containing the splice and taps, seals

shall be provided in that conduit in Class I, Division 1 locations. See (A)(2) above for where these are required.

(D) Cable Seals, Class I, Division 1. In Class I, Division 1 locations that have multiconductor cables installed in conduit, the cable is to be considered a single conductor when the cable is capable of transmitting vapors or gases through its core. Sealing of these cables is covered in (A) above.

If cables have a gas/vapor-tight continuous sheath that is capable of transmitting gases or vapors by means of the cable core, they must be sealed in Division 1 locations. First the jacket and any other coverings must be removed so that sealing compound can completely surround each individual conductor and can also seal the outer jacket.

(E) Cable Seals, Class I, Division 2. The following information is for the location of cable seals in Class I, Division 1 locations:

(1) Cables must be sealed at the point where they enter enclosures that are approved for Class I locations. For the seal fittings, (B)(1) above must be complied with. Even though multiconductor cables are within a gas/vapor-tight continuous sheath that would allow transmission of gases or vapors through the cable core, they must be sealed in fittings approved for the purpose in Division 2 locations after the inside jacket is removed, so that the sealing compound completely surrounds each individual insulated conductor in such a way as to minimize the passage of gases and vapors. Multiconductor cables in conduit shall be sealed as described in (D) above.

(2) If there is a gas/vapor-tight continuous sheath over the cable, and if the cable won't transmit gas or vapor through the cable core in any quantity that exceeds that which is permitted in seal fittings, it shouldn't be required to be sealed except as required in (C)(1) above. The minimum length of such a cable shall be not less than the length that limits the passing of gas/vapor to the rate permitted for seal fittings—.007 cubic feet per hour of air at a water pressure of 6 inches.

Note

See *ANSI/UL 886, Outlet Boxes and Fitting for Use in Hazardous Locations.*

Note

Interstices of the conductor strands are not included in the cable core.

(3) When the sheath of a cable is continuous and is gas/vapor-tight, yet is capable of transmitting gases or vapors through the core of the cable, sealing won't be required except as covered in (E)(1) above, unless the cable is attached to process equipment or to devices that may cause a pressure in excess of 6 inches of water to be exerted at the cable's end. If a greater pressure is transmitted, a sealing, barrier, or other means must be provided to prevent flammables from passing into an unclassified area.

Exception

A cable that has unbroken outer sheath will be allowed to pass in a continuous length to a Class I, Division 2 location without seals.

(4) If a gas/vapor-tight continuous sheath is installed over cable, it shall be sealed at the boundary of the Division 2 and nonhazardous location. This shall be done in such a way that the passage of gas and vapors into a nonhazardous location will be kept to a minimum.

Note

Metal or nonmetallic material may be used for the sheath mentioned in (D) and (E) above.

(F) Drainage

(1) Control Equipment. In control equipment where there is a possibility of liquid accumulating or vapor condensing and becoming trapped in the enclosure, provision shall be made for periodically draining off this liquid, and the draining means shall be of an approved type.

(2) Motors and Generators. The authority enforcing the *Code* has the responsibility of deciding where there is a probability of liquid or condensed vapor accumulating in motors and generators. If judged that there is this accumulation, the conduit systems shall be arranged so as to minimize this possibility and, if a means of permitting periodic draining is judged necessary by the inspector, such means shall be provided at the time of manufacture and shall be an integral part of the machine.

(3) Canned Pumps, Process Connections, and the like. These frequently depend on a single-seal diaphragm or tube to prevent process fluids from entering the electrical conduit system. An additional approved seal or barrier shall be provided with an adequate drain between the seals in such a manner that leaks would be obvious.

501.6: Switches, Circuit Breakers, Motor Controllers, and Fuses

Switches, circuit breakers, motor controllers, and fuses shall conform to the following:

(A) Class I, Division 1. Class I, Division 1 locations shall have the switches, circuit breakers, motor controllers, and fuses, including push buttons, relays, and similar devices, in approved enclosures together with the apparatus that they serve that is approved for Class I locations.

(B) Class I, Division 2. Switches, circuit breakers, motor controllers, and fuses in Class I, Division 2 locations shall conform to the following:

(1) Type Required. Switches, circuit breakers, and motor controllers that are intended to interrupt current in the normal performance of the function for which they are installed shall be approved, along with the enclosures for same, for Class I locations. See Section 501.3(A). Exceptions to this are (a) where the interruption of the current occurs within a chamber hermetically sealed against the entrance of gases and vapors; and (b) where the current-interrupting contacts are oil-immersed and the device is approved for locations of this class and division: 2-inch minimum immersion for power and 1-inch minimum immersion for control.

General-purpose oil-immersed circuit breakers and controllers may not completely confine the arc that might be produced under heavy overloads, so they shall be specifically approved for Class I, Division 2 locations.

(2) Isolating Switches. Isolating switches need not have fuses included, as they are not intended to be opened under load or to interrupt current. Their purpose is to isolate the equipment after the electricity is shut off by other means approved for that purpose, so they may be of the general-purpose type. Most inspectors prefer that they be marked "Don't Open Under Load."

(3) Fuses. Except as provided in (B)(4) below, the following may be used for the protection of motors, appliances, and lamps: (a) If installed in enclosures approved for the purpose of location, cartridge fuses and plug fuses may be used if of the proper voltage for the job; and (b) some fuses have the operating element immersed in oil or other liquid, such as carbon tetrachloride. These may be used or they may be enclosed in chambers hermetically sealed against the entrance of gases or vapors.

(4) Fuses or Circuit Breakers for Overcurrent Protection. When not more than 10 sets of approved enclosed fuses, or not more than 10 circuit breakers that are not intended to be used for switching, are installed for branch-circuit purposes or for feeder protection for only one room area or section of the Class I, Division 2 location, general-purpose enclosures may be used for these fuses or circuit breaker purposes; that is, if they are for protection only of circuits or feeders that are supplying lighting that is maintained in a fixed position.

Please note that this does not cover portable lamps. For the purpose of this part, three fuses on a three-phase system to protect the three ungrounded conductors are considered a set. Also, a single fuse to protect the ungrounded conductor of a single-phase circuit is considered one set. Fuses that conform to (B)(3) above need not be counted in the 10 sets covered in this part. See the *NEC*.

(5) Fuses Internal to Lighting Fixtures. Lighting fixtures may have approved cartridge-type fuses in the fixture as supplementary protection.

501.7: Control Transformers and Resistors

Transformers, impedance coils, and resistors used as or in conjunction with control equipment for motors, generators, and appliances shall conform to the following:

(A) Class I, Division 1. Transformers, impedance coils, and resistors, together with their switching equipment, shall be provided with approved (explosion-proof) enclosures for Class I, Division 1 locations in accordance with Section 501.3(A).

(B) Class I, Division 2. Control transformers and resistors used in Class I, Division 2 locations shall conform to the following:

(1) Switching Mechanisms. When these are used in conjunction with transformers, impedance coils, and resistors, a switching mechanism shall meet the requirements of Section 501.6(B); in other words, shall be approved for Class I locations.

(2) Coils and Windings. If provided with adequate vents to take care of the prompt escape of gases or vapors that may enter the enclosure, transformers, solenoids, and impedance coils may be of the general-purpose type.

(3) Resistors. Resistors shall be enclosed, and the assembly must be approved for Class I locations, unless the resistance is nonvariable and the maximum operating temperature in degrees Celsius does not exceed 80 percent of the temperature that will ignite the gas or vapor involved, or unless it has been tested and approved as being incapable of igniting the gas or vapor.

501.8: Motors and Generators

Motors and generators shall conform to the following:

(A) Class I, Division 1. Motors and generators used in Class I, Division 1 locations shall be (a) explosion-proof and approved for Class I locations; (b) totally enclosed types with positive-pressure ventilation from a source of clean air and the air discharged into a safe area. A totally enclosed motor is not necessarily an explosion-proof motor. The motor and ventilating fan must be interlocked electrically so that the motor can't be started until the fan has purged the motor with air in a quantity of at least 10 times that volume of air in the motor. The interlock must also be arranged to stop the motor if the air source fails; (c) a totally enclosed type, pressurized by a suitable and reliable source of inert gas with interlocking to stop the motor in the event that the gas pressure drops; or (d) designed to be submerged in a liquid that is flammable only when it is vaporized and mixed with air, or in a gas or vapor where the pressure is greater than atmospheric pressure and is flammable only when mixed with air, and the machine is designed to prevent starting until it has been purged with the liquid or gas to get rid of all air, and is also designed so that it will automatically be deenergized when the supply of liquid, gas, or vapor drops and the pressure is reduced to atmospheric pressure.

In (b) and (c), the operating temperature of any external surface of the motor shouldn't exceed 80 percent of the ignition temperature of the gases or vapors involved in the hazardous location. These temperatures are to be determined by ASTM test procedure (Designation D2155-69). Any device used to detect the rise in temperature shall be of an approved type. All auxiliary equipment used for ventilating or pressurizing must also be approved for the location.

(B) **Class I, Division 2.** Motors, generators, and other rotating electrical machinery used in Class I, Division 2 locations shall be approved for Class I locations (explosion-proof) if they have sliding contacts; centrifugal or other switching devices or mechanisms (including motor overcurrent, overloading, and overtemperature devices); or integral resistance devices, either while starting or running, unless the sliding contacts or switching devices and resistance devices are provided for in enclosures approved for the location.

If operated at rated voltage, and the exposed surface of a space heater is installed to prevent condensation of moisture during periods of shutdown, the space heater shouldn't exceed 80 percent of the temperature in degrees Celsius that would ignite the gas or vapors involved.

The above rules don't prohibit the use of open, squirrel cage, nonexplosion-proof motors that don't use brushes or switching mechanisms.

Note

The temperature of internal and external surfaces that may be exposed to flammable atmosphere must be considered.

501.9: Lighting Fixtures

Lighting fixtures must comply with (A) and (B) below.

(A) **Class I, Division 1.** In Class I, Division 1 locations, fixtures for lighting shall conform to the following:

(1) **Approved Fixtures.** All fixtures shall be approved for Class I, Division 1 locations and shall be approved as a complete unit with the maximum wattage of lamps that may be used plainly marked on them. Portable fixtures shall also be approved as a complete unit.

Many times in the field, explosion-proof and dust-tight lighting fixtures are confused. They definitely are not the same. An explosion-proof fixture has to contend primarily with explosions from within, so the glass enclosure must be thick and strong. While the temperature of operation is also a factor, it is not the same as with dust-tight fixtures. In dust-tight units, the operating temperature of the fixture is the important factor, so the glass will be thinner, but the enclosure will be larger so the heat will have more surface to dissipate from. Look for the group letter on the fixture and check with the atmospheric groups that are listed in the first part of Article 500.

(2) Physical Damage. The fixtures shall be protected from physical damage, either by guards or by location. The breaking of a lamp bulb might cause the hot filament to start an explosion or fire.

(3) Pendant Fixtures. Rigid metal conduit or threaded steel intermediate conduit stems shall be used to suspend pendant fixtures. The conduit shall have threaded joints and a set-screw shall be provided to prevent loosening. Fixtures having stems longer than 12 inches may be installed in one of two ways: (a) The stems may be securely supported at a height not to exceed 12 inches above the fixture; these supports shall secure the fixture laterally and shall be of a permanent nature; or (b) an approved, explosion-proof, flexible connector may be used if it is not more than 12 inches below the box or fitting that supports the fixture.

(4) Supports. Support boxes and fittings used to support fixtures shall be approved for the purpose and also be approved for Class I locations.

(B) Class I, Division 2. In Class I, Division 2 locations, lighting fixtures shall conform to the following:

(1) Portable Lighting Equipment. Portable lighting must comply with (A)(1) above. In other words, it shall be approved as a complete unit for Class I locations.

Exception

If portable lighting equipment is mounted on movable stands and flexible cord is used to connect them, as in Section 501.11, they may be mounted in any position if they conform to Section 501.9(B)(2) below.

(2) Fixed Lighting. Fixed lighting fixtures shall be protected from physical damage by guards or by location. If there should be a danger of hot particles falling from lamps or fixtures, they shall be suitably guarded to prevent these hot particles falling into areas of concentrations of flammable gases or vapors. Where the lamps are of a size that might reach an operating temperature of more than 80 percent of the ignition temperature of the gases or vapors involved, as determined by ASTM test procedure (Designation D286-30), the fixtures shall conform to Section 501.9(A)(1).

(3) Pendant Fixtures. This is a repetition of (A)(3) above.

(4) Switches. Section 501.6(B)(1) covers switches that are a part of an assembled fixture or of an individual lampholder.

(5) Starting Equipment. The requirements of Section 501.7(B) shall be met for starting and control equipment used with electric discharge lighting. This section requires that they be in an approved enclosure, or that they not be capable of emitting sparks or arcs that will ignite or explode the gases or vapors in that area.

Exception

Thermal protection installed into a thermally protected ballast for fluorescent lighting must be used if approved for locations in this class and division.

501.10: Utilization Equipment

Utilization equipment, fixed or portable, must conform to the following:

(A) Class I, Division 1. All utilization equipment, including electrically heated and motor-driven equipment, that is used in Class I, Division 1 locations shall be approved for the location.

(B) Class I, Division 2. The following shall be complied with in Class I, Division 2 locations for all utilization equipment:

(1) Heaters. Either (a) or (b) below shall be conformed to for electrically heated utilization equipment: (a) When continuously energized at the maximum rated ambient temperature for the location, heaters shouldn't exceed 80 percent of the ignition temperature in degrees-Celsius rating of the gas or vapor involved on any surface of the heater that is exposed to the gas or vapor. If no controller of temperature is used, the conditions apply if the heater is operated at 120 percent of the rated voltage used; or (b) any heaters used shall be approved for Class I, Division 1 locations.

Exception

See Section 501.8(B), which gives requirements for motor-mounted space heaters that are used to stop condensation.

(2) Motors. Motors in this location are treated the same as other motors, as explained in Section 501.8(B).

(3) Switches, Circuit Breakers, and Fuses. These are covered in Section 501.6(B).

501.11: Flexible Cords, Class I, Divisions 1 and 2

Flexible cord shall be used only for portable lighting fixtures or portable utilization equipment and the supply-circuit fixed portion thereof, and where used: (1) the cord must be the type suitable for extra-hard usage; (2) it shall contain a grounding conductor complying with Section 400.23 in addition to the other conductors; (3) the connection of the supply conductors to terminals shall be done in an approved manner; (4) the cord shall be supported by clamps or other approved means so that there will be no tension on the terminal connections; and (5) wherever the cord enters boxes, fittings, or enclosures that are the explosion-proof type, seals suitable for the purpose shall be used.

Exception

Sections 501.3(B)(6) and 501.4(B) provide some exceptions for cords.

Submersible electric pumps that are provided with means of removing the pump without entering a wet pit are considered portable utilization equipment.

It is permitted to extend a flexible cord between a wet pit and a power source, if the cord is encased in a suitable raceway.

Note

Section 501.13 covers flexible cords that may come in contact with liquids that have a deteriorating effect on insulation.

501.12: Receptacles and Attachment Plugs, Class I, Divisions 1 and 2

Receptacles and plugs shall be of a type approved for Class I locations (explosion-proof), and shall be equipped with a connection for a grounding conductor. The grounding conductor shall make connection first and the other terminals shall be so arranged that, upon insertion or removal, no sparks will result in ignition or explosion, except as provided in Section 501.3(B)(6).

501.13: Conductor Insulation Class I, Divisions 1 and 2

Where condensed vapors or liquids may come in contact with the insulation on conductors or cords, the insulation shall be of an approved type for use in this location or the conductors shall be protected by a lead sheath or other approved means.

TW insulation, as such, is not approved for use where it will be exposed to gasoline. However, there is a TW wire with a nylon cover that is approved, as well as THWN and many others that go by the trade names of the manufacturer. The best method of finding out whether the insulation is approved for contact with any liquid is to refer to the Underwriter's Laboratories listing and check. If it was tested for this use, it will be there.

501.14: Signal, Alarm, Remote-Control, and Communication Systems

Signal, alarm, remote-control, and communication systems shall conform to the following:

- **(A) Class I, Division 1.** All apparatus, equipment, etc., used in Class I, Division 1 locations for signaling systems, alarms, remote-control, and communications shall be approved for Class I locations, irrespective of the voltages that are involved. The wiring methods and sealing, prescribed in Sections 501.4(A) and 501.5(A) and (C), apply to the installation of the same. There are no exceptions.
- **(B) Class I, Division 2.** The following covers what must be complied with for signaling alarm, remote-control, and communication systems, and in Class I, Division 2 locations:
  - **(1) Contacts.** The enclosures for switches, circuit breakers, the make-and-break contacts of push buttons, relays, alarm bells, and horns shall be approved for Class I, Division 1 locations in accordance with Section 501.3(A).

Exception

You may use general-purpose enclosures, provided the interrupting contacts are as follows: (a) immersed in oil; (b) enclosed within a chamber that is hermetically sealed to prevent gases or vapors from entering; or (c) their circuits don't emit sufficient energy to cause ignition of specific atmospheric mixtures.

  - **(2) Resistors and Similar Equipment.** Resistors, resistance devices, thermionic tubes, and rectifiers shall meet the same requirements as covered in Section 501.3(B)(2).
  - **(3) Protectors.** The enclosures for lightning protective devices and for fuses may be of the general-purpose type.
  - **(4) Wiring and Sealing.** All wiring must conform to the requirements for Class I locations covered in Section 501.4(B), and the sealing shall conform to the requirements of Section 501.5(B) and (C).

2005

501.15: Live Parts, Class I, Divisions 1 and 2
No live parts operating at more than 30 volts in dry locations or 15 volts in wet locations are permitted where exposed, except if they are protected according to Sections 500.7(E), 500.7(F), or 500.7(G).

501.16: Grounding, Class I, Divisions 1 and 2
Not only are wiring and equipment required to be grounded in Class I, Division 1 and 2 locations as specified in Article 250, but they must also meet the following additional requirements:

- **(A) Bonding.** The locknut-bushing and double-locknut types of connection shouldn't be depended upon for proper bonding purposes, but bonding jumpers with listed fittings or other approved methods of bonding are required. This method of bonding applies to all intervening raceways, fittings, boxes, enclosures, and the like between the point of grounding in Class I locations and the grounding point at the building disconnecting means. See Section 250.78.
- **(B) Types of Equipment-Grounding Conductors.** If the flexible metal conduit or liquid-tight flexible metal conduit is used as permitted in Section 501.4(B), and if it is to be relied upon for the sole equipment-grounding path, you must use an internal or external bonding jumper paralleled with each flexible conduit, and Section 250.79 shall be complied with.

501.17: Surge Protection, Class I, Divisions 1 and 2
Surge arresters are required, including their installation and connection, in compliance with Article 280. Also, if surge arresters are installed in Class I, Division 1 locations, the enclosure shall be suitable for the location. Metal oxide varistors and other types of sealed devices that are designed for a specific duty can be installed in Class 1, Division 2 locations, in a general-purpose enclosure. Any other types of surge supressors in these areas must be in an enclosure rated for the area.

501.18: Polyphase Branch Circuits
Polyphase circuits that feed equipment in Class 1, Division 1 areas must have separate grounded (neutral) conductors.

Article 502—Class II Locations

502.1: General

In Section 500.6, Class II locations were covered and the coverage included in this classification thoroughly outlined. The general rules of the *Code* apply to all Class II locations, but there are exceptions in this article.

It will be recalled that Class II locations are those areas where dusts are in suspension or otherwise present to the extent that they might be ignitable or explosive. The article covers the supplemental requirements for wiring installations that are necessary to take care of this hazardous condition. In dealing with this article, it should be remembered that most dusts in the proper suspension might become explosive. Not all dusts are covered in the *Code*, and it becomes the responsibility of the inspection authority to judge whether or not these dusts may be a problem and to what extent.

A few definitions that are applicable to these locations are as follows:

In this article *dust-ignition-proof* means that a location will be enclosed in a way that will exclude amounts of dust that might affect performance or rating and that, when installed in compliance with the *NEC*, won't permit arcs, sparks, or heat that might be generated or liberated in the enclosure to cause ignition of suspended or accumulated specified dusts that might occur in the vicinity of the enclosure.

One major problem of Class II locations is temperature. Equipment used in Class II locations must be capable of functioning at full rating without developing surface temperatures high enough to cause excessive dehydration or to cause the dust to carbonize, especially organic dust.

Note

Spontaneous combustion is likely to occur in dusts that are excessively dry or carbonized.

In the above, the answer is in the fact that not only will arcs and sparks cause explosions when dust is in suspension in the proper amounts in the atmosphere, but also the fact that the temperatures of ignition are very low, and when organic dusts are exposed to a high temperature over a prolonged period, the composition of the dust will change and, in most cases, the flashpoint will even be lowered.

Equipment and wiring defined in Article 100 as being explosion-proof won't be required and aren't acceptable in Class II locations, except where specifically approved for Class II locations.

The point here is that the explosion possibility may be taken care of, but the temperature also has to be considered, since many dusts have a lower flashpoint than gases or vapors in Class I locations. *NFPA 497M*, mentioned in other parts concerning hazardous locations, gives the flashpoints and other pertinent facts that will be encountered with dusts. Inspectors and anyone concerned with design and installation of wiring in these areas should be familiar with (or at least know where to find) the facts applicable to the conditions that might prevail.

Where Class II, Group E and F dusts with a resistance less than 105 ohm-centimeters are present in quantities that can be hazardous, they shall be in only Division 1 locations.

502.2: Transformers and Capacitors

The installation of transformers and capacitors shall conform to the following:

(A) **Class II, Division 1.** Transformers and capacitors in Class II, Division 1 locations shall meet the following conditions:

- **(1)** **Containing Liquid That Will Burn.** Sections 450.41 through 450.48 cover transformer vaults. These requirements apply to transformers in Class II locations that contain a liquid that will burn. In addition (a) any openings such as doors, and the like that communicate with the Division 1 locations must have self-closing fire doors, and these are to be on both sides of the wall; they shall be closely fitted so that they can close readily, and suitable seals shall be provided (such as weather stripping), to keep to a minimum the dust that enters into a vault; (b) any ducts or vents shall enter only into the outside air; and (c) suitable pressure-relief openings shall communicate with the outside air and must be provided to relieve any pressures that might occur from an explosion in the vault and communicate this pressure to the outside air away from hazardous locations.
- **(2)** **Not Containing Liquid That Will Burn.** Transformers and capacitors that contain a liquid that won't burn shall be installed in a transformer vault that conforms to the requirements of Sections 450.41 to 450.48 inclusive, or they shall be approved as a complete assembly, including terminal connections, for Class II locations.
- **(3)** **Metal Dusts.** Transformers and capacitors shouldn't be installed in locations in which dust from magnesium, aluminum, aluminum-bronze powders, or other metals of

similar hazardous characteristics may be present. Section 500.6(A)covers electrically conducting dusts from materials such as coal, charcoal, or coke. Even though these are not considered metallic dusts, it will be found that most inspection authorities will, in practically all cases, treat them as metal dusts.

(B) Class II, Division 2. Class II, Division 2 locations are treated as having a lesser hazard than Class II, Division 1 locations, and transformers and capacitors shall conform to the following:

(1) Containing Liquid That Will Burn. Transformers and capacitors containing a liquid that will burn, in Class II, Division 2 locations, shall conform to the requirements for transformer vaults as covered in Sections 450.41 to 450.48 inclusive.

(2) Containing Askarel. Transformers in excess of 25 kVA that contain askarel must (a) have a pressure-relief vent (these naturally must vent outdoors and away from hazardous areas); (b) either have a means of absorbing any gas that might be generated by arcing internally, or the pressure-relief vents must be connected to a chimney or other type of flue that will carry the gases outside the building; and (c) there shall be a space of not less than 6 inches left between the cases of the transformers and combustibles adjacent to it.

(3) Dry-Type Transformers. Vaults shall be used for dry-type transformers, or they shall (a) be enclosed in tight metal enclosures without vents or any other openings (this means that they should be hermetically sealed); and (b) operate at a voltage not to exceed 600 volts, nominal, (including both the primary and secondary voltages).

502.4: Wiring Methods

Wiring methods must comply with (A) and (B) below:

(A) Class II, Division 1. The only two wiring methods approved for Class II, Division 1 locations are threaded rigid conduit, threaded steel intermediate metal conduit, or MI cable with fittings approved for the location and properly supported so that there will be no strain on the fittings.

(1) Fittings and Boxes. Fittings and boxes in which taps, joints, or terminal connections are made shall be designed to minimize the entrance of dust and they shall have threaded bosses and be equipped with telescoping or close-fitting covers, have other effective means to prevent the escape of sparks or burning materials, and have no openings (such as screw holes for attachment) through which, after installation, sparks or burning material might escape, or through which adjacent combustible materials might be ignited.

The above, in summary, states that all fittings and boxes shall be dust-tight and approved for Class II locations. (See Figure 502-1).

2005

(2) Flexible Connections. Flexible connections are often a necessity. When they are, the following types may be used: dust-tight flexible connectors, liquid-tight flexible metal conduit with approved fittings, liquid-tight nonmetallic conduit with approved fittings, or flexible cord approved for extra-hard usage and provided with bushed fittings. Interlocked armor MC cables are permitted with an overall jacket of suitable material and with termination fittings listed for use in these locations.

When using the above methods, some precautions are required, such as a grounding conductor in flexible cords.

If flexible connections are subject to oil or corrosive conditions, the conductors shall be of a type that has insulation for the condition, or a suitable sheath may be used to protect them.

Exception

Where the dusts are of an electrical conducting nature, flexible metal conduit shouldn't be used. Flexible cords shall be provided with dust-tight seals at both ends.

(B) Class II, Division 2. For Class II, Division 2 locations, the following may be used for the wiring methods: rigid metal conduit; intermediate metal conduit; EMT; dust-tight wireways; Type MI, MC, or SNM cables used with approved termination fittings; or Type PLTC or TC cables that are installed in ventilated channel-type cable trays, ladder-type cable trays, or ventilated trough-type cable trays. They shall be laid in a single layer, with a space not less than the diameter of the largest cable left between the cables.

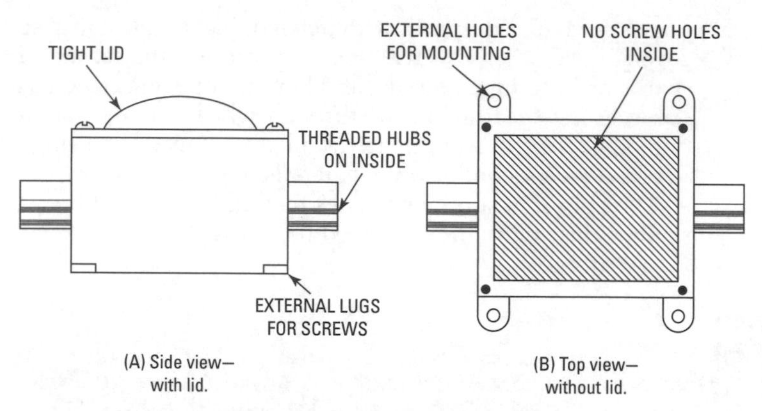

Figure 502-1 Type of boxes for Class II locations.

Exception

You may use any methods that are suitable for ordinary wiring locations, provided that under normal conditions this wiring method can't release sufficient energy to cause ignition of combustible dust fixtures by opening, shorting, or grounding.

- **(1) Wireways, Fittings, and Boxes.** Wireways, their fittings, and boxes provided for taps, joints, or terminal connections must be of a design that will minimize the entrance of dust, and (a) they shall be provided with telescoping or close-fitting covers or some other effective means that will prevent sparks or burning material from escaping; and (b) there shall be no openings in the enclosure such as for attachment with screws that after mounting could allow sparks or burning material to escape or through which nearby ignitable material may be ignited. The screw holes for attachment should be in the metal part of the enclosure external to the inside of the enclosure.
- **(2) Flexible Connections.** Where flexible connections are required, (A)(2) above applies.

502.5: Sealing, Class II, Divisions 1 and 2

In the installation of an enclosure that is dust-ignition-proof and one that is not, there may be communication of dust between the two enclosures. The entrance of dust into the dust-ignition-proof enclosure must be prevented. This may be done by one of the following means: (1) by use of an effective and permanent seal, such as used in Class I locations; (2) by connection to a horizontal raceway that is not less than 10 feet in length; or (3) by means of a vertical raceway not less than 5 feet in length that extends downward from the dust-ignition-proof enclosure. Note this carefully-if the raceway extends downward from the conventional enclosure, it won't meet the requirements.

If a raceway provides communication between an enclosure that must be dust-tight and dust-ignition-proof and an enclosure that is located in an unclassified location, seals won't be required.

Where seals are used they are to be accessible (see Figure 502-2).

502.6: Switches, Circuit Breakers, Motor Controllers, and Fuses

Switches, circuit breakers, motor controllers, and fuses shall conform to the following:

(A) **Class II, Division 1.** In Class II, Division 1 locations, the following will cover switches, circuit breakers, motor controllers, and fuses:

(1) **Type Required.** The emphasis of this part is on any devices that are intended to interrupt a current, such as switches, circuit breakers, motor controllers, and fuses. This also includes push buttons, relays, and similar devices installed in Class II, Division 1 locations. Mention is also made of dusts having an electrical conductive nature.

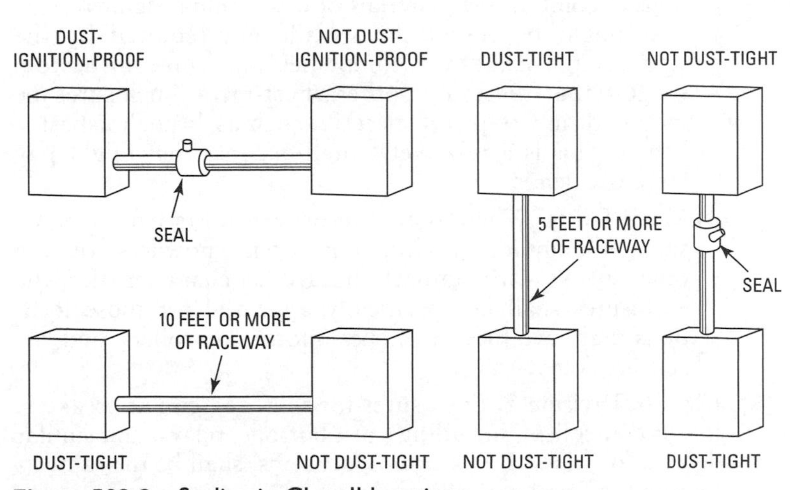

Figure 502-2 Sealing in Class II locations.

Such devices shall be installed in dust-ignition-proof enclosures and shall be approved for Class II locations. In this discussion it should be mentioned that there is an enclosure available for industrial use that is termed a dust-tight enclosure. However, to secure approval, the enclosure must be listed for a hazardous location and classified by a Group number, as covered in Article 500. In this classification are included the following: (a) service and branch-circuit fuses, (b) switches and circuit breakers, and (c) motor controllers; including push-buttons; pilot switches; relays; motor overload protective devices; and switches, fuses, and circuit breakers for the control and protection of lighting and appliance circuits.

(2) Isolating Switches. Isolating switches don't need to contain fuses and are not intended to interrupt current. Their purpose is to isolate the conductors for working on motors and the equipment driven by the motors. When isolating switches are installed in locations that don't have dusts of an electrical-conducting nature, they shall be so designed as to minimize the entrance of dust, and shall (a) have telescoping covers or close-fitting covers or any other means that will prevent the escape of sparks or burning material; and (b) have no holes within the enclosures for attaching enclosures with screws where, after installation, sparks or burning materials could escape, or through which exterior accumulations of combustible materials or dust could be ignited.

It might be stated that, while not required by the *Code*, some insurance companies and some inspectors suggest that these and other sheet-metal enclosures be mounted on fire-proof material such as ¼-inch asbestos board. This is a relatively small expense that might pay large dividends.

(3) Metal Dusts. Where there are present dusts from magnesium, aluminum, or aluminum-bronze powders, or any other metals with similarly hazardous characteristics, the enclosures shall be specifically approved for those locations that have fuses, switches, motor controllers, and circuit breakers.

(B) Class II, Division 2. Enclosures for switches, circuit breakers, motor controllers, including push buttons, relays, and similar devices for Class II, Division 2 locations, shall be the same as required in Section 502.6(A)(2).

502.7: Control Transformers and Resistors

Transformers, solenoids, impedance coils, and resistors used as or in conjunction with control equipment for motors, generators, and appliances must conform to the following:

(A) **Class II, Division 1.** Dust-ignition-proof enclosures approved for Class II, Division 1 locations shall be required for transformers, solenoids, impedance coils, and resistors. There shall be no transformer impedance coil, or resistor installed in locations where dusts from magnesium, aluminum, aluminum-bronze powders or other similar metal dusts are present, unless the enclosures in which they are installed have been specifically approved for such a location.

(B) **Class II, Division 2.** Transformers or resistors in Class II, Division 2 locations must conform to the following:

(1) **Switching Mechanisms.** Section 502.6(A)(2) also applies here for switching mechanisms (including overcurrent devices) associated with control transformers, solenoids, impedance coils, and resistors. That section basically calls for enclosures that are designed to minimize the entrance of dusts and to prevent sparks or burning material from escaping from the enclosure provided in dust-tight enclosure.

(2) **Coils and Windings.** When not located in the same enclosure with the switching mechanism, control transformers, solenoids, and impedance coils are to be mounted in tight metal enclosures and shall have no ventilation openings.

(3) **Resistors.** Resistors become heated when current passes through them, so they must be mounted in dust-ignition-proof enclosures approved for Class II locations. The maximum operating temperature of the resistor shouldn't exceed 120°C (248°F) for nonadjustable resistors and resistors that are a part of an automatically timed starting sequence. In the case where the temperature does not exceed that just specified, the resistors may have enclosures conforming to (B)(2) above, that is, tight metal cases without ventilation openings.

502.8: Motors and Generators

Motors and generators must conform to the following:

(A) **Class II, Division 1.** Motors, generators, and other rotating electrical machinery installed in Class II, Division 1 locations shall be:

(1) Approved for Class II, Division 1 locations.

(2) Totally enclosed and pipe-ventilated, meeting temperature requirements in Section 502.1.

Motors, generators, or rotating electrical machinery located in areas where metal dust such as magnesium, aluminum, aluminum-bronze powders, or similar dusts are present, shouldn't be installed unless they are specifically approved for such locations and are totally enclosed or totally enclosed and fan-cooled.

(B) Class II, Division 2. Motors, generators, and other rotating electrical equipment shall be totally enclosed and nonventilated in Class II, Division 2 locations, totally enclosed if pipe-ventilated, totally fan-cooled or dust-ignition-proof when the maximum full-load external temperature meets the requirements of Section 500.3(D) under normal operation in free air (not dust-blanketed air), and shall have no external openings.

There are exceptions to this that may be granted if the authority that has jurisdiction believes that the accumulations of nonconducting, nonabrasive dust will be moderate. They may also be granted if the machines may be easily reached for routine cleaning and maintenance. If these conditions are met and approval is given, the following may be installed:

(1) Standard open-type machines that don't have any sliding contacts, centrifugal switches, or any other type of switching mechanism, including overcurrent, overloading, and overtemperature devices, or built-in resistance devices.

(2) Standard open-type machines that have contacts, switching arrangements, or resistance devices and are enclosed in a tight metal housing that has no ventilating devices or openings.

(3) Squirrel cage—type textile motors that are self-cleaning.

The *NEC* is a very fair instrument that grants permission in many instances to deviate from the strict rules, provided that we are sure that such deviations won't cause unsafe conditions. We must always consider what might happen after the inspection is made and ask ourselves whether the same conditions will be lived up to.

502.9: Ventilating Piping

Vent pipes for motors, generators, or other rotating electrical machinery, or for enclosures for electrical apparatus, where it is permissible to use vents for these types of enclosures, shall (1) be of a metal not lighter than No. 24 MSG; (2) be of other substantial noncombustible material; (3) lead directly out to clean air outside the building; (4) be screened to prevent the entrance of small animals or birds; (5) be protected from physical damage; and (6) be protected against rust and other corrosion.

In addition to the above, they shall conform to the following:

(A) **Class II, Division 1.** In Class II, Division 1 locations, vent pipes, including their connections to motors or to the dust-ignition-proof enclosures for other equipment or apparatus, shall be dust-tight throughout their length. When metal pipes are used, they shall be (1) riveted (or bolted) and soldered, (2) welded, or (3) rendered dust-tight by some other equally effective means.

(B) **Class II, Division 2.** Ventilating pipes and their connections in Class II, Division 2 locations are to be sufficient to prevent entrance of appreciable quantities of dust into the ventilated equipment or enclosure. They shall also be installed so as to stop sparks, flames, or burning material that could possibly cause ignition of any dust accumulations or any combustibles that are in the vicinity. The metal pipes shall have lock seams and welded joints are permitted, and when necessary to permit some flexibility, tight-fitting slip joints are permitted for the connections to motors.

For metal pipes, the following may be used: lock seams, riveted joints, welded joints, or tight-fitting slip joints where some flexibility is necessary, as at motor connections.

502.10: Utilization Equipment

Utilization equipment, fixed and portable, shall conform to the following:

(A) **Class II, Division 1.** Utilization equipment, fixed and portable, located in Class II, Division 1 locations, also includes electrically heated and motor-driven equipment. Equipment of this type shall be dust-ignition-proof and approved for Class II, Division 1 locations. Where dusts from metals such as magnesium, aluminum, aluminum-bronze powders, or other similar dust are present, the equipment shall be approved for these locations. This will be Group E equipment.

(B) Class II, Division 2. All utilization equipment used in Class II, Division 2 locations must comply with the following:

(1) Heaters. Only approved equipment for Class II locations will be permitted for electrically heated utilization equipment.

Exception

Radiant heating panels that are metal-enclosed are to be dust-tight and marked in accordance with Section 500.3(D).

(2) Motors. The motors used on motor-driven utilization equipment must comply with Section 502.8(B). This section allows some latitude and leaves it to the inspection authority to decide what is involved and the equipment permitted.

(3) Switches, Circuit Breakers, and Fuses. Enclosures used for switches, circuit breakers, and fuses must be approved for being dust-tight.

(4) Transformers, Impedance Coils, and Resistors. These shall conform to Section 502.7(B). There are three parts to this section, and you are advised to read it again.

502.11: Lighting Fixtures

Recall that in Class II locations the ignition of dusts can be caused by sparks, arcs, or heat. Bear this in mind while reading the following:

(A) Class II, Division 1. In Class II, Division 1 locations, lighting fixtures for fixed and portable lighting shall conform to the following:

(1) Approved Fixtures. For Class II locations where ordinary dusts are present, the fixtures shall be approved and plainly marked. In locations where dusts from magnesium, aluminum, aluminum-bronze powders, or similar metal dusts with hazardous characteristics are present, the fixtures shall be approved and so marked.

(2) Physical Damage. Suitable guards or suitable locations shall be provided for fixtures to protect against physical damage. *By location* means that they are to lay mounted high enough that they are not likely to be bumped or hit.

(3) Pendant Fixtures. Pendant fixtures shall be suspended by threaded rigid metal conduit stems, chains with approved fittings, or other approved means, including threaded steel intermediate metal conduit. If the stems are longer than 12 inches, they shall either be rigidly braced laterally at a point not more than 12 inches above the lower end of the stem, or there shall be an approved flexible fitting or flexible connection used, and these shouldn't be mounted more than 12 inches from the attachment to the supporting box or fitting.

The wiring shall be rigid conduit, or, where necessary, a flexible rubber cord that is approved for hard usage may be used if it is sealed properly where attachments are made. The cord shouldn't support the fixture but shall be supported by other suitable means. Where stems are used, a setscrew shall be provided to keep the stem from loosening.

(4) Supports. Lighting fixtures shall be supported by approved boxes, box assemblies, and fittings for Class II locations. Class II boxes and fittings, of course, are to be dust-tight.

(B) Class II, Division 2. Lighting fixtures of Class II, Division 2 locations must comply with the following:

(1) Portable Lighting Equipment. Portable lighting equipment shall be approved for Class II locations and shall be plainly marked as such. Also, the maximum size of lamp that may be used in them shall also be plainly marked.

(2) Fixed Lighting. Fixed lighting shall be approved for Class II locations, except, if not approved, they shall be enclosed to prevent an accumulation of dust on the lamps and to prevent the escape of sparks, burning material, or hot metal. Each fixture shall be plainly marked to indicate the maximum size lamp that may be used in the fixture and, when in use, the maximum temperature of the exposed surface shouldn't exceed 165°C (329°F) under normal use.

(3) Physical Damage. Fixed lighting fixtures must be protected from damage by a guard, or by being in a location where they can't be damaged.

(4) Pendant Fixtures. The requirements for pendant fixtures for Class II, Division 1 locations are practically the same as for pendant fixtures in Class II, Division 1. Please refer to (A)(3) above.

(5) Electric-Discharge Lamps. Section 502.7(B) gives the requirements to be complied with for starting and control equipment for discharge lighting.

502.12: Flexible Cords, Class II, Divisions 1 and 2

Flexible cords used in Class II, Divisions 1 and 2 must conform to the following: (1) The cord must be approved for extra-hard usage; (2) a grounding conductor complying with Section 400.23 shall be enclosed in the cord with the circuit conductors; (3) an approved manner for connecting the terminals to the supply conductors shall be used; (4) clamps or other suitable means of support causing no tension on the terminal connections shall be used; and (5) when flexible cords enter boxes or fittings that must be dust-ignition-proof, suitable seals to prevent the entrance of dust shall be used.

502.13: Receptacles and Attachment Plugs

(A) **Class II, Division 1.** Plugs and receptacles for use in Class II, Division 1 locations shall be provided with a grounding conductor connection for the flexible cord and be dust-ignition-proof-approved for Class II locations.

(B) **Class II, Division 2.** Plugs and receptacles for use in Class II, Division 2 locations shall be provided with a grounding-conductor connection for the flexible cord and be so designed that connection to the supply circuit can't be made or broken while live parts are exposed.

502.14: Signaling, Alarm, Remote-Control, and Communication Systems, Meters, Instruments, and Relays

Refer to Article 800, covering the rules governing the installation of communication circuits, and to the definition in Article 100.

The communication circuits will include telephone, telegraph, fire and burglar alarms, watchman, and sprinkler systems. Signal, alarm, remote-control, and local loudspeaker intercommunication systems shall conform to the following:

(A) **Class II, Division 1.** In this classification, they shall conform to the following:

(1) **Wiring Methods.** In any location where accidental damage or break-down of insulation might cause arcs, sparks, or heating, the wiring method to be used shall be by rigid conduit, intermediate metal conduit, electrical metallic tubing, or type MI cable with approved fittings. The number of conductors to be installed in conduit or EMT is not the same for current-carrying conductors, but is only limited by the 40-percent fill requirement. This is to prevent damage to the insulation when pulling in the conductors.

Flexible cord approved for extra-hard usage may be used where flexibility is required and where it won't be subject to physical damage.

(2) Contacts. Class II, Groups E, F, or G, as required, shall be the types of enclosures for switches, circuit breakers, relays, contactors, and fuses that may interrupt other than voice currents, and for current-breaking contacts for bells, horns, howlers, sirens, and other devices in which sparks or arcs may be produced. An exception to this is if the contacts are immersed in oil or when the interruption of current occurs within a sealed chamber; then the first part need not apply. The purpose of all this is to be certain that any sparks or arcs are isolated so that no ignition or explosion will occur.

(3) Resistors and Similar Equipment. Enclosures shall be approved for Class II locations where there are resistors, transformers, choke coils, rectifiers, thermionic tubes, or any other heat-generating equipment.

Exception

General-purpose enclosures may be used where resistors or similar equipment are immersed in oil or if the chambers enclosing them are so sealed as to prevent the entrance of dust.

(4) Rotation Machinery. Section 502.8(A) shall be compiled with for motors, generators, and other rotational electric machinery.

(5) Combustible Electrically Conductive Dusts. When the dust happens to be of a conductive nature, such as magnesium, aluminum, or aluminum-bronze powders, or metal of other similarly hazardous characteristics, all equipment shall be specifically approved for such conditions.

(6) Metal Dusts. See Section 502.6(A)(3).

(B) Class II, Division 2. The following shall be complied with where signaling, alarm, remote-control, and communication systems and meters, instruments, and relays are stored in Class II, Division 2 locations:

(1) Contacts. Enclosures for contacts in Class II, Division 2 locations shall either conform to the requirements for contacts in Class II, Division 1, as outlined in (A)(2) above, or be enclosed in tight metal enclosures that are designed to minimize the entrance of dust, and shall have telescoping, tight-fitting covers with no openings through which, after installation, sparks or burning materials might escape.

Exception

General-purpose enclosures may be used in circuits that under normal operating conditions don't release sufficient energy to ignite a dust layer.

(2) Transformers and Similar Equipment. The windings of transformers and terminal connections of the transformers, choke coils, and similar equipment shall be enclosed in tight metal enclosures that don't have ventilating openings.

(3) Resistors and Similar Equipment. In the installation of resistors, resistance devices, thermionic tubes, and rectifiers in Class II, Division 2 locations, resistors and similar equipment shall either conform to the requirements in (A)(3) above, or thermionic tubes, nonadjustable resistors, or rectifiers that have a maximum operating temperature that doesn't exceed 120°C (248°F) may be installed in general-purpose type enclosures.

(4) Rotating Machinery. The same requirements apply to rotating machinery of this type as apply to other rotating machinery in Class II, Division 2. See Section 502.8(B).

(5) Wiring Methods. Section 502.4(B) shall be complied with for the wiring methods.

502.15: Live Parts, Class II, Divisions 1 and 2

Live parts in Class II, Division 1 and 2 locations shouldn't be exposed.

502.16: Grounding, Class II, Divisions 1 and 2

In addition to the requirements of Article 250, the following requirements are required in Class II, Division 1 and 2 locations for grounding.

(A) Bonding. Locknut bushings and double-locknut types of contact are not to be the grounding means for bonding purposes, but bonding jumpers used with the proper fittings or other approved types of bondings may be used. This includes the bonding type of bushing bonded to the enclosure.

Recall that the sheet-metal type of bonding clamp is not approved. Approved means of bonding are to be used not only in hazardous locations, but also in intervening raceways, fittings, boxes, enclosures, etc., between hazardous locations and the point of grounding for service equipment.

(B) Types of Equipment-Grounding Conductors. Where, as permitted in Section 502.4, flexible conduit use is permitted, it shall be installed with internal or external bonding jumpers in parallel with each conduit if they comply with Section 250.79.

Exception

With liquid-tight flexible metal conduit 6 feet or less in length, if the fittings are used and the overcurrent protection is limited to 10 amperes or less and the load is not a power utilization type of load, the grounding jumper may be omitted in Class II, Division 2 locations.

502.17: Surge Protection, Class II, Divisions I and 2

Surge arresters, including installation and connection, must comply with Article 280. Also, surge arresters installed in Class II, Division 1 locations must have enclosures suitable for this type of location.

The surge capacitors shall be of a type designed for the purpose (Figure 502-3). Take note that there is overcurrent protection in the capacitor circuit.

502.18: Multiwire Branch Circuits

Multiwire branch circuits are not permitted in Class II, Division 1 areas unless the disconnecting device for the circuit opens all ungrounded conductors simultaneously.

Article 503—Class III Locations

While Class III locations might be considered as having fewer hazards than Class I or Class II locations, the hazards are not to be underestimated. Wherever there is a likelihood of fire or explosion, the hazard must be given the respect that is due to it.

503.1: General

The general wiring and installation requirements of the *Code* apply to Class III locations, as classified in Section 500.7, with the exceptions that are covered in this article.

Class III locations are areas in which fibers and flyings are present. The major point to remember is that the wiring equipment and apparatus shall operate at full-rating without causing excessive dehydration or gradual carbonization of fibers or flyings. Any organic substance that is carbonized or very dry is susceptible to spontaneous combustion (ignition). The maximum surface

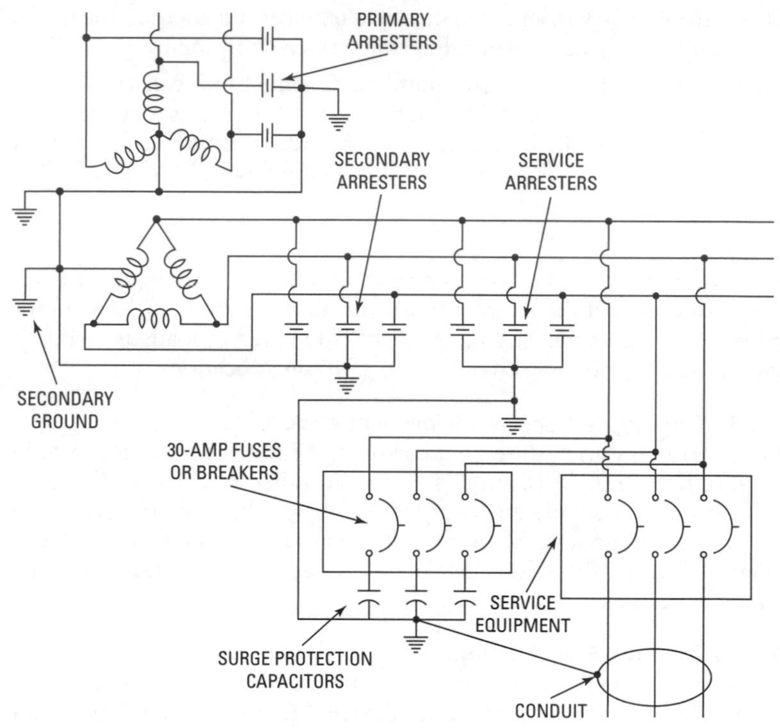

Figure 502-3 Lightning and surge protection.

temperature under operating conditions (1) shouldn't exceed 165°C (329°F) for equipment not subject to overloading, and (2) shouldn't exceed 120°C (248°F) for motors, power transformers, etc. that may be overloaded.

Note

NFPA 505-1999 (ANSI) covers electric trucks.

503.2: Transformers and Capacitors, Class II, Divisions 1 and 2

Refer to Section 502.2(B), which covers transformers and capacitors.

503.3: Wiring Methods

(A) and (B) below cover wiring methods:

(A) **Class III, Division 1.** Wiring methods in Class III, Division 1 shall be rigid metal conduit; rigid nonmetallic conduit; intermediate metal conduit; EMT; dust-tight wireways; and Type

MI, MC, or SNM cables if approved fittings are used at terminations.

(1) Boxes and Fittings. Only dust-tight boxes and fittings shall be used.

(2) Flexible Connections. If it is necessary to have flexible connections in dust-tight locations, you shall use flexible connectors and liquid-tight flexible metal conduits, and, of course, approved fittings are to be used. Extra-hard usage cord that is provided with bushed fittings may be used. An additional conductor for equipment grounding must be provided in the flexible cord, or other approved means may be used for the grounding purpose.

Section 502.4(A)(2) approves dust-tight flexible connectors, flexible metal conduit with approved fittings, or flexible cord approved for extra-hard usage and with approved fittings and seals. Grounding conductors are also required for bonding.

(B) Class III, Division 2. Wiring methods shall be the same as for Class III, Division 2 locations covered in (A) above, with the exception that in sections, compartments, or areas used solely for storage and containing no machinery, open wiring on insulators conforming to the requirements of Article 398 may be used, but only on the condition that protection as required by Section 398.14 be provided where conductors are not run in roof spaces, and the open wiring be well out of reach of sources of physical damage. Section 398.14 provides for the protection of open wiring from physical damage.

503.4: Switches, Circuit Breakers, Motor Controllers, and Fuses, Class III, Divisions 1 and 2

The items in the title of this section, and also push buttons, relays, and similar devices, shall be in dust-tight enclosures.

503.5: Control Transformers and Resistors, Class III, Divisions 1 and 2

The same requirements apply here as in the Class II, Division 2 locations covered in Section 502.7(B). An exception to this is that in Class III, Division 1 locations, if these devices are in the same enclosures with the switching devices of control equipment, and are used only for starting or for short-time duty, the enclosure may have telescoping covers or other effective means for preventing the escape of sparks or burning material as covered in Section 503.1.

503.6: Motors and Generators, Class III, Divisions 1 and 2

Motors, generators, and other rotating machinery located in Class III, Division 1 and 2 locations shall be totally enclosed and nonventilated, totally enclosed and pipe-ventilated, or totally enclosed and fan-cooled.

An exception to the above is that the *Code*-enforcing authority may accept the following if in his or her judgment there is only moderate accumulation of lint and flyings:

(A) Squirrel cage textile motors if they are self-cleaning.

(B) Standard open-type machines that don't have open arcing contacts.

(C) Standard open-type machines, if the switching mechanisms or resistance devices are enclosed within tight housings without any openings for ventilation.

503.7: Ventilating Piping, Class III, Divisions 1 and 2

Ventilating piping in Class III, Division 1 and 2 locations that is used for motors, generators, or any other rotating electric machines, or used for enclosures for electrical equipment, shall be made of metal no lighter than No. 24 MSG or other noncombustible substantial materials, and must comply with the following: (1) Ventilating pipes shall lead directly outside into clean air; (2) the other end of the ventilating pipes shall be screened to prevent the entrance of small animals and birds; and (3) the ventilating pipes shall be protected from physical damage, rusting, or corrosive influences.

These ventilating pipes shall be tight enough for both the pipes and their connections to prevent the entrance of small amounts of fibers or flyings into the ventilated equipment or enclosure and to prevent the escape of sparks, flames, or burning materials that could ignite accumulations of fibers or flyings or other nearby combustible materials. The pipe shall be of the lock-seam type, and riveted or welded joints are to be used. It is permitted to use tight-fitting slip joints where some flexibility is necessary in connections to motors.

503.8: Utilization Equipment, Class III, Divisions 1 and 2

Utilization equipment must conform to the following:

(A) **Heaters.** Only approved electric heater-utilization equipment is allowed in Class III locations.

(B) **Motors.** Motors of motor-driven utilization equipment that is used in Class III locations are subject to the same requirements as motors covered in Section 503.6. When motor-driven utilization equipment is readily movable from one location to another, it must meet the standards required for the most hazardous location in which it is to be used.

(C) Switches, Circuit Breakers, Motor Controllers, and Fuses. Section 503.4 applies to switches, circuit breakers, motor controllers, and fuses, and also to the same items used with utilization equipment.

503.9: Lighting Fixtures, Class III, Division 1 and 2

Lamps shall be installed in fixtures that conform to the following:

(A) Fixed Lighting. There shall be enclosures for fixed lamps and lampholders that (1) are designed to minimize the entrance of fibers and flyings; (2) prevent the escape of sparks, burning materials, or hot metal; (3) are plainly marked for the maximum wattage lamp that can be used; and (4) are capable of being used without exceeding a maximum exposed-surface temperature of 165°C (329°F) under normal operating conditions.

(B) Physical Damage. Fixtures are to be either guarded or located so as not to be subject to physical damage.

(C) Pendant Fixtures. In the suspension of pendant fixtures, the suspension stem shall be threaded rigid metal conduit, threaded intermediate metal conduit, equivalent sizes of threaded metal tubing, or chains that use approved fittings.

If the stem is longer than 12 inches (305 mm), it shall be permanently and effectively supported laterally against displacement, and this bracing shouldn't be more than 12 inches (305 mm) above the lamp fixture, and have an approved flexible connector provided at a distance of not more than 12 inches (305 mm) from the point of attachment to the supporting box or fitting.

(D) Portable Lighting Equipment. Portable lighting equipment shall conform to all of the requirements of (A) above and, in addition, shall be provided with a handle, have a suitable guard for the lamp, and have a lampholder of the unswitched type, with no exposed metal parts, with no means for receiving an attachment plug, and with all exposed noncurrents carrying metal parts grounded.

503.10: Flexible Cords, Class III, Divisions 1 and 2

Flexible cords used in Class III, Division 1 and 2 locations must comply with the following: (1) Cords must be approved for extra-heavy usage; (2) cords must contain a grounding conductor in addition to the conductors of the circuit in compliance with Section 400.23; (3) the connections to terminals or supply conductors must be made in an approved manner; (4) cords must be supported by clamps or other suitable means; and (5) there must be a suitable means to prevent the entrance of fibers or flyings at the point where the cord enters boxes or fittings.

503.11: Receptacles and Attachment Plugs, Class III, Divisions 1 and 2
Receptacles and attachment plugs must be of the grounding type and shall be so designed as to allow only minimal accumulations of fibers or flyings and to prevent sparks of molten particles to escape.

Exception
If, in the opinion of the authority that has jurisdiction, there will be only moderate accumulations of lint or flyings in the vicinity of the receptacle, and if the receptacle is readily accessible for cleaning, general-purpose grounding-type receptacles, if they are mounted so as to minimize the entry of fibers or flyings, may be allowed.

503.12: Signal, Alarm, Remote-Control, and Local Loudspeaker Intercommunication Systems, Class III, Divisions 1 and 2
See the *NEC*.

503.13: Electric Cranes and Hoists, and Similar Equipment, Class III, Divisions 1 and 2
Where traveling cranes and hoists for handling materials and similar installed equipment operate over fibers or accumulations of flyings, they shall conform to the following:

- **(A)** The power supply shall be isolated from all other systems (this may be done by isolation transformers) and shall be equipped with an acceptable ground detector that will give an audible and visual alarm, maintaining the alarm as long as power is supplied to the system and the ground fault remains, and automatically deenergize the contact conductors in the case of a fault to ground.
- **(B)** Contact conductors for cranes and hoists shall be so located or guarded as to be inaccessible to other than authorized persons and protect against accidental contact with foreign objects.
- **(C)** Current collectors shall be located or guarded so as to confine normal sparking and prevent the escape of sparks or hot particles. In order to reduce sparking, there shall be two or more separate contact surfaces for each contact conductor and a reliable means for keeping the contact conductors and current

collectors free of accumulation of dust and of flyings and lint.

(D) Sections 503.4 and 503.5 cover various control equipment. The requirements of these sections apply to control equipment for electric cranes, hoists, and similar equipment in Class III, Division 1 and 2 locations.

503.14: Storage-Battery Charging Equipment, Class II, Divisions 1 and 2

Equipment for the charging of storage batteries shall be located in separate rooms, shall be built or lined with material that is substantially noncombustible and so constructed as to satisfactorily exclude flyings or lint, and shall be well ventilated.

Not only are the lint and flyings a problem, but so are the fumes from the batteries. Proper ventilation will also assist in taking care of these fumes.

2005

503.15: Live Parts, Class III, Divisions 1 and 2

No live parts operating at more than 30 volts in dry locations or 15 volts in wet locations are permitted where exposed, except if they are protected according to Sections 500.7(E), 500.7(F), or 500.7(G).

503.16: Grounding, Class III, Divisions 1 and 2

Wiring and equipment in Class III, Division 1 and 2 locations shall be grounded as specified in Article 250 with the following additional requirements:

(A) **Bonding.** The locknut-bushing and double-locknut types of contact shouldn't be depended upon for bonding purposes, but bonding jumpers with the proper fittings or other approved means of bonding shall be used. This could be bonding-type bushings bonded to the enclosure. This means of bonding must apply to all intervening raceways, fittings, boxes, enclosures, and so on between the Class III locations and the point where the service equipment bonding begins.

(B) **Types of Equipment-Grounding Conductor.** Where flexible metal conduit is used, as permitted in Section 503.3, it shall be installed with internal or external bonding jumpers in parallel with the flexible conduit that meet the requirements of Section 250.79.

Article 504—Intrinsically Safe Systems

504.1: Scope

Intrinsically safe wiring systems for Class I, II, and III locations are covered in this article. An intrinsically safe circuit does not produce sufficient energy to shock, explode, or create fires.

504.2: Equipment Approval

All intrinsically safe equipment must be approved.

504.3: Application of Other Articles

All other parts of the *Code* (except as overridden by this article) apply to intrinsically safe wiring systems.

504.4: Definitions

In this section, the *NEC* gives a number of definitions related to these systems. Refer to the *Code* for specific information.

504.10: Equipment Installation

- **(A)** **Control Drawing.** Intrinsically safe equipment must be installed in accordance with control drawings.
- **(B)** **Location.** Intrinsically safe systems may be installed in hazardous locations, and may use general-purpose enclosures.

504.20: Wiring Methods

Intrinsically safe equipment can be wired with types of wiring that are suitable for nonclassified locations. Sealing must be done as per Section 504.70, and separation must be as per Section 504.30.

504.30: Separation of Intrinsically Safe Conductors

- **(A)** From Nonintrinsically Safe Conductors
  - **(1)** **Open Wiring.** Conductors or cables being used by intrinsically safe systems must be separated from those of nonintrinsically safe systems by at least 2 inches.

Exception

(a) When all of the intrinsically safe conductors are inside of types MC, MI, or SNM cables, or (b) when all of the nonintrinsically safe conductors are in raceways or in MI, MC, or SNM cables whose outer sheath is able to carry the fault current.

  - **(2)** **In Raceways, Cable Trays, and Cables.** Conductors for intrinsically safe circuits are not allowed to share conduits, cable trays, or cables with conductors from nonintrinsically safe circuits.

Exception No. 1
When the conductors are separated by at least 2 inches and secured at this distance by a grounded metal or insulating partition.

Exception No. 2
When either all of the intrinsically safe or all of the nonintrinsically safe conductors are enclosed in metal-sheathed or -clad cables whose sheath is capable of carrying the full ground-fault current.

(3) Within Enclosures. (a) Conductors of intrinsically safe and nonintrinsically safe systems may be in the same enclosure, as long as they are separated by at least 2 inches, or as specified in Section 504.30(A)(2). (b) All conductors must be secured so that a conductor can't come loose from its terminal and touch another terminal.

2005

(B) From Different Intrinsically Safe Conductors. Different intrinsically safe conductors must be in different cables, or must be separated by either a grounded metal shield or insulation on each wire at least 0.01 inches thick. The clearance between wiring terminals must be at least ¼ inch (6 mm), unless this distance is reduced "by the control drawings." Keep in mind that the *Code* is clearly referring to properly engineered and stamped drawings in this case.

504.50: Grounding

(A) Associated Apparatus and Raceways. These items must be grounded as per Article 250.

(B) **Connection to Grounding Electrodes.** When a connection to a grounding electrode is required, the grounding electrode must conform to the requirements of Sections 250.81 and 250.26(C). Section 250.83 can't be used if electrodes specified in Section 250.81 are available.

(C) **Shields.** The shields of shielded conductors or cables must be grounded.

504.60: Bonding

Intrinsically safe equipment in hazardous locations must be bonded as per Sections 250.78, 501.16(A), 502.16(A), and 503.16(A).

504.70: Sealing

All conduits or cables must be sealed according to Sections 501.5 and 502.5.

Exception

It is not required to provide seals for enclosures that contain only intrinsically safe equipment, except as required by Section 501.5(F)(3).

504.80: Identification

All labels must be suitable for the environment in which they are installed.

- **(A)** **Terminals.** Intrinsically safe circuits must be identified at terminal or junction locations. This will prevent the circuits from being interrupted or interfered with during testing and servicing.
- **(B)** **Wiring.** Labels must be placed on all raceways, cable trays, and open wiring that carry intrinsically safe circuits. The labels must be visible after installation, and must be no more than 25 feet (7.62 m) apart.
- **(C)** **Color Coding.** The conductors for intrinsically safe systems can be colored light blue, if light blue is the only color used. (In other words, either all the conductors must be light blue, or there can be no color code at all.)

Article 505—Class I, Zone 0, 1, and 2 Locations

This article covers the zone classification system, which is an alternative to the long-established division classification system. See the *NEC* for details.

Article 510—Hazardous (Classified) Locations—Specific

510.1: Scope

Articles 501, 502, and 503 give the general requirements for Class I, II, and III locations. The provisions of Articles 511 to 517, inclusive, shall apply to occupancies or parts of occupancies that are hazardous because of atmospheric concentrations of hazardous gases, vapors, or liquids, or because of deposits or accumulations of materials that may be readily ignitable.

The intent of these articles is to assist *Code*-enforcing authorities in the classification of areas of locations with respect to hazardous conditions. These hazardous conditions may or may not require construction and equipment conforming to Articles 501, 502, and 503. They also set forth additional requirements that may be necessary in specific hazardous locations.

510.2: General

The general rules of the *Code* shall apply to the installation of electric wiring and equipment for use in the occupancies covered within the scope of Articles 511 to 517, inclusive, and the inspection authority is responsible for judging with respect to the application of specific rules. These include:

Article 511—Commercial Garages, Repair, and Storage
Article 513—Aircraft Hangars
Article 514—Gasoline Dispensing and Service Stations
Article 515—Bulk-Storage Plants
Article 516—Spray Application, Dipping, and Coating Processes
Article 517—Health Care Facilities

It is recommended by the *Code* that the authorities enforcing the *Code* become familiar with the National Fire Protection Association's standards, which apply to occupancies included within the scope of Articles 511 to 517, inclusive. Not only should inspectors be familiar with these codes and standards, but anyone who designs or makes electrical installations should also be familiar with them. Some of the NFPA standards involved are listed in Appendix A of the *NEC*. See also Section 90.3.

There are other standards that will be of great assistance in evaluating the hazards that might be involved, and these may be purchased from the National Fire Protection Association, Batterymarch Park, Quincy, MA 02269, in an eight-volume set that includes all standards. The individual codes may also be purchased.

Article 511—Commercial Garages, Repair, and Storage

511.1: Scope

This article covers occupancies or locations that are used for service and repair in conjunction with self-propelled vehicles (including passenger automobiles, buses, trucks, tractors, and so on) where flammable liquids are used for fuel or power for operation.

Note

See the definition of garage in *NEC* Article 100.

511.2: Locations

Occupancies for flammable fuel that is transferred into vehicle fuel tanks are covered in Article 514. Areas used exclusively for parking garages and parking or storage, and where no repair work is done except exchange of parts and routine maintenance that don't require the use of electrical equipment, open flame, or the use of volatile flammable liquids, are not classified. There shall, however, be adequate ventilation to carry the exhaust fumes away from engines. (These last items are exceptions that one does not ordinarily see.)

Note

Information on parking structures is covered in *NFPA 88B-1997, Standard for Repair Garages*. See the definition of volatile flammable liquid in *NEC* Article 100.

511.3: Class I Locations

This section is for further clarification of the usages classified in Article 500:

(A) Up to a Level of 18 Inches Above the Floor. The entire area of a floor up to a height of 18 inches above the floor is a Class I, Division 2 location, unless the enforcing agency determines that there is mechanical ventilation that will change the air completely four times every hour. Figure 511-1 will assist in a clarification of the above.

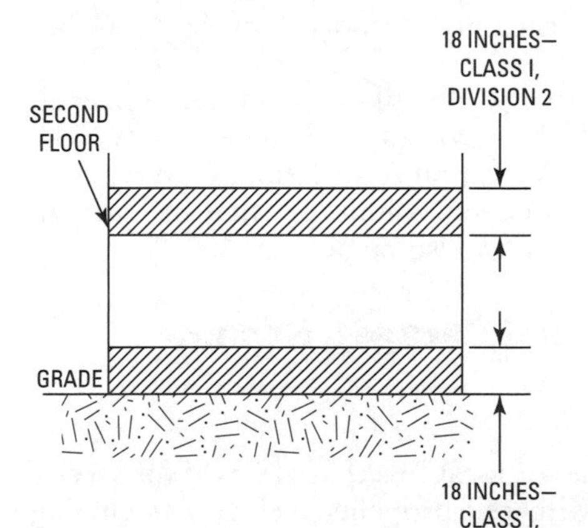

Figure 511-1 Hazardous areas of commercial garages above grade level.

(B) Any Pit or Depression Below Floor Level. Pits or depressions below floor level, that extend up to the floor level shall be considered as Class I, Division 1 locations, except that if the pit or depression is unventilated, the authority enforcing the *Code* may judge the area to be a Class I, Division 2 location (Figure 511-2).

Any pit or depression in which six air changes per hour are exhausted at the floor level of the pit may be judged by the enforcing agency to be a Class I, Division 2 location.

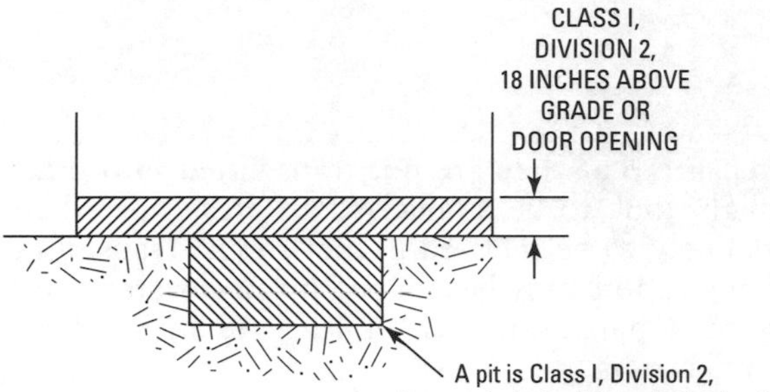

Figure 511-2 Hazardous areas and classification of pits or depressions in garage floors.

In cases where there is adequate and positive ventilation of the floors below grade level, it is up to the inspection authority to classify the areas. They may judge the floors to be Class I, Division 2 up to a height of 18 inches above the floors, even though they are below grade level. It is not mandatory that they judge them this way, however (Figure 511-3).

(C) Areas Adjacent to Defined Locations with Positive Pressure Ventilation. Areas adjacent to Class I, Division 2 locations in which there is no likelihood of hazardous vapors being released (such as stockrooms) are also Class I, Division 2 locations unless the floor of the adjacent area is elevated 18 inches above the floor of the hazardous area, or unless a separation between the two areas is provided by an 18-inch tight curb or partition.

Gasoline vapors are heavier than air and settle to the floor area. Since these vapors settle to the floor, the purpose of this

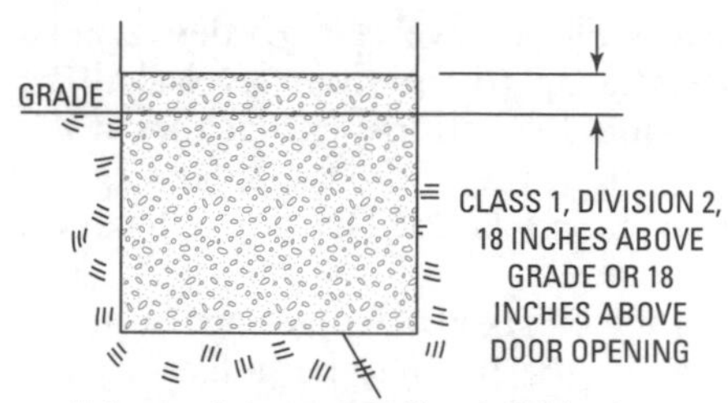

Figure 511-3 Hazardous areas of commercial garages below grade level.

section is to ensure that they are not transmitted into other rooms that might look to be nonhazardous areas since there are no vapors likely to be released in that room. Transmittal of the hazardous vapors may be stopped by an elevated floor, a tight curb, or a partition, each at least 18 inches high, between the two rooms.

(D) Adjacent Areas by Special Permission. Adjacent areas that are well ventilated or have different air pressure or physical spacing and where, in the opinion of the authority that has jurisdiction, no ignition hazards exist, shall be classified as nonhazardous.

(E) Fuel-Dispensing Units. Article 514 describes the regulations for the installation of fuel-dispensing units that handle other than petroleum gas, which is prohibited, in buildings.

If mechanical ventilation is provided in the area where dispensing occurs, the controls shall be interlocked so that the dispenser can't be operated without the ventilation equipment operating as described in Section 500.7(B).

(F) Portable Lighting Equipment. Portable lighting equipment such as extension cords and lamps must have a handle, lampholder, hook, and substantial guard, all attached to the lampholder handle. Exterior surfaces that are likely to come into contact with battery terminals, wiring terminals, or other objects shall be of a nonconducting material or shall be effectively insulated. Lampholders shall be of the unswitched type and shall have no plug-in for attachment plugs. The outer shell shall be of molded composition or other material approved for the purpose. Metal-shell, lined lampholders, either of the switched or unswitched type, shouldn't be used.

Unless the lamp and cord are supported or arranged so that they can't be used in the hazardous areas classified in Section 511.3, they shall be of an approved type for hazardous locations.

511.4: Wiring and Equipment in Class I Locations

In hazardous locations as defined in Section 511.3, all of the requirements of Article 501 shall apply. The location will have to be judged either Class I, Division 1 or Class I, Division 2, and the wiring and equipment will have to conform to the classification.

Raceways embedded in a masonry wall or buried beneath a floor shall be considered to be within the hazardous area above the floor if any connections or extensions lead into or through such areas.

As an example, take the case of a garage at grade level with the doors opening at grade level. The classification would be Class I, Division 2 to a height of 18 inches above the floor. If all of the wiring and equipment is kept above the 18-inch height from the floor, EMT and general-duty equipment and devices can be used. If the wiring goes below the 18-inch point (possibly conduit is run in the floor slab), this part would be Class I, Division 2 and the wiring would have to meet the requirements of this classification, and would have to have seals above the 18-inch height to pass from the hazardous area into the nonhazardous area. Any equipment that is below the 18-inch level will also have to be approved for a Class I, Division 2 location.

511.5: Sealing

Seals must be provided according to Section 501.5. Section 501.5(B)(2) must be applied to both horizontal and vertical boundaries of Class I locations.

511.6: Wiring in Spaces Above Class I Locations

(A) **Fixed Wiring Above Class I Locations.** The only types of wiring permitted above hazardous areas are metal raceways, rigid nonmetallic conduit, electrical nonmetallic tubing, flexible metal conduit, liquid-tight flexible metal or nonmetallic conduits, types PLTC, MI, TC, SNM, or MC cable. Cellular metal floor raceways may be used only to supply ceiling outlets or extensions to areas below the floor; outlets from the same shouldn't be connected above the floor.

(B) **Pendants.** Pendants must be approved for hard usage and the type of service for which they are to be used.

(C) Grounded Conductors. Circuits that supply pendants or portables must have a grounded conductor as covered in Article 200. With receptacles and attachment plug connectors or similar devices, the grounded conductor that carries current to the equipment supplied shall be attached to a polarized prong, and the grounded conductor shall be connected to the screw-shell of the lampholder or to the grounded connection of the utilization equipment supplied.

In explanation, the standard 120-volt receptacle has two different length slots, but the majority of attachment plugs have both prongs the same width. The wider slot of the receptacle connects to the grounded (neutral) conductor of the circuit (the white screw). In garages, plugs that have two different width prongs are to be used. Thus, the attachment plug can only go into the receptacle one way. The neutral of the cord is attached to the wide prong and, if this is an extension cord with a portable lamp, the grounded conductor is to be connected to the screw-shell of the lampholder. Thus, the polarity of the circuit can be maintained (see Figure 511-4). Care must also be taken to connect the grounded terminal of utilization equipment to the grounded conductor of the cord or circuits.

(D) Attachment Plug Receptacles. Attachment plugs and receptacles shall be above the Class I location or shall be of the approved type for Class I locations. From 18 inches down to the floor is the classified area.

511.17: Equipment Above Class I Locations

(A) Arcing Equipment. If equipment is less than 12 feet above the floor level and is capable of producing arcs, sparks, or particles of hot metal such as cutouts, switches, charging panels, generators, motors, and other equipment (excluding receptacles,

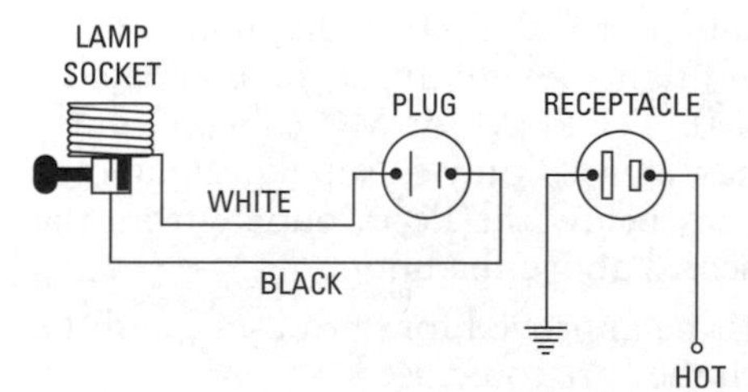

Figure 511-4 Receptacles and attachment plugs.

lamps, and lampholders) and has make-and-break or sliding contacts, they are to be of the totally enclosed type or so constructed as to prevent escape of sparks or hot metal particles.

(B) Fixed Lighting. Lamps and lampholders mounted in a fixed location for lighting and mounted over lanes through which vehicles are ordinarily driven or where the fixtures may be exposed to physical damage, are to be located not less than 12 feet above floor level. An exception to this is that if the fixtures are of a totally enclosed type and are constructed so as to prevent the escape of sparks or hot metal particles, they may be located lower than 12 feet.

511.8: Battery Charging Equipment

Battery chargers and their control equipment, and the batteries that are being charged, are not to be located in areas classified in Section 511.3.

511.9: Electric Vehicle Charging

(A) Connections. When flexible cords and connectors are used for charging, they shall be of the type that has been approved for extra-hard usage, and the ampacity of the conductors must be adequate to handle the charging current.

(B) Connector Design and Location. Connectors that are installed for disconnecting must be designed and installed so that they will disconnect easily at any position of the charging cable. Any live parts shall be guarded to prevent accidental contact. No connector shall be located in a Class I area, which was defined in Section 511.3.

(C) Plug Connections to Vehicles. If plugs are provided for direct connection, then Section 511.3 shall be adhered to so that the connection won't be in a Class I location, and if the cords are supplied from overhead, the lowest point of the plug shall be at least 6 inches above the floor. When the vehicle is equipped with an approved plug that may be readily disconnected and an automatic arrangement is provided for both cord and plug so they are not exposed to physical damage, no other connectors will be required in the cable at the outlet.

511.10: Ground-Fault Circuit Interrupter Protection for Personnel

A ground-fault circuit interrupter shall be installed for the protection of personnel in all 125-volt single-phase 15- and 20-ampere receptacles that are used for electrical connection to diagnostic equipment, electrical hand tools, and where portable lighting devices are used.

Article 513—Aircraft Hangars

513.1: Definition

This shall include occupancies that are used for storage or servicing of aircraft in which gasoline, jet fuels, or other volatile flammable fuels or flammable gases are used. Not included are locations used exclusively for aircraft that have never contained such volatile flammable gases or liquids and aircraft that have been drained or properly purged.

Note

See the definition of volatile flammable liquids in *NEC* Article 100.

513.2: Classification of Location

Classification under Article 500.

(A) **Below Floor Level.** Pits or depressions below the hangar floor level shall be considered as Class I, Division 1 locations, and this classification shall extend to the floor level (Figure 513-1).

(B) **Areas Not Cut Off or Ventilated.** The entire area of the hangar shall be considered as a Class I, Division 2 location to a height of 18 inches above the floor and shall also include adjacent areas into which the hazards may be communicated unless suitably cut off from the hangar as described in (D) below.

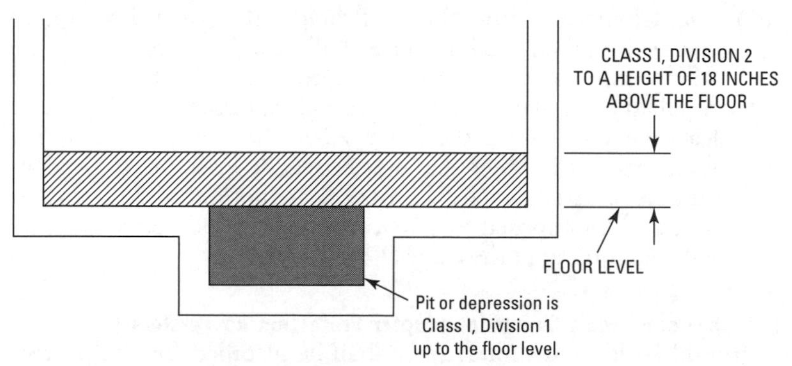

Figure 513-1 Hazardous areas in aircraft hangars.

(C) **Vicinity of Aircraft.** The area immediately adjacent to the aircraft shall be a Class I, Division 2 location. Such an area is defined as being within 5 feet horizontally from aircraft power plants, aircraft fuel tanks, or aircraft structures containing fuel. These locations shall extend from the floor level vertically to a level of 5 feet above the wings and above the engine enclosures.

(D) **Areas Suitably Cut Off and Ventilated.** See (B) above. Any adjacent area into which hazardous vapors are likely to be communicated or released, such as stockrooms and electrical control rooms, shall be a Class I, Division 2 location up to a height of 18 inches. These adjacent areas need not be classified as hazardous if they are effectively cut off from the hangar itself by walls or partitions and are adequately ventilated.

513.3: Wiring and Equipment in Class I Location

All wiring, whether portable or fixed, and all equipment that is located in the hazardous area of an aircraft hangar (hazardous as defined in Section 513.2) shall conform to the provisions of Article 501. If above the floor, the area shall be a Class I, Division 2 location, and if under the floor, the area shall be a Class I, Division 1 location. If any wiring is located in vaults, pits, or ducts, adequate drainage shall be provided, and the wiring shouldn't be placed in the same compartment with any service, with the exception that wiring may be placed in a duct with compressed air. Approved outlets, attachment plugs, and receptacles are to be used in Class I locations, or the design shall be such that they can't be energized while the connection is being made or broken.

513.4: Wiring Not Within Class I Locations

(A) Fixed Wiring. With all of the fixed wiring that is installed in hangars—but not within Class I locations, which are defined in Section 513—the wiring shall be installed in metal raceways, or Type MI, TC, SNM, or MC cable may be used.

Exception

All wiring that is located in nonhazardous areas as defined in Section 513.2(D) must be of a type recognized in Chapter 3.

(B) **Pendants.** If flexible cords are used for pendants, they must be suitable for the location and also approved for hard usage. Each cord must contain an equipment-grounding conductor as well as the current-carrying conductors.

(C) Portable Equipment. Flexible cord may be used for portable utilization equipment and lamps, provided the flexible cord is approved for extra-hard usage. Included in such cord shall be a grounding conductor.

(D) Grounded and Grounding Conductors. When circuits supplying portables and pendants include a grounded conductor (Article 200), they shall have the following devices of the polarized type, and the grounded conductor of the flexible cord shall be connected to the screw-shell of any lampholder or to the grounded terminal of all utilization equipment: receptacles, attachment plugs, connectors, and similar devices.

Grounding continuity that is acceptable shall be provided between fixed raceways and the noncurrent-carrying parts (metallic) of pendant fixtures, portable lamps, and portable and utilization equipment. This may be accomplished by bonding or attaching the grounding conductor to the fixed raceway system.

513.5: Equipment Not Within Class I Locations

(A) Arcing Equipment. In locations other than those described in Section 513.2, equipment that may produce arcs, sparks, or particles of hot metals, such as lamps and lampholders for fixed lighting, cutouts, switches, receptacles, charging panels, generators, motors, or other equipment that has make-and-break or sliding contacts, shall be of the totally enclosed type to prevent the escape of sparks or hot metal particles, or shall be so constructed as to prevent escape of sparks or hot metal particles. In areas described in Section 513.2(D), equipment may be of the general-purpose type.

(B) Lampholders. Metal-shell, fiber-lined-type lampholders may not be used for fixed incandescent lighting.

(C) Portable Lighting Equipment. Only portable lighting equipment that is approved for use within a hangar shall be used.

(D) Portable Equipment. All portable utilization equipment that is or may be used in the hangar shall be of an approved and suitable type for use in Class I, Division 2 locations.

513.6: Stanchions, Rostrums, and Docks

(A) In Class I Locations. Electrical wiring, outlets, and equipment, including lamps, that are attached to or on stanchions, rostrums, or docks that are either located or might be located in a Class I location, as defined in Section 513.2(C), must meet the requirements of Class I, Division 2 locations.

(B) **Not in Class I Location.** Stanchions, rostrums, or docks that are not, or won't likely be, located in the hazardous areas defined in Section 513.2(C) should have wiring and equipment that conform to areas that are classified as hazardous. The wiring for these areas is covered in Section 513.4, and the equipment for these areas is covered in Section 513.5. Exceptions to these requirements are as follows: (1) Wiring and equipment that will be within the 18-inch classification above the floor shall be approved for Class I, Division 2 locations as covered in Section 513.6(A); and (2) receptacles and attachment plugs shall be of the locking type, which won't be readily pulled apart.

(C) **Mobile Type.** Mobile stanchions with electrical equipment that conforms to part (B) above shall carry at least one sign, permanently affixed to the stanchion, that reads, "WARNING—KEEP 5 FEET CLEAR OF AIRCRAFT ENGINES AND FUEL TANK AREAS."

513.7: Sealing

Sealing in Class I locations is covered in Sections 501.5, 501.5(A), and 501.5(B)(2). The requirements of these sections apply to horizontal as well as to vertical boundaries of the hazardous areas. Raceways that are embedded in or under concrete will be in the same hazardous area that is above the floor when any connections lead into or through such areas. This will require seal-offs when connecting to enclosures where there is a possibility of arcs or sparks, or when going from a hazardous to a nonhazardous area.

513.8: Aircraft Electrical Systems

See the *NEC*.

513.9: Aircraft Battery-Charging and Equipment

It is not permitted to charge aircraft batteries while they are installed in the aircraft, if the aircraft is located in the hangar or partially located within the hangar.

Battery chargers and their control equipment, tables, racks, trays, and wiring must conform to the provisions of Article 480 and shouldn't be located in a hazardous area as defined in Section 513.2, but should be located in a separate building or in an area as defined in Section 513.2(D), that is, an area suitably isolated from the hazardous area. Mobile chargers shall have at least one permanently affixed sign reading, "WARNING—KEEP 5 FEET CLEAR OF AIRCRAFT ENGINES AND FUEL TANK AREAS."

513.10: External Power Sources for Energizing Aircraft

(A) Not Less than 18 Inches Above Floor. Aircraft energizers shall be so designed and mounted that all of the electrical equipment and fixed wiring is at least 18 inches above the floor and won't be operated in a hazardous area as defined in Section 513.2(C).

(B) **Marking for Mobile Units.** Mobile energizers shall have at least one permanently affixed sign reading, "WARNING—KEEP 5 FEET CLEAR OF AIRCRAFT ENGINES AND FUEL TANK AREAS."

(C) **Cords.** Flexible cords used for aircraft energizers or ground-support equipment shall be of an approved and an extra-hard-usage type; an equipment-grounding conductor shall also be installed within the cord.

513.11: Mobile Servicing Equipment with Electric Components

(A) General. Mobile servicing equipment that is not suitable for Class I, Division 2 locations shall be so designed and mounted that all fixed wiring and equipment is at least 18 inches above the floor. This equipment includes vacuum cleaners, air compressors, air movers, and similar equipment. None of this equipment shall be operated in the hazardous areas defined in Section 513.2(C). All such equipment shall have at least one permanently affixed sign reading, "WARNING—KEEP 5 FEET CLEAR OF AIRCRAFT ENGINES AND FUEL TANK AREAS."

(B) **Cords and Connectors.** Flexible cords for mobile equipment shall be suitable for the type of service and be approved for extra-hard usage. They shall also include an equipment-grounding conductor. Receptacles and attachment plugs shall be suitable for the type of service and approved for the location in which they are installed. There shall also be a means of connecting the equipment-grounding conductor to the raceway system. This grounding may be accomplished in any approved manner.

(C) **Restricted Use.** No equipment shall be used or operated in areas where maintenance operations are likely to release flammable gases or vapors or flammable liquids, unless the equipment is approved for Class I, Division 2 locations.

513.12: Grounding

All metal raceways and all noncurrent-carrying metal portions of fixed or portable equipment, irrespective of their operating voltages, are to be grounded as covered in Article 250.

Article 514—Gasoline-Dispensing and Service Stations

514.1: Definition

This classification shall include locations where fuel is transferred to the fuel tanks of self-propelled vehicles, or to auxiliary tanks. The fuels referred to are gasoline, other volatile flammable liquids, and liquefied flammable gases. A responsibility is placed on the inspection authority by the following part:

If the authority that has jurisdiction is able to satisfactorily determine that flammable liquids that have a flash point below 38°C (100°F), such as gasoline, won't be handled, he or she may classify such a location as nonhazardous.

Reference is also made to the *NFPA Automotive and Marine Service Station Code, NFPA 30A-2000.*

In practically every gasoline-dispensing location there are also lubritoriums, service rooms, repair rooms, offices, salesrooms, compressor rooms, storage rooms, restrooms, and possibly a furnace room, as well as other areas that might be associated with a filling station. The wiring in these locations shall conform to Articles 510 and 511.

514.2: Class I Locations

Apply Table 514.2 where Class I liquids are stored, handled, or dispensed; it shall be used to delineate and classify service stations. An area beyond an unpierced wall, roof, or other solid partition shouldn't be classified as a Class I location. (Refer to the *NEC* for Table 514.2 and see Section 90.3).

Figures 514-1 and 514-2 show the hazardous area around a gasoline-dispensing pump. Figure 514-3 shows the hazardous area around vapor pipes.

This section was inserted in the *Code* because of the practice of dispensing such liquids as white gasoline, from 55-gallon drums and the like, for special purposes. Please note that this is a hazard.

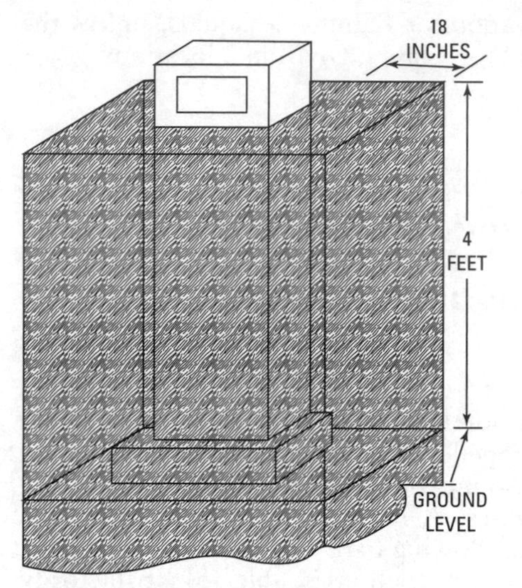

Figure 514-1 Hazardous areas around a gasoline dispenser.

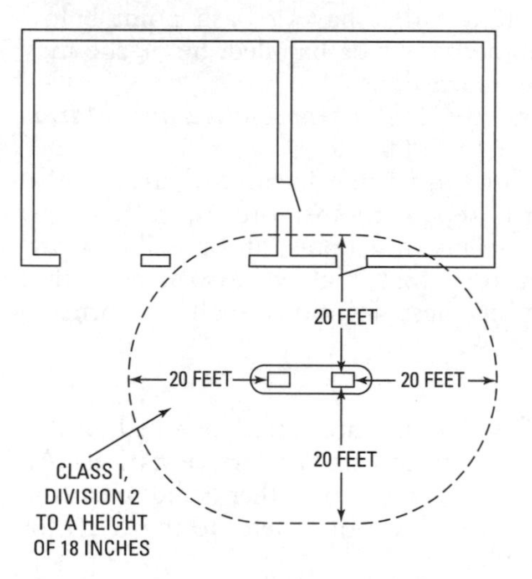

Figure 514-2 Hazardous areas around gasoline service stations.

514.3: Wiring and Equipment Within Class I Locations

Electrical equipment and wiring in Class I, Division 1 and Class I, Division 2 locations, as defined in Section 514.2, must be approved and suitable for these locations. The only wiring methods acceptable (see Article 501) are threaded rigid metal conduit, threaded steel intermediate metal conduit, and Type MI cable. General-purpose devices and equipment are not permitted in these locations.

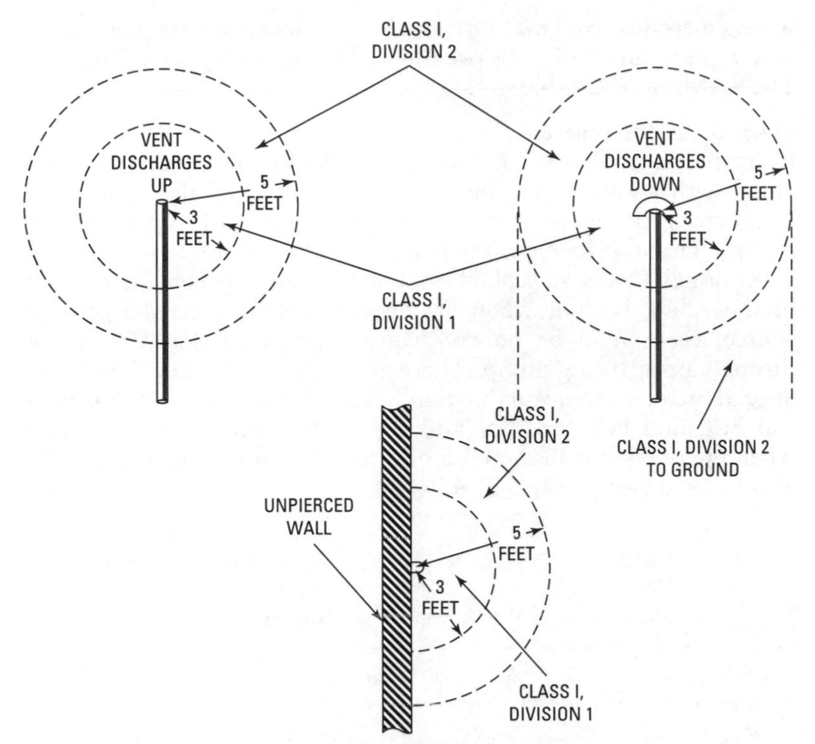

Figure 514-3 Hazardous areas around vent pipes.

Section 501.13 states that conductors used in these locations shall be approved for the location. There are many conductors under various trade names that are approved for the location. These conductors have a nylon outer covering to protect them from gasoline. The Underwriter's Laboratories have a listing on the conductors that have been tested. TW wire, as such, is not approved. Also lead-covered cable may be used for the conductors, but is seldom used because of costs and the difficulty of pulling it into conduit.

Exception
Unless Section 514.8 permits.

514.4: Wiring and Equipment Above Class I Locations
Class I locations are defined in Section 514.2. Sections 511.6 and 511.7 cover wiring methods and equipment for use above Class I locations. These requirements apply to this article also. The only

wiring methods recognized for use in these locations are metal raceway rigid nonmetallic, electrical nonmetallic tubing, and Type MI, TC, SNM, or MC cable.

514.5: Circuit Disconnects

Every circuit leading to or through a dispensing pump must be provided with a switch or other suitable means that disconnects all conductors at the same time from the source of supply to the pumps. This also includes the neutral conductor, if one is used.

Although this is very plain, it is often misinterpreted. The intent of the ruling is that, when the supply is disconnected from the source, there shall be no conductors connected that lead to or through a dispensing pump. There are various means of doing this. Special breakers are available that have a pigtail that ties to the neutral bus, and both the hot and the neutral will be disconnected when the breaker is in the OFF position. A double-pole switch may also be used (see Figure 514-4).

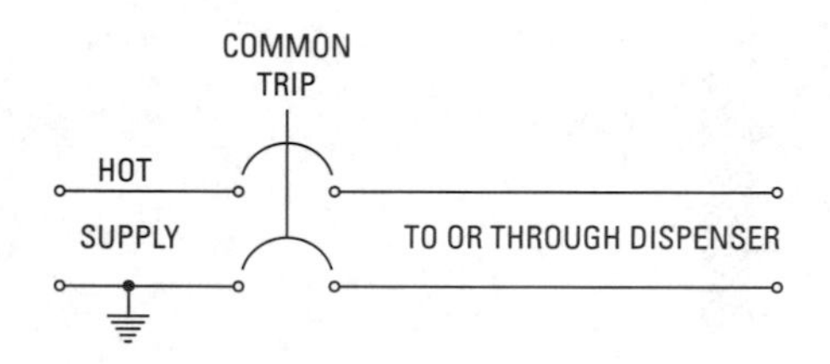

Figure 514-4 Neutral must also be switched to islands.

514.6: Sealing

(A) At Dispenser. A seal shall be provided in every conduit run that enters or leaves a dispenser or any enclosures or cavities that happen to be in direct communication with the dispenser. The sealing shall be the first fitting where the conduit leaves the earth or concrete.

This also is often misinterpreted. Notice the words *first fitting*. This means exactly what it says. There is often an attempt to define a conduit coupling as not being a fitting, but this is not the case—it is a fitting.

All seals are to be readily accessible. Often someone attempts to install a seal at the edge of the hazardous area and bury it in the earth. This is not allowed. It shall be carried into the building or some other place out of the hazardous area and then installed. If, for instance, the seal goes into the service panel, and there is a hazardous area below the panel, the seal must be

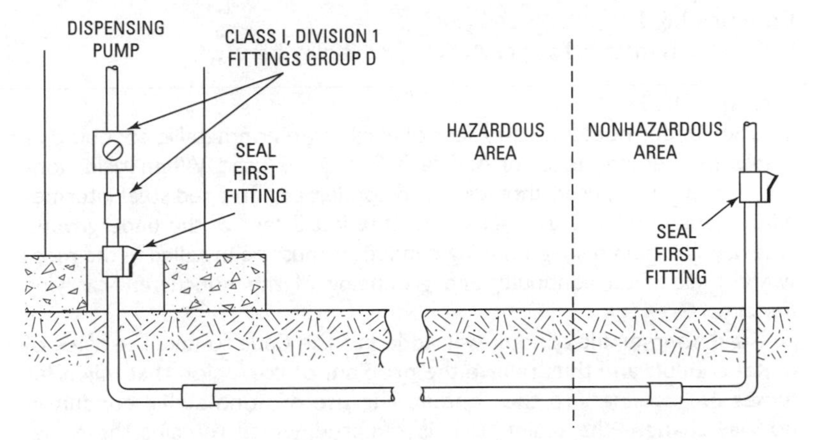

Figure 514-5 Seals required in gasoline service stations.

18 inches or more above the floor, and there can be no fitting of any kind between the floor and the seal (see Figure 514-5).

(B) At Boundary. Additional seals shall be provided as covered in Section 501.5. Section 501.5(A)(4) and 501.5(B)(2) will apply to both horizontal and vertical boundaries between the hazardous and nonhazardous areas.

514.7: Grounding

All metal parts of the dispensing pump, metal raceways, and electrical equipment, regardless of voltage, shall be grounded as provided for in Article 250. This grounding will extend back to the service equipment and service ground. Care should be taken in grounding properly, for an accidental fault could cause a poor ground connection to arc and cause an explosion. See *NEC* Section 501.16.

514.8: Underground Wiring

The wiring installed underground shall be threaded rigid metal conduit or threaded steel intermediate metal conduit (remember that the soil conditions must be such as to prevent electrolysis from damaging the conduits; means should be used to protect the conduits). Any portion of the electrical wiring or equipment that is below the surface of Class I, Division 1 or 2 locations (See Table 514.2) will be considered to be in a Class I, Division 1 location extending to at least the point of emergence above grade. You are also referred to Section 300.5(A), Exception No. 3.

Exception No. 1
If MI cable is installed as per Article 332, it may be used.

Exception No. 2
If buried under not less than 2 feet of earth, rigid nonmetallic conduit that meets the requirements of Article 352 may be used. When rigid nonmetallic conduit is used, threaded rigid conduit or threaded steel intermediate metal conduit must be used for the last 2 feet of the underground raceway; an equipment-grounding conductor must be installed in the raceway for electrical continuity and grounding of the noncurrent-carrying metal parts.

Rigid nonmetallic conduit was added to the *Code* to take the place of metal conduit and thus relieve the problem of corrosion that might let fumes or gasoline into the conduit. The use of nonmetallic conduit in no way changes the sealing that is required, which remains the same. Metal conduit is still to be brought up into the dispenser or the nonhazardous location, as the case might be. This is necessary to prevent physical damage.

Article 515—Bulk Storage Plants

515.1: Definition
This designation shall include locations where gasoline or other volatile flammable liquids are stored in tanks having an aggregate capacity of one carload or more, and from which such products are distributed (usually by tank truck).

515.2: Class I Locations
If Class I liquids are stored, handled, or dispensed, Table 515.2 shall be applied and shall also be used to describe and classify bulk storage plants. The classified area shouldn't extend beyond unpierced walls, roofs, floors, or other solid partitions.

See Table 515.2 in the *NEC* for the requirements. Also refer to Figures 515-1 through 515-5.

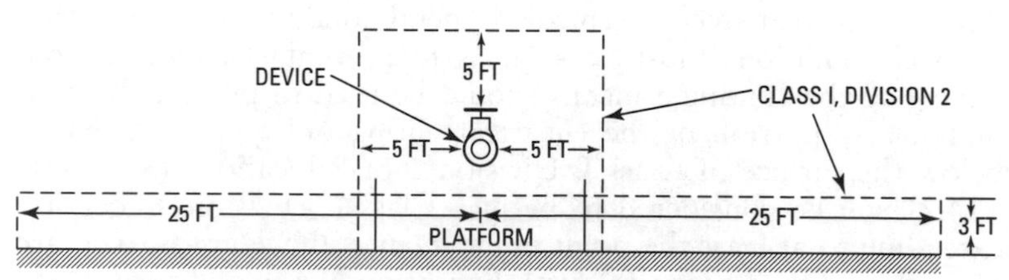

Figure 515-1 Adequately ventilated indoor areas.

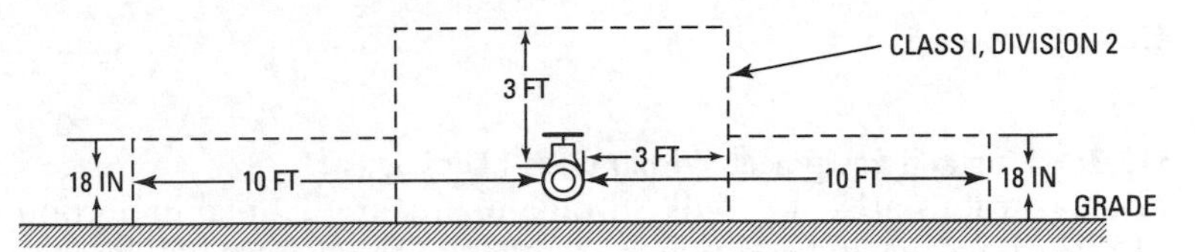

Figure 515-2 Outdoor areas.

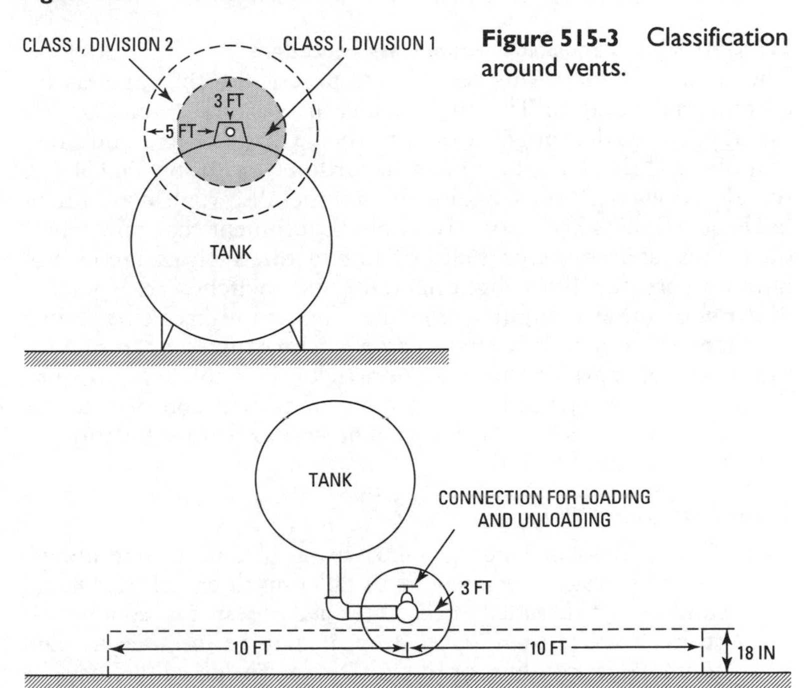

Figure 515-3 Classification around vents.

Figure 515-4 Classification around bottom filler.

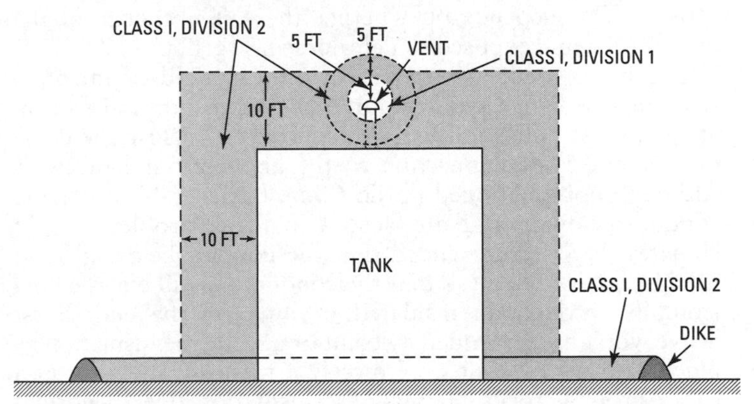

Figure 515-5 Classification around vertical tank.

515.3: Wiring and Equipment Within Class I Locations

The hazardous areas for bulk storage plants are defined in Section 515.2. All wiring and equipment installed in these areas will be subject to the regulations of Article 501.

515.4: Wiring and Equipment above Class I Locations

The requirements of this section are practically the same as for commercial garages. The main concern is sparks, arcs, or hot metal particles that might drop into the hazardous area and cause trouble. All fixed wiring above hazardous locations shall be in metal raceways, PVC schedule-80 nonmetallic rigid conduit, or be Type MI, TC, SNM, or MC cable. Equipment that might produce arcs, sparks, or particles of hot metal, such as lamps and lampholders for fixed lighting, cutouts, switches, receptacles, motors, or other equipment that has make-and-break or sliding contacts, shall be totally enclosed or so constructed as to prevent the escape of sparks or hot metal particles. Portable lamps or utilization equipment shall be approved and shall conform to the provisions of Article 501 for the most hazardous location in which they might be used.

515.5: Underground Wiring

(A) Wiring Method. Underground wiring to and around aboveground storage tanks may be installed in threaded rigid metal conduit or threaded steel intermediate metal conduit. If buried not less than 2 feet deep, it may be installed in rigid nonmetallic conduit or duct. Direct burial cable may be installed if it is approved. One point that might come up here is (B) below; that is, the insulation shall conform to Section 501.13. This depends on whether there is a chance of the insulation being exposed to deteriorating agents.

If cable is used, it must be enclosed in rigid or threaded steel intermediate metal conduit. This protection must begin at the lowest point and will be required to enclose the cable until it makes its connection to the above-ground raceway. Although not mentioned in the *Code*, buried cables entering conduits below-ground are faced with a frost problem in cold climates. It is recommended that the conduit below ground level be in an L shape so that the conductors will emerge horizontally through an insulated bushing on the end. Frost heave won't have a tendency to cut or abrade the insulation as much as if the conduit were merely a vertical run. Also, in a rocky ground, rocks may cut the insulation due to normal

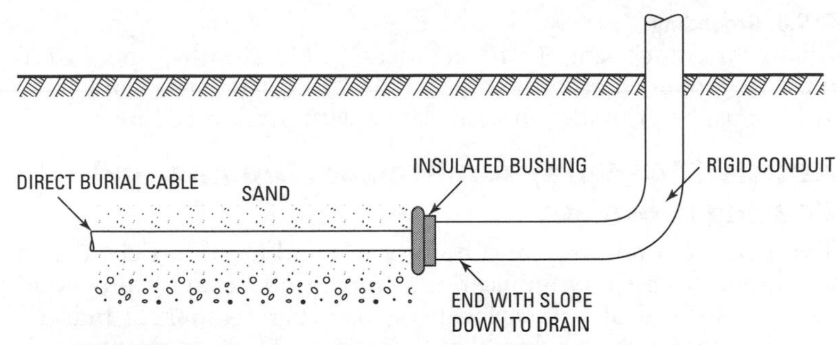

Figure 515-6 Direct burial cable.

pressures and to frost heave. Thus, direct burial cable should be protected below and above by a fine sand bed and covering (see Figure 515-6).

(B) Insulation. Section 501.13 gives the requirements that conductor insulation must meet. Recall that the insulation must be approved for exposure to vapors and liquids.

(C) Nonmetallic Wiring. When rigid nonmetallic conduit or cable that has a nonmetallic sheath is used, an equipment-grounding conductor shall be included to maintain continuity of grounded raceway systems and for noncurrent-carrying parts.

This is standard procedure for practically all wiring in order to provide grounding protection for raceways and equipment where there is no raceway continuity. The grounding continuity must be complete back to the service equipment and the service ground. See *NEC* Section 501.16.

515.6: Sealing

Sealing shall be required in accordance with Section 501.5. Sealing requirements of Section 501.5(A)(4) and 501.5(B)(2) shall apply to both the horizontal and vertical boundaries of the location that is classified as a Class I location. Raceways that are buried under defined hazardous areas will be considered to be in the same classification or area.

515.7: Gasoline Dispensing

When gasoline is dispensed from a bulk-storage plant, all applicable requirements of Article 514, which covers the dispensing of gasoline, must be followed.

515.8: Grounding

All metal components (conduit, metal cable sheaths, and so on) must be grounded according to Article 250. Installations in Class I areas must be grounded in accordance with Section 501.16.

Article 516—Spray Application, Dipping, and Coating Processes

Two definite conditions must be kept in mind in this article. One is the vapor from spray application, dipping and coating processes, and the storage of paints, lacquers, or other flammable finishes, and the other is the residues that accompany finishing processes.

Vapors can be coped with easier than the residues because the residues create a heat problem, as well as the hazard created by the vapors. Bear this in mind while reading this article.

See Appendix A in the *NEC*.

516.1: Scope

This article covers the application of flammable liquids, combustible liquids, and combustible powders by spray operations done regularly at temperatures that exceed their flash points. This may be done by dipping, coating, spraying, or other means.

Note

NFPA 33-2000 (ANSI) gives further information covering these processes, including fire protection, the position of warning signs, and maintenance. The NFPA code called *Spray Application Using Flammable and Combustible Materials, NFPA 34-2000 (ANSI)*, covers dipping and coating processes using flammable or combustible liquids. Additional information on ventilation is covered in *NFPA 91-1999, Blower and Exhaust Systems, Dust, Stock, and Vapor Removal and Conveying*.

516.2: Classification of Location

The classification is based on dangerous quantities, flammable vapors, combustible mists, dusts, residues, and deposits.

- **(A)** **Class I or Class II, Division 1 Locations.** The following outlines the spaces that are to be considered Class I or Class II, Division 1 locations, as applicable for the conditions involved:
  - **(1)** Unless specific provision is made in Section 516.3(D), the interiors of spray booths and rooms are classified.
  - **(2)** The interior of exhaust ducts, because they accumulate residues.

(3) Any other areas that are in the direct path of the spray operation.

(4) When dipping or coating operations are involved, the space that is within 5 feet in any direction from the vapor source, including a distance of 5 feet to the floor. The vapor source shall be considered the liquid surface of the dip tank, the surface of the wetted drain boards, and the dipped object surface. The classified area shall be from the drain-board surface or the wetted surface of the drain board, and shall extend from these surfaces to the floor.

(5) The whole of any pit that is within 25 feet of the vapor surface and extends beyond that distance will be classified Class I, Division 1 unless there is a vapor-stop provided.

(B) Class I or Class II, Division 2 Locations. The areas covered below are considered Class I or Class II, Division 2, whichever classification is applicable:

(1) When open spraying is used, all space outside, but within 20 feet horizontally and 10 feet vertically, is a Class I, Division 1 location as defined in Section 516.2(A), if it is not separated by partitions.

(2) When spray and operations are conducted under a closed top that has an open-front spray booth, a space shown in Figure 2 in this section of the *NEC*, and also a space within 3 feet in any direction from openings not including the open front or face.

The following, as can be seen from Figure 2, shall be Class I or Class II, Division 2, extending from the open face or front of the spray booth in conformance to the following: (a) If the spraying equipment is interlocked with the ventilating equipment so that it is impossible to spray if the ventilating equipment is not operating first, the space 5 feet out from the open face of a spray booth, and as shown in Figure 2A shall be designated Class I or Class II, Division 2. See the *NEC* for Figure 2A, which shows a situation where ventilation is interlocked with spray equipment. (b) Should the spraying operation not be interlocked with the ventilating equipment, so that the spraying can be done without the ventilating equipment being in operation, the space designated Class I or Class II, Division 2 shall extend 10 feet from the open face or front of a spray booth, which is shown in Figure 2B. See the *NEC* for Figure 2B, which shows a situation where ventilation is not interlocked with spray equipment.

(3) Spray booths that have an open top 3 feet above the booth and 3 feet from other booth openings shall be considered as Class I or Class II, Division 2.

(4) When spray operations are confined to a spray booth or room, any space 3 feet in all directions from any openings into the spray area are to be considered Class I or Class II, Division 2, as may be seen in Figure 3. See the *NEC* for Figure 3.

(5) Figure 4 illustrates dip tanks and drain boards, and other hazardous operations; all space beyond the limits for Class I, Division 1 and within 8 feet are defined in (A)(4) as being the vapor source. In addition, the vapor source that was defined in (A)(4) and shown in Figure 4 shall cover all space to the floor to 3 feet above the floor, and it shall stand horizontally 25 feet from the vapor source.

(C) **Adjacent Locations.** If locations adjacent to Class I or Class II locations are cut off from them by tight partitions and have no communicating openings where hazardous vapors are likely to be released, they shall be classified as nonhazardous locations.

(D) **Nonhazardous Locations.** Locations utilizing drying, curing, or fusion apparatus may be judged nonhazardous by the authority enforcing the *Code* if (1) there is adequate positive mechanical ventilation provided to prevent the accumulation of concentrations of flammable vapors, and (2) there are interlocks provided to deenergize all electrical equipment (other than Class I equipment) in the event of a failure of the ventilating equipment.

Note

See *NFPA 86-2003 (ANSI), Ovens and Furnaces, Design, Location, and Equipment.*

516.3: Wiring and Equipment in Class I Locations

(A) **Wiring and Equipment—Vapors.** All electrical wiring and equipment located in a Class I location (where vapor only is contained and not residues), which was defined in Section 516.2, must comply with provisions of Article 501.

Article 501 covers Class I, Division 1 and 2 locations. Remember that such areas are in Group D. Group D covers gases and vapors but not deposits and residues. It will be found that no equipment is approved for use in spray booths for protection involving deposits and residues. For example, a motor may be approved for a Class I, Division 1, Group D location; but the motor can't be mounted in an exhaust duct from a spray booth because of the deposits and residues from the spraying operation, which are highly hazardous from a temperature standpoint. In addition, residues will deposit on the motor and cause a hazard. The same is true of lighting fixtures. A belt used to drive an exhaust fan creates a static electricity problem that must be remembered in designing the system. The NFPA states that the belt shall be enclosed in a nonferrous enclosure and that the fan shall be of nonferrous material. Even if the motor is mounted externally, it is possible to have gases or vapors come into contact with the motor, so it still must be approved for a Group D location.

(B) Wiring and Equipment—Vapors and Residues. Unless specifically approved for readily ignitable deposits and flammable vapor locations, no electrical equipment shall be installed or used in hazardous areas where it can be subjected to accumulations of readily ignitable deposits or residues, or subject to spontaneous heating and ignition of some residues, the likelihood of which will be greatly increased by a rise in temperature. Type MI cable and wiring in threaded rigid metal conduit, or threaded steel intermediate metal conduit, may be used in such locations.

(C) Illumination. Because there seem to be no lighting fixtures approved for deposits and residues, illumination of readily ignitable areas through panels of glass or other approved transparent or translucent material is permissible only where (1) fixed lighting units are used as the source of illumination; (2) the panel effectively isolates the hazardous area from the area in which the lighting unit is located (Figure 516-1); (3) the lighting unit is approved specifically for a Class I, Division 1, Group D location; (4) the panel is of a material or is so protected that breakage will be unlikely; and (5) the arrangement is such that the normal accumulation of hazardous residues on the surface of the panel won't be raised to a dangerous temperature by radiation or conduction from the illumination source (Figure 516-1). The inspection authority must judge the hazard of the area in which lighting is mounted behind glass.

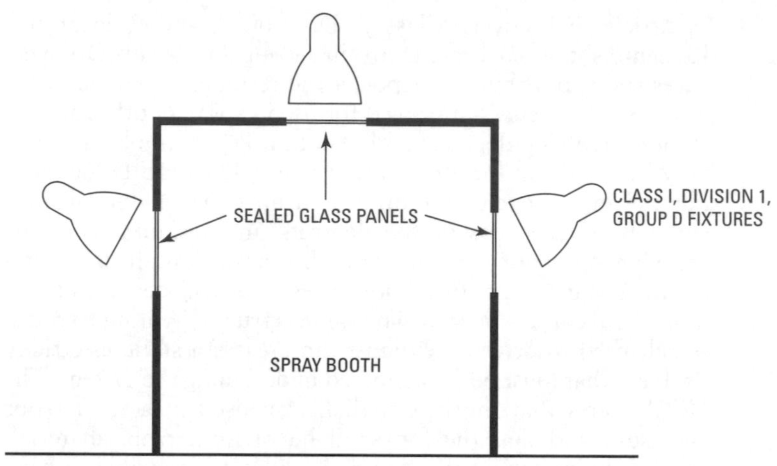

Figure 516-1 Approved lighting for spray booths.

(D) Portable Equipment. Portable lamps and utilization equipment shouldn't be used in hazardous areas during the finishing operations. They may be used during cleaning or repair operations, but they shall be approved for Class I locations, and all of the metal noncurrent-carrying parts shall be grounded.

Exception No. 1

If portable lamps are required in spaces not readily illuminated by the fixed lighting in the spraying area, the portable lamps shall be a listed and approved type for Class I, Division 1 locations where there will be readily ignitable residues.

Exception No. 2

Where portable electric drying equipment is used for automobile finishing spray booths and the following requirements are met: (1) Its electrical connections are not located in the spray enclosure during spraying operations; (2) electrical equipment within 18 inches of the floor is approved for Class I, Division 2 locations; (3) the drying apparatus has all the metal parts electrically bonded and grounded; and (4) interlocks are provided that will prevent the operation of spray equipment during the time that the drying equipment is within the spraying location; this will allow a three-minute purge of the area before the drying equipment is energized, and also shuts off drying apparatus if the ventilating system fails.

(E) **Electrostatic Equipment.** Only as provided in Section 516.4 will electrostatic spraying or detearing equipment be installed.

516.4: Fixed Electrostatic Equipment

For the requirements of this section, refer to the *NEC*.

516.5: Electrostatic Hand Spraying Equipment

For the requirements of this section, refer to the *NEC*.

516.6: Powder Coating

See the *NEC*.

516.7: Wiring and Equipment Above Class I and II Locations

(A) Wiring. Fixed wiring above Class I and II locations shall be in metal raceways, rigid nonmetallic conduit, electrical nonmetallic tubing, or Types MC, MI, TC, or SNM cable. Cellular metal floor raceways will be permitted in the floor only to supply ceiling outlets or extensions to areas below the floor of the spray area or Class I or II location. Such floor raceways are not to have any connections that lead to or pass through Class I or II locations above the floor unless suitable seals are provided.

(B) **Equipment.** Equipment that may produce arcs, sparks, or hot metal particles, such as lamps, lampholders used for fixed lighting, cutouts, switches, receptacles, or any other type of electrical equipment that includes make-and-break or sliding contacts, if installed above Class I or II locations or above locations where freshly finished goods are to be handled, must be of the totally enclosed type, or the construction shall be such that it will be impossible for escaping sparks or hot metal particles to enter Class I or II locations.

516.8: Grounding

All metal raceways and noncurrent-carrying parts of either fixed or portable equipment shall be grounded as provided in Article 250, regardless of the voltage at which they are operated.

Article 517—Health Care Facilities

I. General

517.1: Scope

This article covers construction and installation of electrical wiring in health care facilities. See Appendix A in the *NEC*.

Note
Veterinary facilities are not included in this article.

Note
Refer to the information that concerns the performance, testing, and maintenance requirements for the health care facilities.

517.2: Definitions

This section gives necessary definitions that come into play with essential electrical systems for health care facilities.

Alternate Power Source. This may be one or more generator sets or battery systems, if the battery system is permitted for the location. The alternate power sources are intended to provide power during interruption of normal electrical service by the public utility electric service, and consist of generators on the premises.

Anesthetizing Location. These are areas in health care facilities that have been designated for the administration of any flammable or nonflammable inhalation of any anesthetic agent in the course of examinations or treatments required for relative analgesia.

Critical Branch. This is a branch of an emergency system that may be feeders or branch circuits supplying energy for illumination, any special power circuits, and receptacles that are selected to serve these areas whose functioning is essential to patient care, and they are connected to an alternate power source by means of one or more transfer switches that energize from the temporary power source when the normal power source is interrupted.

Electrical Life-Support Equipment. Any electrically powered equipment that must have continuous operation to maintain the patient's life.

Emergency System. Feeders to branch circuits that conform to Article 700 and are intended to supply power from an alternate source to a limited number of designated functions that are vital for the protection of life and for the patient's safety and must operate within 10 seconds of the interruption of the normal power source.

Equipment System. This consists of feeders or branch circuits arranged for delay manual or automatic connection to the power source. This ordinarily serves three-phase power.

Essential Electrical System. A system supplying the alternate power source, all distribution systems, and ancillary equipment that have been designed to ensure electrical power continuity to designated areas and functions of a health care facility when the normal power source is disrupted. It shall also be designed to minimize disruption of power in the internal wiring system.

Exposed Conductive Surfaces. These surfaces are capable of carrying electrical current and are unprotected, not enclosed, and not guarded, permitting personal contact. Painting, anodizing, and similar coatings are not considered suitable insulation, unless they have been listed for that purpose.

Flammable Anesthetics. Gases or vapors such as fluroxene, cyclopropane, divinyl ether, ethyl chloride, ethyl ether, and ethylene that can form flammable or explosive mixtures when combined with air, oxygen, or reducing gases such as nitrous oxide.

Flammable Anesthetizing Location. This is any area that has been designated to be used for the administration of any flammable inhalation anesthetic agents that will be used during normal examinations or treatment.

Hazard Current. A set of connections in the isolated power system where the total current could flow through a low impedance should it be connected between either the ground or an isolated conductor.

> FAULT HAZARD CURRENT: The isolated systems that are hazardous current when all devices are connected, except the line isolation monitor.
>
> MONITOR HAZARD CURRENT: The line isolation monitor alone and its hazardous current.
>
> TOTAL HAZARD CURRENT: The combined hazardous current in the preceding two definitions.

Health Care Facilities. These are buildings or parts of buildings that are not limited to occupancies such as hospitals, nursing homes, residential care facilities, facilities for supervisory care, clinics, doctors' and dentists' offices, and ambulatory health care facilities that may be mobile or fixed.

Hospital. A building or a portion of a building that is used for medical, psychiatric, obstetrical, or surgical care and is open and maintained on a 24-hour basis and has facilities for four or more inpatients. For the purpose of this article, hospitals include general hospitals, mental hospitals, tuberculosis hospitals, children's hospitals, and places providing inpatient care.

Isolated Power System. This consists of an isolation transformer or its equivalent and a monitor showing that it is isolated, and the circuit conductors are not to be grounded.

Isolation Transformer. Such a transformer consists of a primary and secondary winding that are physically separated, and that by induction couples the secondary to the feeders that are used to energize the primary winding.

Life Safety Branch. This consists of heaters and branch circuits that meet the requirements of Article 700 and the intent of which is to provide power needs adequate to secure the power source to the patients and personnel, and that shall automatically be connected to the alternate source of power upon interruption of the normal supply of power.

Limited Care Facility. A building or part of a building that is used continually for the care of four or more persons who are not capable of caring for themselves.

Line Isolation Monitor. This is a test instrument that is designed to give a continual check of the balanced and unbalanced impedance from each line of an isolated circuit to ground. It shall be equipped with a built-in test circuit that allows the alarm without adding leakage-current hazard.

Note

This was formerly known as a ground contact indicator.

Nursing Home. A building or portion thereof for lodging, boarding, and nursing care on a 24-hour basis for four or more persons, who because of their incapacity may be unable to provide for their own needs and safety unless aided by another person. Nursing homes, when used in this *Code*, include nursing and convalescent homes in which there are skilled nursing facilities and immediate care facilities, and infirmary areas or homes for the aged.

Nurses' Stations. The location of a group of nurses that serve the needs of patients; also a point at which patient' calls are received and from which nurses are dispatched, where the nurses' notes are written and inpatient charts are maintained, and also where any medication is prepared to be distributed to patients. If more than one location is provided as a nursing unit, all these are to be included as part of the nurses' station.

Patient Equipment Grounding Point. In order to eliminate electromagnetic interference problems in the grounding of electrical appliances serving patients, a jack or terminal bus is to serve as the collection point for redundant grounding of electric appliances.

Patient Vicinity. Patients are normally cared for in what is termed a patient vicinity. This is the space likely to be contacted by the patient or the attending person. A typical patient room shall contain space for the bed and its terminal location, and the space beyond the bed shall be not less than 6 feet horizontally and extend vertically not less than 7½ feet from the floor.

Psychiatric Hospital. This is a building that is used exclusively to give psychiatric care to four or more patients on a 24-hour basis.

Reference Grounding Point. The ground bus of a panelboard or isolated power system that supplies power to a patient care area.

Residential Custodial-Care Facility. A building or part of one that provides boarding or lodging for four or more persons who are incapable of taking care of themselves because of age or physical or mental handicap. Included here are facilities for taking care of the aged, nurseries (custodial care for children under six years of age), and for caring for the mentally handicapped. If they don't provide lodging and board and day care facilities for the occupants, they are not to be classified as residential custodial-care places.

Room Bonding Point. One or more grounding terminals are points that provide a place to ground exposed metal or conductive building circuits in a room.

Selected Receptacles. A minimal number of receptacles for use with appliances that under normal conditions are required for tasks in the room, and that are likely to be used for the emergency treatment of patients.

Task Illumination. This is the minimum lighting required to conduct the necessary tasks in a given area. It also includes safe access to exits and supplies and equipment.

Therapeutic High-Frequency Diathermy Equipment. This is therapeutic induction and dielectric heating equipment operating high-frequency diathermy equipment.

Wet Location, Health Care Facility. An area for patients that is normally subjected to water on the floor, or where routine dousing or drenching is required for housekeeping procedures. Places where liquids are accidentally spilled don't constitute wet locations.

517.3: General

The requirements that you will find under Parts III, IV, and V apply not only to a building with a single function, but also to respective forms of occupancy within a building used for multifunctions (for example, a doctor's room located in a residential facility for custodial care would be required to meet the provisions of Part III).

II. Wiring Design and Protection

517.10: Applicability

Part II applies to all health care facilities. Iinstallations in these facilities that provide for patient care, services, or equipment covered in other parts of this article shall also be required to comply with that part.

517.11: General Construction/Construction Criteria

This article outlines installation methods that are intended to avoid hazards by keeping only low-potential differences between exposed conductive surfaces.

517.12: Wiring Methods

Wiring methods, except as modified in this article, shall comply with the appropriate requirements of Chapters 1 through 4 of the *NEC*. Wiring methods not permitted are found in Articles 362, 334, and 352, and shouldn't be used in those portions of health care facilities that are designated for patient care areas.

517.13: Grounding of Receptacles and Fixed Electrical Equipment

In inpatient care areas, the grounding terminal of receptacles, all noncurrent-carrying surfaces that are conductive, and fixed electrical equipment that could possibly become energized and are subject to personnel contact, operating at over 100 volts, are to be grounded by an insulated copper conductor. Sizing the grounding conductor shall be done in accordance with Table 250.95, and it shall also be installed in metal raceways with the branch circuit supplying the fixed equipment or receptacles. All branch circuits that supply patient care areas must be provided with a ground path for fault currents by any approved grounding means, as defined in Section 250.91(B).

Exception No. 1

For Type MC, MI, or AC cable where an insulated grounding conductor is used, metal raceways are not required.

Exception No. 2

Metal faceplates may be grounded. This may be done by means of metal mounting screw(s) securing the metal faceplate to a grounded outlet box or some device used for grounding.

517.14: Panelboard Bonding

In normal and essential electrical system panelboards, the equipment-grounding terminal must be bonded with an insulated continuous conductor, and this conductor shall be smaller than No. 10 AWG copper. The conductor can be broken only where it terminates in a grounding bus.

517.16: Receptacles with Insulated Grounding Terminals

If receptacles have insulated grounding terminals as permitted by Section 250.74, they must be identified so that the identification is still visible after installation.

Note

Care is very important in specifying a system that uses insulated receptacle grounds, because the impedance of the grounding is controlled by the grounding wires, and no benefit is derived functionally from parallel grounding paths.

517.17: Ground-Fault Protection

(A) Feeders. If ground-fault protection is provided to cover the operation of service disconnecting means as covered in Section 230.95, an additional step or ground-fault protection shall be provided at the next level of feeders from the original feeder, thus downstream from the load. This ground-fault protection consists of overcurrent devices and current transformers. Other equivalent protection that will cause the feeder to be disconnected may be used.

(B) **Selectivity.** Ground-fault protection for the service and disconnecting means shall be selective to the extent that such device and the feeder, and not the service device, will open if a ground-fault occurs on the load side of the feeder device. There shall be a six-cycle minimum differential between the service and the feeder tripping of the ground fault. The time of operation of the disconnecting device must be considered when selecting time spread between these two bands, so as to achieve 100-percent selectivity.

Note

Refer to *NEC* Section 230.95 and the fine-print note there covering the transfer of alternative sources of protection.

(C) **Testing.** When equipment ground-fault protection is installed, each performance level must be tested to see that it conforms to (B) above.

517.18: General Care Areas

(A) Patient Bed Location Branch Circuits. Every patient bed must be supplied by at least two branch circuits. One of the circuits must come from a normal system panelboard. All branch circuits of the normal system have to come out of the same panelboard.

Exception No. 1

Special purpose branch circuits are not required to all come from the same panelboard.

Exception No. 2

Clinics, outpatient facilities, offices, etc. are accepted, as long as they meet the requirements of the exceptions to Section 517.10.

(B) **Patient Bed Location Receptacles.** At least 4 hospital grade receptacles must be provided near each hospital bed. These receptacles must be grounded with a copper wire, according to Table 250.95. It is not intended that all existing nonhospital-grade receptacles be immediately replaced.

Exception No. 1

Rehabilitation and similar hospitals, which conform to the exceptions of Section 517.10.

Exception No. 2

Psychiatric security rooms are not required to have any receptacles.

(C) **Pediatric and Psychiatric Locations.** Fifteen- or 20-amp, 125-volt receptacles in these areas must be of the tamper-proof type.

517.19: Critical Care Areas

(A) Patient Bed Location Branch Circuits. Every patient bed must be supplied by at least two branch circuits. At least one circuit must be from the emergency system (and be marked, showing that it is from the emergency system, and must also show the panel name and circuit number), and at least one must come from the same panelboard, and all emergency circuits must come from the same panelboard.

Exception

Special-purpose branch circuits are not required to come from the same panelboard.

(B) **Patient Bed Location Receptacles.** At least six receptacles are required at each patient bed location. These receptacles (which may be either single or duplex receptacles) must be hospital-grade devices, and must be grounded with an insulated copper conductor.

(C) **Grounding and Bonding, Patient Vicinity.** In some hospitals, a special grounding point with several grounding jacks is used to ground the various pieces of equipment. When these are used, there must be connections made between the grounding point and the grounding terminals of the various receptacles, made with a No. 10 or larger copper wire.

(D) **Panelboard Grounding.** When feeders are run to panelboards in metal raceway, Type MC cable, or Type MI cable (which is commonly done), one of the following grounding techniques must be used at each termination:

- **(1)** When double-locknut connections are used with metal raceway, grounding must be ensured by installing a grounding bushing and a copper wire jumper. The jumper must be sized according to Section 250.95.
- **(2)** Grounding may be ensured by connecting the raceway or cable to the enclosure with threaded hubs.
- **(3)** Other approved techniques may be used, such as bonding locknuts or bushings, etc. Note that approved means "acceptable to the authority that has jurisdiction." Usually this means that the inspector says it is satisfactory.

(E) **Additional Protective Techniques in Critical Care Areas.** Isolated power systems are allowed to supply critical care areas. When used in this way, they must

(1) Be approved.

(2) Be properly designed for the purpose.

(3) Meet the requirements of Section 517.160.

Audio and visual monitors for the isolated systems can be located at the nurses' station that serves the area where the system is installed.

(F) **Isolated Power System Grounding.** The grounding conductor of these systems (when the first fault current is limited) can be run outside of the enclosure that houses the power conductors. The author considers it a good trade practice to keep the power conductors and grounding conductors together whenever possible.

(G) **Special-Purpose Receptacle Grounding.** The requirements for special-purpose receptacles are somewhat different from those for other patient receptacles. The grounds for these receptacles must be taken to the branch circuit's reference grounding point. This ground can be run according to part (F) of this section if it is supplied from an isolated, ungrounded system.

517.20: Wet Locations

(1) All equipment and receptacles in wet areas must be protected by ground-fault interrupters. However, there are times when the interruption of power can't be tolerated due to a greater danger to patients in the area. In these instances, the power must be supplied by an isolated power system.

(2) When isolated power systems are used, they must be listed for the purpose and must conform to the requirements of Section 517.160.

Exception

Branch circuits that supply only listed therapeutic or diagnostic equipment can be supplied from the normal, grounded service. In such cases, all conductive surfaces of the machines must be grounded, and the grounded and isolated circuits must be kept in separate raceways.

III. Essential Electrical System

517.25: Scope

Clinics, medical and dental offices, outpatient facilities, nursing homes, residential custodial-care facilities, hospitals, and other health care facilities that serve patients must have an essential electrical system. This system must be able to supply enough power to maintain essential lighting and power in the case of an interruption of the normal power service.

517.30: Essential Electrical Systems for Hospitals

(A) Applicability. This segment of the Code (Part III, Sections 517.30 through 517.35) applies to hospitals that require essential electrical systems. Facilities covered by Sections 517.40 and 517.50 are exempted.

(B) General

(1) The essential electrical systems this section covers are actually made up of two separate systems—the emergency system and the equipment system. These systems must automatically activate when normal power is lost for any reason.

(2) The emergency system must be limited to circuits that are essential to maintaining life and safety. There are two parts of this emergency system: the life safety branch and the critical branch.

(3) The equipment system should supply power to major pieces of electrical equipment that are essential for either hospital operations or patient care.

(4) Each branch of the essential electric system must be served by at least one transfer switch. This is shown in *NEC* Diagrams 517.30(1) and 517.30(2), which you should carefully review. One transfer switch is allowed to serve more than one branch of the essential electrical system, up to 150 kVA. The exact number of transfer switches for a facility must be based on a good engineering design, considering load, switch design, switch reliability, and so on.

(C) Wiring Requirements

(1) **Separation from Other Circuits.** The life safety and critical branches of the emergency system are not allowed to be installed in a common raceway, enclosure, or box with any other wiring system, except in the following circumstances: (a) in transfer switches; (b) in exit or emergency lighting fixtures that are supplied by two sources; (c) in a junction box feeding a fixture such as in (b) of this list.

The wiring of the equipment system may share raceways and the like with other wiring systems.

(2) Isolated Power Systems. Isolated power systems that are installed in anesthetizing locations or in special environments must be supplied by an individual, dedicated circuit.

(3) Mechanical Protection of the Emergency System. For mechanical protection, the wiring of an emergency system must be installed in metal raceways.

Exception No. 1
Cords of appliances and other pieces of equipment connected to the emergency system.

Exception No. 2
The secondary circuits of communication or signaling systems that are supplied by transformers don't have to be enclosed in metal raceways, except as required by other sections of the *Code*. (See Chapters 7 and 8.)

Exception No. 3
Schedule-840 PVC is permitted, except for branch circuits that serve patient care areas.

2005

Exception No. 4
Schedule-40 PVC conduit, jacketed interlocked Type MC cable, or EMT may be used if encased in 2 inches of concrete. However, these wiring methods may not be used for branch circuits in patient care areas.

(D) Capacity of Systems. Essential electrical systems must be sufficient to supply any demand placed on them. This generally means that they must be over-engineered to provide more than enough capacity under any conditions. This is extremely important, because if they fail to meet the demand placed upon them, death of patients in the facility could result.

517.31: Emergency System

The functions of patient care that depend on lighting or appliances that may be connected to the emergency system are divided into two required branches: the life safety branch and the critical branch, descriptions of which appear in Sections 517.32 and 517.33.

It is essential that the branches of the emergency system be installed and so connected to the alternate source of power that all functions necessary for them to perform as specified herein for the necessary emergency system are automatically put into operation within 10 seconds after the interruption of the normal source of power.

517.32: Life Safety Branch

The following lighting, receptacles, and equipment shall be supplied from the life safety branch:

(A) **Illumination of Means of Egress.** The means of exit such as corridors, passageways, stairways, and landings at exits shall have illumination provided. Switching the patient corridor lighting in hospitals from general illumination to circuits for night illumination is permitted, but only one of two circuits can be selected, and it shall be made impossible for both circuits to be turned on at the same time.

Note

Life Safety Code, NFPA 101-2000 (ANSI), Section 5.10 has additional information.

(B) **Exit Signs.** Exit lights and exit direction lights shall have sources of normal power and also the alternate source of power available.

Note

Life Safety Code, NFPA 101-2000 (ANSI), Section 5-11, has additional information.

(C) **Alarm and Alerting Systems.** Life safety alarm and alert systems that include:

(1) Both automatic fire alarms and manually operated station.

(2) Sprinkler systems that have electric water-flow alarm devices in the system.

(3) Devices that will detect products of combustion and operate automatically in the case of fire or smoke.

Note

See Life Safety Code, NFPA 101-2000 (ANSI), Sections 12-1 and 12-2.

(4) Piping for nonflammable medical gases where alarms are required for such systems.

Note
For additional information on nonflammable medical gas systems, see *NFPA 56F*.

(D) **Communication Systems.** In hospitals, systems that are used to communicate instructions during emergency conditions.

(E) **Generator Set Location.** There shall be task illumination and receptacles that are selected for maintenance work and the generator set location.

(F) **Elevator Cab Lighting, Control, and Signal Systems.** Only those functions listed between (A) and (F) may be connected to the life safety branch.

517.33: Critical Branch

(A) Task Illumination and Selected Receptacles.

(1) All receptacles, fixed equipment, and task illumination in anesthetizing locations.

(2) In special environments, the isolated power systems.

(3) Selected receptacles and illumination for patient care areas in:

See (a) through (g) in the *NEC*.

Refer to the *NEC* for (4) through (8), including (a) through (i).

(9) Special power circuits needed for effective hospital operation, such as additional task illumination, receptacles, and special power circuits needed for effective hospital operation. Also single-phase fractional-horsepower motors used for exhaust fans and interlocked with the three-phase motors, may be connected to the critical branch circuit.

(B) **Subdivision of the Critical Branch.** Two or more circuits may be divided from the critical power branch.

Note
Analyzing the consequences is very important when supplying an area with only critical branch-circuit power when failure occurs between the transfer switch and that area. It may be appropriate to supply normal and critical power by means of separate transfer switches.

517.34: Equipment System Connection to Alternate Power Source

Refer to the *NEC* for this, including (A)(1) through (A)(3), the fine-print note, (B)(1), and the exception, which includes (a) through (c). Also included under (B) are the following types of equipment: elevator(s) that are selected for the transfer of patients, surgical and obstetrical, to the ground floors when normal power has been interrupted [See the *NEC*, Section 517.46 (B)(2), "Elevator Service"]; there shall be supply and exhaust ventilating systems in surgical and obstetrical delivery suites, in intensive care units, isolating rooms and emergency treatment spaces, and those areas constructed specifically for infection-control laboratory fume hoods [see the *NEC* for (4) through (8)].

517.35: Source of Power

It is recommended that this section be thoroughly covered in the *NEC*. Basically, essential electrical systems shall have two sources of power available (minimum). One may be the normal source and the other may be the alternate source(s) for use when the normal power is interrupted, or may be a generator set(s) driven by a prime mover and located on the premises. Where normal power consists of generating unit(s) on the premises, the alternate source may be another generating set(s) or an external utility source.

Care must be exercised in the location of equipment to protect it from floods, fires, or icing. The electrical characteristics of the normal and emergency sources of power shall be compatible. Refer to the *NEC*.

517.40: Essential Electrical Systems for Nursing Homes and Residential Custodial-Care Facilities

- **(A) Applicability.** The requirements given in Sections 517.40(C) through 517.44 apply to nursing homes and residential custodial-care facilities. However, such facilities that comply fully with Section 517.10, Exception 3 are exempted.
- **(B) Inpatient Hospital Care Facilities.** Sections 517.30 through 517.35 apply to nursing homes and residential custodial-care facilities that provide hospital-type care for inpatients.
- **(C) Facilities Contiguous with Hospitals.** When a nursing home or residential care facility is part of a hospital, it may use the hospital's essential supply system.

517.41: Essential Electrical Systems

- **(A) General.** The essential electrical system in nursing homes and residential custodial-care facilities shall consist of two separate branches, namely, the life safety branch and the critical branch. Both branches shall be capable of supplying a limited amount of lighting and power service, which is essential to protect life and ensure effective operation of the institution should normally supplied electrical service to said institution be interrupted for any reason. The installation and connections of essential electrical systems shall automatically take over if normal power fails and automatically be disconnected when normal power service is restored.

(B) **Transfer Switches.** The reliability, design, or load consideration shall determine the number of transfer switches used. There shall be one or more transfer switches in each branch of the essential electrical system. See the *NEC* for Diagrams 517.44(1) and 517.44(2). If the maximum demand does not exceed 150 kVA, one transfer switch may serve one or more branches of the facility of the electrical system.

Note

The *NEC* refers you to *Health Care Facilities, NEPA 99 (ANSI)*; for the description of transfer switching operations, it refers you to Section 8-4.5.2; the features of automatic transfer switches are also covered in Section 8-2.2.4; and you are referred in Section 8-2.2.4 for nonautomatic transfer switches.

(C) Capacity of System. The capacity of the essential electrical system shall be large enough so that all the requirements can be served at the same time.

(D) **Separation from Other Circuits.** The life safety branches must be kept entirely separate from the regular wiring and equipment and shall never enter the same raceways, boxes, or cabinets with other wiring except in the following:

(1) By necessity, the two must meet in transfer switches.

(2) Two sources may supply exit and emergency lighting fixtures.

(3) Two sources may be in a common junction box that is attached to emergency-exit lighting fixtures.

The wiring of the critical branch circuits will be permitted to occupy the same raceways or cabinets if they are not part of the life safety branch.

517.42: Automatic Connection to Life Safety Branch

The life safety branch shall be installed in connection with the alternate source of power supply in such a manner that all functions specified herein will be automatically restored to operation within 10 seconds after the interruption of the normal source of power. The following lighting, receptacles, and equipment shall be capable of being supplied from the life safety branch:

Note

The *NEC* defines the life safety branch as emergency systems; refer to *Health Care Facilities, NFPA 99 (ANSI)*.

(A) **Illumination of Means of Egress.** Means of egress shall be provided with illumination as necessary in passageways, stairways, landings, exit doors, and all ways of approach to exits. A means providing switching for patient corridor lighting from general-illumination circuits will be permitted, but only if only one circuit can be used at a given time.

Note

Refer to *Life Safety Code, NFPA 101-2000 (ANSI)*, Section 5.10.

(B) **Exit signs.** Exit directional signs and exit signs.

Note

You are referred to Sections 5.11 of *NFPA 101*, which is referred to in the fine-print note to (A).

(C) **Alarm and Alerting Systems.** Alarm and alerting systems, including:

(1) Manually operated fire alarm stations, water flow alarm systems that are electrically connected with sprinkler systems, and alarms for fire or smoke that automatically detect products of combustion.

Note

Again the fine print: *Life Safety Code, NFPA, 101-2000 (ANSI)*, Sections 7-6, 12-3.4, and 12-3.5.

(2) Piping of nonflammable medical gases for alarms that must be used on them.

Note

The *NEC* refers to the *Standard for Nonflammable Medical Gas Systems, NFPA 56F (ANSI)*.

(D) **Communication Systems.** Where used for issuing instructions in emergency communication systems.

(E) Dining and Recreation Areas. Sufficient illumination to exits from dining and recreational areas.

(F) Generator Set Location. Illumination and selected receptacles in the generator location.

(G) Elevator Cab Lighting, Control, and Communication Systems. Elevator cab lighting, control, and communication systems may be connected to the nearby power systems.

Only functions listed in (A) through (G) shall be connected to the life safety branch.

517.43: Connection to Critical Branch

The critical branch shall be so installed and connected with the alternate source of power that equipment listed in Section 517.43(A) shall be automatically connected at appropriate intervals of time lag to put the life safety branch into operation. Equipment listed in Section 517.43(B) shall also be arranged either by delayed automatic action or by manual operation.

(A) Delayed Automatic Connection. Listed below is the equipment that shall be connected to the critical branch, and it is to be arranged for automatic delayed connection to the critical branch.

- **(1)** Task illumination and receptacles that are in inpatient care areas: (a) areas where medication is prepared; (b) dispensing areas of pharmacies; (c) unless there is adequate lighting from corridor illumination, nurses' stations in this classification.
- **(2)** Other equipment, including sump pumps required for the safe operation of major apparatus, and the control systems, including alarms.

(B) Delayed Automatic or Manual Connection. Listed below is the equipment that shall be connected to the critical branch. It shall be arranged for either delayed automatic or manual connection to the alternate power source:

- **(1)** Any heat equipment that provides heat for the patient's room.

Exception

Under the following conditions, heating the patient's room during disruption of the normal electrical source is not required: (a) The design temperature outside is higher than 20°F (−6.7°C); (b) If the design temperature is lower than 20°F (−6.7°C), and there is a selected room(s) provided to take care of all the needs of a confined patient, then only such room(s) need to be heated; or (c) Where two sources of normal power can serve the facility as described in Section 517.44(C), fine-print note.

(2) Where disruption of power would result in elevators stopping between floors, the critical branch must have a throw-over facility that will temporarily allow the operation of any elevator so that passengers may be released. Lighting-, control-, and signal-system requirements are covered in Section 517.42(G).

(3) More illumination receptacles and other equipment will be permitted to be connected only to the vertical branch.

517.44: Sources of Power

(A) Two Independent Sources of Power. Essential electrical systems must have a minimum of two independent sources of power: normal power that ordinarily supplies the entire electrical system and one or more alternate power sources to be used in the event of disruption of the normal power.

(B) **Alternate Source of Power.** A generator(s) that is driven by a prime mover(s) located on the premises may constitute the alternate source of power.

Exception No. 1
If the normal electrical supply is generated on the premises, the alternate source of power may be either an external source from utility service or another generator.

Exception No. 2
Nursing homes and residential custodial-care facilities that meet all the requirements of the exception to Section 517.40 may use a battery system or self-contained battery that is a part of the equipment.

(C) **Location of Essential Electrical System Components.** Locations for the essential electrical system shall be carefully analyzed so that the effects of interruptions due to natural forces will be minimized (for example, storms, floods, earthquakes, or hazards created by structures adjacent thereto or other activities). Careful consideration must also be given to normal electrical supplies that may be interrupted by similar causes, as well as disruptions of normal electrical services that may be caused by internal wiring or equipment failure.

Note

Facilities whose normal source of power is obtained from two or more central stations experience electrical service reliability that is greater than that of facilities whose normal supplies of power are served from only a single source. Such a source of electrical power consists of power supplied from two or more electrical generators or two or more electrical services supplied from separate utility distribution networks that have local power in the input sources and that are arranged so as to provide mechanical and electrical separation. This is so that a fault between the facilities and the generating source won't be likely to cause interruption of more than one of the service feeder facilities.

517.50: Essential Electrical Systems for Clinics, Medical and Dental Offices, Outpatient Facilities, and Other Health Care Facilities Not Covered in Sections 517.30 and 517.40

(A) Acceptability. The requirements given in this section apply to facilities that are both described in this section and in which anesthetics are inhaled, or where patients need electrical life support equipment.

(B) **Connections.** The essential electrical system must supply power for the following:

- **(1)** Lighting that is necessary for the safety of life, and which is required for the completion of procedures already underway when normal power is lost.
- **(2)** All equipment used for anesthesia and resuscitation, including any alarm or alerting devices.
- **(3)** All life support equipment.

(C) **Alternate Source of Power**

- **(1)** **Power Source.** The alternate power source is to be specifically designed for this purpose. It may be the generator, the battery system, or a self-contained battery that is a part of the equipment being used.
- **(2)** **System Capacity.** The alternate source of power shall be entirely independent of the normal source of power and separate there from. It shall also have the capacity to maintain the connected loads for at least 1½ hours after loss of the normal power.
- **(3)** **System Operation.** This alternate source of power shall be arranged so that if the normal power source fails, the alternate source of power shall be on the line in 10 seconds or less.

Note

It is recommended that you see *Health Care Facilities, NFPA 99-1984 (ANSI)*, Section 8-5.5.2 for description of the switching operation with engine generator sets, or Section 8-5.5.3 for description of transfer switch operation with battery systems.

IV. Inhalation Anesthetizing Locations

517.60: Anesthetizing Location Certification

(A) Hazardous (Classified) Location

(1) Any location where flammable anesthetics are used is considered a Class I, Division 1 area between the floor and a level 5 feet above the floor.

(2) Any location where flammable anesthetics are stored is considered a Class I, Division 1 location from floor to ceiling.

(B) **Other Than Hazardous (Classified) Location.** Locations where only nonflammable anesthetics are used are not considered classified locations.

517.61: Wiring and Equipment

(A) Within Hazardous Anesthetizing Locations

(1) Circuits in these areas must be supplied by an isolated power system, except as allowed in Section 517.160.

(2) Isolated power equipment must comply with Part VII of this Article, and must be listed.

(3) Wiring in the hazardous locations identified in Section 517.60 and operating at more than 10 volts must comply with Sections 501.1 through 501.16(B).

(4) When a box, fitting, or the like is only partially located in a hazardous area, the entire item must be treated as though it were in a classified area.

(5) Receptacles used in these areas must be rated for Class 1, Group C locations, and must have provision for grounding connections.

(6) Cords that connect equipment in these areas must follow the requirements of Table 400.4, and must have a grounding conductor.

(7) Flexible cords must be provided with a storage device of some type, and must not be bent to a radius of less than 3 inches.

(B) **Above Hazardous Anesthetizing Locations**

(1) All wiring above the hazardous locations specified in Section 517.60 must be in rigid metal conduit, EMT, intermediate metal conduit, MI cable, or MC cable that has a continuous gas or vapor-tight shield.

(2) Any equipment that may cause sparks, arcs, or particles of hot metal must be totally enclosed, or designed not to allow the escape of these arcs, sparks, and so on.

Exception

Wall-mounted receptacles that are above the hazardous location areas for anesthetizing location don't have to be totally enclosed or guarded.

(3) Surgical lighting fixtures (or other lighting fixtures) that conform to Section 501.9(B).

Exception No. 1

The temperature restrictions of 501.9(B) don't have to be applied.

Exception No. 2

Pendant switches, or switches that are built in to other electrical items, can't be lowered into the hazardous area don't have to be explosion-proof.

(4) Seals must be provided in accordance with Section 501.5.

(5) All receptacles and plugs above these hazardous locations must be listed for hospital use.

(6) Receptacles and plugs that are rated for 250 volts and for connecting 50- and 60-amp medical equipment must be of the two-pole, three-wire design, with a grounding contact connected to an equipment-grounding wire. 60-amp receptacles must be designed to accept either the 50- or 60-amp plug, and the 50-amp receptacle must be designed to accept only a 50-amp plug.

(C) **Other Than Hazardous Anesthetizing Locations.**

(1) Wiring in these locations must be installed in a metal raceway system or a metal-clad cable assembly that is listed as an approved grounding means [see Section 250.91(B)]. Type MC cables can be used if they have an outer metal jacket that is an approved grounding means.

Exception

Pendant receptacle cords of type SJO or better more than 6 feet from the floor.

(2) Receptacles and plugs in these areas must be listed for hospital use.

(3) Receptacles and plugs that are rated for 250 volts and for connecting 50- and 60-amp medical equipment must be of the two-pole, three-wire design, with a grounding contact connected to an equipment-grounding wire. Sixty-amp receptacles must be designed to accept either the 50- or 60-amp plug, and the 50-amp receptacle must be designed to accept only a 50-amp plug.

517.62: Grounding

In anesthetizing locations, it is required that all metallic raceways be grounded, as well as metal-sheathed cables and all noncurrent-carrying parts that are conductive portions of either portable or fixed electrical equipment.

Exception

When equipment operates at not over 10 volts between the conductors, it shouldn't be required to be grounded.

517.63: Grounded Power Systems in Anesthetizing Locations

(A) General Purpose Lighting Circuit. A separate lighting circuit must be installed to each room. The power for this circuit is to come from the normal power service.

Exception

When connected to an alternate source (see Section 700.12) that is separate from the source that supplies the emergency system.

(B) Branch Circuit Wiring. Branch circuits located in nonhazardous locations, or over hazardous locations, and supplying fixed medical equipment can be supplied from a single-phase or three-phase grounded normal service, as long as:

(1) Isolated and grounded wiring conductors don't occupy the same raceway.

(2) All conductive surfaces of the equipment are grounded.

(3) Equipment is at least 8 feet from the floor, or outside of the anesthetizing location. (Except X-ray tubes, and the leads to the tubes.)

(4) Switches for the grounded circuit are not installed in the hazardous location. Items (3) and (4) apply only to hazardous locations.

(C) **Fixed Lighting Branch Circuits.** These fixtures can be supplied by a grounded normal branch circuit, provided that:

(1) They are at least 8 feet above the floor.

(2) Their surfaces are grounded.

(3) Wiring for these fixtures and wiring for isolated system circuits don't occupy the same raceways.

(4) Switches are located above the hazardous locations, and are wall-mounted.

(D) **Remote Control Stations.** Wall-mounted remote control stations are allowed in anesthetizing locations, provided they operate at 24 volts or less.

(E) **Location of Isolated Power Systems.** These systems (this section refers to the main equipment) and their feeders can be located in anesthetizing areas, as long as they are above the location, or not in a hazardous area.

(F) **Circuits in Anesthetizing Locations.** All circuits in these areas (except as permitted above) must be isolated from all other distribution systems.

517.64: Low-Voltage Equipment and Instruments

This section refers to requirements for the manufacture of equipment. For specific information, refer to the *Code*.

V. X-Ray Equipment

There is nothing in this part that can be construed as specifying safeguards from useful beam or stray X-ray radiation.

Note

Several classes of X-ray equipment are regulated under Public Law 90-602, which covers radiation, safety, and performance requirements. The enforcement of such regulations comes under the Department of Health and Human Services.

517.71: Connection to Supply Circuit

(A) Fixed and Stationary Equipment. The general requirements of this Code shall be used for fixed and stationary X-ray equipment.

Exception

X-ray equipment supplied or rated at 30 amperes or less may be supplied if extra-hard-usage cord is used and suitable plug and attachments are also used.

(B) **Portable, Mobile, and Transportable Equipment.** Individual branch circuits are not required for portable, mobile, and transportable medical X-ray equipment if it does not require over 60 amperes for operation.

(C) **Over 600-Volt Supply.** If X-rays are operated at over 600 volts, they must comply with Article 710.

517.72: Disconnecting Means

(A) **Capacity.** A disconnecting means that has adequate capacity to provide at least 50 percent of the input required for momentary operation rating or 100 percent of the input required for long-term rating of the X-ray equipment, whichever is greater, shall be provided for disconnecting the supply circuit.

(B) **Location.** A disconnecting means shall be provided that is operable from a location that is easily accessible from the X-ray control.

(C) **Portable Equipment.** If portable equipment operates at 30 amperes or less and 120 volts, a grounding-type attachment plug and receptacle, with a rating that is appropriate for the load it is to carry, may be used as a disconnecting means.

517.73: Rating of Supply Conductors and Overcurrent Protection

(A) Diagnostic Equipment

(1) The ampacity of the branch-circuit conductors, and also the rating of the overcurrent devices protecting the branch circuit, may not be less than 50 percent when the diagnostic equipment is used only momentarily, or a 100-percent rating if used for a long time, whichever is greater.

(2) The ampacity of feeder conductors and the current rating of protective devices that supply two or more branch circuits supplying X-ray units, if the X-ray units are used only for momentary usage, shouldn't be less than 50 percent of the largest unit, 25 percent of the next largest unit, and 10 percent of the demand of each additional unit. Where simultaneous X-ray examinations are undertaken by X-ray units, the ampacity of the supply conductors and the overcurrent protective devices shall be 100 percent of the momentary rating of each X-ray unit.

Note

The minimum ampacity of branch circuits and feeder conductors from the circuit shall also be governed by the voltage regulation requirements. The manufacturer usually supplies specific information: minimum transformer and conductor sizes, the rating of the disconnecting means, and overcurrent protection.

(B) **Therapeutic Equipment.** The current rating shall be not less than the 100 percent of the rating of medical X-ray therapy equipment.

Note

The ampacity of branch circuit conductors and the ratings of overcurrent protection devices and disconnecting means for the protection of X-ray equipment are in most cases supplied by the manufacturer supplying the specific installation.

517.74: Control Circuit Conductors

(A) **Number of Conductors in Raceway.** Section 300.17 covers the number of control circuit conductors that may be installed in a raceway.

(B) **Minimum Size of Conductors.** As specified in Section 725.16, sizes No. 18 or No. 16 fixture wires are required. Flexible cords may be permitted for the circuits from control and operation of the X-ray and its auxiliary equipment if they are protected by 20-ampere or less overcurrent devices.

517.75: Equipment Installations

The equipment for use with new X-ray equipment for installation and for all used or reconditioned X-ray equipment that is moved and reinstalled at a new location shall be of an approved type.

517.76: Transformers and Capacitors

Transformers and capacitors that are a part of X-ray equipment shouldn't be required to comply with Articles 450 and 460.

Capacitors shall either be grounded or mounted in insulating material.

517.77: Installation of High-Tension X-Ray Cables

These cables (provided they have grounded shields) can be run in cable trays or cable troughs, along the power and control conductors of X-ray equipment. No barriers are required.

517.78: Guarding and Grounding

(A) High-Voltage Parts. All high-voltage parts of X-rays, including X-ray tubes, are to be mounted in grounded enclosures. You may use air, oil, or gas or other suitable insulation media to insulate the high-voltage components from the grounded enclosure. High-voltage shielding cables shall be used for connecting high-voltage equipment to X-ray tubes and any other high-voltage components that are required.

(B) Low-Voltage Cables. Oil-resistant insulation is required for low-voltage cables that are connected to oil-filled units when they are not completely sealed. This includes transformers, condensers, oil coolers, and high-voltage switches.

(C) Noncurrent-Carrying Metal Parts. Metal parts of X-rays and associated noncurrent-carrying equipment, such as controls, tables, X-ray tube supports, transformer tanks, shielded cables, X-ray tube heads, and the like must be grounded in a manner specified in Article 250, which may be modified by Section 517.11(A) and (B) under the criteria set forth in Section 517.81 covering critical care areas.

Exception

Battery-operated equipment is an exception.

VI. Communications, Signaling Systems, Data Systems, Fire Protective Signaling Systems, and Systems Less Than 120 Volts, Nominal

517.80: Patient Care Areas

All low-voltage systems must be insulated, isolated, and grounded according to the same requirements as the electrical distribution systems for patient care areas.

517.81: Other-Than-Patient-Care Areas

Installations in these areas must be according to Articles 725, 760, and 800.

517.82: Signal Transmission Between Appliances

(A) General. Signal transmission systems must prevent hazardous grounding interconnection between the appliances.

(B) Common Signal Grounding Wire. These wires are allowed to be run between appliances that are all in the patient area, and are all served by the same reference grounding point.

VII. Isolated Power Systems

517.160: Isolated Power Systems

(A) Installations.

(1) Every power circuit that is within or partially within an anesthetizing location must be isolated from any distribution system other than that supplying anesthetizing locations. Every isolated power circuit must be controlled by a switch for disconnecting and shall disconnect each isolated conductor. The isolating circuits shall be supplied from one or more isolating transformers that have no electrical connection between the primary and secondary windings, or they may be supplied by motor generator sets, or by suitable isolated batteries.

(2) The primaries of isolating transformers shouldn't operate at voltages exceeding 600 volts between conductors and shall be protected by overcurrent protection of proper size. The secondary of isolated systems shouldn't exceed 600 volts between conductors of every circuit, and all conductors, such as isolated secondaries, shall be ungrounded. They shall also be provided with an approved overcurrent device of the proper rating in each conductor. If circuits are supplied directly from batteries or from a motor generator set, the conductors shall be ungrounded and overcurrent protection shall be provided in the same fashion as transformer-supplied secondary circuits. When an electrostatic seal is present, it shall be connected to the reference grounding point.

(3) The isolating transformers, motor generator set, batteries and battery charges, and primary and/or secondary overcurrent protection devices may not be installed in a hazardous (classified) location. The requirements of Section 501.4 shall be used when the isolating secondary circuit wiring is extended into the hazardous anesthetizing location.

(4) An isolated branch circuit that supplies an anesthetizing location won't be allowed to supply any other location. The insulation on the conductors on the secondary side of the isolated power supply must have a dielectric constant of 3.5 or less. If pulling-compound increases the dielectric constant, it shouldn't be used on the secondary conductors of the isolated power supply.

(5) The identification of isolated conductors must be as follows: See the *NEC*.

(B) Line Isolation Monitor

(1) A continual-operating line isolation monitor that indicates possible leakage or fault currents from any of the isolated conductors to ground shall be used in addition to the usual overcurrent protective devices that are required in isolated power systems. The design of this monitor shall be such that a green signal lamp is conspicuously visible to persons in the anesthetizing location. It remains lighted if the system is properly isolated from ground and there is a red signal lamp and also an audible warning signal, which can be remote if desired. These will be energized when the total hazardous current (which could result from resistive or capacitive current leakage) emanating from either isolated conductor to ground reaches a value of 5 milliamperes or slightly less when the line voltage conditions are nominal. The line isolation monitor shouldn't act where fault hazard current is less than 3.7 milliamperes. The isolating line alarm monitor is not to sound when the total hazardous current is less than 5 milliamperes.

Exception

It may be designed for operation at a lower threshold of total hazardous current. A line isolation monitor for such a system may be approved, with the provision that the fault hazard current may be reduced, but not to less than 35 percent of the value of the hazardous current of the above section, and the monitor hazard current may also be reduced to not more than 50 percent of the alarm threshold value of the total hazard current as above.

Note

The systems in the section above provide little electric safety, and are to be used only in special applications.

(2) The internal impedance of a line isolation monitor shall be designed so that when properly connected, the internal current can flow through the isolation monitor, and if any point of the system is grounded, the current shall be 1 milliampere.

Exception

Low-impedance type of line isolation monitor shall be permitted so that current through the monitor, when any point of the isolation system is grounded, won't exceed twice the alarm threshold value, and this shall be for a time period not exceeding 5 milliseconds.

Note

If the reduction of the monitor hazard current results in increased "not alarm" threshold hazard current, it will increase the circuit capacity.

(3) An ammeter that is calibrated for the total current of the system (contribution of the faulthazard current plus monitor hazard current) shall be mounted in a place that is plainly visible on the line isolation monitor, with the "alarm on" ampere zone located at approximately the center of the ammeter scale and visible to persons in the anesthetizing locations.

Exception

A line isolation monitor may be more than one unit that has a sensing section carried by cable to a separate display panel section where the alarm and/or text functions may be located.

Article 518—Places of Assembly

518.1: Scope

The coverage of this article includes all buildings or portions of buildings or structures that are designed and intended to be used for the assembly of 100 or more persons.

518.2: General Classifications

See the *NEC* for a list of places that could be used for the assembly of 100 or more persons.

Any room or space used for occupancy by less than 100 persons in a building used for other occupancy, even though incidental to such occupancy, shall be classed as part of the other occupancy, and shall be subject to the provisions applicable thereto.

If said building, part of building, or other structure has a projection booth or there is a stage platform where theatrical or musical productions are performed, whether these stages are fixed or portable, the wiring methods of the equipment and all the area and wiring used in the production, if not permanently installed, must comply with Article 520.

Note

It is very important that you use the *NFPA Life Safety Code, 101-1985 (ANSI)*, which contains the methods for determining how many occupants may be allowed in an area of assembly.

518.3: Other Articles

(A) **Hazardous (Classified) Areas.** Hazardous (classified) areas located in any assembly place of occupancy must be installed in accordance with Article 500, which covers hazardous (classified) locations.

(B) **Temporary Wiring.** In exhibition halls used for display booths, as in trade shows, any temporary wiring must be installed in accordance with *NEC* Article 305, "Temporary Wiring," except that approved flexible cables and cords will be permitted to be laid on the floors where they can be protected from contact by the general public.

(C) **Emergency Systems.** Article 700, "Emergency Systems," must be followed in the installation for control of emergency systems.

518.4: Wiring Methods

Wiring methods of the fixed type must be metal raceways, nonmetallic raceways if encased in not less than 2 inches of concrete, or Type MI or MC cable.

Exception No. 1

If the applicable building code does not require the buildings or portions thereof to be fire-rated construction by the applicable building code, nonmetallic-sheathed cable, EMT, Type AC cable, or rigid nonmetallic conduit may be installed in those buildings or portions thereof.

Exception No. 2

As covered in Article 640, "Sound Reproduction and Similar Equipment," in Article 800, "Communication Circuits," and in Article 725 for Class 2 and Class 3 remote-control and signaling circuits, and in Article 760 for fire protection signaling circuits.

Note

Refer to the *NEC.*

518.5: Supply

Portable switchboards and power distribution systems can be supplied only from listed outlets that are sufficient in voltage and current. These outlets must have listed circuit breakers or fuses in accessible locations, and must have provisions for the connection of grounding conductors. The neutral of any feeder supplying three-phase, four-wire dimming systems must be rated as a current-carrying conductor. Permanent outlets that supply portable switchboards must be marked with the available fault current.

Article 520—Theaters and Similar Locations

I. General

520.1: Scope

This article covers buildings or portions of buildings that are used for motion-picture projection, dramatic presentations, musicals, or similar purposes. It also applies to areas that incorporate assembly areas for motion picture and television studios.

520.3: Motion-Picture Projectors
See the *NEC.*

520.4: Sound Reproduction
See the *NEC.*

520.5: Wiring Methods
The fixed wiring method must be metal raceways, Type MI cable, or Type MC cable. Nonmetallic raceways that are encased in a minimum of 2 inches of concrete may be used.

Exception No. 1
See Article 640 for sound reproduction, Article 800 for communication circuits, and Article 725 for Class 2 and Class 3 remote-control and signaling circuits; and for fire protection signaling circuits, see Article 760.

Exception No. 2
Switchboards of the portable type, stage-set lighting, stage effects, and other wiring that is not maintained in a fixed location but is portable, may be connected by flexible cords and cables as provided elsewhere in this article. The use of uninsulated staples or nails for securing these cords and cables in place won't be permitted.

Exception No. 3
Nonmetallic-sheathed cables, AC cables, EMT, and rigid nonmetallic conduit can be used in areas that don't require fire-rated construction by the applicable building code.

520.6: Number of Conductors in Raceway
Table 1 of Chapter 9 governs the number of conductors for border or stage-pocket circuits, or for remote-control conductors, which may be installed in any metal conduit, rigid nonmetallic conduit, or electrical metallic tubing. For auxiliary gutters or wireways, the sum of the cross-sectional areas of all of the conductors contained therein shouldn't exceed 20 percent of the interior cross-sectional area of the gutter or raceway. The limitation of thirty conductors covered in Sections 376.5 and 366.5 does not apply.

520.7: Enclosing and Guarding Live Parts
See the *NEC.*

520.8: Emergency Systems
See the *NEC.*

520.9: Branch Circuits
A branch circuit of any size that supplies receptacles can be used to supply stage lighting. The circuit or receptacles can't be rated lower than the branch circuit rating.

520.10: Portable Equipment

Indoor lighting equipment marked "Suitable for Damp Locations" may be used out of doors, but only during fair weather and when kept away from the general public.

II. Fixed Stage Switchboard

520.21: Dead Front

See the *NEC.*

520.22: Guarding Back of Switchboard

See the *NEC.*

520.23: Control and Overcurrent Protection of Receptacle Circuits

See the *NEC.*

520.24: Metal Hood

See the *NEC.*

520.25: Dimmers

(A) through (D) below give the requirements for dimmers:

- **(A) Disconnection and Overcurrent Protection.** When dimmers are installed in ungrounded conductors, the overcurrent for each dimmer shouldn't exceed 125 percent of the dimmer rating, and the dimmers shall be disconnected from all ungrounded conductors when the master or individual switch or circuit breaker supplying such dimmer is in an open position.
- **(B) Resistance or Reactor-Type Dimmers.** It is permissible that resistance or reactor-type dimmers be placed in the grounded neutral conductor, as long as they don't open the circuit. Resistance or series reactance dimmers may be placed in either the grounded or ungrounded conductor of the circuit.

  Where designed to open either the supply circuit to the dimmer, or the circuit that is controlled by it, the dimmer shall comply with Section 380.1, which requires that no switch shall open the grounded conductor unless the grounded conductor and ungrounded conductor or conductors are simultaneously opened.
- **(C) Autotransformer-Type Dimmers.** Autotransformer-type dimmers shall be supplied by a source that does not exceed 150 volts, and the input and output grounds shall be common, as provided for in Section 210.9.
- **(D) Solid-State-Type Dimmers.** Circuits not exceeding 150 volts between two conductors may be used for solid-state dimmers,

subject to the dimmer that has approval for higher-voltage operation. Where the grounded conductor supplies a dimmer, it must be common to the input and output circuits, and the dimmer chassis is to be connected to the equipment-grounding conductor.

Note

Refer to Section 210.9 on circuits derived from autotransformers.

520.26: Type of Switchboard

One or all of the following may be used for stage switchboards:

(A) **Manual.** Handles mechanically linked to the control devices, such as dimmers and switches.

(B) **Remotely Controlled.** A pilot-type control console or panel is used to operate the devices that are electrically operated. The pilot control panels must be either part of the switchboard in another location.

(C) **Intermediate.** A switchboard that has interconnection is a secondary switchboard (patch panel), or it is a panelboard that is remote to the primary stage switchboard. Overcurrent protection shall be contained therein. If the branch circuit overcurrent protection is provided in the dimmer panel, the intermediate switchboard shall be permitted.

520.27: Stage Switchboard Feeders

The feeder used to supply the stage switchboards shall be one of the following:

(A) **Single Feeder.** A single feeder with a single disconnect. The neutral must be considered a current-carrying conductor for three-phase, four-wire systems.

(B) **Multiple Feeders to Intermediate Stage Switchboard (Patch Panel).** As long as they are part of a single system, any number of feeders are allowed. See text of *NEC* for neutral sizing.

(C) **Separate Feeders to Single Primary Stage Switchboard (Dimmer Bank).** Each such feeder must have a separate disconnect. The dimmer bank must be labeled, stating the number and location of disconnecting means. Neutrals for three-phase, four-wire systems shall be considered current-carrying conductors.

III. Stage Equipment—Fixed

520.41: Circuit Loads

Branch circuits that supply footlights, border lights, and proscenium sidelights shouldn't carry a load that exceeds 20 amperes. Where circuits have heavy-duty lampholders, an exception is made to provide for these, but the conditions in Article 210 that cover heavy-duty lampholders must be complied with.

520.42: Conductor Insulation

When wiring with connectors used for foot, border, proscenium, or portable strip-lighting fixtures and any connector strips used with these, it is required that the installation be suitable for the temperature encountered. However, the conductors will be rated at not less 125°C (257°F). Drops from connector strips rated with 90°C-wire and 6 inches or less of wire extending into the connector strip may be installed without applying *NEC* Note 8 for the ampacity table for 0 through 2000 volts.

Note

Table 310.13 covers conductor types.

520.43: Footlights

See the *NEC*. Notice in (B) that disappearing footlights shall be arranged so as to automatically disconnect from the circuit when they are in the closed position. Otherwise, they would present a great fire hazard if energized while in the closed recess.

520.44: Borders and Proscenium Sidelights

(A) General. General border and proscenium lights shall be (1) constructed as specified in Section 520.43; (2) suitably stayed and supported; and (3) designed so that reflector flanges or other suitable guards are provided to protect the lamps from contact with other scenery or combustible material and give mechanical protection from injury that could be caused by accidental contact with the above.

(B) Cables for Border Lights. Only types of cable that are listed for extra-hard usage can be used to supply border lights. For ampacities of these cords, refer to Table 520.44. These flexible cords shall be suitably supported and used only where flexible conductors are necessary.

520.45: Receptacles

Receptacles shall be rated in amperes when used on stages for electrical equipment and fixtures.

Articles 310 and 400 shall be complied with for conductors supplying receptacles.

520.46: Connector Strips, Drop Boxes, and Stage Pockets

Receptacles shall comply with Section 520.45 when intended for use with portable stage-lighting equipment and shall be mounted in suitable pockets or enclosures.

520.47: Lamps in Scene Docks

When lamps are used in scene docks, their location shall be such that they are protected from physical damage, and an air space is to be provided of not less than 2 inches between any combustible material and the lamps.

520.48: Curtain Motors

The conditions in (A) through (F) below are to be adhered to if curtain motors have brushes or sliding contacts:

- **(A)** **Types.** They shall be enclosed-pipe-ventilated, totally enclosed, or enclosed-fan-cooled type.
- **(B)** **Separate Rooms or Housings.** They may be installed in separate rooms or housings, both of which are built of noncombustible material, and are to be constructed so as to exclude flyings or lint. There shall also be proper ventilation from a source of clean air.
- **(C)** **Solid Metal Covers.** Motors with brushes or sliding contacts at the end of a motor are to be enclosed in solid metal covers.
- **(D)** **Tight Metal Housings.** They may be enclosed in substantial housings that are tight and made of metal where there are brushes or sliding contacts.
- **(E)** **Upper- and Lower-Half Enclosures.** The upper half of the brushes or sliding contacts at the end of the motor may be enclosed by means of a wire screen, or it may be perforated metal, but solid metal covers shall be used for the lower half.
- **(F)** **Wire Screens or Perforated Metal.** They may have screens that are perforated metal installed at the commutator of the brush end, provided no dimension of openings in the wire screen or perforated metal exceeds 0.05 inch, irrespective of the shape of the openings or of the material that is used to cover the openings.

520.49: Flue Damper Control

If stage flue dampers are opened by means of an electrical device, the circuit that operates the devices will normally be closed, and it is to be controlled by means of at least two switches for the operation. One of these switches is to be placed at the electrician's station, and the authority that has jurisdiction will determine where the other switch is to be located. The operating devices shall be designed for the full voltage of the circuit to which it is connected; no resistance shall be inserted. The control device is to be located in the loft above the scenery, and it shall be enclosed in a tight metal enclosure that has a self-closing door.

IV. Portable Switchboards on Stage

520.50: Road-Show Connection Panel (A Type of Patch Panel)

This is a panel designed for a road show with connection to portable stage switchboards and portable lighting outlets by using supplementary circuits permanently installed. The panel, the supplementary circuits, and any outlets must meet the requirements of (A) through (D) below:

- **(A)** **Load Circuits.** Grounding and polarized inlets of current and voltage rating that match the fixed-load receptacle shall be connected only to circuits that meet the requirements of this.
- **(B)** **Circuit Transfer.** If circuits are transferred between fixed and portable switchboards, they must have the line and neutral transferred at the same time.
- **(C)** **Overcurrent Protection.** The devices that supply supplementary circuits shall be protected by branch-circuit overcurrent protective devices. Each supplementary circuit within the road-show connection panel and the theater is to be protected by branch-circuit overcurrent protective devices, the ampacity of which must be suitable and installed within the road-show connection panel.
- **(D)** **Enclosure.** Article 408 covers the construction of the enclosures of panels.

520.51: Supply

Only outlets of sufficient voltage and ampere rating are to be used on portable switchboards. Only externally operable, enclosed fused switches or circuit breakers are to be used for these outlets, and they shall be mounted either on the stage or at the permanent switchboard in locations that are readily accessible from the stage floor. Provision shall be made for equipment-grounding connections. If the neutral of a three-phase, four-wire system is used on a dimmer-circuit system, the neutral shall be considered to be a current-carrying conductor.

520.52: Overcurrent Protection

Circuits originating from portable switchboards that directly supply equipment that contains incandescent lamps of not over 300 watts are to be protected by overcurrent protection devices that have a rating or setting not to exceed 20 amperes. If overcurrent protection complies with Article 210, circuits for lampholders of over 300 watts will be permitted.

520.53: Construction and Feeders

Refer to the *NEC.*

V. Stage Equipment—Portable

For Sections 520.61 through 520.68, refer to the *NEC.*

VI. Dressing Rooms

Refer to the *NEC.*

VII. Grounding

520.81: Grounding

You must ground all metal raceways and metal-sheathed cables. Metal frames and enclosures of all equipment—including border lights and portable light fixtures—shall be grounded, and the grounding shall comply with Article 250.

Article 525—Carnivals, Circuses, Fairs, and Similar Events

I. General Requirements

525.1 Scope.

This article concerns temporary installations of carnivals, fairs, and so on. It includes wiring in or on all structures installed in these places.

525.3 Other Articles.

Article 525 takes precedence over other articleson the NEC regarding wiring installed in these areas. The exceptions to this are:

Artricles 518 and 520 apply to permanent structures.
Article 640 applies to audio equipment.
Article 680 applies to pools, fountains, and similar installations.

525.5 Overhead Conductor Clearances.

(A) Vertical Clearances. The requirements of Section 225.18 apply.

(B) Clearance To Rides and Attractions. Wiring under 600 volts must be kept at least 15 feet (4.5 m) in any direction from rides and attractions, wiring that supplies power to the rides or attractions excepted. Wiring over 600 volts must not be run above rides and attractions, and may not be within 15 feet (4.5 m) of them horizontally.

525.6 Protection of Electrical Equipment.
Where subject to physical damage, wiring must be protected.

II. Power Sources

525.10 Separately Derived Systems
Generators must comply with Article 445. Transformers must comply with Article 450 and with Sections 240.4(A), (B)(3), (C), and 250.30.

525.11 Services.
Services must comply with Article 230 and Sections 525.11(A) and (B).

(A) Guarding. Service equipment must not be accessible to unqualified people, except if it is locked.

(B) Mounting and Location. Service equipment must be securely-mounted. If not weatherproof it must not be exposed to the weather.

III. Wiring Methods

525.20 Wiring Methods.

(A) Type. Cables or cords, if used, must be listed for extra hard use. If not exposed to any physical damage, cords or cables can be of the hard use type. (Not extra-hard use.) Cords used outdoors must be sunlight resistant and listed for wet locations. Extra-hard useage cords and cables may be installed as permanent wiring methods of rides, but only ifthey are not exposed to physical damage.

(B) Single Conductors. Only sizes of conductors #2 AWG or larger can be installed as single conductors.

(C) Open Conductors. Open conductors are not permitted, except as part of a festoon lighting system installed per Article 225.

(D) Splices. No splices are permitted between boxes or fittings.

(E) Cord Connectors. Cord connectors must not be exposed to public foot traffic, and must be listed for wet locations if laid on the ground.

(F) Support. Wiring may not be supported by a ride or similar structure, unless it is designed for the support of wiring.

(G) Protection. Cables must be laid in such a way as to prevent people tripping over them. Nonconductive mats may be used for this purpose. Also, cords and cables may be buried, and do not have to comply with the requirements of Section 300.5.

(H) Boxes and Fittings. A box or fitting must be used for every connection or outlet.

525.21 Rides, Tents and Concessions.

(A) Disconnecting Means. Every ride or concession must have its own disconnect. (Which may be a fused switch or circuit breaker.) This disconnect must be within 6 feet (1.8 m) of the-operator's station, and within sight. Disconnects must be lockable if accessible to the public. A shunt trip switch may substitute for a disconnect within 6 feet if it is installed in the operator's panel.

(B) Portable Wiring Inside tents and Concessions. Wiring for lighting in these areas must be securely installed and protected where exposed to damage. All lamps must be protected. That is, open bulbs may not be used.

525.22 Portable Distribution or Termination Boxes.

Boxes must be designed so that no live parts are exposed. If installed outdoors, they must be weatherproof, and may not be installed within 6 inches of the ground.

Busbars must be properly rated for the equipment they supply, and must be fitted with proper terminals for any conductors that terminate on them.

Receptacles must have overcurrent protection installed inside their box. They must be properly rated.

Any single-pole connectors must comply with Section 530.22

525.23 GFCI Protection For Personnel.

All standard receptacles (125-volt, 15- or 20-ampere) must have GFCI protection if they are used by personnel. GFCI devices must be installed in attachement plug orin supply cords, within 12 inches

(30 mm) or the plug. Egress lighting must be connected to the load side of a GFCI device.

GFCIs are not required for appliances that would malfunction with a GFCI.

Other types of receptacles may either have GFCI protection, or must be accompanied by a written grounding conductor management procedure, per Section 527.6.

IV. Grounding and Bonding

525.30 Equipment Bonding.

The following must be bonded to the grounding system: Metal raceways,metal-sheathed cable, metal electrical equipment enclosures, metal parts of rides, concessions, tents, and other equipment.

525.32 Grounding Conductor Continuity Assurance.

The continuity of the griounding system must be verified every time portable electric equipment is connected. This means that a tester must be used before the cord is plugged-in.

Article 527—Temporary Wiring

527.1: Scope

Temporary electrical lighting and wiring methods are covered here. The type of wiring used for temporary wiring is sometimes of lower quality than may be required for a permanent wiring.

This gives the inspection authorities that have jurisdiction something to use for decisions.

527.2: All Wiring Installations

All other requirements in this *Code* for permanent wiring installations apply except as Article 527 may modify them.

527.3: Time Constraints

(A) **During the Period of Construction.** This fundamentally pertains to temporary electrical wiring and lighting installations that are required and permitted while construction, remodeling, maintenance, repair, or demolition is taking place.

Since its evolution, OSHA, in addition to the authority that has jurisdiction, has had a great deal to say about the safety of temporary wiring. Everyone working in this field has seen very dangerous practices in temporary wiring.

(B) **Ninety Days.** Christmas decorations come under temporary wiring for decorative lighting, as do events such as carnivals.

The time limit permitted for this type of temporary wiring shouldn't exceed 90 days.

(C) Emergencies and Tests. Emergencies and tests or experiments that cover development work are included under temporary electrical wiring installations.

(D) Removal. When construction is completed or any other type of temporary wiring is installed, it shall be removed immediately. For instance, wiring for Christmas lighting shouldn't be permitted to stay year-after-year.

527.4: General

(A) Services. Article 230 must be followed on temporary services.

(B) Feeders. Temporary feeders must conform to Article 240. They must also originate in an approved method for the distribution of current. Cord or cable assemblies may be used if they are of a type shown in Table 400.4 for hard usage or extra-hard usage. If the voltage does not exceed 150 volts to ground and the conductors are not subjected to physical damage, they may be run in open wiring, but to meet this requirement they must be supported at intervals of 10 feet or less.

Exception
As specified in Section 527.3(C).

(C) Branch Circuits. All branch circuits must originate from an approved power outlet or panelboard, and must be protected according to Article 240. For voltages that don't exceed 150 volts to ground, open wiring may be used, provided it is supported every 10 feet (3.05 m) or less. No open wiring may be laid on a floor. Cable assemblies or multiconductor cords may be used, provided they are suitable for the intended use (see Table 400.4).

The three preceding parts are a far cry from the temporary wiring so often used on construction sites. If the overcurrent panels and the like are exposed to the elements, the same types of equipment shall be used as would be used under similar conditions on permanent wiring.

(D) Receptacles. Grounding-type receptacles only shall be used. All branch circuits shall contain an equipment-grounding conductor. A grounded metal raceway or grounded metal cable will serve as the equipment-grounding conductor; otherwise, a separate grounding conductor shall be run. Irrespective of the method of grounding conductor as outlined, every receptacle shall be electrically connected to the equipment-grounding conductor. Any branch circuit on temporary construction wiring or lighting shouldn't have any receptacle installed on it. If the circuit is a multiwire circuit, receptacles are not to be connected to the same ungrounded conductor as the temporary lighting.

(E) Disconnecting Means. Disconnecting means that are suitable for use, such as switches and plug connectors, must be installed so as to disconnect permanently all ungrounded conductors of each of the temporary circuits. The disconnecting means shall disconnect all ungrounded conductors at the same time at the power outlet or panelboard where the branch circuit originated. Approved handle ties will be permitted.

(F) Lamp Protection. Lighting fixtures on temporary wiring for illumination must be protected from accidental contact by persons or breakage. To do this, a suitable fixture or a suitable guard is to be used.

If ordinary brass-shell, paper, line sockets, or any other metal-cased sockets are used, the metal encasing shall be grounded.

(G) Splices. On construction sites, boxes are not required for splices or junction connections. This applies to circuit conductors or multiconductor cord or cable assemblies or open conductors. Refer to the *NEC* for Sections 110.14(B) and 400.9. If a change is made to a raceway system or cable system that is metal-clad or sheathed, a box, conduit body, or terminal fitting with a separate bushed hole for each conductor must be used.

(H) Protection from Accidental Damage. All cords or cables must be protected where accidental damage may occur. Any projections or sharp corners make accidental damage likely, and shall be avoided. When cords or cables pass through doorways or any other place where they could be readily damaged, additional protection is required over the cord or cable.

(I) Securing. Cables that enter boxes and panels that contain device terminations must be secured to the boxes and panels with appropriate fittings.

(J) Carnival Locations. Feeders and branch circuits must be terminated in approved enclosures.

527.6: Ground-Fault Protection for Personnel

To comply with (A) or (B), below-ground-fault protection for personnel shall be installed on construction sites.

(A) Ground-Fault Circuit Interrupters. Fifteen-, 20-, and 30-ampere receptacles for 125-volt single-phase, if these outlets are not a part of the permanent building or structure and are for the use of the construction employees, shall have GFCI protection for persons.

Exception No. 1
If a 5-kW or less portable generator is vehicle mounted, two-wire, single-phase does not require GFCI protection—if the circuit conductors are insulated from the generator frame and on any other grounded surface.

Exception No. 2
GFCI protective devices can be installed on cordsets, but only in industrial buildings where the electrical system is supervised by qualified personnel.

(B) Use of Other Outlets. See the *NEC.*

Author's Note
This section is quite necessary, as extension cords and portable equipment used on construction sites usually receive rather severe usage and can become dangerous to personnel.

527.7: Guarding
To prevent access to other than authorized and qualified personnel, any temporary wiring over 600 volts, nominal, must have suitable fencing or barriers or some other effective means to prevent access by unqualified persons.

Article 530—Motion Picture and Television Studios and Similar Locations

For details on these very specialized installation requirements, refer to the *Code*.

Article 540—Motion-Picture Projectors

This article is seldom used by the trade. Therefore, refer to the *NEC* for answers to any questions involving motion-picture projectors.

Article 545—Manufactured Building

I. General

545.1: Scope
This article covers requirements for a manufactured building and/or building components as herein defined.

545.2: Other Articles

Wherever requirements of other articles of this *Code* and Article 545 differ, the requirements of Article 545 shall apply.

545.3: Definitions

There are a number of definitions in this section, for which reference to the *NEC* should be made, but since manufactured buildings are becoming a way of life, they will be mentioned here.

See the *NEC* definitions of the following: manufactured building, building component, building system, and closed construction.

545.4: Wiring Methods

(A) Methods Permitted. The wiring systems specifically intended and listed for manufactured buildings shall be permitted with listed fittings and with fittings listed for manufactured buildings. Otherwise, all raceway and cable-wiring methods included in the *Code* shall apply.

(B) Securing Cables. In closed construction, cables may be secured only at cabinets, boxes, or fittings, provided No. 10 AWG or smaller conductors are used, and they must be protected against physical damage. Requirements for this protection shall be found in Section 300.4.

Manufactured buildings will in many instances require inspection by the authority that has jurisdiction in the area in which they are being installed. It's a good idea to check beforehand to find out local regulations covering such buildings.

545.5: Service-Entrance Conductors

This is one problem that must be controlled because of the electric utility line locations. The requirements of Article 230 must be met for service-entrance conductors, which must be routed from the service equipment to the point of attachment to the service. Reference is made to Section 310.10 for temperature of conductors.

545.6: Installation of Service-Entrance Conductors

From the preceding section one may readily see why the service-entrance conductors are installed after the erection of the building.

Exception

Except where the point of attachment is known prior to manufacture.

545.7: Service Equipment Location

The service disconnecting means may be installed either inside or outside the building in a readily accessible location. It must also be near the point of entrance of the service conductors.

545.8: Protection of Conductors and Equipment

Manufactured buildings are usually moved from the place of manufacture in parts or as one unit and are thus subjected to various strains and weather conditions: The protection of exposed conductors and equipment must be provided during the process of manufacture, in transit, and during the construction of the building.

545.9: Outlet Boxes

(A) Other Dimensions. Outlet boxes of dimensions other than those required in Table 314.6(A) will be permitted if they are identified, tested, and listed to standards applicable.

(B) Not Over 100 Cubic Inches. Outlet boxes not over 100 cubic inches in size shall be mounted in closed construction affixed with approved anchors and clamps to provide rigid and secure installation.

545.10: Receptacle or Switch with Integral Enclosure

Receptacles and switches that are part of an enclosure and mounting means may be used where they have been identified, tested, and listed for installation for the standards that apply to them. Check with UL, NEMA, or ANSI.

545.11: Bonding and Grounding

Bonding and grounding are as important in manufactured buildings as in any other building. Panels and/or components of the building that come prewired must provide bonding and/or grounding in accordance with Article 250, Parts V, VI, and VII, if it is likely that these parts may become energized.

545.12: Grounding Electrode Conductor

This is to be no different from the requirements of Part X of Article 250.

545.13: Component Interconnection

When intended to be concealed at time of assembly, fittings and conductors, if they have been identified, tested, and listed for the standards to which they are applicable, will be permitted for on-site connection to modules or other building components. The fittings and connectors must be equal to the wiring method used in insulation, temperature rise, and withstanding of fault current, and shall be capable of withstanding vibrations or any minor motions that occur in the components of the manufactured building.

In a number of cases, manufactured buildings have concerned inspection authorities and some require inspection at the manufacturing site before being moved, as well as after set in place.

Article 547—Agricultural Buildings

547.1: Scope

This article applies to agricultural buildings or parts of buildings that are covered in (A) and (B) below:

(A) **Excessive Dust and Dust with Water.** This covers agricultural buildings where excessive dust or dust with water accumulate. These buildings contain areas where poultry- and livestock-confinement systems are environmentally controlled, where feed dust, and litter dust, including mineral feed particles, may accumulate, and also enclosed areas of similar nature.

(B) **Corrosive Atmosphere.** Agricultural buildings in which a corrosive atmosphere exists are covered here. Such buildings include areas (1) where poultry and animal excrement may cause corrosive vapors; (2) where corrosive materials may combine with water; (3) that are damp or wet, because periodic washing is required for cleaning and for sanitation by the use of water and cleaning agents; or (4) where similar conditions exist.

547.2: Definitions

Distribution Point. The source of electrical supply to farm buildings and associated dwellings that are under a single management.

Equipotential Plane. An area which is provided with a very strong grounding and bonding system so that the entire area has an equal potential to ground.

547.3: Other Articles

In agricultural buildings where conditions such as were covered in the previous section occur, the wiring installations may be made using applicable articles of the *Code*.

The above calls to the author's mind a turkey building where general-purpose service-entrance equipment and a mast service was installed. It was only a very short time until shorts and grounds occurred in the service equipment. This was due to the high humidity rising up the mast and condensing, causing water to run down on the circuit breakers. This was stopped by installing duct-seal at both the top and bottom of the mast.

547.4 Surface Temperatures

When electrical equipment and devices are installed in compliance with provisions of this article, the installation shall be in such a manner that the equipment will be able to work at its full rating and not develop surface temperatures in excess of normal safe operation of the equipment or device.

2005

547.5: Wiring Methods

(A) **Wiring Systems.** With agricultural buildings described in Section 547.1(A) and (B), you may use Type UF, NMC, or SNM cable, copper-Type SE cables, Type MC cables, rigid nonmetallic conduit, liquid-tight flexible nonmetallic conduit, or other cables or raceways that will be suitable for the location, and if approved terminal fittings are used. For areas described in Section 547.1(A), the wiring methods of Section 502, Part II may be used. Securing of cables within 8 inches of each cabinet, box, or fitting is required.

(B) Mounting. Cables must be supported within 8 inches (200 mm) of each box, fitting, or cabinet in agricultural buildings covered by this article. However, the ¼ inch (6 mm) air gap required by Section 300.6(C) is waived in these buildings.

(C) Equipment Enclosures, Boxes, Conduit Bodies, and Fittings. Enclosures used in the following conditions demand special attention:

- **(1) Excessive Dust.** These enclosures must be designed to minimize the entrance of dust, and may not have openings through which dust could enter, even for mounting screws. In other words, they must be specially designed for the purpose.
- **(2) Damp or Wet Locations.** These enclosures must be designed or properly placed to minimize the entrance of moisture. In wet locations, the enclosures must be listed, and equipment enclosures must be weatherproof.
- **(3) Corrosive Atmosphere.** These enclosures must be have corrosion resistance appropriate for the environments in which they are installed. See Table 490.91 for enclosure designations.

(D) Flexible Connections. Flexible connections are permitted when necessary, and must be made with dust-tight flexible connectors, liquid-tight conduit, or hard-usage cords. All connectors and fittings must be identified for the purpose.

(E) Physical Protection. All wiring and equipment must be suitably protected.

(F) Separate Equipment-Grounding Conductor. A separate copper equipment-grounding conductor is required for these installations. While metal conduit systems are permitted to serve as their own equipment-grounding conduictor in most situations, that is not the case here. Also, for an underground run, the equipment-grounding conductor must be insulated.

2005

(G) Receptacles. All 125-volt, single-phase, 15- and 20-ampere general-purpose receptacles must be protected with a ground-fault circuit interrupter if they are installed in the following locations:

(1) Areas having an equipotential plane.

(2) Outdoors.

(3) In damp or wet locations.

(4) In dirt confinement areas. [See Section 547.10(B).]

Equipment whose use is incompatible with ground-fault protection is excepted from this requirement.

Other circuits providing power in dirt confinement areas to metal equipment that could be energized and that is exposed to livestock must be protected by a ground-fault circuit interrupter. This GFCI protection is not required for circuits where GFCI protection for equipment that is incompatible with such protection. That is, safety or emergency devices (or, conceivably, others), whose deenergization would make the area less safe are not to be protected by GFCIs.

Note

If raceway systems are exposed to widely varying temperatures, you should refer to Sections 300.7 and 352.9 for regulations governing their installation.

547.6: Switches, Circuit Breakers, Controllers, and Fuses

When the parts in the title of this section, including pushbuttons, relays, and similar devices, are used in Section 547.1(A) and (B), they are to be provided with weatherproof, corrosion-resistant enclosure. They shall be designed to minimize the entrance of dust, water, and corrosive elements, and shall be equipped with close-fitting lids or telescoping covers.

547.7: Motors

Motors or other operational machinery must be equipped so that they are totally enclosed, or must be designed to keep the entrance of dust, moisture, or corrosive particles to a minimum.

547.8: Lighting Fixtures

When installed in agricultural buildings covered in Section 547.1, lighting fixtures installed must meet the following requirements:

- **(A)** **Minimize the Entrance of Dust.** Design of the installed lighting fixtures must be such as to keep the entrance to a minimum of dust, foreign material, corrosive material, and moisture.
- **(B)** **Exposed to Physical Damage.** If exposed to physical damage, the lighting fixture must have suitable guards for protection.
- **(C)** **Exposed to Water.** If they are exposed to water and condensation and cleaning or water solutions used to clean the building, lighting fixtures shall be watertight.

547.10: Grounding, Bonding, and Equipotential Plane

- (A) **Grounding and Bonding.** Article 250 must be complied with for grounding and bonding.

Note

Section 250.21 covers current that would be objectionable passing over the grounding conductors.

Exception No. 1

If all the following conditions are met, it won't be required to have a main bonding jumper at a distribution panelboard when the buildings house livestock or poultry: (a) The same owner owns the buildings' and premises' wiring; (b) run with the supply conductors is an equipment-grounding conductor that is the same size as the supply conductors; (c) there is a service disconnecting means at the point of distribution to these buildings; (d) the service-entrance equipment and equipment-grounding conductor are bonded together; and (e) an approved grounding electrode is provided that is also connected to the equipment-grounding conductor in the distribution panelboard.

Exception No. 2

This exception allows the metal interior water-piping system to be bonded to the service enclosure with an approved impedance device, as long as the following conditions are met: (a) The impedance device has a short-circuit withstand rating of at least 10,000 amps; (b) an insulated copper-bonding conductor is used that is No. 8 AWG or larger, and has no splices; (c) the bonding jumper is installed in a raceway; and (d) the bonding conductor is connected to the metal piping system by a pressure connector that is listed and suitable for the conditions, or by an exothermic weld.

(B) **Concrete Embedded Elements.** Either wire mesh or other conductive elements that are provided in the concrete floor of locations where animals are confined shall provide a plane of equipotential, and it shall be bonded to the grounding electrode system. The bonding conductor shall be only copper that is insulated, covered, or bare, and shouldn't be smaller than No. 8 AWG. The bonding of the equipment-grounding conductor shall be connected to the wire mesh or conductive elements by means of pressure connectors or clamps. These must be of brass, copper, copper alloy, or equally substantial approved means.

Equipotential Plane: An area where wire mesh or other conductive elements are embedded in the concrete and bonded to all adjacent equipment, structures, or surfaces that are conductive and are connected to the grounding system. This must prevent a difference of potential from occurring in the plane.

Note

If wire mesh or other conductive grid embedded in concrete floor or platform is bonded to the electrical grounding system, should livestock make contact between the connected floor or platform and the equivalent or metal structure, they will be less likely to be exposed to a potential that will change the behavior or the productivity of the animals.

(C) **Separate Equipment-Grounding Conductor.** As described in Section 547.1(A) and (B) and covering the agricultural buildings described there, any noncurrent-carrying equipment, raceways, or other enclosures that might accidentally be grounded must be grounded by a copper equipment-grounding conductor. This shall be installed to the equipment and the building disconnection means. If this equipment-grounding conductor is installed underground, it must be insulated or covered.

Note

When grounding electrode systems have lower resistances than those required in Article 250, Part IX, this will also reduce any potential differences within the livestock areas.

(D) **Water Pumps and Metal Well Casings.** The frame and motor of any water pump must be grounded as required by Section 430.142, unless the unit is specifically designed to be fully insulated from ground. When submersible pumps are installed in metal casings, they must be bonded to the pump circuit's equipment-grounding conductor (green wire) or to the grounding bus of the panel that supplies power to the circuit.

Article 550—Mobile Homes and Mobile Home Parks

I. General

550.1: Scope

This article covers electrical conductors and equipment that are installed in or on mobile homes and conductors from the mobile homes to a supply of electricity, and includes the installation of electrical wiring fixtures, equipment, and other portions related to electrical installations within the mobile-home park up to the mobile home. If service-entrance conductors don't exist, then the mobile-home service equipment shall be used.

Don't let this confuse you, as the service equipment, as will be seen in Section 550.4, shall be located adjacent to, and not in or on, the mobile home.

Mobile homes have caused much discussion by *Code* panels and inspection authorities in general. They are built and sold to the public as a complete unit with the walls in place. Thus, local inspection authorities can't really check to see that the wiring meets the *Code* requirements. Underwriter's Laboratories have a labeling service that is available to mobile-home manufacturers and the mobile homes come from such factories with the label on them. However, these seem to be in the minority. Some inspection authorities demand an inspection before the units are allowed in their jurisdictional area. Others find that some manufacturers ask for local inspection and label their product accordingly to show that a local inspection has been made, feeling that the public is entitled to purchase a mobile home with adequate and safe wiring. These local labels are often accepted by other inspection authorities.

Another problem that faces inspection authorities is that so often the wheels are removed and the mobile home is set on a permanent foundation. Local laws will govern whether it is still considered a mobile home or not. In many localities it may stay a mobile home, regardless. In other localities, it may stay a mobile home as long as a license is purchased annually for a mobile home.

There is also a popular type of mobile home that is built in two or more mobile sections that are hauled to the location and bolted together to become a large home.

These problems can't be covered by the *National Electrical Code*; they are problems to be answered by local rulings and laws.

550.2: Definitions

There are a number of definitions included that cover mobile homes. They won't be repeated, so refer to the *NEC*; however, the headings are given here:

(1) Appliance, Fixed.
(2) Appliances, Portable.
(3) Appliance, Stationary.
(4) Distribution Panelboard.
(5) Feeder Assembly.
(6) Laundry Area.
(7) Mobile Home.
(8) Mobile Home Accessory Building or Structure.
(9) Mobile Home Lot.
(10) Mobile Home Park.
(11) Mobile Home Service Equipment.
(12) Park Electrical Wiring Systems.

As stated previously, the following is given: Basically, the local authority enforcing the *Code* is interested that:

(1) The electrical panel in a mobile home is basically a feeder panel, and that the neutral bus is isolated from the panel enclosure and the equipment grounding. This must be taken care of by an extra conductor for grounding purposes only from the source of supply through the cord or cords supplying the home. See Figures 550-1 and 550-2.

(2) If there are two cords supplying the home, the two panels must be treated as entirely separate panels and not interconnected.

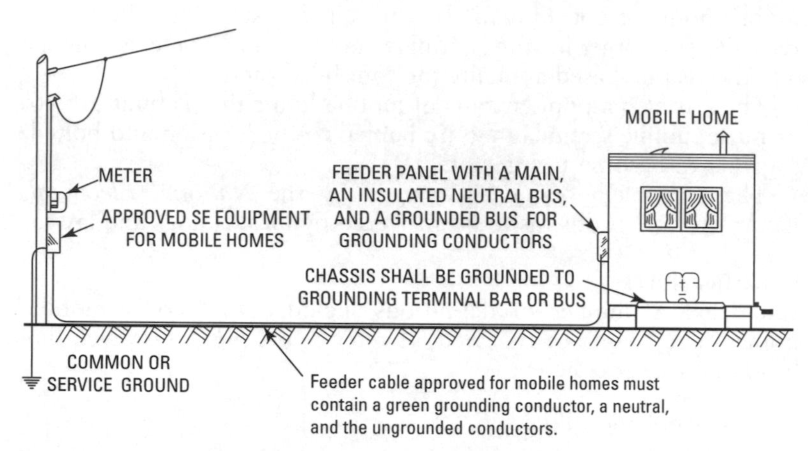

Figure 550-1 Pole-mounted service-entrance equipment for mobile home use with the feeder cable above ground.

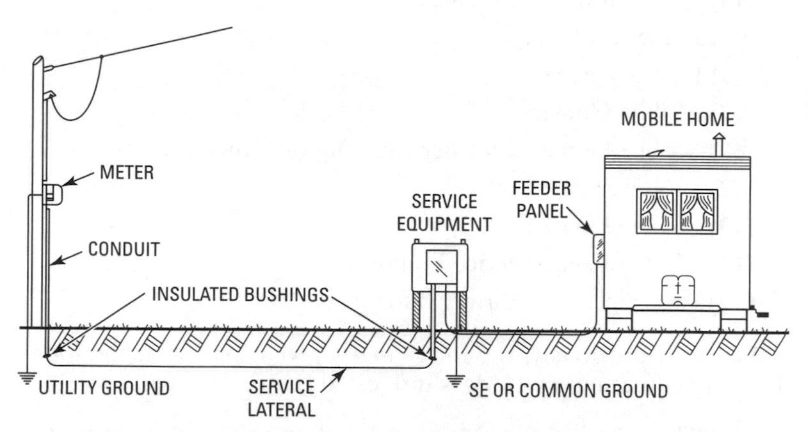

Figure 550-2 Underground service lateral for use with a mobile home.

(3) Supply cord or cords shall be no longer than 36½ feet and no less than 20 feet in length.

(4) Where the calculated load exceeds 100 amperes, or where a permanent feeder is used as a supply, four permanently installed conductors, one being identified as the grounding conductor as required by the *Code*, must be used. See Figures 550-3 and 550-4.

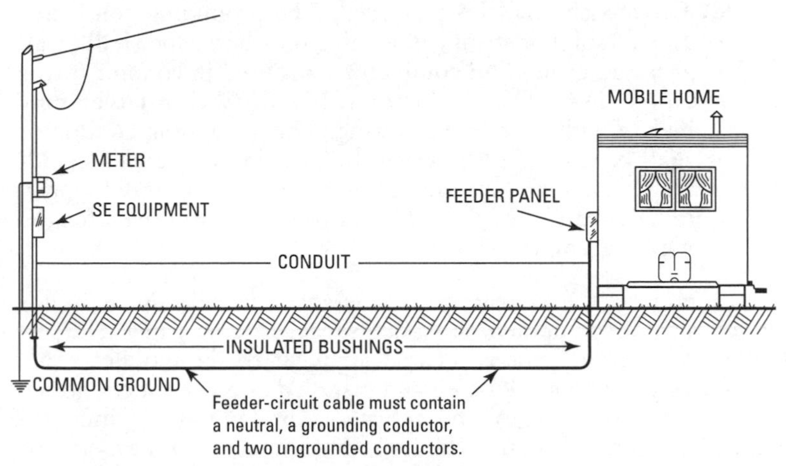

Figure 550-3 Pole-mounted service-entrance equipment for mobile home use with the feeder cable buried.

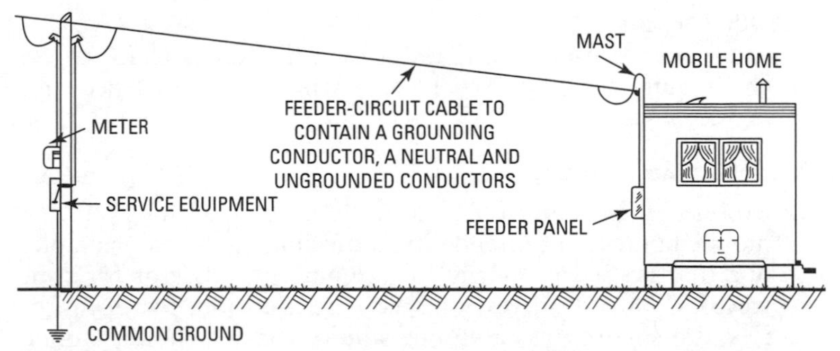

Figure 550-4 An overhead feeder-cable installation to supply power to a mobile home.

(5) Grounding of both electrical and nonelectrical metal parts in a mobile home is through connection to a grounding bus in the mobile home distribution panel. The grounding bus is grounded through the green-colored conductor in the supply cord or the feeder wiring to the service ground in the service-entrance equipment located adjacent to the mobile home location. Note this last part. Neither the frame of the mobile home nor the frame of any appliance may be connected to the neutral conductor in the mobile home.

(6) The chassis shall be grounded. The grounding conductor may be solid or stranded, insulated or bare, and shall be an armored grounding conductor or routed in conduit if it is No. 8 AWG. The conductor, if No. 6 AWG or larger, may be run without metal covering. The grounding conductor shall be connected between the distribution panel grounding terminal and a terminal on the chassis. Grounding terminals shall be of the solderless type and approved for the wire size employed.

In summary, the distribution panel of a mobile home is not service-entrance equipment. It is, in essence, a feeder panel. The service-entrance equipment is located adjacent to the mobile home. When a mobile home is located and installed as a permanent home, then the picture changes and the necessary alterations must be made in the wiring system to meet the requirements of the *Code* for a permanent home.

Note under definition of mobile home that it is designed to be used as a dwelling unit(s) without a permanent foundation. This rules out a mobile home as a prefabricated or modular home.

550.3: Other Articles

If requirements of other articles of the *NEC* differ from Article 550, the requirements of Article 550 will take precedence and apply.

550.4: General Requirements

(A) Mobile Home Not Intended As a Dwelling Unit. When a mobile home is not intended as a dwelling unit—for example, one that is used primarily for sleeping purposes or for contractors' on-site offices for a construction job, for dormitories, dressing rooms in mobile studios, banks, clinics, mobile stores, or for exhibiting and demonstrating merchandise or machinery—it won't be required to meet the provisions of this article pertaining to the number of capacity of the circuits required. All other applicable requirements in this article must be followed if a mobile home of the type mentioned above is provided with an electrical installation that will be energized from a 120-volt or 120/240-volt AC power system. If a different voltage is required, either by available power supply systems or design, adjustments must be made to meet the requirements of other articles and sections covering the voltage used.

(B) **In Other Than Mobile Home Parks.** Mobile homes, if they are installed in other than regular mobile home parks, must comply with the provisions of this article.

(C) **Connection to Wiring System.** The provisions of this article apply to mobile homes that will be connected to a wiring system rated 120/240 volts, nominal, three-wire AC, with a grounded neutral.

(D) **Listed or Labeled.** The electrical materials, including devices, appliances, fittings, and other equipment, must be listed or labeled. This must be done by a qualified testing agency, and the installation and connection of the same shall be done in an approved manner.

Refer to the *NEC* for complete coverage of this article.

II. Mobile Homes

550.5: Power Supply

(A) **Service Equipment.** The service equipment for a mobile home must be located adjacent to the mobile home and shouldn't be mounted in or on the mobile home. The power supply to the mobile home shall be a feeder assembly from the adjacent service equipment, and it shall consist of not more than one listed 50-ampere mobile home power-supply cord that has a molded cap that is part of the cord on the permanently installed feeder.

Exception

If the mobile home has been factory-equipped with gas- or oil-fired central-heating equipment and cooking appliances, the mobile home may be supplied with a power-supply cord that is listed and rated at 40 amperes.

(B) **Power-Supply Cord.** If a power-supply cord goes to the mobile home, it shall be permanently attached to the distribution panel or a junction box that is a permanently connected distribution panelboard, and the free end of this cord is to be terminated with an attachment plug cap.

Cords that use adapters and pigtail-end extension cords and similar items shouldn't be shipped with a mobile home or attached thereto.

A clamp or other device that is suitable to afford strain-relief for the cord shall be used at the distribution panelboard knockout. The purpose of this is to prevent strain to the terminals when the power cord is handled in a normal manner.

The cord must be a listed type and include four conductors, one of which is to be identified by a continuous green color or a continuous green color with one or more yellow stripes. This conductor shall be used for the grounding conductor.

(C) **Attachment Plug Cap.** The attachment plug cap shall be a three-pole, four-wire, grounding type that is rated at 50 amperes, 125/250 volts, such as is shown in the *NEC* in Figure 550.5(C), and that is intended for use with 50-ampere, 125/250-volt receptacles. The attachment plug shall be molded of butyl rubber, neoprene, or other materials that have been found to be suitable for this purpose, and the plug shall be molded to the flexible cord in such a manner that it adheres tightly to the cord where the attachment plug enters the receptacle cap. When a right-angle cap is required, the "configuration must be such that the grounding member is furthest from the cord."

Note

Complete details may be seen in the *NEC* in Figure 550.5(C). For the balance of this section, refer to the *NEC*.

550.6 Through 550.15

These sections apply mainly to the manufacture of mobile homes. You may refer to the *Code* for extract requirements.

III. Services and Feeders

550.21: Distribution System

The electrical power to mobile home lots must be 120/240 volts, single-phase. For the purposes of Part III of this article, when the service to the park is over 240 volts, the transformers and secondary distribution panelboards should be treated as service equipment.

550.22: Minimum Allowable Demand Factors

Electrical wiring systems for mobile home parks must be calculated based on the larger of:

(A) 16,000 volt-amperes for each lot.

(B) The load calculated according to Section 550.13 for the largest typical mobile home that each lot can accept. It is allowed to calculate the feeder or service load based upon Table 550.22. Note that no demand factors are allowed, except as stated in this *Code*.

Service and feeder conductors may be sized according to Article 310, Note 3, or the notes to the ampacity tables for 0 to 2000 volts.

550.23: Mobile Home Service Equipment

(A) **Service Equipment.** The service equipment for a mobile home must be located within sight and within 50 feet (9.14 m) of the mobile home it serves. It may not be mounted in or on the mobile home itself.

Exception

The service equipment can be located in other parts of the premises (not adjacent to the mobile home), as long as all of the following conditions are met: (a) A disconnecting means for the service equipment is located within sight and within 50 feet of the mobile home; (b) a grounding electrode is installed at the disconnecting means (see Part VIII of Article 250); and (c) a grounding electrode conductor (see Part X of Article 250) connects the grounding electrode to the equipment-grounding terminal of the disconnecting means.

Section 250.24's provisions may be applied at the disconnecting means.

(B) **Rating.** Service equipment for mobile homes must be rated at no less than 100 amps. Power outlets that are a part of mobile-home service equipment can be rated up to 50 amps, and must have overcurrent protection accordingly. These power outlets must be configured according to Figure 550.5(C) of the *NEC*.

(C) **Additional Outside Electrical Equipment.** The service equipment must also be provided with some means for connecting an accessory building or structure, etc. by a fixed wiring method.

(D) **Additional Receptacles.** Additional receptacles for equipment outside the mobile home are permitted, but all such 125-volt, 15- or 20-ampere receptacles must be protected by a listed ground-fault interrupter.

(E) **Mounting Height.** An outdoor disconnecting means for a mobile home must be installed so that the bottom of its enclosure is no less than 2 feet from the ground, and that the center of the handle (in its highest position) is not more than 6½ feet from the ground or platform.

(F) **Grounded.** Every mobile home service must be grounded according to Article 250.

550.24: Feeder

(A) **Feeder Conductors.** Mobile home feeder conductors may be either a listed cord (factory-installed in accordance with Section 550.5(B)), or a permanently installed group of four color-coded wires. The wires may be marked in either the factory or in the field, but must comply with Section 310.12. It is not allowed to mark the equipment-grounding (green) conductor by merely stripping the insulation.

(B) **Adequate Feeder Capacity.** Feeder conductors must be rated at least 100 amps, and must be sufficient for the load they will carry.

Exception

A mobile home feeder between the service equipment and the disconnecting means does not need to include an equipment-grounding conductor if the neutral is grounded in accordance with Section 250.24(A).

Articles 551—Recreational Vehicles and Recreational Vehicle Parks

Part I applies to wiring systems used in the vehicles. Part II applies to recreational-vehicle park wiring systems and includes type of distribution, receptacles, load calculations, lot service equipment, and installation requirements for overhead and underground conductors.

It appears that any authority that has jurisdiction over the enforcement of the *National Electrical Code* has a very comprehensive guide for inspection and enforcement of the *NEC* in regard to recreational vehicles and parks. See the *NEC* for coverage of this article.

Article 553—Floating Buildings

See the *NEC*.

Article 555—Marinas and Boatyards

These areas have the possibility of becoming very hazardous from an electrical aspect. When wiring marinas and boatyards, be sure that you follow the prescribed methods outlined in the *NEC*.

Chapter 6

Special Equipment

Article 600—Electric Signs and Outline Lighting

While electric sign work is in many ways similar to other types of electrical work, it is a specialty trade. Most municipalities have different proficiency exams for electrical contractors and sign contractors. This article reiterates that all sign installations must follow the established requirements for other electrical installations, and also gives specific requirements for sign installations. For specific details, review this article in the *NEC*.

Article 604—Manufactured Wiring Systems

604.1: Scope

This article applies to wiring modules that are factory-made, and field-installed as branch circuitry. They can also be used for remote-control circuits, signaling circuits, or communications circuits. These subassemblies may only be installed in accessible locations, such as above suspended ceilings.

604.2: Definition

A *manufactured wiring system* is a system of component parts assembled in a factory that can't be inspected in the field without damage to the assembly.

604.3: Other Articles

All other requirements of the *Code* apply to manufactured wiring systems, except as modified by this article.

604.4: Uses Permitted

These systems can be installed in dry, accessible locations. They can also be used in plenums and spaces used for environmental air when listed for the purpose and installed according to Section 300.22.

Exception

In concealed spaces, one end of a tapped cable can extend into hollow wall spaces to be terminated in switches and receptacles. They may also be used out of doors if listed for such use.

604.5: Uses Not Permitted

These systems can't be used when their conductors or cables are limited by Articles 320 and 330.

604.6: Construction

(A) Cable or Conduit Types.

(1) When cable is used for the wiring modules, it can be either Type AC cable or Type MC cable. The conductors in the cable must be No. 10 or larger, and insulated for at least 600 volts. The grounding conductor can be bare.

(2) If conduit is used for the wiring modules, it must be flexible metal conduit. The conductor requirements are the same as those in (1) above.

Exception No. 1

This exception applies to both (1) and (2) above: A fixture tap no longer than 6 feet, connecting a single fixture, can contain conductors as small as No. 18 AWG.

Exception No. 2

This section applies to both (1) and (2) above: Conductors smaller than No. 12 can be used for listed communications, signaling, or remote control circuits.

(3) Each section must be clearly marked, identifying the type of cable or conduit.

(B) **Receptacles and Connectors.** These must be of a locking type, polarized, identified for the purpose, and must be part of a listed assembly.

(C) **Other Component Parts.** All other component parts must be listed for their intended use.

604.7: Unused Outlets

All unused outlets must be capped.

Article 605—Office Furnishings (Consisting of Lighting Accessories and Wired Partitions)

605.1: Scope

Lighting accessories, electrical equipment, and the wiring used to connect same, which may be installed in relocatable wired partitions, are covered by this article.

Office areas use many relocatable partitions so that areas of work may be relocated as business conditions change.

605.2: General

Only identified and suitable wiring systems shall be used to provide power for lighting accessories and appliances where supplied by wired partitions, and these partitions shouldn't extend from the floor to the ceiling.

- **(A)** Use. As provided by this article, these assemblies shall be installed and used only as covered here.
- **(B)** **Other Articles.** All other articles of the *Code* apply, unless modified by this article.
- **(C)** **Hazardous (Classified) Locations.** When they are used in hazardous (classified) locations, manufactured wiring systems shall conform to Articles 500 through 517, in addition to the requirements of this article.

605.3: Wireways

The installation of conductors and their connections is to be contained within wiring channels that are made of metal or other material that is suitable for the conditions of use. There shall be no projections or other conditions that might damage the insulation of the conductors.

Wireways for this purpose often contain a divider so that telephone wire and electrical conductors may be run in the same wireway but are entirely separated. Drop ceilings are usually used and connections to the regular wiring system above the ceiling are used to connect the wireway, but only approved wiring methods shall be used for this connection.

Note

Flexible cords are not to be used as conductors in these partitions.

605.4: Partition Interconnections

The electrical connection between partitions shall be flexible assembly identified for use with wired partitions.

Exception

It will be permitted to use flexible cords between partitions if all the following conditions are met: (a) The cords are extra-hard usage type, (b) the partitions are mechanically contiguous; (c) the flexible cord in no case is longer than 2 feet; it may be up to 2 feet long if necessary for maximum positioning of the partitions; and (d) the cord is terminated with an attachment plug- and cord-connector, and provision is made for strain relief.

605.5: Lighting Accessories

The following conditions must be met, and the lighting equipment must be listed and identified for use with wired partitions.

- **(A) Support.** There shall be means for secure attachment or support provided.
- **(B) Connection.** If cord-and-plug connections are provided, the cord that is used for this application and its length shall be suitable, but the length shouldn't exceed 9 feet in length. Cords no smaller than No. 18 AWG shall contain an equipment-grounding conductor, and the cord shall be of hard-usage type. Any connections by other means must be identified and suitable for the conditions for which it is used.
- **(C) Receptacle Outlet.** There shall be no convenience receptacles included in a lighting fixture.

605.6: Fixed-Type Partitions

If partitions are wired and are secured to the building surface, they must be permanently connected to the building surface, and one of the means covered in Chapter 3 shall be used for wires.

605.7: Free-Standing-Type Partitions

One of the wiring methods covered in Chapter 3 may use partitions that are of the free-standing type and will be permanently connected to the building electrical system.

605.8: Free-Standing-Type Partitions, Cord- and Plug-Connected

Individual partitions that are of a free-standing type, or groups of individual partitions that are electrically connected, mechanically contiguous, and don't exceed 30 feet when assembled, may be connected to the electrical system of the building by means of a single flexible cord and plug, provided all the following conditions are met:

- **(A) Flexible Power Supply Cord.** The flexible supply for the power must be extra-hard-usage-type cord and shall be No. 12 AWG or larger conductors and include an equipment-grounding conductor, and the length of the flexible cord connection shouldn't exceed 2 feet.
- **(B) Receptacle Supplying Power.** Receptacle(s) are to be supplied with power by a separate circuit that is not used to supply any other loads, only the panels themselves, and the receptacle shouldn't be over 12 inches from the panel to which it is connected.

(C) **Receptacle Outlets, Maximum.** Individual partitions and groups of partitions that are interconnected shouldn't be allowed to contain more than thirteen 15-ampere, 125-volt receptacle outlets.

(D) **Multiwire Circuits Not Permitted.** Individual panels or groups of panels that are interconnected shouldn't be permitted to have multiwire circuits installed in them.

Note

You are referred to Section 210.4, which covers circuits that supply partitions in Sections 605.6 and 605.7.

Article 610—Cranes and Hoists

Refer to the *NEC.*

Article 620—Elevators, Dumbwaiters, Escalators, and Moving Walks

Refer to the *NEC.*

Article 630—Electric Welders

Refer to the *NEC.*

Article 640—Sound-Recording and Similar Equipment

Refer to the *NEC.*

Article 645—Information Technology Equipment

645.1: Scope

This article applies to the various types of data-processing equipment, interconnecting wiring, and the grounding of this equipment.

645.2: Special Requirements for Electronic Computer/Data Processing Equipment Room

This article applies, provided all of the following conditions are met:

(1) Disconnecting means according to Section 645.10 are provided.

(2) A separate HVAC system is dedicated for data-processing equipment and is separated from all other areas of occupancy. If fire/smoke dampers are used at the point of entry to the data processing area, HVAC systems that serve other areas can also serve data-processing locations.

(3) Only listed data-processing equipment is used.

(4) The data-processing area is occupied only by personnel necessary for the operation of the data-processing equipment.

(5) The data-processing room has fire-resistant-rated floors, walls, and ceilings with protected openings.

(6) The construction of the area conforms to the applicable building code.

645.5: Supply Circuits and Interconnecting Cables

(A) Branch Circuit Conductors. Branch circuit conductors that supply more than one piece of data processing equipment must have an ampacity of at least 125 percent of the total connected load.

(B) Connecting Cables. A data-processing system can be connected to a branch circuit by any of the following means:

- **(1)** Data processing cables and attachment plug caps.
- **(2)** Flexible cord with an attachment plug cap.
- **(3)** A cord-set assembly, which must be protected if run on a floor.

(C) Interconnecting Cables. Data-processing units can be interconnected by listed cables and assemblies. When run on a floor, these cables must be protected.

(D) Under Raised Floors. Power cables, communications cables, connecting cables, interconnecting cables, and receptacles associated with data-processing equipment can be installed under a raised floor, provided the following conditions are met:

- **(1)** The raised floor must be accessible, and of a suitable construction.
- **(2)** The branch-circuit conductors to receptacles or field-wired equipment must be in rigid metal conduit, intermediate metal conduit, EMT, rigid nonmetallic conduit, metal wireway, surface metal raceway (with metal cover), flexible metal conduit, liquid-tight flexible metal or nonmetallic conduit, MI cable, MC cable, or AC cable. The supply conductors must comply with Section 300.11.
- **(3)** The ventilation in the underfloor area can be used only for the data-processing equipment and area.
- **(4)** Cable openings in the raised floor must protect the cables from abrasion and minimize the passage of debris.

(5) Cable types other than those listed in (2) above must be listed as DP cable with adequate fire-resistant qualities for use under computer-room floors.

2005

(6) Abandoned cables must be removed. The *Code* considers cables as abandoned if they are not terminated in a connector and are not tagged for future use.

Exception No. 1
When connecting cables are enclosed in conduits or raceways.

Exception No. 2
Interconnecting cables listed with equipment made before July 1, 1994, can be reinstalled with that equipment.

Exception No. 3
This exception lists other acceptable cable types. See the *NEC.*

(7) The areas under raised floors can't be used for storage, and unused cables must be removed from these areas.

(E) **Securing In Place.** All cables, boxes, connectors, etc. that are part of data processing equipment are not required to be secured in place.

645.6: Cables Not in Computer Room
Cables that extend out from computer rooms must conform to the requirements of this article.

645.7: Penetrations
Penetrations of fire-rated room boundaries must comply with Section 300.21.

645.10: Disconnecting Means
There must be a disconnecting means for all electronic equipment in the data-processing rooms. A similar disconnecting means must be provided for the dedicated HVAC system that serves the area. Disconnecting the power to this HVAC unit must cause all of the fire/smoke dampers to close. These disconnects must be grouped together, identified, and installed next to the primary exit and

entrance doors. It is allowable to combine these two disconnects into one unit.

Exception

Installations that qualify under the requirements of Article 685.

645.11: Uninterruptible Power Supplies (UPS)

These systems and their supply and output circuits must comply with Section 645.10 when installed in data-processing rooms. The disconnecting means must also disconnect the battery from its load.

Exception No. 1

Installations that qualify under the requirements of Article 685.

Exception No. 2

A disconnect according to Section 645.10 is not required for batteries rated 10 amp-hours or less inside data-processing equipment, or for batteries that are inside UPS systems and are rated 750 volt-amperes or less.

645.15: Grounding

Data-processing equipment must either be grounded according to Article 250, or double insulated. For the purpose of applying Section 250.20(D), power systems derived within the various pieces of equipment that supply other parts of the system through receptacles or cable assemblies are not considered separately derived systems.

645.16: Marking

Each piece of data-processing equipment that connects to a branch circuit must be marked, showing the manufacturer's name, the operating voltage, frequency, and maximum load in amps.

Note

For the use of isolated grounding systems in these areas, refer to Section 250.74.

Article 650—Organs

Refer to the *NEC*.

Article 660—X-Ray Equipment

Refer to the *NEC*.

Article 665—Induction and Dielectric Heating Equipment

Refer to the *NEC*.

Article 668—Electrolytic Cells

Refer to the *NEC*.

Article 669—Electroplating

Refer to the *NEC*.

Article 670—Industrial Machinery

Refer to the *NEC*.

Article 675—Electrically Driven and Controlled Irrigation Machines

Most of these machines pivot in a circle from a source of irrigation water, usually supplied from a pump drawing water from an irrigation ditch or a ground-water well supplying water from a pumping source.

Some are electrically propelled, and some are water-propelled, but electrically controlled, so that they may be kept in line over rough ground, but any that the author has seen use electrical controls at the very least; many are driven by electric motors.

Refer to the *NEC*.

Article 680—Swimming Pools, Fountains, and Similar Installations

I. General

Due to the large usage of underwater lighting and other electrical equipment in conjunction with swimming pools, this has become a very important article in the *Code*. The water and conductivity of wet surfaces around the pool have become a very serious hazard, and in designing or installing electrical installations in such locations, special precautions should be taken at all times.

680.1: Scope

This article covers the construction and installation of wiring for equipment that will be either in or adjacent to all swimming, wading, therapeutic, and decorated pools, fountains, hot tubs, spas, and hydromassage bathtubs, and this will be irrespective of whether they are of a permanent nature or can be stored; it also pertains to the metal auxiliary equipment. This will cover pumps, filters, ladders, and any other similar equipment in or near the water.

Note

The term *pool* as used in this article includes swimming, wading, and permanently installed therapeutic pools. *Fountains* as used in this article includes fountains, ornamental pools, display pools, and reflection pools.

680.2: Definitions

The *NEC* covers definitions in this section. The following definitions are covered:

(1) Cord- and Plug-Connected Lighting Assembly.

(2) Dry-Niche Lighting Fixture.

(3) Forming Shell.

(4) Hydro-Massage Bathtub.

(5) No-Niche Lighting Fixture.

(6) Permanently Installed Decorative Fountains and Reflection Pools.

(7) Permanently Installed Swimming, Wading, and Therapeutic Pools.

(8) Pool Cover, Electrically Operated.

(9) Spa or Hot Tub.

(10) Storable Swimming or Wading Pools.

(11) Wet-Niche Lighting Fixtures.

680.3: Other Articles

The requirements of Chapters 1 to 4, inclusive, are applicable to the wiring of swimming pools with the exception of the modifications in this article, which are necessary due to the hazards involved.

Note

You are referred to Section 314.13 for the junction boxes, Section 352.3 for rigid nonmetallic conduit, and Article 720 for low-voltage lighting.

680.4: Approval of Equipment

All electrical equipment that may be installed in the water or the walls or decks of pools, fountains, and similar installations must be installed according to this article. Underwriter's Laboratories' approval is practically universally accepted for this equipment.

680.5: Ground-Fault Circuit Interrupters

(A) **Transformers.** Transformers and their enclosures that supply fixtures shall be identified for the purpose. The transformers of the two-winding type with a grounded metal barrier between the primary and secondary windings shall be used. Check the green UL listing book to be certain that the one you use has been approved.

(B) **Ground-Fault Circuit Interrupters.** Self-contained units may be used for ground-fault circuit interruption, or circuit-breaker types may be used, or receptacle types, or any other type that might be approved for such usage.

(C) **Wiring.** All conductors on the load side of ground-fault circuit interrupters or transformers shall be kept entirely separate from all other wiring and equipment. This is in reference to the equipment used to comply with Section 680.20(A)(1).

Exception No. 1
Ground-fault circuit interrupters may be installed in a panelboard even though it contains some circuits that are not protected by ground-fault circuit interrupters.

Exception No. 2
If the supply conductors are of the feed through, receptacle type, you will be permitted to install them in the same enclosure ground-fault circuit interrupters.

Exception No. 3
The load side conductors from a ground-fault circuit interrupter may occupy conduit, boxes, or enclosures where conductors in those items are also protected by ground-fault circuit interrupters.

680.6: Receptacles, Lighting Fixtures, Lighting Outlets, and Switching Devices

(A) Receptacles.

(1) Any receptacles on the property must be mounted at least 10 feet away from the inside walls of a pool or fountain.(See Figure 680-1.)

Exception
Receptacle(s) installed to supply power to water-pump motor(s) for a pool that is permanently installed, as permitted in Section 680.7, may be installed between 5 and 10 feet from the inside walls of the pool. They shall be single receptacles and of the locking type and grounding type. It is also required that they be protected by GFCIs.

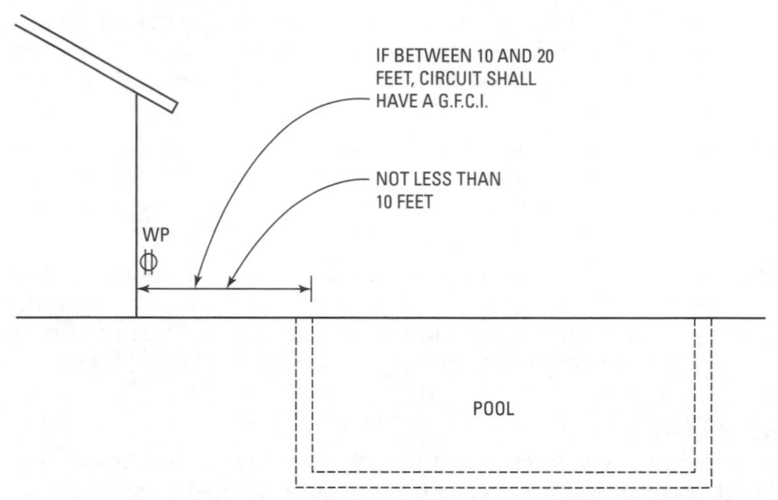

Figure 680-1 Location and protection of all outside receptacles.

(2) At dwelling unit(s) with a permanently installed pool, there shall be at least one 125-volt convenience receptacle, and this is to be located at a distance of at least 10 feet and not more than 20 feet from the inside of the wall of the pool. It may be no more than 6½ feet above the walking level.

(3) Ground-fault circuit interrupters, as covered in Section 210.8(A)(3), will be required for lighting fixtures and outlets that are located within 20 feet of the inside walls of the pool.

Note

The distance is determined for the above dimensions by measuring the shortest path of a supply cord of an appliance that will be connected to the receptacle. This path is followed without piercing a floor, wall, ceiling, or doorway that has either hinged or sliding doors, any window openings, or any other effective barrier.

Note

Section 400.8 prohibits cord wiring where it is run through a window, doorway, or similar opening.

(B) Lighting Fixtures and Lighting Outlets.

(1) No lighting fixtures, lighting outlets, or ceiling fans are to be installed over the pool or over the area extending 5 feet in a horizontal direction from the inside walls of the pool unless they are a minimum of 12 feet above the maximum water level.

Exception No. 1
If there are existing lighting fixtures or outlets that are located less than 5 feet from the inside walls of the pool in a horizontal direction, they shall be at least 5 feet above the surface of the maximum level of the water and must be rigidly attached to the existing structure.

Exception No. 2
The limitation of Section 680.6(B)(1) doesn't apply to indoor pool areas if all the following conditions are met: (a) They are the totally enclosed type of fixture; (b) a GFCI is installed in the branch circuit that is supplying the fixture(s); and (c) the distance from the bottom of the fixture to the maximum water level is not less than 7½ feet.

(2) Where lighting fixtures and lighting outlets are installed in the area extending between 5 feet and 10 feet from the inside walls of the pool, a GFCI shall be used unless they are installed 5 feet above the maximum water level and are rigidly and securely attached to the structure adjacent to or enclosing the pool.

(3) As set forth in Section 680.7, any cord- and plug-connected lighting fixtures shall meet these specifications, as would any other plug- and cord-connected equipment when installed within 6 feet at any point measured radially on the surface of the water.

(C) Switching Devices. Any switching devices on the property shall be located a minimum of 5 feet from the inside walls of the pool, unless they are separated from the pool by a solid fence or other permanent type of barrier.

680.7: Cord- and Plug-Connected Equipment
Whether fixed or stationary, any equipment rated 20 amperes or less, not including underwater lighting installed permanently in the pool, may be connected by a flexible cord to aid in disconnection and removal for maintenance and repair. For other than storable pools, the flexible cord shouldn't exceed 3 feet in length, the cord must contain a copper equipment-grounding conductor that isn't smaller than No. 12, and a grounding-type attachment plug shall be used.

Note

For flexible cords used for connection, see Section 680.25(E).

680.8: Overhead Conductor Clearances

The following parts of pools are not to be placed under existing service-drop conductors or any other open-type overhead wiring; nor shall such type of wiring be installed above any of the following: (1) pools, including the area extending 10 feet horizontally from the inside wall of the pool; (2) structures designed for diving; or (3) platforms, towers, or stands used for observation.

Exception No. 1

Structures covered in (1) through (3) above will be permitted under supply lines or service drops owned, operated, and maintained by a utility- if clearances are provided as indicated by the table here in the *NEC*, the fine-print note, and Figure 680-8, Exception No. 1.

Exception No. 2

Communication conductors owned, operated, and maintained by a utility; community antenna system coaxial cables that comply with Article 820; and the supporting messengers for the cable may be at a height not less than 10 feet above swimming and wading pools, diving structures, diving stands, towers, or platforms.

Note

For clearances not covered by this section, you are referred to Sections 225.18 and 225.19.

680.9: Electric Pool Water Heaters

Electric water heaters for pools shall have the heating elements subdivided so that the load on each element is not over 48 amperes, and the protection for that circuit shouldn't be more than 60 amperes.

The branch-circuit conductor's ampacity and the rating or setting of the overcurrent devices shall be not less than 125 percent of the total rating on the nameplate.

680.10: Underground Wiring Location

Any underground wiring going to the pool may not be installed under the pool or the area extending 5 feet horizontally from the inside wall of the pool.

Exception No. 1

Wiring that is necessary for the operation of the pool equipment, as permitted by this article, may be permitted within this area.

Exception No. 2

If, due to space limitations, the wiring can't be limited to within 5 feet or more from the pool, this wiring may be in rigid metal conduit, intermediate metal conduit, or nonmetallic raceway systems. The minimum burial depths of these conduits must be as follows:

- **(a)** Rigid or Intermediate Metal Conduit—6 inches
- **(b)** Rigid Nonmetallic Conduit (approved for direct burial without concrete encasement) or other types of approved raceways—18 inches

680.11: Equipment Rooms and Pits

It is not permitted to install electrical equipment in rooms or pits that are not provided with adequate drainage to prevent water accumulation under normal operation, or for filter maintenance.

680.12: Disconnecting Means

Disconnecting means must be accessible. They must also be within sight of the pool and at least 5 feet away from the pool.

II. Permanently Installed Pools

680.20: Underwater Lighting Fixtures

(A) General. Paragraphs (A) through (C) in this section apply to underwater lighting fixtures that are below the normal water level in the pool.

(1) Lighting fixtures that are supplied either from a branch circuit or by a transformer that complies with the requirements of Section 680.5(A) are to be designed so that, when the fixture is properly installed without a GFCI, there will be no likelihood of shock conditions with any probable combinations of fault conditions when the pool is in normal use. Relamping is not covered here.

There shall also be a GFCI installed in the branch circuit that supplies the fixtures that operate at more than 15 volts between conductors so that there is no possibility of a shock hazard during relamping. The GFCI shall be installed is such a way that there is no likely shock hazard with any fault-condition combination involving a person in a conductive path from any ungrounded part of the branch circuit or the fixture to ground.

This requirement shall be complied with by the use of an approved underwater lighting fixture and with the installation of a GFCI in the branch circuit.

(2) No lighting fixtures shall be installed that require over 150 volts between supply-circuit conductors.

(3) Underwater lighting fixtures are to be installed with the top of the fixture at least 18 inches below normal water level of the pool. If the lighting fixture faces upward, it shall be properly guarded to prevent contact by a person (see Figure 680-2).

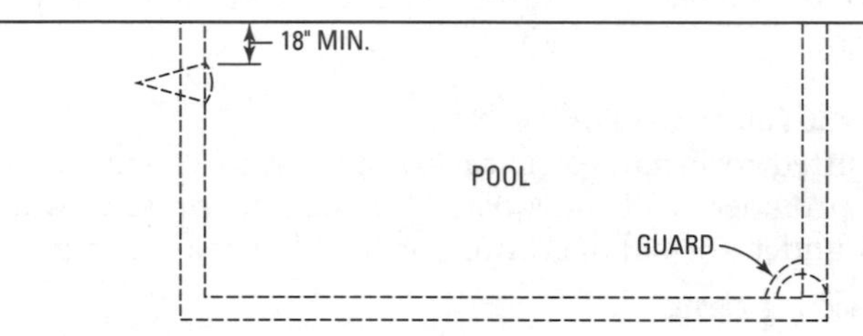

Figure 680-2 Installation of fixtures below water level.

Exception

Lighting fixtures shall be permitted if they have been identified for use at a depth of not less than 4 inches below the water level of the pool.

(4) Submerged lighting fixtures that depend upon submergence for proper operation must be inherently protected against hazardous overheating when not submerged.

(B) Wet-Niche Fixtures.

(1) An approved forming shell of metal shall be installed for mounting the wet-niche underwater fixtures, and this shall be equipped with threads for entries of conduit.

Conduit shall run from the forming shell to a suitable junction box or other enclosure located as covered in Section 680.21. The conduit shall be rigid metal conduit, intermediate metal conduit, or rigid nonmetallic conduit.

Metal conduit shall be made of brass or other corrosive-resistant metal.

If rigid nonmetallic conduit is used and insulated, copper conductor of size No. 8 AWG must be installed in the nonmetallic conduit, and it shall have provisions for terminating in the forming shell, junction box, transformer, or GFCI enclosure. The No. 8 conductor shall be terminated in the forming shell, and it shall be encapsulated or covered by a listed potting compound, the purpose of which is to protect such connection from any possible deteriorating effect of the pool water. Metal parts of the lighting fixture and forming shell in contact with the water in the pool must be of brass or other approved corrosion-resistant metal.

(2) The end of a flexible-cord jacket and the flexible-cord conductor terminations within a fixture must be covered with, or encapsulated in, a suitable potting compound; this is to prevent the entry of water through the fixture, the cord, or its conductors. In addition thereto, the grounding connection within a fixture shall be similarly treated to protect such connection from any deteriorating effects that the pool water might have in the event of water entering into the fixture.

(3) The fixture must be bonded and secured to the forming shell by means of a positive locking device that ensures there will be low-resistance contact and that requires a tool to remove the fixture for the forming shell.

(C) Dry-Niche Fixtures. Dry-niche lighting fixtures shall be provided with (1) some provision for draining water, and (2) a means for connecting the equipment-grounding conductor for each conduit entry.

Rigid metal conduit, intermediate metal conduit, or rigid nonmetallic conduit, that has been approved must be installed from the fixture to the service equipment or panelboard. A junction box is not required, but if one is used, it shouldn't be required to be elevated or located as specified in Section 680.21(A)(4), provided the fixture is specifically identified for the purpose.

Junction boxes mounted above the grade of finished walkways around the pool shouldn't be located in the walkway unless afforded additional protection such as location under diving boards, adjacent to fixed structures, or the like.

The purpose of this ruling is easy to understand. Junction boxes shall be so located as not to cause a stumbling hazard. The area around the pool is always wet and slippery—if installed without protection, the boxes might be a great hazard and likely to cause falls.

(D) No-Niche Fixtures. These fixtures must be supplied by a transformer [the transformer must conform to the requirements of Section 680.5(A)], and must:

(1) Have no exposed metal parts.
(2) Operate at 15 volts or under.
(3) Have a polymeric (plastic) impact-resistant body and lens.
(4) Be listed for the purpose.

Exception

If installed in or on buildings, EMT may be used to protect conductors.

680.21: Junction Boxes and Enclosures for Transformers or Ground-Fault Circuit Interrupters

(A) **Junction Boxes.** When the conduit is connected directly to the forming shell, the following requirements must be met:

(1) The junction box connected to the conduit must have threaded hubs, bosses, or nonmetallic hubs listed for the purpose.

(2) The junction box shall be made of copper, brass, suitable plastic, or any other material that is corrosion resistant and approved for the purpose.

(3) The electrical continuity must be maintained between every metal conduit and the grounding terminals. This shall be by means of copper, brass, or any other approved noncorrosive metal that is part of the box.

(4) Junction boxes are to be located not less than 8 inches measured from the bottom of the box to the ground level, pool deck, or maximum water level. The greater of these distances is to be used, and the location shall be not less than 4 feet from the inside wall of the pool unless separated from the pool by a solid fence, wall, or other permanent barrier (see Figure 680-3).

Exception

When lighting systems operate at 15 volts or less, a flush junction box on the deck may be used if (a) prevention of moisture entering the box may be accomplished by using an approved potting compound to fill the box; and (b) the flush deck box is not less than 4 feet from the inside edge of the pool.

(B) **Other Enclosures.** Enclosures such as for a transformer, GFCI, or similar device that connects to a conduit extending directly to the forming shell shall be:

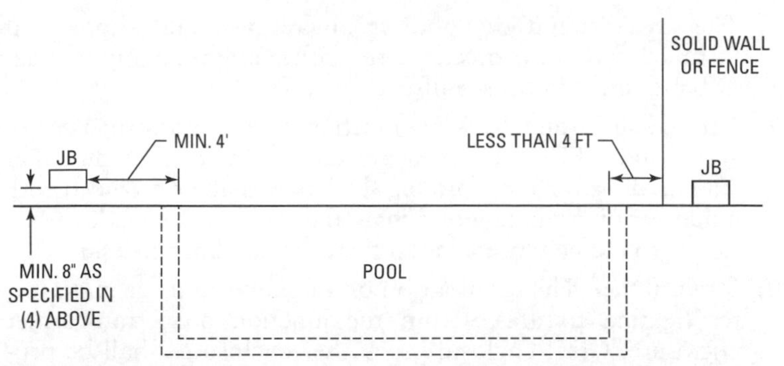

Figure 680-3 Junction box placement where walls or fence are used.

(1) See (A)(1) of this section.

(2) Provided with an approved seal, such as duct seal, at the conduit connection; this prevents circulation of air between conduit and enclosures.

(3) See (A)(4), but replace the word box with enclosure.

(4) See (A)(3), but replace the word box with enclosure. (See Figure 680-4.)

(C) Protection. Junction boxes and enclosures mounted above the finished walkway grade around the pool shouldn't be located in the walkway unless protected in addition by being, for example, located under diving boards or adjacent to fixed structures.

The purpose of this ruling is easy to understand. Junction boxes shall be so located as not to cause a stumbling hazard.

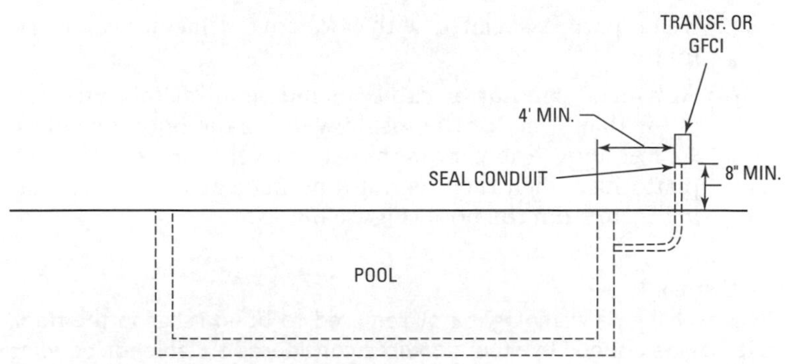

Figure 680-4 Showing conduit seal.

The area around the pool is always wet and slippery; if installed without protection, the boxes might be a great hazard and likely to cause falls.

(D) **Grounding Terminals.** When junction boxes, transformer enclosures, and GFCI enclosures are connected to a conduit that extends directly to the forming shell, they shall be provided with a number of grounding terminals and there shall be at least one more grounding terminal than there are conduit entries.

(E) **Strain Relief.** The termination of a flexible cord for underwater lighting fixtures within the junction box, transformer enclosure, GFCI enclosure, or other enclosures shall be provided with some means of approved strain relief.

680.22: Bonding

The intent of this subsection, in requiring No. 8 or larger solid copper bonding conductors, is to point out that the conductor is not required to be extended or attached to any remote panelboard, servicing equipment, or any electrode. It is employed only to eliminate voltage gradient that could occur in the pool area as prescribed.

(A) **Bonded Parts.** The following parts shall be bonded together:

- **(1)** All metal parts used in the pool structure, including reinforcing metal of the pool shell, coping stones, and deck.
- **(2)** All forming shells.
- **(3)** All metal fittings within or attached to the pool structure.
- **(4)** Metal parts of the electrical system equipment that are associated with the pool water circulating system, and including the pump motors.
- **(5)** Metal parts associated with pool covers, including electric motors.
- **(6)** Any metal conduit or cable, including all metal parts that are within 5 feet of the inside walls of the pool, or within 12 feet above the maximum water level of the pool or any platforms, when there is not a permanent barrier separating them from the pool (Figure 680-5).

Exception No. 1

Special welding or clamping is not required to bond rebar in the pool, but it may be bonded by steel tie wires considered suitable for bonding reinforcing steel. Such tie wires must be made up tight.

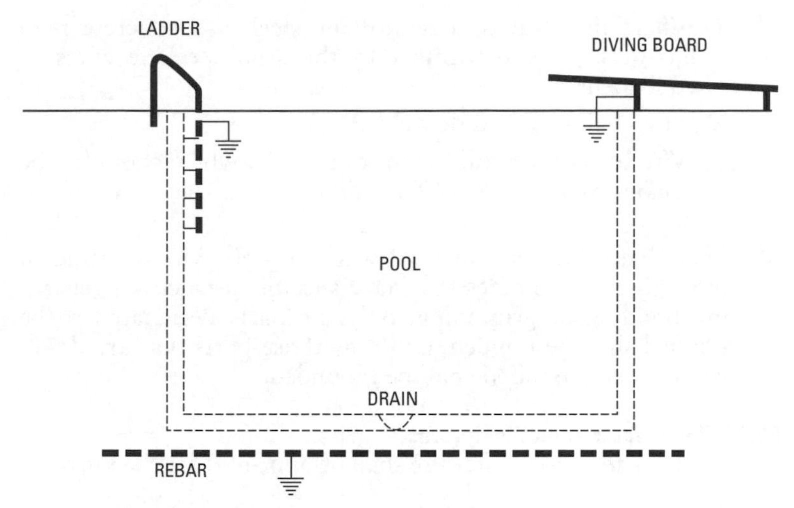

Figure 680-5 Bonding all metal equipment together by using solid copper wire.

Exception No. 2
Metal parts that are not more than 4 inches in dimension and don't penetrate the interior of the pool structure more than 1 inch are not required to be bonded.

Exception No. 3
Bonded or welded metal structural parts of walls with reinforcing steel will be permitted as a common bonding grid for non-electrical connection. They can be made as specified in Section 250.8.

Exception No. 4
Listed pump motors that are double insulated are not required to be bonded. These motors must be clearly marked as double insulated by the manufacturer.

(B) Common Bonding Grid. All these parts are to be of common bonding grid connected with solid copper conductor that is insulated, covered, or bare, and shouldn't be smaller than No. 8 AWG. The connection may be made by pressure connectors or clamps made of brass, stainless steel, copper, or copper alloy. The common bonding grid shall be as follows:

(1) Where the structural reinforcing steel of a concrete pool has been bonded together by the usual steel tie wires or equivalent.

(2) If the wall is bolted or welded.

(3) With the copper solid wire connection, which shouldn't be smaller than No. 8 AWG, it may be insulated, covered, or bare.

(C) **Pool Water Heaters.** In pool water heaters with a rating of more than 50 amperes that have specific instructions regarding bonding or grounding, only the parts designated to be bonded shall be bonded, and only those parts that are designated to be grounded shall be grounded.

680.23: Underwater Audio Equipment

Audio systems for underwater use shall be approved for the purpose.

(A) **Speakers.** They are to be mounted in an approved metal forming shell, and the front shall be enclosed by a captive metal screen, or the equivalent, that is bonded to the forming shell by means of a positive locking device so that a low-resistance contact is ensured and a special tool must open the installation to service the speaker. The forming shell must be recessed in the wall or the floor of the pool.

(B) **Wiring Methods.** Rigid metal conduit, or intermediate metal conduit that is made of brass or other identified corrosive-resistant metal, or rigid nonmetallic conduit may be used to extend from the speaker enclosure forming shell to a suitable junction box or other enclosure. Section 680.21 covers this. If rigid nonmetallic conduit is used, No. 8 AWG copper conductor that is insulated must be installed in the conduit and provisions shall be made for terminations in the forming shell and the junction box. In the forming shell the termination of the No. 8 conductor shall be covered or encapsulated in a suitable potting compound to protect against possible deterioration of the terminal by the pool water.

(C) **Forming Shell and Metal Screen.** Both the forming shell and the metal screen shall be of brass or other approved metal that is corrosion-resistant.

680.24: Grounding

Grounding is required for the following equipment: (1) wet-niche lighting fixtures underwater; (2) dry-niche lighting fixtures installed underwater; (3) any equipment located within 5 feet of the inside of the pool wall; (4) all equipment that is associated with the recirculation of the pool water; (5) junction boxes; (6) transformer enclosures; (7) GFCIs; and (8) panelboard that is not part of the service-entrance equipment, but supplies any electrical equipment associated with the pool.

680.25: Methods of Grounding

(A) General. The following provisions apply to the grounding of underwater lighting fixtures, junction boxes, metal transformer enclosures, panelboards, motors, and all other electrical equipment and enclosures.

(B) **Pool Lighting Fixtures and Related Equipment.** See the *NEC*, starting with (B) of this section. See Figures 680-6 and 680-7.

680.26: Electrically Operated Pool Covers

(A) Motors and Controllers. These devices must be at least 5 feet from the inside wall of the pool, unless separated by a permanent barrier. Any electric motors installed below grade must be the totally enclosed type.

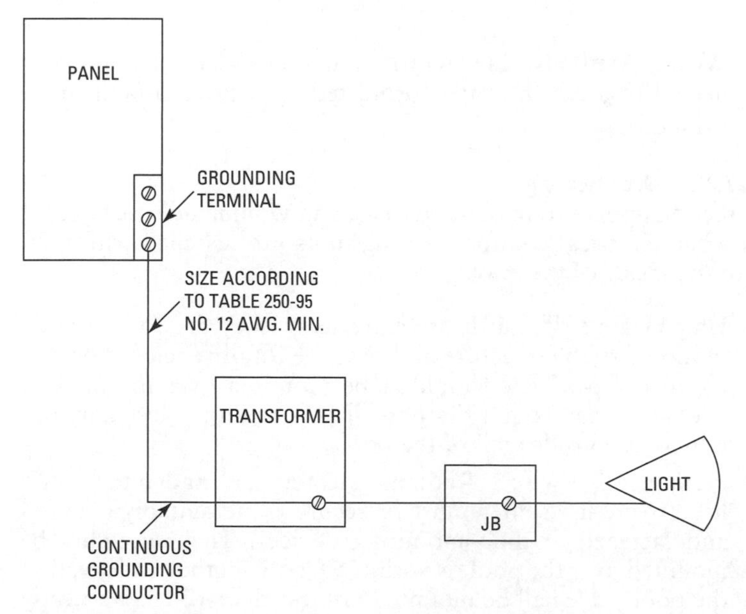

Figure 680-6 Grounding all electrical equipment to the panelboard.

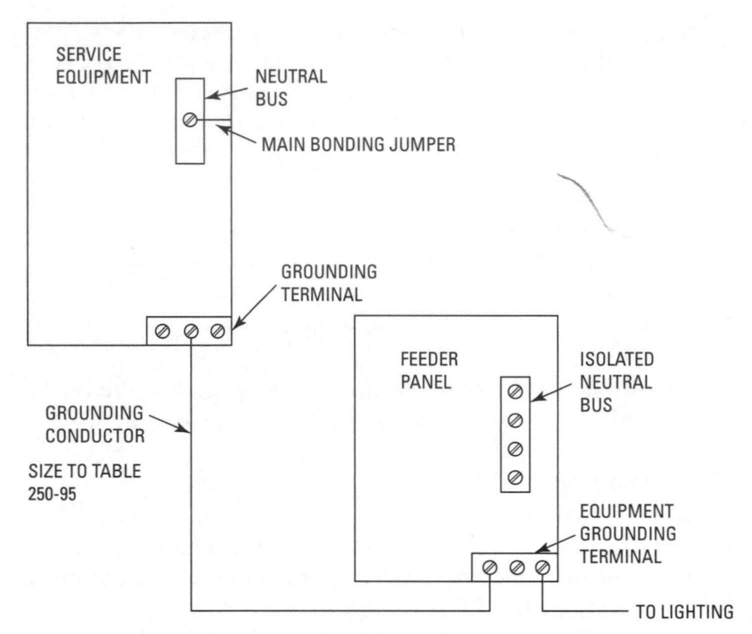

Figure 680-7 Grounding feeder panel to panelboard.

(B) Wiring Methods. The motor and controller must be connected to a circuit that is protected by a ground-fault interrupter.

680.27: Deck-Area Heating

This section applies to pool deck areas, and will include the covered pool when electrical comfort heating units are installed within 20 feet of the inside of the pool.

(A) Unit Heaters. If unit heaters are used, they are to be rigidly mounted to the structure and must be totally enclosed or of a guarded type. They shouldn't be mounted over the pool or over the area around the pool that is within 5 feet horizontally of the inside wall of the pool.

(B) Permanently Wired Radiant Heaters. If radiant electric heaters are used, they must be securely and suitably guarded and fastened to their mounting devices. They shouldn't be mounted over the pool or within 5 feet from the inside wall of the pool, and shall be mounted not less than 12 feet vertically above the pool deck unless approval for other type of mounting is given.

(C) Radiant Heat Cables Not Permitted. No radiant heaters are permitted to be mounted in or under the deck.

III. Storable Pools

680.30: Pumps

Any cord-connected pool filter pump must be double-insulated (or the equivalent), and only the internal metal parts that don't carry current may be grounded. An equipment-grounding conductor and receptacle-and-cap assembly with a grounding prong can be used to accomplish this internal grounding.

680.31: Ground-Fault Circuit Interrupters Required

All electrical equipment (including cords) used with storable pools must be protected with ground-fault interrupters.

680.32: Lighting Fixtures

Lighting fixtures that are installed in the walls of storable pools must be cord- and plug-connected and must:

(1) Have no exposed parts.

(2) Have a lamp that operates at 15 volts or less.

(3) Have an impact-resistant polymeric (plastic) body, lens, and transformer enclosure.

(4) Have a transformer with a primary rating of no more than 150 volts, and conforming to Section 680.5(A).

(5) Be listed for the purpose.

IV. Spas and Hot Tubs

680.40: Outdoor Installation

The provisions of Parts I and IV cover a spa or hot tub installed outdoors.

Exception No. 1

This allows the exemption covered in Section 680.22 for metal bands or hoops used to secure the wooden staves.

Exception No. 2

You may use a cord-and-plug connection if its package is listed for such use, but the cord shall be no longer than 15 feet and shall be protected by GFCI.

Exception No. 3
Metal-to-metal bonding on a common frame shall be permitted.

680.41: Indoor Installations
Spas and hot tubs that are installed indoors must meet the requirements of this part, and the wiring connections to the same shall comply with Chapter 3.

Exception
Listed packaged units that are rated 20 amperes or less may be cord- and plug-connected to facilitate removal and disconnection for maintenance and repair.

(A) Receptacles.

(1) Receptacles that are mounted or are located on the property shall be at least 5 feet from the inside of the spa or the hot tub.

(2) GFCI shall be used to protect 120-volt receptacles that are located within 10 feet of the inside wall of the spa or hot tub.

Note
The distances covered above are to be the shortest path that the supply cord for the appliance to be connected to the receptacle would follow. Piercing a floor, wall, or ceiling of a building or other effective barrier is not permitted.

(3) GFCIs must be used on receptacles that provide the power for the hot tub or spa.

(B) **Lighting Fixtures and Lighting Outlets.**

(1) Any lighting fixtures or outlets over the spa or hot tub or that are within 5 feet of the inside wall of the spa or hot tub are to be a minimum of 7½ feet above the maximum level of the water, and shall be protected by GFCIs.

Exception No. 1
If lighting fixtures, lighting outlets, and ceiling fans are located a minimum of 12 feet above the maximum water level, they shouldn't require a GFCI for protection.

Exception No. 2
Lighting fixtures meeting (a) or (b) below and protected by a GFCI, may be installed less than 7½ feet over a spa or hot tub: (a) They must have a glass or plastic lens and shall be recessed, and the trim shall be of a

nonmetallic material or electrically-isolated metal trim that may be used in wet locations. (b) If they are surface-mounted, fixtures shall be equipped with a glass or plastic globe and nonmetallic body recognized for use in wet locations.

(2) The provisions of Part II of this article must be complied with for underwater lighting fixtures.

(C) **Wall Switches.** Wall switches must be mounted at least 5 feet away from the inside of the spa or hot tub.

(D) **Bonding.** Listed below are the parts that shall be bonded together:

(1) All metal fittings, when attached to or within the spa or hot tub structure.

(2) All metal parts of any of the electrical equipment used with the spa or hot tub for water circulating systems, and including pump motors.

(3) If not separated from the spa by a suitable barrier, metal conduit and piping that are within 5 feet horizontally from the inside of the spa or hot tub.

(4) Unless a permanent barrier is installed, all metal surfaces that are within 5 feet of the spa or hot tub.

(5) Any electrical device or controls that are not used in association with the hot tub or spa are to be maintained at a minimum distance of 5 feet from such units, or shall be bonded to the spa or hot tub system.

(E) **Methods of Bonding.** Any of the following means may be used for bonding all metal parts associated with the spa or hot tub: the interconnection of threaded metal piping and fittings, metal-to-metal mounting when on a common base or frame, or the use of a copper bonding jumper that isn't smaller than No. 8 solid and may be insulated, covered, or bare.

(F) **Grounding.** Grounding is required on the following:

(1) Any electrical equipment that might be located within 5 feet of the inside of the spa or hot tub.

(2) The electrical equipment used with a circulation system of the spa or hot tub.

(G) **Methods of Grounding.**

(1) Article 250 must be complied with for grounding all electrical equipment, and methods covered in Chapter 3 shall be used for connecting wiring.

(2) If equipment is connected by means of a flexible cord, the equipment-grounding conductor in the cord shall be connected to the metal part of the assembly.

(H) Electric Water Heaters. Spa or hot tub electric water heaters must be listed and have the heating elements subdivided into loads that won't exceed 48 amperes, and they shall be protected by not more than 60-ampere overcurrent devices.

The branch-circuit conductor's ampacity and the rating or setting of the overcurrent protection devices must be not less than 125 percent of the total load listed on the nameplate.

V. Fountains

680.50: General

Water fountains are defined in Section 680.4, and Part V applies to water fountains that have water common to the pool.

Exception

Part V does not cover self-contained portable fountains that are no larger than 5 feet in any dimension.

680.51: Lighting Fixtures, Submersible Pumps, and Other Submersible Equipment

(A) Ground-Fault Circuit Interrupter. A GFCI must be installed in the branch circuit that supplies fountain equipment.

Exception

If a circuit of 15 volts or less is supplied by a transformer that complies with Section 680.5(A), a GFCI won't be required for equipment operation. It will be recalled that this type of transformer, which has two windings and a grounded metal barrier between the windings, must also be an approved type.

(B) Operating Voltage. Lighting fixtures shall be installed for operation at 150 volts or less between conductors. Submersible pumps and any other submersible equipment shall operate at 300 volts or less between conductors.

(C) Lighting Fixture Lenses. Lighting fixtures shall be installed so that the top of the lens is below the normal water level of the fountain, unless they are approved for above-water-level locations. A lighting fixture facing upward must have the lens properly and adequately guarded to prevent contact by persons.

(D) Overheating Protection. A low-water cutoff or other approved means if the water level drops below the normal level shall be provided for electrical equipment that depends upon submersion for protection against overheating.

(E) Wiring. Equipment shall be provided with threaded conduit entries or with flexible cords. The exposed cord may not be over 10 feet in length. If the cords extend beyond the fountain perimeter, they are to be enclosed in approved wiring enclosures. Any metal parts or equipment that make contact with the water are to be of brass or other corrosion-resistant metal.

(F) Servicing. All equipment used in the fountain shall be capable of being removed from the water for relamping or any normal maintenance. Fixtures may not be permanently embedded in the fountain structure so that the water level has to be lowered or the fountain drained when the relamping or maintenance is required.

(G) Stability. Equipment must be inherently stable or securely fastened in place.

680.52: Junction Boxes and Other Enclosures

(A) General. When junction boxes or other enclosures are installed for other than underwater installation, they must comply with Section 680.21(A)(1)–(A)(3), and (B), (C), and (D) of the same section.

(B) Underwater Junction Boxes and Other Underwater Enclosures. Junction boxes and other underwater enclosures immersed in water or exposed to water spray shall comply with the following: (1) They shall be equipped with provisions for threaded conduit entries or compression glands or seals for cord entry; (2) they shall be copper, brass, or other approved corrosion-resistant material; (3) they shall be located below the water level in the fountain wall or floor. An approved potting compound shall be used to fill the box to prevent the entry of moisture; and (4) when the junction box is supported only by the conduit, the conduit shall be of copper, brass, or other approved corrosion-resistant metal. When the box is fed by nonmetallic conduit, it shall have additional supports and fasteners of copper, brass, or other approved corrosion-resistant material. The box must be firmly attached to the supports or directly to the fountain surface and bonded as required.

Note

Refer to Section 314.13, which covers this type of enclosure.

680.53: Bonding

All metal piping associated with the fountain must be bonded to the equipment-grounding conductor of the branch-circuit supply power to the fountain.

Note

The sizing of these equipment-grounding conductors is covered in Section 250.95.

680.54: Grounding

The following equipment must be grounded: (1) all electrical equipment mounted within the fountain or within 5 feet of the inside wall of the fountain; (2) all equipment that is a part of the recirculating system of the fountain water; and (3) panelboards that are not a part of the service equipment that supply power to any of the equipment involved with the fountain.

680.55: Methods of Grounding

- **(A) Applied Provisions.** Section 680.25 applies, but this excludes paragraph (E) of that section.
- **(B) Supplied by a Flexible Cord.** Any equipment that is supplied by flexible cords must have noncurrent-carrying exposed metal parts grounded, and this shall be by means of an insulated, copper equipment-grounding conductor that is integral to the wiring of the cord. This grounding conductor must be connected to a grounding terminal in the supply junction box, transformer enclosure, or any other enclosure.

680.56: Cord- and Plug-Connected Equipment

- **(A) Ground-Fault Circuit Interrupter.** GFCIs are required on all electrical equipment, and include power supplied by cords.
- **(B) Cord Type.** Type SO or ST flexible cords shall be used when they are exposed to water.
- **(C) Sealing.** Suitable potting compound must be used to prevent water entering the equipment through the cord or its

conductors; this shall be done by covers or potting compound wherever the flexible cord jacket and flexible cord conductor terminate in equipment. Also, in addition to the above, the equipment-grounding conductor is to be similarly treated to protect it from possible deterioration from water entering the equipment.

(D) **Terminations.** Permanent connection of flexible cord may be used, except that grounding-type receptacles and attachment plugs shall be permitted for aiding in the removal or disconnection for repair and maintenance of equipment not located in any water-containing part of a fountain.

VI. Therapeutic Pools and Tubs in Health Care Facilities

680.60: General

See Section 517 for definition of health care facilities. Part VI is intended to cover pools and tubs in health care facilities. Portable therapeutic appliances must comply with Article 422.

680.61: Permanently Installed Therapeutic Pools

Parts I and II of this article must be complied with for therapeutic pools that are constructed in or on the ground or that may be located in a building in a manner so that the pool can't be readily taken apart.

Exception

If lighting fixtures are of the totally enclosed type, the limitation put on them in *NEC* Section 680.6(B)(1) won't be applicable.

680.62: Therapeutic Tubs (Hydrotherapeutic Tanks)

If patients are treated in therapeutic tanks that are not easily moved from one place to another in the normal sequence of use, or if they are fastened or otherwise secured to a particular location, including associated piping systems, they shall comply with this part.

(A) **Ground-Fault Circuit Interrupter.** GFCIs shall be used for the protection of all therapeutic equipment.

Exception

Compliance with Section 250.45 is required for portable therapeutic appliances.

(B) Bonding. You must bond the parts listed below. See the *NEC* for (1) through (5).

(C) Methods of Bonding. The metal parts that are associated with therapeutic tubs are to be bonded by methods as follows: the interconnection of threaded metal piping and fittings; metal-to-metal contact mounting on a common base or frame; connections made by suitable metal clamps; or by solid copper bonding jumpers, insulated, covered, or bare that are not smaller than No. 8 AWG.

(D) Grounding. The equipment below shall be grounded:

- **(1)** Any electrical equipment located 5 feet or less from the inside of the tub.
- **(2)** All electrical equipment that in any way is associated with the circulating system in the tub.

(E) Methods of Grounding.

- **(1)** Article 250 spells out how electrical equipment shall be grounded, and the connections are covered in Chapter 3.
- **(2)** With fixed metal parts of the assembly that are supplied by flexible cord, the grounding conductor of the cord shall be used for the grounding purpose.

(F) Receptacles. Any receptacle that is within 5 feet of a therapeutic tub must be protected by a GFCI.

680.63: Lighting Fixtures

Totally enclosed lighting fixtures are required when installed near therapeutic tub area.

VII. Hydromassage Bathtubs

680.70: Protection

A GFCI shall be used to supply hydromassage bathtubs and their associated electrical components.

680.71: Other Electric Equipment

The requirements of Chapters 1 through 4 in the *NEC* cover the installation of equipment in a bathroom applying to fixed lighting, switches, receptacles, and other electrical equipment that is located in the same room and area, but may not be directly associated with a hydromassage bathtub.

Article 685—Integrated Electrical Systems

I. General

685.1: Scope

Integrated electrical systems, but not unit equipment that is not integrated with other unit equipment, is covered by this article. In an integrated system, it is essential that an orderly shutdown of the equipment occur in industrial operations where one unit of the operation depends on other unit(s) for their proper operation. Thus the following conditions must be met: (1) In order to keep hazards to personnel and damage to the equipment low, an orderly shutdown is required; (2) there must be qualified persons to service the system and maintain and supervise it properly; and (3) the authority that has jurisdiction is assured that effective safety guards are acceptable.

685.2: Application of Other Articles

In the *Code*, other articles contain information that apply to an orderly shutdown in addition to the information contained in this article or are modifications of them.

See the *NEC* for a list of applicable sections for integrated systems.

II. Orderly Shutdown

685.10: Location of Overcurrent Devices in or on Premises

Overcurrent devices that are necessary for the operation of integrated circuits may be installed at heights that are accessible for security of operation to nonqualified personnel.

685.12: Direct-Current System Grounding

Two-wire DC circuits will be permitted to be ungrounded.

685.14: Ungrounded Control Circuits

For the purpose of control, operational continuity of service is required. If the voltage of the control circuits is 150 volts or less, they may be supplied from separately derived systems and shouldn't be required to be grounded.

Article 690—Solar Photovoltaic Systems

Solar photovoltaic systems are not exactly a new method of power generation, but much progress is being made. This technology has been used for many years in photolight meters to measure light intensity by photoelectric cells. Streetlights, yard lights, and the like have been turned on and off by the same means, that is, by a cell that gives off a voltage when exposed to light.

As with the beginning of all new technology, photovoltaic cells are being improvised, and numbers of these cells are connected in series, parallel, or series and parallel to produce larger current supplies and higher voltages. The current produced is direct current, which, in turn, in most cases must be changed to alternating current and electronically changed to 60 hertz.

See the *NEC*.

Chapter 7

Special Conditions

Article 700—Emergency Systems

I. General

There are three articles on emergency systems; so in applying the regulations that cover emergency systems, be certain you have the applicable article(s): Article 700, 701, or 702.

700.1: Scope

This article applies to design for electrical safety, installation, operation, and maintenance. The specific application is emergency systems made up of circuits and equipment, the intent of which is to supply, distribute, and control electricity for illumination and/or power that will be required if the normal service of electricity is interrupted.

Emergency systems are those systems legally required and classed as emergency by municipal, state, federal, or any other codes, or any other governmental agencies under whose jurisdiction they may come. The intent is that these emergency systems automatically supply electricity for illumination and power to areas designated to require emergency power. If the normal supply of power is interrupted for any reason at all, then the emergency system takes over to distribute and control power and illumination that is required for safety to human life.

Note

Further information on the installation of emergency systems is covered in Article 517.

Note

More information on performance and maintenance required of emergency systems can be found in *Health Care Facilities, NFPA 99-2002 (ANSI)*.

Note

Emergency systems are quite often installed in places of assembly to provide illumination in the event of a normal power outage so that there will be a means of safe exit and panic control in those buildings, which may be occupied by a large number of persons, such as hotels, theaters, sports arenas, health care facilities, and so on. Emergency systems may supply power for ventilation that may be essential for sustaining life, for fire protection and alarm systems, elevators, fire pumps, public safety communications, or industrial processes where interruption of current could cause serious life, safety, or health hazards.

Note
See *Life Safety Code, NFPA 101-2000 (ANSI)* for specifications where emergency lighting is required for safety.

Note
Further information on emergency and standby power systems and their performance can be found in *Emergency and Standby Power Systems, NFPA 110-2002.*

Reference is made to *NFPA Life Safety Code (NFPA No. 101)* for specification of locations where emergency lighting is considered essential to life safety.

700.2: Application of Other Articles

The requirements of the *National Electrical Code* as covered elsewhere in the *Code* are applicable, except where modified by this article.

700.3: Equipment Approval

Approval of all equipment is required when used on emergency systems.

700.4: Tests and Maintenance

(A) **Conduct and Witness Test.** The authority that has jurisdiction should inspect the emergency system upon installation, and periodically thereafter.

(B) **Tested Periodically.** In order to ensure proper operation of an emergency system, the authority that has jurisdiction shall require periodic tests to be conducted on the system to ensure its maintenance and proper operating condition.

(C) **Battery System Maintenance.** The authority that has jurisdiction shall require periodic maintenance on battery systems and unit equipment, including batteries used for starting or ignition of auxiliary engines.

(D) **Written Record.** Records of tests are to be kept in writing, and shall cover maintenance and operation of same.

(E) **Testing Under Load.** This requires a means of testing lighting and power emergency systems while they are under actual maximum load conditions.

700.5: Capacity

(A) Capacity and Rating. The capacity of the emergency system shall be adequate to handle the requirements of all the equipment to be operated simultaneously. The equipment must be sufficient to handle the available fault current.

(B) Selective Load Pickup and Load Shedding. The alternate power source may supply emergency service, legally required standby, and other optional standby system loads where automatic selection of pickup and shedding of power is provided as needed to ensure that adequate power is available to (1) emergency circuits; (2) legally required standby circuits; and (3) the optional standby circuit; and it shall be done in this order of priority.

Note

The test requirements of Section 700.4(B) are covered for the load sharing operation to satisfy test requirements when all conditions are met as covered in Section 700.4.

A portable or temporary alternate source of power must be available whenever the emergency generator has to be taken out of service for major maintenance.

700.6: Transfer Equipment

The transfer equipment must be automatic and identified for emergency service, or approved by the authority that has jurisdiction. The transfer equipment shall be so designed and installed that accidental transfers—interconnecting the normal and the emergency source—cannot result in any operation of the transfer equipment. You are referred to Section 230.83.

A means for isolating the transfer switch is permitted. If isolation switches are provided, accidental parallel operations shall be avoided.

700.7: Signals

For the following purposes, audible and visible signaling devices shall be used where practical:

(A) Derangement. To give warning of a derangement or nonfunctioning of the emergency or auxiliary system.

(B) Carrying Load. To indicate if the batteries or generator are ready to carry the load should it become necessary for the emergency or auxiliary system to take over.

(C) Not Functioning. To indicate if the battery charger is functioning properly. The batteries and their condition should also be checked—this will come under the periodic checks required to be taken and written up in the log that is kept of the auxiliary or emergency system.

(D) **Ground Fault.** In solidly grounded wye emergency systems that are more than 150 volts to ground, protective devices rated at 1000 amperes or more are used to indicate a ground fault. The sensor to indicate the ground-fault signal device shall be mounted ahead of the disconnecting means for the emergency source, and the maximum setting of this sensing device shall be for a ground-fault current of 1200 amperes. Instructions shall be issued for the required action to take in case indicated ground fault shall be located at or near the sensor location.

The *Code* refers you to *NFPA 110-2002 (ANSI)* for signals for generator sets.

700.8: Signs

A sign shall be posted at the service-entrance equipment indicating the type and location of the on-site emergency power source.

II. Circuit Wiring

700.9: Wiring, Emergency System

(A) **Identification.** All boxes and enclosures that contain emergency circuits are to be marked so that they will be easily identified as being a part of the emergency circuit.

(B) **Wiring.** Emergency source wiring, including its means of disconnecting overcurrent protection supplying the emergency load, is to be kept entirely separate from all other wiring and equipment, raceways, cables, and cabinets that contain other than emergency wiring. These systems must be located and designed to avoid damage due to vandalism.

Exception No. 1

Wiring in transfer equipment enclosures.

Exception No. 2

Exit for emergency lighting fixtures that are supplied from two sources.

Exception No. 3

Common junction boxes for exit lights and emergency lighting when supplied by two sources of power.

Exception No. 4
Two or more circuits supplied by the same emergency power may be run in the same raceway, box, cable, or cabinet.

Exception No. 5
In a common junction box attached to unit equipment that contains only the branch circuit supplying that equipment and the emergency circuit supplied by the unit equipment.

III. Sources of Power

700.12: General Requirements

The same types of emergency systems might not be suitable in all cases. The conditions must be evaluated as to whether the emergency system will be needed for a long period of time or a short period of time, and how much capacity the emergency system must have to supply the emergency demands. For example, the exit lights in a theater may be needed for only a period long enough to ensure lighting while the theater is being emptied. In the case of interruption of service to a hospital, whether from within or without, the emergency system might be required to furnish a large amount of power for a long period of time. Each particular condition requires a thorough evaluation of the possible needs, the type of system, and its capacity, and so on, in addition to the requirements of the *Code*.

Whether the emergency system supplies only one building or a group of buildings, it is required that the emergency service be in operation not to exceed 10 seconds after the failure of the normal source of power.

The emergency supply system, in addition to meeting the requirements of this section, shall also be permitted to consist of one or more of the types of systems that are covered in (A) through (E) below. Unit equipment that is covered in Section 700.12(F) shall satisfy the requirements of this article.

In selecting an emergency source of power, the occupancy and the type of service required must be considered: whether it is for short duration, as for the evacuation of a theater, or for longer durations, supplying light and power for indefinite periods of failure of power from either inside or outside the building.

When designing and installing emergency systems, care must be taken to avoid all hazards that could result and cause a total failure due to floods, fires, icing, or vandalism.

Each case of designing the emergency system must be individually evaluated as to the conditions and requirement for the particular installation.

(A) Storage Battery. Storage batteries may be used for a source of emergency power supply in some systems if they are suitably rated and have capacity to supply and maintain total load for a period of 1½ hours without the voltage falling below 87½ percent of normal.

Whether the batteries are acid or alkali, they shall be designed and constructed to meet the requirements of the emergency service, and shall be compatible with the charger used to recharge them.

Automotive-type batteries are not to be used. If the battery container is of a sealed type, it does not have to be transparent. With the lead-acid–type battery that requires water additions, the container must be transparent or translucent. It must provide an automatic battery charger to keep the batteries fully charged.

(B) Generator Set.

(1) A generator is supplied by a prime mover that will be accepted by the authority that has jurisdiction, and it is sized as covered in Section 700.5. It must have automatic starting of the prime mover when the normal source of power fails, and must have a transfer switch for all electrical equipment in the emergency circuit. To prevent immediate retransfer in cases of short-term restoring of the normal source of power, a time-delay feature allowing for a 15-minute setting shall be provided.

(2) Internal combustion engines used for prime movers must have an on-site fuel supply that will function at full demand for not less than 2 hours of operation.

(3) Prime movers shouldn't rely solely upon public utility gas systems for the fuel supply. Automatic transferring means shall be provided for transferring from one fuel supply to another when a dual fuel supply is used.

Exception

When acceptable to the authority that has jurisdiction, it shall be permitted to use other than on-site fuels when there is a low probability of the failure of both the on-site fuel delivery system and the power from the outside electrical utility company.

(4) If a storage battery is used for control or signal power or as a means of starting prime movers, it must be suitable for that type of service and shall be equipped with an automatic charging means in the generator set.

(5) When an emergency generator requires more than 10 seconds to develop power, an auxiliary power supply will be acceptable to energize the emergency system until the regular generator is capable of picking up the load.

(C) Uninterruptible Power Supplies. Uninterruptible power supplies providing power for the emergency system must comply with the applicable provisions of Section 700.12(A) and (B).

(D) Separate Service. If acceptable to the authority that has jurisdiction, a second service will be permitted. This separate service shall comply with Article 230 and shall have a separate surface drop or lateral that is widely separated, both electrically and physically, from the normal service, in order to minimize the simultaneous interruption of the supply of services.

(E) Connection Ahead of Service Disconnecting Means. If acceptable to the authority that has jurisdiction, connections may be made at, but not within, the main service disconnecting means. The emergency service must be sufficiently separated from the normal service disconnection means so as to prevent simultaneous interruption of the service when there is an occurrence within the building or group of buildings.

Note

Equipment that may be connected ahead of the disconnecting means is covered in Section 230.82.

(F) Unit Equipment. Unit equipment for individual emergency illumination shall consist of (1) a rechargeable battery, often a unit system with a triple charger and lamps attached to the charger and hung in various locations; (2) a means of charging the battery; (3) provisions for attaching one or more lamps on equipment, and it is also permitted to have terminals for connecting lamps in remote locations; and (4) a relay device designed to energize the lamps immediately upon failure of the supply to the unit equipment. The batteries must be suitable for maintaining the supply at not less than 87½ percent of the normal battery voltage for the total load of lamps attached thereto and for a period of at least 1½ hours, or the equipment shall supply and maintain at least 60 percent of the emergency lighting for a period of at least 1½ hours. The supply batteries, whether alkali- or acid-type, shall be designed and constructed to meet the requirements of the emergency service.

This unit equipment is to be permanently installed and not portable, and all the wiring to the unit shall be done by wiring methods covered in Chapter 3. Flexible cord may be used if it does not exceed 3 feet in length. The branch circuit supplying the emergency units shall be the same branch circuit that supplies the normal lighting in that area, and it shall be connected ahead of any local switches controlling regular lighting. Emergency illumination fixtures, which are to obtain power from the unit equipment, are to be wired to the unit equipment by one of the wiring methods covered in Section 700.9 and Article 300.

Exception

In a separate area that is uninterrupted and is supplied by a minimum of three lighting circuits, you may install a branch circuit for the unit equipment provided it and the normal lighting circuits originate from the same panelboard, but this separate branch circuit shall be provided with a lock-on feature.

IV. Emergency Circuits for Lighting and Power

700.15: Loads on Emergency Branch Circuits

No loads, except those specifically required for the emergency service, shall be connected to or supplied from the emergency supply. Emergency equipment is not intended to take care of the entire load unless designed for this purpose.

700.16: Emergency Illumination

Emergency illumination includes means of egress lighting and auxiliary illumination required for the safe evacuation of a building or buildings. This illumination should also supply sufficient light for the purposes for which it is intended. Many new structures are being built without provisions for illumination from the outside daylight. This condition makes it even more important to have an emergency source of illumination to supply sufficient lighting for the evacuation of the building or buildings. This has been experienced in outages that covered a large area.

Emergency lighting systems shall be so designed and installed that the failure of an individual lighting element, which could be the burnout of a bulb, won't leave in total darkness any space required to have emergency illumination. The burning out of an individual lamp won't, in most cases, cause total interruption of the lighting source in any one area since most areas are supplied by more than one lamp, but there are installations where this is possible.

There are battery-operated self-contained units that have a trickle charger to keep the battery fully charged. If these are installed, they should be on the emergency supply circuit so that if a breaker should trip or a fuse blow on the regular circuit, the battery could still charge and the unit be operable until the malfunction is corrected.

700.17: Circuits for Emergency Lighting

Emergency lighting branch circuits must be installed and supplied from a source of power complying with Section 700.12, if the normal power supply is interrupted. These installations shall supply one of the following: (1) an emergency lighting supply that is independent of the general lighting supply, and has provisions for automatically transferring emergency light if the general lighting supply is interrupted, or (2) two or more complete and separate systems with independent power supply, each system supplying enough current for the purpose of emergency lighting. If both lighting systems are used for regular lighting purposes and both systems are kept lighted, there shall be means provided for automatically energizing one system if the other system fails. Either or both systems will be permitted for automatic energizing of either system upon failure of the other. Both or either of the systems shall be part of the general lighting system of the projected occupancy if circuits supplying lights for emergency illumination are installed as permitted by other sections of this article.

700.18: Circuits for Emergency Power

With branch circuits supplying equipment classed as emergency, there shall be an emergency supply source to which this load will automatically transfer in the event of loss of the normal power supply.

V. Control Emergency Lighting Circuits

700.20: Switch Requirements

The switch or switches that control emergency circuits are to be accessible only to qualified persons or persons authorized to have control of these circuits.

Exception No. 1

When two or more single-throw switches are connected in parallel and control a single circuit, at least one of these switches shall be accessible to authorized persons. This means that switches for control of emergency lighting may be paralleled but, if they are, one must be accessible only to authorized persons.

Exception No. 2

There may be additional switches in addition to the one that is controlled by authorized persons, but they shall be arranged so that the unauthorized person may put the lighting into operation but won't be able to disconnect it.

Switches connected in series, or three- and four-way switches shouldn't be used.

700.21: Switch Location

All manual switches that control emergency circuits must be easily accessible to the people responsible for their actuation. In places of assembly, such a switch must be located in the lobby. Never can a control switch be placed in a projection booth or on a stage.

Exception

If multiple switches are used, a manual switch may be installed in these locations, but it must be designed so that it can energize the circuit, but not deenergize it.

700.22: Exterior Lights

Lights mounted outside a building that are not required to be turned on when there is sufficient illumination by daylight may be actuated by an automatic light-actuating device after dark.

VI. Overcurrent Protection

700.25: Accessibility

Only authorized persons shall have access to overcurrent devices in emergency circuits. This will prevent tampering or interference with the operation of the emergency circuits.

Note

The reliability of emergency systems can be ensured if the fuses and circuits provided for overcurrent protection are coordinated so that selective clearance of fault currents is ensured.

700.26: Ground-Fault Protection of Equipment

The alternate source that supplies emergency systems is not required to have ground-fault protection of equipment according to Section 700.7(D).

Article 701—Legally Required Standby Systems

I. General

701.1: Scope

This article covers the electrical safety and design, installation, operation, and maintenance of standby systems consisting of circuits and equipment intended to supply distributing and controlling electricity to facilities that are legally required to have standby power to cover illumination and/or power should there be an interruption in the normal source of power.

The systems in this article involve only those that are permanently installed in their entirety and that also include the normal power source.

Note

See *Health Care Facilities, NFPA 99-2002 (ANSI)* for additional information.

Note

See *Emergency and Standby Power Systems, NFPA 110-2002* for further information covering the performance of emergency standby systems.

701.2: Legally Required Standby Systems

Legally required standby systems are those required by municipal, state, federal, or by other codes, or any governmental agency that has jurisdiction over the same, the intent of which is to supply power to selected loads (other than those classified as emergency systems) in the event that the normal power source fails.

Note

Typical installations of legally required standby systems are for operation to serve loads such as heating, refrigerator, communication, ventilation and smoke removal, sewage disposal, and industrial processes systems, which, if normal operation of the normal power supply fails, could create hazards or hinder rescue or fire-fighting operations.

Many industrial buildings must be designed for eliminating dust entrance and thus may not have windows. Therefore, some emergency lighting is required for such things as exit lights and lighting exit corridors.

701.3: Application of Other Articles

Except as modified by this article, all other articles of the *Code* apply.

701.4: Equipment Approval

All equipment shall be listed or approved by this article. This includes approval by the authority that has jurisdiction.

701.5: Tests and Maintenance for Legally Required Standby Systems

(A) Conduct or Witness Test. The authority that has jurisdiction shall conduct or witness a test when the installation is completed.

(B) Tested Periodically. Systems are to be periodically inspected on a schedule approved by the authority that has jurisdiction. This is to ensure their maintenance in proper condition for operation.

(C) Battery Systems Maintenance. When batteries are used for starting or igniting the prime movers, periodic maintenance is to be required by the authority that has jurisdiction.

(D) Written Record. A permanent written record must be kept of all these tests and maintenance.

(E) Testing Under Load. Legally required standby systems shall have means provided for testing under load conditions.

This gives the authority that has jurisdiction the right to make repeated inspections to ensure compliance with this article.

701.6: Capacity and Rating

A legally required standby system shall have adequate capacity and rating for supplying all equipment intended to be operated at one time.

The alternate power source may be permitted to supply legally required standby and optional standby system loads when these loads are automatically picked up for load shedding so as to ensure power to the legally required standby circuits.

701.7: Transfer Equipment

Automatic transfer equipment shall be identified or approved for standby use by the authority that has jurisdiction. The automatic transfer systems equipment must be so designed that no accidental interconnection of the normal and alternate sources of supply will occur in the operation of any transfer equipment.

Again, the authority that has jurisdiction must approve this transfer equipment.

Isolation equipment may be used to isolate the transfer equipment. If isolation equipment is used, inadvertent parallel operation shall be avoided.

701.8: Signals

Where possible, audible and visual signal devices are to be provided for the following purposes:

(A) Derangement. To notify any derangement of the standby source.

(B) Carrying Load. To show that the standby source is carrying the load.

(C) **Not Functioning.** Should the battery charger fail to function, it will cause an alarm.

Note

See *Emergency and Standby Power Systems, NFPA 110 (ANSI)*, which covers signals for generator sets.

701.9: Signs

Signs shall be located at the service entrance to indicate the type and grounding location of the legally required on-site standby power source.

II. Circuit Wiring

701.10: Legally Required Standby Systems

Legally required standby service may occupy the same raceways, cables, boxes, and cabinets as other general wiring.

III. Sources of Power

Much of this section is very similar to Section 700.12 in the preceding article.

Refer to the *NEC* for Section 701.11.

IV. Overcurrent Protection

This is the same as Sections 700.25 and 700.26.

701.15: Accessibility

Only authorized persons shall have access to overcurrent devices in emergency circuits. This will prevent tampering or interference with the operation of the emergency circuits.

Note

The reliability of emergency systems can ensure that the fuses and circuits provided for overcurrent protection are coordinated so that selective clearance of fault currents is ensured.

701.17: Ground-Fault Protection of Equipment

The alternate source that supplies emergency systems is not required to have ground-fault protection of equipment.

Article 702—Optional Standby Systems

I. General

702.1: Scope

Optional standby installation operations are covered in this article.

This covers only those standby systems that are permanently installed in place, and also includes the prime mover.

702.2: Optional Standby Systems

Optional standby systems are intended only for the protection of business or property and do not include places where life safety is dependent on performance of the system. They may be operated manually or automatically.

With brownouts and blackouts over the country, many individuals, and especially farms and milking operations, may have a standby source of power to eliminate losses from power outages.

Note

This type of system is typically installed to provide an alternate source of electrical power for facilities such as industrial and commercial buildings, farms and residences, heating or refrigeration systems, data processing and communication systems, and industrial processes, which, if stopped during a power outage, could cause serious interruption to the process or damage to the product, and so on.

702.3: Application of Other Articles

As with most special conditions, other articles of the *Code* apply unless specifically exempted by this article.

702.4: Equipment Approval

All equipment shall be listed or approved for this article. This includes approval by the authority that has jurisdiction.

702.5: Capacity

Optional standby systems shall have capacity and rating adequate to supply all the equipment intended to be operated at one time.

Note

They are needed to supply loads selected by the user.

702.6: Transfer Equipment

Transfer equipment need only be suitable for its intended use, and designed and installed so as to prevent accidental connection with normal or alternate sources of power.

Again, the authority that has jurisdiction must approve this transfer equipment.

702.7: Signals

For the following purposes, audible or visible signals may be installed:

- **(A) Derangement.** To indicate derangement of the optional source of power.
- **(B) Carrying.** To indicate that the optional source of power is carrying the load.

702.8: Signs

A sign must be placed at the service-entrance equipment and grounding location stating the type and location of the optional power source.

II. Circuit Wiring

702.9: Wiring Optional Standby Systems

The wiring from the optional standby equipment may be in the same raceways, cables, boxes, and cabinets as other general wiring.

Article 705—Interconnected Electric Power Production Sources

705.1: Scope

This article covers one or more sources of power that will operate in parallel with the primary source(s) of power.

Note

The primary source of power could be a utility supply or such other sources as on-site electric power source(s).

705.2: Definition

The following definition applies for the purpose of this article:

> **Interactive System.** Any source of electric power system that is operating in parallel and is also capable of delivering power to the primary source of the supply system.

705.3: Other Articles

Interconnected systems shall comply not only with this article, but also with applicable requirements of the following articles: See the *NEC*.

705.10: Directory

A permanent plaque or directory must be installed at every electric service and also at all the electrical power production sources that can be interconnected, and it shall denote all sources of electrical power on the premises.

Exception

If a large number of these production services are involved, they shall be designated by groups.

Since interconnected electrical power sources are rarely encountered in most wiring installation, refer to the *NEC* for the balance of this article.

Article 720—Circuits and Equipment Operating at Less Than 50 Volts

This article covers the installations that operate at less than 50 volts, either direct current or alternating current, with the exception of low-voltage systems below 50 volts, as covered in Articles 650, 725, and 760. Refer to the *NEC*.

Article 725—Class 1, Class 2, Class 3, and Class 4 Remote-Control, Signaling, and Power-Limited Circuits

I. General

725.1: Scope

This article covers remote-control, signaling, and power-limited circuits that are not part of a device or appliance. Because these circuits use far less power (and are therefore generally less dangerous) than the circuits covered in the first parts of the *Code*, they have alternate requirements.

725.2: Location and Other Articles

These circuits and associated equipment must comply with the following *Code* sections and articles:

- **(A)** Section 300.21, regarding the spread of fire or the products of combustion.
- **(B)** Section 300.22, regarding these circuits when installed in ducts, plenums, or other spaces used for environmental air.
- **(C)** Articles 500 through 516 and 517, Part IV, regarding installations in hazardous locations.
- **(D)** Article 318, which covers when these circuits are installed in cable trays.
- **(E)** Article 430, Part VI, which covers when these circuits are tapped from the load side of the motor branch circuit protective device. See Section 430.72(A).

725.3: Classifications

A remote-control, signaling, or power-limited circuit is considered to be the portion of the wiring system between the load side of the overcurrent device or power-limited supply and all connected equipment. These circuits must be Class 1, Class 2, or Class 3. The classes of circuits are as follows:

(A) **Class 1 Circuits.** These are circuits that comply with Part II of this article, and have voltage and power (measured in watts or VA) limits according to Section 725.31.

(B) **Class 2 and Class 3 Circuits.** These are circuits that comply with Part III of this article, and have their voltage and power limited according to Section 725.31.

725.4: Safety-Control Equipment

Remote-control circuits that supply safety-control equipment must be classified as Class 1 if the equipment failing would cause a fire hazard or other hazard to life. Such equipment as controls for heating or air-conditioning equipment is not to be considered safety-control equipment.

725.5: Communications Cables

Class 1 circuits cannot be run in the same cable as communications circuits. Class 2 and Class 3 circuits can be run in the same cable as communications circuits, but when so installed, they must be considered communications circuits, and must comply with the requirements of Article 800.

Exception

Cables specifically designed to be used this way can be used, with the Class 2 and 3 conductors retaining their Class 2 and 3 ratings.

725.6: Access to Electrical Equipment behind Panels Designed to Allow Access

Access to equipment cannot be obstructed by an accumulation of wires or cables.

II. Class 1 Circuits

725.11: Power Limitations for Class 1 Circuits

(A) **Class 1 Power-Limited Circuits.** The sources that supply these circuits must have a rated output of no more than 30 volts and 1000 volt-amperes (*volt-amperes*, abbreviated VA, are essentially the same thing as watts). When power sources other than transformers are used to supply these circuits, the sources must be protected by overcurrent devices that are rated at no more than 167 percent of the rated current (volt-amps divided by rated voltage). The overcurrent device may be a part of the power supply, and cannot be interchangeable with overcurrent devices of a higher rating.

(1) **Transformers.** When transformers are used to supply these circuits, they must comply with Article 450.

(2) **Other Power Sources.** Power sources for these circuits other than transformers can have a maximum power output of 2500 VA. [Note that in 725.11(A) the *rated* power is covered, and in this subsection, the *maximum possible output* is discussed.] Also, the maximum possible voltage and the maximum possible current should be supplied by each other, and the total can be no more than 10,000. These ratings must be calculated based on there being no overcurrent devices in the circuit, which won't occur in a proper installation. This set of requirements allows for a worst possible case scenario.

(B) Class 1 Remote-Control and Signaling Circuits. These circuits may operate on up to 600 volts, and the power output of the source does not have to be limited.

725.12: Overcurrent Protection

Conductors No. 14 or larger must be protected according to Section 310.15, but derating factors cannot be applied. No. 18 conductors must be protected at no more than 7 amperes, and No. 16 conductors must be protected at no more than 10 amperes.

Exception No. 1

Where different levels of protection are *required* (or permitted) by other parts of the *Code*.

Exception No. 2

Class 1 conductors that are supplied by two-wire transformer secondaries can be supplied with overcurrent protection only on the primary side of the transformer. See Section 450.3.

Exception No. 3

Class 1 circuit conductors that are No. 14 or larger and tapped from the load side of the overcurrent device of controlled power circuits are required only to have ground-fault and short-circuit protection. If the rating of the branch-circuit protective device is not more than three times the rating of the circuit conductors, no other protection is required.

725.13: Location of Overcurrent Devices

These devices must be located at the point of supply.

Exception No. 1

When the overcurrent protective device for a larger conductor also protects a smaller conductor.

Exception No. 2
When protection is provided according to Section 725.12.

725.14: Wiring Methods
Class 1 circuits must be installed according to the requirements of Chapter 3 of the *Code*.

Exception No. 1
As allowed by Sections 725.15 through 725.17.

Exception No. 2
When other articles permit or allow other methods.

725.15: Conductors of Different Circuits in the Same Cable, Enclosure, or Raceway
Class 1 circuits can occupy the same raceways, cables, and so on, as long as all the conductors are insulated for the maximum voltage of any conductor. Note that in this subsection we are referring to conductors of different circuits, not conductors of different systems. Power-supply and Class 1 circuits can be run in the same cable, raceway, and so on, only when they are associated.

Exception No. 1
In control centers.

Exception No. 2
Underground conductors in a manhole, as long as one of the following conditions is met: (a) The conductors are in a metal-enclosed or UF cable; (b) the conductors are separated from the power supply conductors by a fixed nonconductor (such as plastic tubing) in addition to the insulation of the conductor; or (c) the conductors are separated from the power supply conductors by mounting on racks or the like.

725.16: Conductors

(A) **Sizes and Use.** Conductors of No. 16 or No. 18 can be used, as long as they do not exceed the ampacities given in Section 402.5. They must be installed in a listed raceway, enclosure, or cable. Conductors larger than No. 16 must comply with Section 310.15, and flexible cords must comply with Article 400.

(B) **Insulation.** Insulation must be rated at least 600 volts. Wires larger than No. 16 must comply with Article 310, and wires No. 16 or No. 18 must have one of the following types of insulation: RFH-2, RFHH-2, RFHH-3, FFH-2, TF, TFF, TFN, TFFN, PF, PFF, PGF, PGFF, PTF, PTFF, SF-2, SFF-2, PAF, PAFF, ZF, ZFF, KF-2, KFF-2, or other types listed for such use.

725.17: Number of Conductors in Cable Trays and Raceway, and Derating

(A) **Class 1 Circuit Conductors.** When only Class 1 conductors are in a raceway, the allowable number of conductors is calculated according to Section 300.17. The derating factors of *NEC* Article 310, Note 8(a) for "Notes to Ampacity Tables 0–2000 volts," can be applied only to conductors that carry continuous loads that are more than 10 percent of their total rated current.

(B) **Power-Supply Conductors and Class 1 Circuit Conductors.** Where Section 725.15 allows power-supply and Class 1 conductors in the same raceway, cable, and so on, the number of conductors must be determined according to Section 300.17. The derating factors mentioned in (A) of this section can apply only as follows:

(1) To all conductors, when the Class 1 conductors carry continuous loads exceeding 10 percent of their rated load, and where there are more than three conductors.

(2) To only the power-supply conductors when there are more than three such conductors, and when the Class 1 conductors do not have continuous loads of more than 10 percent of their rated current.

(C) **Class 1 Circuit Conductors in Cable Trays.** For these installations, the requirements of Section 392.9 through 392.11 must be met.

725.18: Physical Protection

Where damage is likely, all conductors must be installed in rigid metal conduit, intermediate metal conduit, rigid nonmetallic conduit, EMT, MI cable, MC cable, or must be protected from damage by some other effective means.

725.19: Circuits Extending Beyond One Building

Such circuits that run aerially must meet the requirements of Article 225.

725.20: Grounding

Grounding of these circuits must be in accordance with Article 250.

III. Class 2 and Class 3 Circuits

725.31: Power Limitations of Class 2 and Class 3 Circuits

The power for Class 2 or Class 3 circuits must be either inherently limited, or limited by a combination of power source and overcurrent limitations. [See Table 725-31(A) and 725-31(B).]

725.32: Interconnection of Power Supplies
These power supplies cannot be paralleled or interconnected unless listed for such use.

725.34: Marking
Class 2 or 3 power supplies must be clearly marked.

725.35: Overcurrent Protection
When overcurrent protection is required, the devices cannot be interchangeable with devices of higher ratings. The device may be part of the power supply.

725.36: Location of Overcurrent Devices
These devices must be installed at the point of supply.

725.37: Wiring Methods on Supply Side
Such wiring must comply with Chapter 3 of the *Code*. Transformers and like devices that are supplied by light or power circuits must be protected at no more than 20 amperes.

Exception
Input leads of transformers can be as small as No. 18 if they are no more than 12 inches long.

725.38: Circuit Conductors Extending Beyond One Building
When these conductors are extended out of a building, and are subject to accidental contact with other systems that operate at over 300 volts to ground, they must meet the requirements of Sections 800.10, 800.12, 800.13, 800.30, and 800.32.

725.39: Grounding
Grounding of these circuits must be in accordance with Article 250.

725.40: Fire Resistance of Cables Within Buildings
All Class 2 and 3 cables must be listed as resistant to the spread of flames, according to Sections 725.50 and 725.51.

725.50 and 725.51
These sections give the listing and marking requirements for Class 2 and 3 and PLTC cables. These sections pertain mostly to the manufacturers of these cables. For details, refer to the *Code*.

725.52: Wiring Materials and Methods on Load Side
These conductors must comply with Section 725.50 and (A) and (B) below:

(A) Separation from Electric Light, Power, and Class I Circuit Conductors.

(1) **Open Conductors.** Class 2 or 3 conductors must be separated by at least 2 inches from the other conductors mentioned above.

Exception No. 1
When either the light, power, or Class 1 conductors, or the Class 2 or 3 conductors are enclosed in raceway, metal-sheathed, metal-clad, nonmetallic-sheathed, or UF cables.

Exception No. 2
Where the conductors of the differing systems are separated by a fixed insulator in addition to the insulation of the conductors, such as a porcelain or plastic tube.

(2) **In Cable, Cable Trays, Enclosures, and Raceways.** Class 2 or 3 conductors cannot be installed in any of these locations with the other systems mentioned in this section.

Exception No. 1
Where they are separated by a barrier.

Exception No. 2
In enclosures and the like, when Class 1 conductors enter only to connect equipment to which the Class 2 and 3 circuits are also connected.

Exception No. 3
Underground conductors in manholes, when one of the following conditions is met: (a) The power of Class 1 conductors is in UF or metal-enclosed cable; (b) the conductors are separated by a fixed insulator, in addition to the insulation of the conductors; or (c) the conductors are firmly mounted on racks or the like.

Exception No. 4
As allowed by Section 780.6, and when installed according to Article 780.

(3) **In Hoistways.** In hoistways, Class 2 or 3 conductors must be installed in rigid metal conduit, rigid nonmetallic conduit, intermediate metal conduit, or EMT.

Exception
As allowed according to Section 620.21, Exceptions 1 and 2.

(4) **In Shafts.** In these locations, Class 2 and 3 conductors must be separated by at least 2 inches from other systems.

(5) **Support of Conductors.** Raceways cannot be used to support Class 2 or 3 conductors, except as allowed by Section 300.11(B).

(B) Conductors of Different Circuits in the Same Cable, Enclosure, or Raceway.

(1) Two or More Class 2 Circuits. This is permitted, as long as every conductor is insulated sufficiently for the highest voltage of any conductor present.

(2) Two or More Class 3 Circuits. Conductors of two or more Class 3 circuits are allowed to share the same raceway, cable, and so on.

(3) Class 2 Circuits with Class 3 Circuits. This is permitted, as long as the insulation of the Class 2 conductors is equivalent to the insulation required for Class 3 conductors.

(4) Class 2 or Class 3 Circuits with Other Circuits. Class 2 or 3 circuits are allowed to be in the same raceway or enclosure as other circuits, provided they are in a jacketed cable of one of the following types: (a) Power-limited power signaling systems, according to Article 760; (b) optical fiber cables, according to Article 770; (c) communications circuits, according to Article 800; or (d) community antenna television systems and radio distribution systems, according to Article 820.

725.53: Applications of Listed Class 2, Class 3, and PLTC Cables

This section explains where each type of cable may be used. But because this information will also be found on the marking of any such cable, it won't be reiterated here.

Article 760—Fire Protective Signaling Systems

I. Scope and General

This article was new with the 1975 edition of the *NEC* and it won't be necessary to repeat it here, but your attention should be called to Section 760.1.

760.1: Scope

Covered in this article is equipment for fire alarms and fire-alarm signal systems that operate at 600 volts, nominal or less.

Note

Refer to the following NFPA Codes:

NFPA 71—Central Station Signaling Systems

NFPA 72A—Local Protective Signaling Systems

NFPA 72B—Auxiliary Protective Signaling Systems

NFPA 72C—Remote Station Protective Signaling Systems

NFPA 72D—Proprietary Protective Signaling Systems
NFPA 72E—Automatic Fire Detectors
NFPA 74—Household Fire Warning Equipment

Note
Article 725 defines Class 1, 2, and 3 circuits.

The preceding standards should be purchased from the NFPA by anyone involved with the installation or maintenance of fire-protection signaling systems. See the *NEC*.

Article 770—Optical Fiber Cables and Raceways

Optical fiber cables are being used with great success for control, signaling, and communication purposes. Optical fiber cables get involved with electrical cables, or are sometimes pulled in raceway with electric cables, so they definitely belong in the *NEC*.

770.1: Scope

This article applies to the installation of optical fiber cables along with electrical conductors. It does not cover the construction or installation of optical fiber cables, except in installations that are covered in this article.

770.2: Other Articles

(A) and (B) below cover equipment:

- **(A)** You are referred to Section 300.21, which covers the spread of fire or products of combustion.
- **(B)** Section 300.22 covers the installation of optical fiber cables in plenums, ducts, or any other air-handling equipment.

Exception to (B)
See Section 770.6(C) for an exception to (B).

770.3: Optical Fiber Cables

By the transmission of light through optical fiber cables, they can be used for control, signaling, and communication.

770.4: Types

There are three types of optical fiber cable:

- **(A)** **Nonconductive.** These cables do not contain any metallic members or other materials that conduct electricity.

(B) Conductive. These cables are noncurrent carrying, but have conductive members such as metallic strength members and metallic vapor barriers.

(C) Composite. These carry optical fiber cables and also current-carrying conductors; they may be classified as electrical cables according to the type of electrical conductors.

770.5: Optical Fibers and Electrical Conductors

(A) With Conductors for Electric Light, Power, or Class 1 Circuits. Optical fiber will be permitted in the same composite cable for use in electric lighting, power, or Class 1 circuits provided they operate at 600 volts or less, and only where optical fibers and the electrical conductors are associated. Nonconductive optical fiber cables may occupy the same raceway or cable tray as conductors for light and power or Class 1 circuits rated 600 volts or less. Conductors in composite optical fiber cables may not be in the same raceway as cable tray for electrical light, or power, or Class 1 circuits.

Nonconductive optical fiber cables shouldn't be allowed to occupy the same cabinet, panel, or any other enclosure with electrical terminations for electrical light, and power, or Class 1 circuits.

Exception No. 1

If nonconductive optical fiber cables are functionally associated with the light, power, or Class 1 circuits, they may occupy the same cabinet, panel, outlet box, or similar enclosure.

Exception No. 2

If nonconductive optical fiber cables are installed in either factory- or field-assembly control centers, the optical fiber cables may occupy the same panel, outlet box, or similar enclosure.

Exception No. 3

In industrial locations where maintenance and supervision ensures that only qualified persons will be servicing the installation, nonconductive optical fiber cables may be with circuits exceeding 600 volts.

Section 300.17 covers the installation of optical fiber cables in raceways.

(B) With Other Conductors. If any of the following are complied with, optical fiber cables shall be permitted in the same cable, and conductive and nonconductive fiber cables shall be permitted in the same raceway, cable tray, or enclosure with other conductors:

(1) If in compliance with Article 725, Class 2 and 3 remote control, signaling, and power-limiting circuits may be installed.

(2) If in compliance with Article 760, signaling systems that are power-limited for fire-protective use are permitted.

(3) If in compliance with Article 800, communication circuits are permitted.

(4) If in compliance with Article 820, distribution systems for community antenna for television and radio are permitted.

(C) Grounding. Noncurrent-carrying conductive parts of optical fiber cables shall be grounded and meet the requirements of Article 250.

770.6: Fire Resistance of Cables

2005

(A) Wiring Within Buildings. When installed in buildings, optical fiber cables shall be Type OFC or OFN and be listed as resistive to the spread of fire. If optical fiber cables run vertically in a shaft, Section 770.6(B) applies, and if cables are installed in ducts, plenums, or other air-handling spaces, Section 770.6(C) applies. Types OFCR and OPNR cable are listed for use in vertical runs as covered in Section 770.6(B). If Types OFCP and OFNP cables are listed for use in ducts, plenums, or any other air-handling space and comply with Section 770.6(E), they may be used as requirements of this section.

Optical fiber raceways are also permitted. These must be listed raceways, installed only in areas they are listed for, and must be installed according to the requirements of Article 362.

Note

UL 1581, known as the vertical-tray flame test, is one method of checking the spread of fire for optical fiber cables.

Exception No. 1

If optical fiber cables are enclosed in raceways or noncombustible tubing, they may be used in the above locations.

Exception No. 2
If the cable does not exceed 10 feet in nonconcealed spaces.

(B) In Vertical Runs. Optical fiber cables shall (1) be Type OFCR or OFNR when run vertically in a shaft, and (2) listed as having fire-resistant characteristics that are capable of preventing the spread of fire from floor to floor. Types OFCP and OFNP cable, which are listed for use in ducts, plenums, and other air-handling areas as covered in Section 770.6(C), shall meet the requirements of this section.

Exception
They are permitted if encased in noncombustible tubing, or the shaft in which they are located is fireproof shaft having firestops at each floor.

(C) In Ducts, Plenums, and Other Air-Handling Spaces. Optical fiber cables shall comply with Section 300.22 as to installation methods where they are installed in ducts, plenums, or other spaces used to handle environmental air.

Exception
Types OFCP and OFNP optical fiber cable that have been listed as having adequate fire resistance and a low smoke-producing characteristic will be permitted for ducts and plenums as described in Section 300.22(B) and other spaces used for environmental air as described in Section 300.22(C).

Note
One method of establishing low-smoke-producing cables to a value that is acceptable is covered in *NFPA 262*, which tests to a maximum of 0.5 peak optical density and a maximum optical density of 0.15. Another similar fire-resistance test for cables is covered in the *NFPA 262* test, and can be defined as having a maximum allowable flame travel distance of 5 feet.

770.7: Grounding of Entrance Cables
Where the noncurrent-carrying members of optical fiber cable are exposed to hostile contact with light or power conductors, they are to be grounded or interrupted as close as possible to the entrance by using an insulating joint or equivalent device.

The point of emergence from an exterior wall, concrete floor slab, or from a rigid metal conduit or an intermediate metal conduit, grounded in accordance with Article 250, shall be considered for this section the point of entrance.

770.8: Cable Marking

See *NEC* Table 770-8, "Cable Markings," for the marking of listed optical fiber cables.

Article 780—Closed-Loop and Programmed Power Distribution

This article was added to the *Code* to allow for the Smart House wiring system, which is being developed by the National Association of Home Builders. This system uses microprocessors to control power to each outlet. The receptacles and devices that plug in actually communicate with each other. Special cables and receptacles are used. The receptacles can be used for not only 120-volt power equipment, but also for plugging in telephones, security devices, or televisions.

While this article has been part of the *Code* since 1987, these systems are not yet available.

Chapter 8

Communications Systems

Article 800—Communication Circuits

I. General

2005

800.1: Scope

This article covers communications circuits, such as telephone systems, and outside wiring for fire and burglar alarm systems. It also covers computer networks, but not computers (covered in Article 645).

800.2: Definitions

General definitions are contained in Article 100 of the *NEC*. The definitions in Article 800 are specific to communication systems. The important definitions in this section of the *Code* are:

> **Abandoned Communications Cable.** Cables that are installed but unterminated and untagged.
>
> **Block.** A city block.
>
> **Point of Entrance.** The point at which a cable enters a building, such as through a wall or floor.
>
> **Premises.** User-owned land and buildings on the user side of the point of demarcation. (The *point of demarcation* is where ownership of the system changes from the utility to the user. Usually this is some type of box or panel.)

800.3: Hybrid Power and Communications Cables

Hybrid cables contain both communications circuits and power circuits under the same cable jacket. They are covered by Section 780.6 for use in closed-loop and programmed power applications, and in Section 800.51(I) for other applications. Note also that there is a similar type of cable used for optical fiber, which is called a *composite* cable. This is covered by Section 770.5(C).

800.4: Equipment

All equipment connected to a communications network must be listed for the purpose and the installation must comply with Section 110.3. Test equipment is exempt from this requirement.

800.5: Access to Equipment behind Panels Designed to Allow Access

Access to electrical equipment cannot be obstructed by accumulated communication cables. This includes the areas above suspended ceilings.

As communications cables have proliferated, they have frequently overloaded electrical rooms and similar areas. This is specifically prohibited by the *Code*.

800.6: Mechanical Execution of Work

All communications circuits must be installed in a neat and workmanlike manner. (See commentary on Section 110.12.)

This section gives specific standards for "workmanship," requiring that cables be supported by methods that will not cause damage. Section 300.4 also applies to the protection of cables from physical damage.

800.8: Hazardous (Classified) Locations

Any communications circuits installed in hazardous areas (as defined in Article 500) must comply with all applicable requirements of Chapter Five of the *NEC*.

II. Conductors Outside and Entering Buildings

800.10: Overhead Communications Wires and Cables

(A) **On Poles and In-Span.** When communications and power conductors are installed on the same pole, four rules apply:

- **(1)** The communications conductors must be installed below the power conductors. While an exception is made for necessities, this is a basic safety issue; workers should not be exposed to the danger of electric power cables unless necessary. This is especially important when the workers are from the communications industry and are not fully educated in dealing with the elevated voltages of power systems.
- **(2)** Communications conductors may not be installed on cross-arms.
- **(3)** The climbing-space through communications cables must meet the requirements of Section 225.14(D).

(4) When communications conductors are run below and parallel to electrical service drops, they must be separated by 12 inches (300 mm) at all points.

(B) Above Roofs. Communications conductors must have a vertical clearance of at least 8 feet (2.5 m) above relatively flat roofs. Conductors above auxiliary buildings such as garages are exempt from this rule. Above a roof with a slope of 4⁄12 (4-inch rise over 12-inches horizontally), a clearance of 3 feet (900 mm) is permitted. If the conductor run terminates at a through-the-roof raceway (usually a service riser), clearance can be reduced to 18 inches (450 mm) for the last 4 feet (1.2 m) of a run above a roof overhang.

800.11: Underground Circuits Entering Buildings

(A) With Electric Light or Power Conductors. If underground communications conductors are installed in a raceway, handhole, or manhole that also contains power wiring, Class 1, or nonpower-limited fire alarm conductors, the communications equipment must be separated from the power equipment by brick, tile, or concrete partitions. The *Code* permits, but does not specify, other suitable barriers.

(B) Underground Block Distribution. When an entire communications circuit is contained in a single city block, the protection mandated by Sections 800.12(A) and 800.12(C) will not be required.

800.12: Circuits Requiring Primary Protectors

Communications circuits that require primary protectors (see Section 800.30) must comply with the following:

(A) Insulation, Wires, and Cables. Unshielded cables that run from the last outdoor support to the protector must be listed for the purpose and must have an ampacity meeting the requirements of either Section 800.30(A)(1)(b) or Section 800.30(A)(1)(c). Note that the run described in this section [Section 800.12]is the typical communications service drop to a building.

(B) On Buildings. When run parallel on the side of a building, communications conductors must be kept at least 4 inches (100 mm) away from open power conductors. This means that communications conductors do not require separation from power conductors run inside of a conduit, only from open wires or cables. Conductors that might be exposed to accidental contact with power conductors, such as the drop described in Section 800.12(A), may not be directly attached to the building, but must be attached to glass or porcelain insulators. See the exception to this section in the *NEC* for an unusual case where separation from the building is not required.

(C) **Entering Buildings.** When the primary protector is located inside the building, special attention must be paid to avoid damage from high voltages (whether from contact with power lines or from the effects of lightning). In these cases, insulating bushings, metal-sheathed cables, or metal raceways are required where the communications conductors enter the building. This is not required if the wall is entirely of masonry, or for some unusual installations involving shielded cables and the omission of fuses. [See this section in the *NEC* and Section 800.30(A)(1) for details.] Through-the-wall passages must have an upward slope toward the inside, so that water will drip out of the building rather than into it. If this is not possible, drip loops must be provided on the outside of the building. Raceways must be grounded and fitted with a service head.

800.13: Lightning Conductors

Communications conductors must be kept at least 6-feet (1.8 m) from lightning-protection system conductors.

Author's Note

This is necessary to prevent flashovers (arcs) from the lightning conductor to the communications conductors in the event of a direct lightning strike. The *Code* does excuse this requirement if it is "impractical," but the author recommends that you follow it if at all possible. Though very rare, deaths have occurred from such events.

III. Protection

800.30: Protective Devices

Protectors are surge-suppression devices, generally simple MOVs (metal-oxide varistors), which divert excess voltages directly to ground, and away from the system they protect. Telephone companies have long used these to protect their systems from power surges, which are of three specific origins:

- Accidental contact with power lines.
- Direct lightning strikes to the system. (Obviously an MOV alone would never be able to handle a direct lightning strike, but at some distance it would protect a single building from a strike to the communications system.)
- Induced voltages from nearby lightning strikes.

(A) Application. Protectors must be installed on every circuit, overhead or underground, that is not contained in a single block. They must also be used on any circuit (even if contained in a single block) that could be exposed to power conductors operating at more than 300 volts to ground. Except in areas not exposed to significant numbers of lightning strikes (the *Code* describes this in a Fine Print Note, but only mentions "areas along the Pacific coast"), circuits run between buildings must have protectors.

(1) Fuseless Primary Protectors. This type of protector may be used:

(a) If the conductors enter a building in shielded cables and the conductors will melt if currents exceed the rating of the protector.

(b) If the conductors begin in shielded cables and extend to other buildings. In addition, the conductors must melt if currents exceed the rating of the protector.

(c) If the conductors begin in unshielded cables and extend to other buildings. In addition, the conductors must melt if currents exceed the rating of the protector and the protector must be listed for this use.

(d) Where communications conductors go from an underground run to an aerial service drop to a building.

(2) Fused Primary Protectors. This type of protector is required where the conditions above do not exist.

(B) Location. Much like the requirements for mail service disconnects, protectors must be located as close as possible to the point of entrance. The *NEC* requires that protectors for mobile homes be within sight and within 30 feet (9 m) of the exterior wall of the mobile home, or at the disconnect.

(C) Hazardous (Classified) Locations. Protectors are not permitted in hazardous locations. See *NEC* Sections 501.14, 502.14, and 503.12 for exceptions.

800.33: Cable Grounding

Metal sheaths on communications cables must be either grounded or interrupted as close as possible to the point of entrance. A suitable insulating joint must be used.

IV. Grounding Methods

800.40: Cable and Primary Protector Grounding

(A) Grounding Conductor. These conductors must (1) have suitable insulation, (2) be a minimum of No. 14 copper (or some other corrosion-resistant conductor), and (3) be as short as possible, not exceeding 20 feet (6 m). The NEC does list one exception to this rule for one-and two-family dwellings only. Also, these conductors must be run in as straight a line as possible, and must be protected from physical damage. If metal raceway is used to provide protection, it must be bonded to the conductor, or to the terminal at which the conductor is attached.

(B) Electrode.

- **(1) In Buildings or Structures with Grounding Means.** If a building has a grounding system (usually the for the power system), the following are considered suitable electrodes for connecting the grounding conductor:
  - **(a)** The power grounding system (See Section 250.50.)
  - **(b)** A grounded interior metal water pipe, within 5 feet of its entrance.
  - **(c)** A service enclosure for the power system. (See Section 250.94.)
  - **(d)** A metal service raceway.
  - **(e)** The grounding electrode conductor, or to this conductor's metal enclosure.
  - **(f)** The grounding electrode conductor, as per Section 250.32.
- **(2) In Buildings or Structures without Grounding Means.** Where there is no grounding system, a separate grounding electrode must be installed. Use the same methods here as for power-system grounding. Refer to Sections 250.52(A), 250.70, and 250.60 for these requirements.

800.41: Primary Protector Grounding and Bonding at Mobile Homes

(A) **Grounding.** If there is no mobile home service equipment within 30 feet (9 m) of the mobile home, and within sight, then a separate grounding electrode must be installed. Use the same methods here as you would for power system grounding. You may refer to Sections 250.52(A), 250.70, and 250.60 for these requirements.

(B) **Bonding.** If there is not service equipment or disconnecting means per Section 800.41(A) above, or if the mobile home is cord- and plug-connected, the following is required: The metal frame or grounding terminal of the mobile home must be bonded to the protector bonding terminal or to the grounding electrode. A No. 12 copper conductor shall be the minimum size used.

V. Communications Wires and Cables within Buildings

800.48: Raceways for Communications Wires and Cables

If raceways are used for communications cables, they must be installed in accordance with the other requirements of the *NEC*.

Exception

Conduit fill requirements do not apply. But the author still recommends them, as over-full conduits pose installation difficulties and make future changes very difficult. Special communications raceways are permitted, but must be installed according to the requirements of Article 362.

800.50: Listing, Marking, and Installation of Communications Wires and Cables

Communications cables used inside of buildings must be listed for their use. They must be applied correctly. Only riser cables may be used in risers; only plenum cables may be used in plenums; cables must be limited to the voltages they are rated for, and so on.

Cables entering a building from outside, and continuously contained in a metal raceway are not considered to be inside the building.

Listed outdoor cables are permitted to extend up to 50 feet (15 m) into a building if they run directly to an enclosure or to a protector. Refer to the *NEC* for the manufacturing requirements of various cable types.

800.52: Installation of Communications Wires, Cables, and Equipment

(A) Separation From Other Conductors.

(1) In Raceways, Boxes, and Cables. Communications conductors may share raceways and enclosures with:

(a) Class 2, Class 3, remote-control, signaling, and power-limited circuits. (See Article 725.)

(b) Power-limited fire alarm circuits. (See Article 760.)

(c) Optical fiber cables, conductive or nonconductive. (See Article 770.)

(d) CATV cables. (See Article 820.)

(e) Low-power network broadband conductors. (See Article 830.)

Power conductors and Class 1 circuits may not be run with communications conductors. Exceptions are made where the two systems are separated by a barrier, or where power conductors are present only as necessary to provide power to communications equipment. In these cases, a minimum separation of ¼ inch (6 mm) must be maintained. The *Code* also makes an exception for elevator cables. (See Section 620.36.)

(2) Other Applications. Communications conductors must be kept at least 2 inches (50 mm) from power conductors, Class 1 conductors, nonpower-limited conductors for fire alarms, or medium-power network broadband circuits. Exceptions are made if the systems are separated by a barrier, or if the power conductors are present only as necessary to provide power to communications equipment. In these cases, a minimum separation of ¼ inch (6 mm) must be maintained.

(B) Spread of Fire or Products of Combustion. The jackets and insulation of communications cables can spread combustion, and can produce toxic gases when burned. Because of this, fire-stopping is required around penetrations of fire-rated walls, floors, partitions, and ceilings. Accessible abandoned conductors must be removed.

(C) Equipment in Other Spaces Used for Environmental Air. All of the requirements of Section 300.22(C) apply here. Smoke and fumes from cables burning in environmental air can move through a building very quickly, and can poison the building's occupants.

(D) **Cable Trays.** Communications cables may be installed in cable trays, but must be Types MPP, MPR, MPG, MP, CMP, CMR, CMG, or CM. As always, cables must be properly listed for the areas in which they are installed.

(E) **Support of Conductors.** Communications cables must not be supported by other raceways. An exception is made for overhead drops to a service-raceway riser.

Article 810—Radio and Television Equipment

This article covers radio- and television-receiving equipment and amateur (ham) radio-transmitting and -receiving equipment, but not antennas or equipment that are used for coupling carrier current to power conductors. Refer to the *Code* for exact requirements.

Article 820—Community Antenna Television and Radio Distribution Systems

Community Antenna Television (CATV) systems started decades ago to provide television signals to communities that could not receive broadcast stations, either because of distance or due to shadow areas where the signal was too weak. Community antennas were installed at a remote location (usually a hill-top), and signals from it were fed to the homes in the area. Later, when technology developed, modern cable television systems came into use, based on the older CATV technology. Article 820 applies to both types of systems, and to all coaxial cable installations.

I. General

820.3: Locations and Other Articles

CATV circuits and equipment must comply with other sections of the *NEC*. The following are specified:

- The spread of fire and smoke must be avoided, per Section 300.21.
- Cables and equipment in air-handling spaces must meet the requirements of Section 300.22. [Also refer to Section 820.53.(A)]
- All equipment must be suitable for the areas where it is installed. (See Section 110.3.)
- The wiring methods of Article 830 may be substituted for those of Article 820.

820.4: Access to Electrical Equipment behind Panels Designed to Allow Access
Access to electrical equipment cannot be obstructed by accumulated communication cables. This includes the areas above suspended ceilings.

As coaxial cables proliferate, they frequently overload electrical rooms and similar areas. This is specifically prohibited by the *Code*.

820.6: Mechanical Execution of Work
All communications circuits must be installed in a neat and workmanlike manner. (See commentary on Section 110.12.)

This section gives specific standards for "workmanship," requiring that cables be supported by methods that will not damage cables. Section 300.4 also applies to the protection of cables from physical damage.

II. Outdoor Cables Entering Buildings

820.10: Outdoor Cables

(A) On Poles. CATV cables must be installed below the power conductors. While the Code makes an exception for unavoidable situations, this is a basic safety issue; workers should not be exposed to the danger of electric power cables. This is especially important because workers from the communications industry may not be fully trained to deal with the elevated voltages of power systems. Communications conductors may not be installed on cross-arms.

(B) **Lead-in Clearance.** Lead-in (aerial drop) conductors must be kept away from power, Class 1, or nonpower-limited conductors. The *Code* states that "where proximity cannot be avoided," this clearance can be reduced to 12 inches (300 mm), but must remain a full 40 inches (1.02 m) from the pole.

(C) **On Masts.** Aerial cable can attach to above-the-roof raceways that do not contain or support power wiring.

(D) **Above Roofs.** CATV conductors must have a vertical clearance of at least 8 feet (2.5 m) above relatively flat roofs. Conductors above auxiliary buildings such as garages are excepted from this rule. Above a roof with a slope of 4/12 (a 4-inch rise over 12 inches horizontally), a clearance of 3 feet (900 mm) is permitted. For the last 4 feet (1.2 m) of a run above a roof overhang, clearance can be reduced to 18 inches (450 mm). In this last case, the conductor run must terminate at a through-the-roof raceway (usually a service riser).

(E) Between Buildings. These cables must be suitable for the purpose, and physically strong enough to withstand loads that may be placed on them. They may also be attached to messenger wires , which may be necessary for withstanding loads.

(F) On Buildings. When run parallel on the side of a building, CATV conductors must be kept at least 4 inches (100 mm) away from open power conductors. This means that communications conductors do not require separation from power conductors run inside of a conduit, only from open wires or cables. These cables must be kept away from other communications conductors to avoid interference. They must also be separated at least 6 feet (1.8 m) from lightning-protection conductors, unless impractical. The author recommends that you maintain the 6-foot separation if possible.

820.11: Entering Buildings

(A) Underground Systems. If underground CATV conductors are installed in a raceway, hand-hole, or manhole that also contains power wiring, Class 1, or non-power-limited fire alarm conductors, the CATV conductors must be separated from the other wiring with a suitable barrier.

(B) Direct-Buried Cables and Raceways. Direct-buried CATV cables must be separated at least 12 inches (300 mm) from power or Class 1 circuits.

Exception

Where either the power conductors or CATV cables are contained in metal raceways or cables.

III. Protection

820.33: Grounding of Outer Conductive Shield

The outer shields of CATV cables must be grounded as close as possible to the point of entrance. Grounding located at mobile-home service equipment must be within 30 feet (9 m) of the mobile home and within sight of the mobile home. If the above is followed, no other protection is required. Alternatively, a protective device may be used, provided that it does not interrupt the grounding system of the building.

IV. Grounding Methods

820.40: Cable Grounding

(A) Grounding Conductor. These conductors must (1) have suitable insulation; (2) be a minimum of No. 14 copper (or some other corrosion-resistant conductor); (3) be as short as possible, not exceeding 20 feet (6 m); (4) run in as straight a line as possible; and (5) be protected from physical damage. If metal raceway is used to provide protection, it must be bonded to the conductor, or to the terminal at which the conductor is attached.

(B) Electrode.

(1) **In Buildings or Structures With Grounding Means.** If a building has a grounding system (usually for the power system), the following are considered suitable electrodes for connecting the grounding conductor:

(a) The power grounding system (See Section 250.50.)

(b) A grounded interior metal water pipe, within 5 feet of its entrance.

(c) A power service enclosure. (See Section 250.94.)

(d) A metal service raceway.

(e) The grounding electrode conductor, or to this conductor's metal enclosure.

(f) The grounding electrode conductor per Section 250.32.

(2) **In Buildings or Structures Without Grounding Means.** Where there is no grounding system, a separate grounding electrode must be installed. Use the same methods here as for power-system grounding. Refer to *NEC* Sections 250.52(A), 250.70, and 250.60 for these requirements.

(C) **Electrode Connection.** Connections must comply with Section 250.70. Thus the requirements here are the same as for power systems.

(D) **Bonding of Electrodes.** A No. 6 or larger bonding jumper must be installed between the antenna-grounding electrode and the power-grounding electrode system. See Section 820.42 for special requirements for mobile homes.

820.41: Equipment Grounding

Unpowered equipment and conductors on the coaxial system are considered grounded if they are connected to the coaxial outer shield.

820.42: Bonding and Grounding at Mobile Homes

(A) **Grounding.** If there is no mobile-home service equipment within 30 feet (9 m) of the mobile home, and within sight of the mobile home, then a separate grounding electrode must be installed. Use the same methods here as for power-system grounding. Refer to *NEC* Sections 250.52(A), 250.70, and 250.60 for these requirements.

(B) Bonding. If the service equipment and disconnecting means do not conform to Section 800.41(A) above, of if the mobile home is connected via cord-and-plug, the following is required: The metal frame or grounding terminal of the mobile home must be bonded to the protector bonding terminal or the grounding electrode. A No. 12 copper conductor shall be the minimum size used.

V. Cables Within Buildings

820.50: Listing Marking, and Installation of Coaxial Cables

CATV cables used inside of buildings must be listed for their use. They must be applied correctly. Only riser cables may be used in risers; only plenum cables may be used in plenums; cables must be limited to the voltages they are rated for, and so on.

Cables entering a building from outside, and continuously contained in a metal raceway are not considered to be inside the building.

Listed outdoor cables are permitted to extend up to 50 feet (15 m) into a building if they run directly to an enclosure or to a protector.

Refer to the *NEC* for the manufacturing requirements of various cable types.

820.52: Installation of Cables and Equipment

(A) Separation From Other Conductors

(1) In Raceways, Boxes, and Cables. CATV conductors may share raceways and enclosures with:

(a) Class 2, Class 3, remote-control, signaling, and power-limited circuits. (See Article 725.)

(b) Power-limited fire alarm circuits. (See Article 760.)

(c) Optical fiber cables, conductive or non-conductive. (See Article 770.)

(d) Low-power network broadband conductors. (See Article 830.)

Power conductors and Class 1 circuits may not be run with communications conductors. Exceptions are made if the two systems are separated by a barrier, or in cases where power conductors are present only as necessary to provide power to communications equipment. In these cases, a minimum separation of ¼ inch (6 mm) must be maintained. There is also an exception made for elevator cables. (See Section 620.36.)

(2) Other Applications. Communications conductors must be kept at least 2 inches (50 mm) from power conductors, Class 1 conductors, non-power limited conductors for fire alarms, or medium-power network broadband circuits. Exceptions are made if the systems are separated by a barrier, or if the power conductors are present only as necessary to provide power to communications equipment. In these cases, a minimum separation of ¼ inch (6 mm) must be maintained.

(B) Equipment in Other Spaces Used For Environmental Air. All of the requirements of Section 300.22(C) apply here. Smoke and fumes from cables burning in environmental air can move through a building very quickly, and can poison the building's occupants

(C) Hybrid power and Coaxial Cabling. Section 780.6 applies to these cables.

(D) Support of Conductors. CATV cables must not be supported by other raceways. An exception is made for overhead drops to a service raceway riser.

Article 830—Network-Powered Broadband Communications Systems

This is a new type of cabling system that transmits not only data, but also low levels of power. A typical cable would contain coaxial, twisted pair, and optical fiber, all under the same sheath. While some proponents expect such systems to be used widely in the future, they are little used now. See the *NEC* for details.

Chapter 9

Tables and Examples

Chapter 9 of the *NEC* contains a large group of tables with technical information. Annexes A through F contain further tables and other reference material. If you need information on allowable numbers of conductors in various types of raceways, conductor properties, engineering data on alternative ampacity calculations, and so on, this is where it may be found.

Index

D

E

F

P

R

S

T